Computation with Linear Algebraic Groups

MONOGRAPHS AND RESEARCH NOTES IN MATHEMATICS

Published Titles

Actions and Invariants of Algebraic Groups, Second Edition, Walter Ferrer Santos and Alvaro Rittatore

Analytical Methods for Kolmogorov Equations, Second Edition, Luca Lorenzi

Application of Fuzzy Logic to Social Choice Theory, John N. Mordeson, Davender S. Malik and Terry D. Clark

Blow-up Patterns for Higher-Order: Nonlinear Parabolic, Hyperbolic Dispersion and Schrödinger Equations, Victor A. Galaktionov, Enzo L. Mitidieri, and Stanislav Pohozaev

Bounds for Determinants of Linear Operators and Their Applications, Michael Gil′

Complex Analysis: Conformal Inequalities and the Bieberbach Conjecture, Prem K. Kythe

Computation with Linear Algebraic Groups, Willem Adriaan de Graaf

Computational Aspects of Polynomial Identities: Volume I, Kemer's Theorems, 2nd Edition Alexei Kanel-Belov, Yakov Karasik, and Louis Halle Rowen

A Concise Introduction to Geometric Numerical Integration, Fernando Casas and Sergio Blanes

Cremona Groups and Icosahedron, Ivan Cheltsov and Constantin Shramov

Delay Differential Evolutions Subjected to Nonlocal Initial Conditions Monica-Dana Burlică, Mihai Necula, Daniela Roşu, and Ioan I. Vrabie

Diagram Genus, Generators, and Applications, Alexander Stoimenow

Difference Equations: Theory, Applications and Advanced Topics, Third Edition Ronald E. Mickens

Dictionary of Inequalities, Second Edition, Peter Bullen

Elements of Quasigroup Theory and Applications, Victor Shcherbacov

Finite Element Methods for Eigenvalue Problems, Jiguang Sun and Aihui Zhou

Introduction to Abelian Model Structures and Gorenstein Homological Dimensions Marco A. Pérez

Iterative Methods without Inversion, Anatoly Galperin

Iterative Optimization in Inverse Problems, Charles L. Byrne

Line Integral Methods for Conservative Problems, Luigi Brugnano and Felice Iavernaro

Lineability: The Search for Linearity in Mathematics, Richard M. Aron, Luis Bernal González, Daniel M. Pellegrino, and Juan B. Seoane Sepúlveda

Modeling and Inverse Problems in the Presence of Uncertainty, H. T. Banks, Shuhua Hu, and W. Clayton Thompson

Published Titles Continued

Monomial Algebras, Second Edition, Rafael H. Villarreal

Nonlinear Functional Analysis in Banach Spaces and Banach Algebras: Fixed Point Theory Under Weak Topology for Nonlinear Operators and Block Operator Matrices with Applications, Aref Jeribi and Bilel Krichen

Partial Differential Equations with Variable Exponents: Variational Methods and Qualitative Analysis, Vicenţiu D. Rădulescu and Dušan D. Repovš

A Practical Guide to Geometric Regulation for Distributed Parameter Systems Eugenio Aulisa and David Gilliam

Reconstruction from Integral Data, Victor Palamodov

Signal Processing: A Mathematical Approach, *Second Edition*, Charles L. Byrne

Sinusoids: Theory and Technological Applications, Prem K. Kythe

Special Integrals of Gradshteyn and Ryzhik: the Proofs – Volume I, Victor H. Moll

Special Integrals of Gradshteyn and Ryzhik: the Proofs – Volume II, Victor H. Moll

Stochastic Cauchy Problems in Infinite Dimensions: Generalized and Regularized Solutions, Irina V. Melnikova

Submanifolds and Holonomy, Second Edition, Jürgen Berndt, Sergio Console, and Carlos Enrique Olmos

Symmetry and Quantum Mechanics, Scott Corry

The Truth Value Algebra of Type-2 Fuzzy Sets: Order Convolutions of Functions on the Unit Interval, John Harding, Carol Walker, and Elbert Walker

Forthcoming Titles

Groups, Designs, and Linear Algebra, Donald L. Kreher

Handbook of the Tutte Polynomial, Joanna Anthony Ellis-Monaghan and Iain Moffat

Microlocal Analysis on R^n and on NonCompact Manifolds, Sandro Coriasco

Practical Guide to Geometric Regulation for Distributed Parameter Systems, Eugenio Aulisa and David S. Gilliam

MONOGRAPHS AND RESEARCH NOTES IN MATHEMATICS

Computation with Linear Algebraic Groups

Willem Adriaan de Graaf

CRC Press
Taylor & Francis Group
Boca Raton London New York

CRC Press is an imprint of the
Taylor & Francis Group, an informa business
A CHAPMAN & HALL BOOK

CRC Press
Taylor & Francis Group
6000 Broken Sound Parkway NW, Suite 300
Boca Raton, FL 33487-2742

International Standard Book Number-13: 978-1-4987-2290-2 (Hardback)

Visit the Taylor & Francis Web site at
http://www.taylorandfrancis.com

and the CRC Press Web site at
http://www.crcpress.com

MIX
Paper from responsible sources
FSC www.fsc.org FSC® C013056

Printed and bound in Great Britain by
TJ International Ltd, Padstow, Cornwall

Contents

Preface **xiii**

1 Closed Sets in Affine Space **1**

1.1 Closed sets in affine space 1
1.1.1 Affine space and polynomial maps 1
1.1.2 Closed sets 2
1.1.3 Closed sets and ideals 2
1.1.4 Coordinate ring and regular maps 4
1.1.5 Irreducible closed sets 4
1.1.6 Products of closed sets 5
1.2 Tangent space 6
1.2.1 Dual numbers and tangent vectors 6
1.2.2 Differentials 7
1.2.3 Tangent space of a product of closed sets 10
1.3 Dimension 10
1.3.1 Specializations and generic points 13
1.4 Dominant maps 14
1.5 Gröbner bases 17
1.6 Elimination 20
1.7 Dimension of ideals 21
1.8 Ideal quotients 23
1.9 Radical of an ideal 23
1.10 Notes 26

2 Lie Algebras **27**

2.1 Basic constructions 27
2.1.1 Algebras 27
2.1.2 Homomorphisms and representations 28
2.1.3 Structure constants 29
2.1.4 Centre, centralizer and normalizer 30
2.2 Jordan decomposition of a linear transformation 31
2.3 Derivations 33
2.4 Nilpotency 33
2.4.1 Engel's theorem 34

2.4.2 Nilradicals 35
2.5 Cartan subalgebras 35
2.5.1 Primary decomposition 35
2.5.2 Fitting decomposition 36
2.5.3 Cartan subalgebras of Lie algebras 37
2.5.4 Weights and roots 39
2.6 Solvability 40
2.6.1 Lie's theorem 41
2.6.2 Cartan's criterion for solvability 42
2.6.3 Computing the solvable radical 43
2.6.4 Finding a non-nilpotent element 43
2.6.5 Computing the nilradical 44
2.7 Semisimple Lie algebras 45
2.7.1 Derivations and the Jordan decomposition 47
2.7.2 Levi's theorem 48
2.8 Root systems 49
2.8.1 Cartan matrices 50
2.8.2 Root systems 52
2.8.3 The Weyl group 56
2.9 Classification of simple Lie algebras 62
2.9.1 Representations of $\mathfrak{sl}(2, k)$ 62
2.9.2 From Lie algebra to root system 63
2.9.3 Canonical generators and isomorphisms 67
2.9.4 Structure constants of semisimple Lie algebras 68
2.10 Universal enveloping algebras 71
2.10.1 Poincaré-Birkhoff-Witt theorem 71
2.10.2 Left ideals of universal enveloping algebras 72
2.10.3 Integral forms 74
2.11 Representations of semisimple Lie algebras 78
2.11.1 Highest weight modules 79
2.11.2 Classification of irreducible modules 81
2.11.3 Path model 82
2.11.4 Constructing irreducible modules 87
2.12 Reductive Lie algebras 90
2.13 The Jacobson-Morozov theorem 92
2.14 Notes 93

3 Linear Algebraic Groups: Basic Constructions 95

3.1 Definition and first properties 95
3.2 Connected components 97
3.3 Semidirect products 100
3.4 The Lie algebra of an algebraic group 101
3.5 Subgroups and subalgebras 104
3.6 Examples 105

3.7 Morphisms and representations 109
3.8 Adjoint representation 110
3.9 Characters and diagonalizable groups 112
3.10 Jordan decomposition 115
3.10.1 In the Lie algebra of an algebraic group 116
3.10.2 In an algebraic group 117
3.11 The unipotent radical 118
3.12 Algebraic groups acting on closed sets 118
3.13 Specification of an algebraic group 119
3.14 Notes 120

4 Algebraic Groups and Their Lie Algebras 121

4.1 $G(\delta)$ 121
4.2 The Lie correspondence 123
4.3 Algebraic Lie algebras 127
4.3.1 The algebraic hull 127
4.3.2 Unipotent groups 131
4.3.3 The structure of algebraic Lie algebras 133
4.4 Computing the algebraic hull 138
4.5 Computing defining polynomials for an algebraic group from its Lie algebra 140
4.5.1 Unipotent case 141
4.5.2 Diagonalizable case 144
4.5.3 General case 147
4.6 The algebraic group generated by a given set 149
4.6.1 Unipotent case 150
4.6.2 Semisimple case 150
4.6.3 The algorithm 151
4.7 Generators, centralizers and normalizers 154
4.7.1 Generating sets 154
4.7.2 The centralizer of an algebraic subgroup 155
4.7.3 The normalizer of an algebraic subgroup 157
4.8 Orbit closures 158
4.9 Notes 160

5 Semisimple Algebraic Groups 161

5.1 Groups defined by certain generators and relations 162
5.2 Chevalley groups 166
5.2.1 Admissible lattices 167
5.2.2 Chevalley's commutator formula 169
5.2.3 Chevalley groups 172
5.2.4 Bruhat decomposition 177
5.2.5 Presentation of G 180

5.3 Semisimple algebraic groups 182
5.4 Root data 190
5.5 Irreducible representations in positive characteristic 192
5.6 Multiplication in a semisimple algebraic group 198
5.7 From matrix to word 204
5.8 Reductive algebraic groups 212
5.9 Regular subalgebras of semisimple Lie algebras 216
5.10 Notes 222

6 Generators of Arithmetic Groups 223

6.1 Brief introduction to arithmetic groups 223
6.2 Algorithms for lattices 229
6.3 Arithmetic subgroups of unipotent groups 232
6.3.1 $\mathfrak{T}$-groups 232
6.3.2 The derived representation 234
6.3.3 The algorithm 238
6.4 Arithmetic subgroups of tori 240
6.4.1 Unit groups of orders in toral matrix algebras 241
6.4.2 Generators of arithmetic subgroups of tori 245
6.5 Notes 247

7 Invariants of Algebraic Groups 249

7.1 The Derksen ideal 250
7.2 The invariant field 252
7.3 Computing invariants of reductive groups 254
7.3.1 Reynolds operator 254
7.3.2 Computing generators of the invariant ring 256
7.4 The nullcone 259
7.4.1 Characteristics 262
7.4.2 Stratification of the nullcone 272
7.4.3 Computing characteristics of the strata 274
7.5 Notes 278

8 Nilpotent Orbits 279

8.1 Weighted Dynkin diagrams 280
8.2 Computing representatives and weighted Dynkin diagrams . 285
8.3 θ-groups 288
8.3.1 Cyclic gradings of $\mathfrak{g}$ 289
8.3.2 The nilpotent orbits of a θ-group 292
8.3.3 Carrier algebras 294
8.4 Listing the nilpotent orbits of a θ-group 300
8.4.1 Using $\mathfrak{sl}_2$-triples 301

8.4.2 Using carrier algebras 302
8.5 Closures of nilpotent orbits 304
8.5.1 Deciding emptiness 307
8.5.2 Traversing a W_0-orbit 308
8.6 Notes 310

Bibliography **311**

Index of Symbols **323**

Index of Terminology **325**

Preface

Since the middle of the 20th century, many books on algebraic groups have appeared. The main aim of these works is to present parts of the theory of algebraic groups in an elegant way. But little attention is paid to the problems that arise when one wants to investigate a concrete algebraic group. The kinds of questions that come up in such a situtation serve as the inspiration of this book, which aims at describing algorithms for working with linear algebraic groups. As such, this book is part of the literature on computational group theory. This is a subject that was initiated in the 1960's, driven by the development of computer technology (although its origins can be traced back much earlier). In computational group theory mostly finitely generated groups are considered, as they can conveniently be defined in ways suitable for computation (that is, by a finite list of generators). However, there is a growing body of literature on computation with non-discrete groups, like algebraic groups and Lie groups. In this book we focus on the former (a thorough account of algorithms for Lie groups would require a separate book).

In order to obtain a reasonably self-contained narrative, parts of the theory of algebraic groups and directly related structures, such as closed sets in affine space and Lie algebras, are contained in this book. The most immediate (sometimes also called "naive" or "non-intrinsic") definition of the concept of algebraic group is used, that is, algebraic groups are matrix groups given as zero loci of sets of polynomials in the matrix entries. This has two advantages. Firstly, the definition can be grasped immediately, without much study. Secondly, it leads to an immediate way to describe an algebraic group by a finite amount of data, namely a set of polynomials, which at the same time lends itself to doing computations. However, the main disadvantage is that this definition makes it difficult to define what a quotient of algebraic groups is. Nowadays, a standard way to develop the theory of algebraic groups ([Bor91], [Hum75], [Spr98]) uses the construction of the quotient variety of an algebraic group by a subgroup quite heaviliy. Here we follow paths similar to those followed in the books by Chevalley ([Che51], [Che55b]), meaning that Lie algebras play a more important role, which in turn limits parts of the theory to characteristic 0. This approach to the theory is mirrored by the algorithms, which occasionally make good use of the Lie algebra of an algebraic group, and therefore, on those occasions, are also limited to algebraic groups of characteristic 0.

Here we are not concerned with the exact definition of the notion of algorithm. We take the position that we know an algorithm when we see one. Also we do not consider the complexity of algorithms as they very often are bad. Indeed, quite a few algorithms use Gröbner bases, and the complexity of the algorithms to compute the latter is known to be doubly exponential. There are other algorithms that need to solve number theoretic problems, for which the algorithms also have excessive complexity. Instead of giving complexity analyses, for a number of algorithms we give examples, and report the running times that the computations took. This gives an idea of the practical possibilities and limitations of these algorithms. The timings have been obtained using the computer algebra systems GAP4 ([GAP16]), MAGMA ([BCP97]) and SINGULAR ([DGPS15]). As it is not our objective to compare these systems, in most examples we do not mention which system has been used.

We describe algorithms roughly in two ways. Firstly, we often describe an algorithm explicitly by a list of steps to be executed. Secondly, a number of algorithms are described more implicitly by text. In those cases it is straightforward to obtain the list of steps from the text, and adding such a list to the text would not add much to the clarity of the exposition.

We now give a very brief summary of the contents of this book. Chapters 1 and 2 contain background material on, respectively, affine algebraic sets and Lie algebras, along with algorithms to work with them. Going into much detail on these topics would take us too far afield, so the style of these chapters is that of an overview rather than a detailed exposition. In Chapter 3 the main concept of this book, the linear algebraic groups, is defined, and some basic constructions are discussed. Some algorithms that deal with algebraic groups given by a set of defining polynomials are discussed. The remaining chapters deal with specific topics. Chapter 4 covers the correspondence between algebraic groups and Lie algebras in characteristic 0. Chapter 5 is devoted to semisimple algebraic groups. In Chapter 6 we look at the problem of determining a finite set of generators of an arithmetic group. Chapter 7 is concerned with invariant theory of (reductive) algebraic groups. The last chapter deals with nilpotent orbits in related settings: a semisimple algebraic group acting on its Lie algebra, and a so-called θ-group.

I thank everyone who, directly or indirectly, helped me to write this book. In particular I am grateful for the suggestions of Alla Detinko, Bettina Eick and Jim Humphreys. I thank the staff at Taylor & Francis Group, personified by Sunil Nair, for their help with the preparation of the manuscript. Finally, I thank Trenitalia, on whose trains large parts of the book were written.

Willem de Graaf

Chapter 1

Closed Sets in Affine Space

In the first part of this chapter we give an overview of some notions and results regarding the Zariski topology in a vector space over an algebraically closed field (which is called an affine space). One of the main points of the Zariski topology is that it translates certain topological notions into properties of ideals of polynomial rings. For example, a closed set is irreducible if and only if its vanishing ideal is prime. This makes it a very useful tool in the study of algebraic groups. In this context the tangent space is of special importance, as for algebraic groups this is the Lie algebra that will play a pivotal role throughout this book.

The second part is concerned with algorithms for working with closed sets in affine space. The main cornerstone of all these algorithms is the concept of Gröbner bases. We give a brief introduction of Gröbner bases, the algorithm to compute them, and some applications that are particularly useful when dealing with algebraic groups.

In some places in the literature the term "affine variety" is used for what we call a "closed set". We have avoided the term "variety" here, as in many other places this is defined in a much more conceptual way, i.e., as a set satisfying certain axioms rather than as a zero locus of some polynomials. In order to avoid confusion, we have therefore decided to stick to the term "closed set".

Throughout the chapter we let K be an algebraically closed field. Not necessarily algebraically closed fields will be denoted $k, k', \ldots$.

1.1 Closed sets in affine space

1.1.1 Affine space and polynomial maps

We write $\mathbb{A}^n$ for the space K^n. It is called the n-dimensional affine space. The symbol $\mathbb{A}^n$ is used rather than just K^n to emphasize that we are studying the geometry of the space, and are not mainly concerned with linear algebra. For $v \in \mathbb{A}^n$ we let v_i denote its i-th coordinate, so $v = (v_1, \ldots, v_n)$.

Let $K[x_1, \ldots, x_n]$ be the polynomial ring in n indeterminates. We view $f \in K[x_1, \ldots, x_n]$ as a function $f : \mathbb{A}^n \to K$ by $f(v) = f(v_1, \ldots, v_n)$. This is called a *polynomial function*. The set of polynomial functions is denoted by

$K[\mathbb{A}^n]$, and called the coordinate ring of $\mathbb{A}^n$. The difference between $K[\mathbb{A}^n]$ and $K[x_1,\dots,x_n]$ is mainly the point of view: elements of the former are functions, and elements of the latter are polynomials. In our treatment this difference will often be forgotten, when, for example, we write that a certain polynomial is an element of $K[\mathbb{A}^n]$.

1.1.2 Closed sets

A set $X \subset \mathbb{A}^n$ is called *closed* (or *algebraic*) if there is a set of polynomials $A \subset K[x_1,\dots,x_n]$ with $X = \{v \in \mathbb{A}^n \mid f(v) = 0 \text{ for all } f \in A\}$.

A set is called *open* if it is the complement of a closed set. More generally, if $X \subset \mathbb{A}^n$ is closed, then a $Y \subset X$ is called open in X if $X \setminus Y$ is a closed set.

The intersection of any number of closed sets is again closed. Indeed, if $A_i \subset K[x_1,\dots,x_n]$ defines the closed set X_i, then $\cup A_i$ defines $\cap X_i$. Also, the union of two closed sets is again closed. Let X_1, X_2 respectively be defined by $A_1 = \{f_1,\dots,f_s\}$ and $A_2 = \{g_1,\dots,g_t\}$. Then $X_1 \cup X_2$ is defined by $\{f_i g_j \mid 1 \le i \le s, 1 \le j \le t\}$. In particular we see that a union of a finite number of closed sets is closed. This means that the collection of closed sets in $\mathbb{A}^n$ defines a topology on $\mathbb{A}^n$. This is called the *Zariski topology.*

Let $U \subset \mathbb{A}^n$ be any set. Then the intersection of the closed sets containing U is again closed. It is the smallest closed set containing U, and called the *closure* of U, denoted $\overline{U}$. A subset U of a closed set X is said to be *dense* in X if $\overline{U} = X$. It is clear that any polynomial which is zero on a dense subset of X must be zero on X.

Example 1.1.1 Let $M_{2,3}(K)$ denote the set of 2×3 matrices over K. Let $R_1(K) \subset M_{2,3}(K)$ be the set of all such matrices of rank at most 1. Note that

$$a = \begin{pmatrix} a_{11} & a_{12} & a_{13} \\ a_{21} & a_{22} & a_{23} \end{pmatrix}$$

lies in $R_1(K)$ if and only if all of its 2×2-minors vanish, in other words, if $a_{11}a_{22} - a_{12}a_{21} = a_{11}a_{23} - a_{13}a_{21} = a_{12}a_{23} - a_{13}a_{22} = 0$. It follows that $R_1(K)$ is closed in $M_{2,3}(K)$.

1.1.3 Closed sets and ideals

Note that the ideal generated by $A \subset K[\mathbb{A}^n]$ defines the same closed set as A. So it makes sense to only consider closed sets defined by ideals. Let $U \subset \mathbb{A}^n$, and set $\mathcal{I}(U) = \{f \in K[\mathbb{A}^n] \mid f(u) = 0 \text{ for all } u \in U\}$. Then $\mathcal{I}(U)$ is an ideal of $K[\mathbb{A}^n]$. It is called the *vanishing ideal* of U. So if U is a closed set, then $\mathcal{I}(U)$ is the largest set of polynomials defining it. Conversely, if $I \subset K[\mathbb{A}^n]$ is an ideal then we construct the closed set $\mathcal{V}(I) = \{v \in \mathbb{A}^n \mid f(v) = 0 \text{ for all } f \in I\}$.

The operations $\mathcal{V}$ and $\mathcal{I}$ invert inclusions. For example, if $U_1 \subset U_2$ then $\mathcal{I}(U_1) \supset \mathcal{I}(U_2)$. Therefore, for a given $U \subset V$ the set $\mathcal{V}(\mathcal{I}(U))$ is the smallest closed set containing U, i.e., it is the closure of U.

But $\mathcal{I}$ and $\mathcal{V}$ are not mutual inverses. For a trivial example consider $\mathbb{A}^1$ and let I be generated by x^2, then $\mathcal{I}(\mathcal{V}(I)) \neq I$. To get a precise idea about the correspondence between ideals and closed sets, the key is the next theorem. To formulate it we need the *radical of an ideal* I which is defined as

$$\sqrt{I} = \{f \in K[x_1, \ldots, x_n] \mid \text{there is an } m > 0 \text{ with } f^m \in I\}.$$

Theorem 1.1.2 (Hilbert's Nullstellensatz) *Let $I \subset K[\mathbb{A}^n]$ be an ideal. Then*

(i) $\mathcal{V}(I) = \emptyset$ *if and only if* $I = K[\mathbb{A}^n]$,

(ii) $\mathcal{I}(\mathcal{V}(I)) = \sqrt{I}$.

The two statements of the theorem are equivalent. The first one says that if $I \neq K[\mathbb{A}^n]$, then $\mathcal{V}(I)$ actually has points. The second affirms that $\mathcal{V}$ and $\mathcal{I}$ are mutual inverses if we restrict ourselves to the set of radical ideals (i.e., ideals I with $I = \sqrt{I}$). We note that the theorem is obviously false if the base field is not algebraically closed. The proof of the theorem is contained in many books, for example, [CLO15].

Another important theorem, also due to Hilbert, is the following. Once more we refer to [CLO15] for a proof of it.

Theorem 1.1.3 (Hilbert's basis theorem) *Every ideal of $k[x_1, \ldots, x_n]$ is finitely generated.*

It follows that a closed set can always be defined by a finite set of polynomials. We note also that this theorem implies that there is no infinite series of strictly increasing ideals in $k[x_1, \ldots, x_n]$. Therefore, there cannot be an infinite strictly decreasing sequence of closed sets in $\mathbb{A}^n$ either.

Let $k \subset K$ be a subfield. A closed set $X \subset \mathbb{A}^n$ is said to be *defined* over k if the ideal $\mathcal{I}(X) \subset K[\mathbb{A}^n]$ is generated by elements of $k[x_1, \ldots, x_n]$. Also, X is called *k-closed* if X is the set of common zeros of a set of elements of $k[x_1, \ldots, x_n]$. It is clear that a closed set defined over k is also k-closed. The reverse implication does not always hold.

Example 1.1.4 Let the notation be as in Example 1.1.1. Let $K[M_{2,3}(K)] = K[x_{11}, x_{12}, \ldots, x_{23}]$ denote the coordinate ring of $M_{2,3}(K)$. Set $f_1 = x_{11}x_{22} - x_{12}x_{21}$, $f_2 = x_{11}x_{23} - x_{13}x_{21}$, $f_3 = x_{12}x_{23} - x_{13}x_{22}$. Let $I \subset K[M_{2,3}(K)]$ be the ideal generated by f_1, f_2, f_3. Then $R_1(K) = \mathcal{V}(I)$. However, it is not immediately clear that $I = \mathcal{I}(R_1(K))$. By Theorem 1.1.2 this is equivalent to $I = \sqrt{I}$. We will describe algorithms for computing the radical of an ideal, making it possible to decide this question.

1.1.4 Coordinate ring and regular maps

Let $X \subset \mathbb{A}^n$ be closed; then we consider polynomial maps from X to K. These, by definition, are the restrictions of elements of $K[\mathbb{A}^n]$ to X. The ring of polynomial maps $X \to K$ is denoted $K[X]$. It is called the *coordinate ring* of X, and it is straightforward to see that $K[X] \cong K[x_1, \dots, x_n]/\mathcal{I}(X)$. For an element $f \in K[\mathbb{A}^n]$ we write $\bar{f}$ for its restriction to X. In particular, $K[X]$ is generated (as an algebra over K) by $\bar{x}_1, \dots, \bar{x}_n$.

Let $Y \subset \mathbb{A}^m$ be closed, and $h_1, \dots, h_m \in K[X]$ be such that $(h_1(v), \dots, h_m(v)) \in Y$ for all $v \in X$. Then $h : X \to Y$ defined by $h(v) = (h_1(v), \dots, h_m(v))$ is called a *regular map.* If the h_i have coefficients in a subfield $k \subset K$, then h is said to be defined over k.

Let $h : X \to Y$ be a regular map; then we define a function $h^* : K[Y] \to K[X]$ by $h^*(f)(v) = f(h(v))$. This map is called the *comorphism* of h. It is a homomorphism of algebras. Furthermore, it is clear that it is injective if and only if $h(X)$ is dense in Y.

A regular map $h : X \to Y$ that has a regular inverse is called an *isomorphism.* The next lemma contains some useful observations on the behaviour of closed sets under isomorphism.

Lemma 1.1.5 *Let $h : X \to Y$ be an isomorphism of closed sets. Then*

1. *h maps closed and open sets in X to closed, respectively open, sets in Y.*

2. *If $X' \subset X$ is dense, then $h(X') \subset Y$ is dense. In particular, for $U \subset X$ we have $h(\overline{U}) = \overline{(f(U))}$.*

Proof. Let $Z \subset X$ be a closed set defined by the polynomials $f_1, \dots, f_r$. Then $h(Z)$ is the closed set defined by the polynomials $(h^{-1})^*(f_i)$.

For the second part, set $Y' = \overline{h(X')}$. Suppose that $Y' \neq Y$; then $Y \setminus Y'$ is non-empty and open, and hence $h^{-1}(Y \setminus Y')$ is non-empty and open in X. But a non-empty open set and a dense set have a point in common (otherwise the dense set would be contained in the complement of the open set, which is a closed set strictly contained in X). Let $u \in X' \cap h^{-1}(Y \setminus Y')$. Then on the one hand $h(u) \in Y \setminus Y'$ and on the other hand $h(u) \in Y'$, a contradiction. Now we restrict h to an isomorphism $h : \overline{U} \to h(\overline{U})$. It follows that $h(U)$ is dense in $h(\overline{U})$. □

1.1.5 Irreducible closed sets

A closed set $X \subset \mathbb{A}^n$ is called *reducible* if it is the union of two other closed sets; otherwise X is irreducible.

Let $X \subset \mathbb{A}^n$ be closed. If it is reducible, then we can write $X = X_1 \cup X_2$, where $X_1, X_2 \subset X$ are two proper non-empty closed subsets. With X_1, X_2 we can continue this process. It has to terminate; otherwise we can construct an infinite strictly decreasing sequence of closed sets $X \supset Y_1 \supset Y_2 \supset \cdots$, which is not possible in view of the remarks following Theorem 1.1.3. It follows that X is a finite union of irreducible closed subsets, $X = X_1 \cup \cdots \cup X_r$. If there are $i \neq j$ such that $X_i \subset X_j$, then we discard X_i. We then obtain a union $X = X_1 \cup \cdots \cup X_s$ such that $X_i \not\subset X_j$ for $i \neq j$. It is not difficult to see that up to order these X_i are uniquely determined ([Sha94], I.3.1, Theorem 2). They are called the *irreducible components* of X.

A closed set X is irreducible if and only if $\mathcal{I}(X)$ is a prime ideal. Indeed, if $X = X_1 \cup X_2$ with the X_i closed, then we can find polynomials f, g such that $f \in \mathcal{I}(X_1)$ and $f \notin \mathcal{I}(X_2)$, $g \in \mathcal{I}(X_2)$, but $g \notin \mathcal{I}(X_1)$. Then $fg \in \mathcal{I}(X_1 \cup X_2)$ but neither f nor g is in this ideal. Therefore it is not prime. Conversely, suppose that X is irreducible, and let f_1, f_2 be such that $f_1 f_2 \in \mathcal{I}(X)$. Set $X_i = X \cap \mathcal{V}(f_i)$. Then $X = X_1 \cup X_2$. Hence $X = X_1$ or $X = X_2$. The first possibility implies that $f_1 \in \mathcal{I}(X)$, the second that $f_2 \in \mathcal{I}(X)$.

Lemma 1.1.6 *Let $f : \mathbb{A}^n \to \mathbb{A}^m$ be a regular map. Let $X \subset \mathbb{A}^n$ be an irreducible closed set. Then $\overline{f(X)}$ is irreducible as well.*

Proof. Suppose that $\overline{f(X)} = Y_1 \cup Y_2$, with Y_i closed. Set $X_i = X \cap f^{-1}(Y_i)$. The inverse image of a closed set is closed. So the X_i are closed, but $X = X_1 \cup X_2$. Hence $X = X_1$ (say). Then $f(X) = f(X_1) \subset Y_1$ and $\overline{f(X)} = Y_1$. □

Lemma 1.1.7 *Let $X \subset \mathbb{A}^n$ be closed and irreducible. Let $U \subset X$ be nonempty and open in X. Then U is dense in X.*

Proof. Suppose that $\overline{U} \neq X$. Then there is a polynomial $p \in K[X]$ which is zero on U, but does not vanish on X. But $X \setminus U$ is closed, and hence there is a $q \in K[X]$ such that q vanishes on $X \setminus U$ but not on X. Hence pq is zero on X, i.e., $pq \in \mathcal{I}(X)$. Since X is irreducible, either p or q must lie in $\mathcal{I}(X)$ but that is impossible. □

A closed set is called *connected* if it is not the union of two disjoint proper closed subsets. If $X \subset \mathbb{A}^n$ is irreducible then it is necessarily connected. The converse is not true: consider $X = \{(u, v) \in \mathbb{A}^2 \mid uv = 0\}$.

1.1.6 Products of closed sets

We identify the product $\mathbb{A}^n \times \mathbb{A}^m$ with $\mathbb{A}^{n+m}$. We write $K[\mathbb{A}^n \times \mathbb{A}^m] = K[x_1, \ldots, x_n, y_1, \ldots, y_m]$ and we have natural inclusions $K[\mathbb{A}^n], K[\mathbb{A}^m] \subset K[\mathbb{A}^n \times \mathbb{A}^m]$.

Let $X \subset \mathbb{A}^n$, $Y \subset \mathbb{A}^m$ be two closed sets. If X is defined by the equations $f_i(v) = 0$, and Y by $g_j(w) = 0$, then $X \times Y$ is defined by the equations $f_i(v, w) = g_j(v, w) = 0$ (where now f_i, g_j are viewed as elements of $K[\mathbb{A}^n \times \mathbb{A}^m]$). It follows that $X \times Y$ is a closed set.

Lemma 1.1.8 *We have* $K[X \times Y] \cong K[X] \otimes K[Y]$.

Proof. Define a map $\mu : K[X] \otimes K[Y] \to K[X \times Y]$ by $\mu(\sum_i f_i \otimes g_i) = \sum_i f_i g_i$. This is an isomorphism as shown in [Sha94], I.2.2, Example 4. □

Theorem 1.1.9 *Let* $X \subset \mathbb{A}^n$, $Y \subset \mathbb{A}^m$ *be irreducible closed sets. Then* $X \times Y$ *in* $\mathbb{A}^{m+n}$ *is irreducible as well.*

Proof. (Outline; for more details see [Sha94], I.3.1, Theorem 3.) Suppose that $X \times Y = Z_1 \cup Z_2$, with the Z_i closed. For a fixed $v \in X$ the set $v \times Y$ is closed and irreducible. But also $v \times Y = (v \times Y \cap Z_1) \cup (v \times Y \cap Z_2)$. Hence $v \times Y$ is contained in Z_1 or in Z_2. Set $X_i = \{v \in X \mid v \times Y \subset Z_i\}$. Then X_i is closed, and $X = X_1 \cup X_2$. So we have $X = X_1$ (say). But then $X \times Y = Z_1$. □

1.2 Tangent space

In this section we define the tangent space to a closed set at a given point. This of the highest importance for the study of algebraic groups, as the tangent space at the identity is the Lie algebra of such a group. The Lie algebra is not only of theoretical importance: many algorithms make good use of it. Here we first define the tangent space in a rather abstract way, using the algebra of dual numbers. Then we show that by differentiating the generators of the vanishing ideal of the closed set, the tangent space can be identified with a subspace of K^n. The approach using dual numbers can be rather useful when dealing with algebraic groups, as the polynomials defining them are not always given explicitly. Example 3.6.4 shows an algebraic group for which we can easily get information about its Lie algebra, without writing the polynomials that define the group.

1.2.1 Dual numbers and tangent vectors

The algebra of *dual numbers* over K is $\mathcal{D} = \{a + b\varepsilon \mid a, b \in K,\ \varepsilon^2 = 0\}$. It is a 2-dimensional associative algebra over K, with multiplication $(a+b\varepsilon)(c+d\varepsilon) = ac + (ad + bc)\varepsilon$.

Let $X \subset \mathbb{A}^n$ be a closed set, with $v \in X$. Then a *close point* to v on X is a K-algebra homomorphism $\varphi : K[X] \to \mathcal{D}$ with $\varphi(f) = f(v) + \lambda(f)\varepsilon$, where $\lambda : K[X] \to K$ is a K-linear map.

As in Section 1.1.4, for $f \in K[\mathbb{A}^n]$, we denote its restriction to X by $\bar{f}$.

For $v \in X$ we let $\mathfrak{m}_v(X) \subset K[X]$ be the ideal consisting of all $f \in K[X]$ with $f(v) = 0$. If it is clear which closed set we mean then we also write $\mathfrak{m}_v$. It is generated by $\bar{x}_i - v_i$ for $1 \leq i \leq n$ (see Section 1.1.4 for the notation). Hence the quotient $K[X]/\mathfrak{m}_v$ is isomorphic to K, and therefore $\mathfrak{m}_v$ is a maximal ideal. We let $\mathfrak{m}_v^2$ be the ideal of $K[X]$ generated by all products of elements of $\mathfrak{m}_v$.

Proposition 1.2.1 *Let $\lambda : K[X] \to K$ be a K-linear map and define $\varphi : K[X] \to \mathcal{D}$ by $\varphi(f) = f(v) + \lambda(f)\varepsilon$. Then the following are equivalent:*

(i) *φ is a close point to v on X.*

(ii) *$\lambda(fg) = f(v)\lambda(g) + g(v)\lambda(f)$ for all $f, g \in K[X]$.*

(iii) *$\lambda(K) = 0$ and $\lambda(\mathfrak{m}_v^2) = 0$.*

Proof. First we note that $\varphi(fg) = \varphi(f)\varphi(g)$ is the same as $\lambda(fg) = f(v)\lambda(g) + g(v)\lambda(f)$. This shows that the first two assertions are equivalent.

Suppose that φ is a close point to v on X. Since φ is a K-algebra homomorphism, $\varphi(\alpha) = \alpha$ for all $\alpha \in K$. This is equivalent to $\lambda(\alpha) = 0$. Moreover, if both $f, g \in \mathfrak{m}_v$ then by (ii), $\lambda(fg) = 0$. We conclude that $\lambda(\mathfrak{m}_v^2) = 0$.

Conversely, suppose that $\lambda(K) = 0$ and $\lambda(\mathfrak{m}_v^2) = 0$. Let $f, g \in K[X]$, then

$$\begin{aligned}\lambda(fg) &= \lambda(g(v)f + f(v)g + (f - f(v))(g - g(v))) \\ &= g(v)\lambda(f) + f(v)\lambda(g)\end{aligned}$$

(since $(f - f(v))(g - g(v)) \in \mathfrak{m}_v^2$, and $\lambda(f(v)g(v)) = 0$). So φ is a close point to v on X. □

A linear map $\lambda : K[X] \to K$ with Proposition 1.2.1(ii) is called a *tangent vector* to X at v. We let $T_v(X)$ be the space of all tangent vectors to X at v. This space is called the *tangent space* to X at v. It is also denoted T_v, if there can be no confusion about the closed set to which v belongs. By Proposition 1.2.1 we see that $T_v \cong (\mathfrak{m}_v/\mathfrak{m}_v^2)^*$ (the dual space of $\mathfrak{m}_v/\mathfrak{m}_v^2$).

We remark that a $\lambda \in T_v(X)$ is completely determined by the values $\lambda(\bar{x}_i)$ for $1 \leq i \leq n$. Indeed, any $f \in K[X]$ can be written $f = a_0 + \sum_{i=1}^n a_i(\bar{x}_i - v_i) + h$, where $a_i \in K$ and $h \in \mathfrak{m}_v^2$. Furthermore, $\lambda(\bar{x}_i - v_i) = \lambda(\bar{x}_i)$. Hence $\lambda(f) = \sum_i a_i \lambda(\bar{x}_i)$.

1.2.2 Differentials

Let $X \subset \mathbb{A}^n$ and $Y \subset \mathbb{A}^m$ be closed sets, and $\sigma : X \to Y$ a regular map. Let $v \in X$, and $w = \sigma(v) \in Y$. Let φ be a close point to v on X, and define $\psi : K[Y] \to K$ by $\psi(f) = \varphi(\sigma^*(f))$. Then ψ is a close point to w on Y. Let $\lambda_\varphi, \lambda_\psi$ be the elements of T_v, T_w corresponding to respectively φ, ψ. Then

$$\lambda_\psi(f) = \lambda_\varphi(\sigma^*(f)).$$

On this basis we have the following definition.

Definition 1.2.2 *The linear map* $\mathrm{d}_v\sigma : T_v \to T_w$ *defined by* $\mathrm{d}_v\sigma(\lambda)(f) = \lambda(\sigma^*(f))$ *is called the* differential *of* σ *at* v.

Lemma 1.2.3 *Let* $X \subset \mathbb{A}^n$, $Y \subset \mathbb{A}^m$, $Z \subset \mathbb{A}^l$ *be closed sets, and* $\sigma : X \to Y$, $\pi : Y \to Z$ *regular maps. Let* $u \in X$ *and* $v = \sigma(u) \in Y$. *Then* $\mathrm{d}_u(\pi \circ \sigma) = \mathrm{d}_v\pi \circ \mathrm{d}_u\sigma$.

Proof. Note that $(\pi \circ \sigma)^* = \sigma^* \circ \pi^*$. This, together with the definition of the differential, implies the statement. □

Corollary 1.2.4 *Let* $X \subset \mathbb{A}^n$, $Y \subset \mathbb{A}^m$ *be closed sets. Let* $\sigma : X \to Y$ *be a regular map, and suppose that it has a regular inverse. Let* $v \in X$ *and* $w = \sigma(v)$. *Then* $\mathrm{d}_v\sigma : T_v \to T_w$ *is an isomorphism.*

Proof. The differential of the identity on X is the identity on T_v. Hence Lemma 1.2.3 implies that $\mathrm{d}_w\sigma^{-1}$ is the inverse of $\mathrm{d}_v\sigma$. □

Now consider the closed set $\mathbb{A}^n$ (i.e., the whole space). Let $v \in \mathbb{A}^n$, and set $\mathfrak{m}_v = \mathfrak{m}_v(\mathbb{A}^n)$. The $x_i - v_i + \mathfrak{m}_v^2$ form a basis of $\mathfrak{m}_v/\mathfrak{m}_v^2$. Hence for $(a_1, \dots, a_n) \in K^n$ there is a unique $\lambda \in T_v(\mathbb{A}^n)$ with $\lambda(x_i) = \lambda(x_i - v_i) = a_i$. It follows that $T_v(\mathbb{A}^n) \cong K^n$.

Proposition 1.2.5 *Let* $X \subset \mathbb{A}^n$ *be a closed set, and* $v \in X$. *Let* $\sigma : X \to \mathbb{A}^n$ *be the inclusion map. Then* $\mathrm{d}_v\sigma$ *maps* $T_v(X)$ *injectively into* $T_v(\mathbb{A}^n)$. *This identifies* $T_v(X)$ *with a subspace of* $T_v(\mathbb{A}^n)$. *Moreover, for* $\lambda \in T_v(\mathbb{A}^n)$ *we have* $\lambda \in T_v(X)$ *if and only if* $\lambda(f) = 0$ *for all* $f \in \mathcal{I}(X)$. *In fact, it is enough to have this condition for a set of generators of* $\mathcal{I}(X)$.

Proof. Note that $\sigma^* : K[\mathbb{A}^n] \to K[X]$ is the restriction map, and hence it is surjective. It maps $\mathfrak{m}_v(\mathbb{A}^n)$ surjectively onto $\mathfrak{m}_v(X)$. Let $\mu \in T_v(X)$. Then $\mathrm{d}_v\sigma(\mu) = 0$ is the same as $\mu(\sigma^*(f)) = 0$ for all $f \in \mathfrak{m}_v(\mathbb{A}^n)$. But then $\mu(f) = 0$ for all $f \in \mathfrak{m}_v(X)$, so that $\mu = 0$. It follows that $\mathrm{d}_v\sigma$ is injective.

Let $\lambda \in T_v(\mathbb{A}^n)$. Suppose that $\lambda \in \mathrm{d}_v\sigma(T_v(X))$, i.e., $\lambda = \mathrm{d}_v\sigma(\mu)$ for a certain $\mu \in T_v(X)$. Let $f \in \mathcal{I}(X)$; then $f \in \mathfrak{m}_v(\mathbb{A}^n)$ so that $\lambda(f) = \mu(\sigma^*(f)) = 0$ because $\sigma^*(f) = 0$. Conversely, suppose that $\lambda(f) = 0$ for all $f \in \mathcal{I}(X)$. Define $\mu : K[X] \to K$ by $\mu(\sigma^*(h)) = \lambda(h)$ for $h \in K[\mathbb{A}^n]$. This is well defined since $\sigma^*(h) = \sigma^*(h')$ implies that $h - h' \in \mathcal{I}(X)$ and $\lambda(h - h') = 0$. Let $f \in \mathfrak{m}_v(X)^2$, i.e., $f = \sum_i g_i g_i'$ for certain $g_i, g_i' \in \mathfrak{m}_v(X)$. Let $h_i, h_i' \in \mathfrak{m}_v(\mathbb{A}^n)$ be such that $\sigma^*(h_i) = g_i$, $\sigma^*(h_i') = g_i'$. Then $\mu(f) = \lambda(\sum_i h_i h_i') = 0$. Hence $\mu \in T_v(X)$ by Proposition 1.2.1, and $\lambda = \mathrm{d}_v\sigma(\mu)$ by construction.

Let $f_1, \dots, f_r \in K[\mathbb{A}^n]$ be generators of $\mathcal{I}(X)$. Let $\lambda \in T_v(\mathbb{A}^n)$ be such that $\lambda(f_i) = 0$ for $1 \le i \le r$. Let $h \in \mathcal{I}(X)$ and write $h = \sum_i g_i f_i$, for certain $g_i \in K[\mathbb{A}^n]$. Then $g_i - g_i(v)$ and f_i lie in $\mathfrak{m}_v(\mathbb{A}^n)$. Therefore $\lambda((g_i - g_i(v))f_i) = 0$, or $\lambda(g_i f_i) = g_i(v)\lambda(f_i)$. It follows that $\lambda(h) = \sum_i g_i(v)\lambda(f_i) = 0$. By the above we conclude that $\lambda \in \mathrm{d}_v\sigma(T_v(X))$. □

Let $X \subset \mathbb{A}^n$ be a closed set and $v \in X$. By the previous proposition, $T_v(X)$ can be identified with a subspace of $T_v(\mathbb{A}^n)$, which in turn can be identified with K^n. Now we describe how to determine the subspace of K^n identified with $T_v(X)$.

Let $(a_1, \ldots, a_n) \in K^n$ and let $\lambda \in T_v(\mathbb{A}^n)$ be such that $\lambda(x_i) = a_i$. Then $\varphi(x_i) = v_i + a_i\varepsilon$ defines the close point to v on $\mathbb{A}^n$ that corresponds to λ. Let $F \in K[\mathbb{A}^n]$, then using the Taylor expansion of F, and the fact that $\varepsilon^2 = 0$ we obtain

$$\begin{aligned}\varphi(F(x_1, \ldots, x_n)) &= F(\varphi(x_1), \ldots, \varphi(x_n)) = F(v_1 + a_1\varepsilon, \ldots, v_n + a_n\varepsilon)\\ &= F(v_1, \ldots, v_n) + \left(\sum_{i=1}^{n} \frac{\partial F}{\partial x_i}(v)a_i\right)\varepsilon.\end{aligned} \tag{1.1}$$

But also we have that $\varphi(F) = F(v) + \lambda(F)\varepsilon$. In view of Proposition 1.2.5 we have proved the following result.

Proposition 1.2.6 *Let $X \subset \mathbb{A}^n$ be a closed set. Let $f_1, \ldots, f_r \in K[\mathbb{A}^n]$ be generators of $\mathcal{I}(X)$. Identify $T_v(\mathbb{A}^n)$ with K^n. Then $T_v(X)$ corresponds to the subspace consisting of $(a_1, \ldots, a_n)$ such that $\sum_{i=1}^n \frac{\partial f_j}{\partial x_i}(v)a_i = 0$ for $1 \le j \le r$. Under this correspondence a $\lambda \in T_v(X)$ corresponds to $(a_1, \ldots, a_n) \in K^n$ where $\lambda(\bar{x}_i) = a_i$.*

We note that this last condition is also equivalent to $f_j(v_1 + a_1\varepsilon, \ldots, v_n + a_n\varepsilon) = 0$ for $1 \le j \le r$. Sometimes this yields a convenient method for computing the tangent space (when we know generators of $\mathcal{I}(X)$).

Now we turn to the problem of determining the differential of a regular map. Let $Y \subset \mathbb{A}^m$ be closed. Let $\sigma : X \to Y$ be a regular map given by $\sigma(v) = (\sigma_1(v), \ldots, \sigma_m(v))$, where $\sigma_i \in K[X]$. Write $K[y_1, \ldots, y_m]$ for the coordinate ring of $\mathbb{A}^m$. Observe that $\sigma^*(y_j) = \sigma_j$. Let φ be a close point to v on X and write $\varphi(\bar{x}_i) = v_i + a_i\varepsilon$. Then φ corresponds to $\lambda \in T_v(X)$ given by $\lambda(\bar{x}_i) = a_i$. Write $w = \sigma(v)$, and let ψ be the close point to w on Y defined by $\psi(f) = \varphi(\sigma^*(f))$. Let $h_j \in K[x_1, \ldots, x_n]$ be such that $\bar{h}_j = \sigma_j$, $1 \le j \le m$. Then $\psi(\bar{y}_j) = \varphi(\sigma_j) = h_j(v_1 + a_1\varepsilon, \ldots, v_n + a_n\varepsilon) = h_j(v) + (\sum_{i=1}^n \frac{\partial h_j}{\partial x_i}(v)a_i)\varepsilon$, by (1.1). We conclude that $\mathrm{d}_v\sigma(\lambda)$ is the element μ of $T_w(Y)$ given by

$$\mu(\bar{y}_j) = \sum_{i=1}^{n} \tfrac{\partial h_j}{\partial x_i}(v)a_i. \tag{1.2}$$

(Note that this does not depend on the choice of $h_j \in K[x_1, \ldots, x_n]$, in view of Proposition 1.2.6.) Equivalently, if we write $\mu(\bar{y}_j) = b_j$, then the coefficients b_j are determined by the relation

$$\sigma(v + (a_1, \ldots, a_n)\varepsilon) = \sigma(v) + (b_1, \ldots, b_m)\varepsilon. \tag{1.3}$$

Example 1.2.7 Let the notation be as in Examples 1.1.1 and 1.1.4. It can be shown that the ideal generated by f_1, f_2, f_3 is radical so that $I = \mathcal{I}(R_1(K))$. Write a as in Example 1.1.1, and suppose $a \in R_1(K)$. Then the equations determining $T_a(R_1(K))$, obtained by differentiating the f_i, are

$$a_{22}x_{11} - a_{21}x_{12} - a_{12}x_{21} + a_{11}x_{22} = 0$$
$$a_{23}x_{11} - a_{21}x_{13} - a_{13}x_{21} + a_{11}x_{23} = 0$$
$$a_{23}x_{12} - a_{22}x_{13} - a_{13}x_{22} + a_{12}x_{23} = 0.$$

As $a \in R_1(K)$ there is an $\alpha \in K$, $\alpha \neq 0$ such that $(a_{21}, a_{22}, a_{23}) = \alpha(a_{11}, a_{12}, a_{13})$. Substituting this, one sees that the matrix of the linear equations above has rank 2 if not all a_{ij} are zero (in the latter case the rank is 0). Hence $\dim T_a(R_1(K)) = 4$, unless a is zero, in which case $\dim T_a(R_1(K)) = 6$.

1.2.3 Tangent space of a product of closed sets

Let X, Y be as in Section 1.1.6. From Lemma 1.1.8 we recall that $K[X \times Y] \cong K[X] \otimes K[Y]$, where the isomorphism is given by $f \otimes g \mapsto fg$. In the sequel we will identify $K[X \times Y]$ and $K[X] \otimes K[Y]$. Let $\pi_1 : X \times Y \to X$ be the projection onto X, i.e., $\pi_1(v, w) = v$. Similarly π_2 will be the projection onto Y. Then $\pi_1^*(f) = f \otimes 1$, and $\pi_2^*(g) = 1 \otimes g$ for $f \in K[X], g \in K[Y]$.

Proposition 1.2.8 *Let $v \in X$, $w \in Y$. Then $\lambda \mapsto (\mathrm{d}_{(v,w)}\pi_1(\lambda), \mathrm{d}_{(v,w)}\pi_2(\lambda))$ is an isomorphism $T_{(v,w)} \to T_v \oplus T_w$.*

Proof. Let $\lambda \in T_{(v,w)}$ and φ the corresponding close point to (v, w). Suppose that λ is mapped to zero. Then $\mathrm{d}_{(v,w)}\pi_1(\lambda) = 0$, or equivalently, $\lambda(f \otimes 1) = 0$ for all $f \in \mathfrak{m}_v$ (note that this implies that $f \otimes 1 \in \mathfrak{m}_{(v,w)}$). Then $\varphi(f \otimes 1) = f(v)$ for all $f \in K[X]$. Similarly we have $\varphi(1 \otimes g) = g(w)$ for all $g \in K[Y]$. Hence $\varphi(h) = h(v, w)$ for all $h \in K[X \times Y]$. This means that $\lambda = 0$ and we have shown that the map is injective.

Now consider the map $\sigma : X \to X \times Y$, $\sigma(v') = (v', w)$. Let $\mu \in T_v$ and $\lambda = \mathrm{d}_v\sigma(\mu) \in T_{(v,w)}$. We have that $\pi_1 \circ \sigma$ is the identity on X. So using Lemma 1.2.3 we infer $\mathrm{d}_{(v,w)}\pi_1(\lambda) = \mu$. Also, $\pi_2 \circ \sigma : X \to Y$ is the constant map $v' \mapsto w$. So for $g \in \mathfrak{m}_w$ we have $(\pi_2 \circ \sigma)^*(g) = 0$. Hence $\mathrm{d}_v(\pi_2 \circ \sigma)(\mu) = 0$, and therefore $\mathrm{d}_{(v,w)}\pi_2(\lambda) = 0$. We see that λ is mapped to $(\mu, 0)$. Similarly, any $(0, \nu)$, for $\nu \in T_w$, lies in the image of the map. So it is surjective as well. □

1.3 Dimension

In this section we look at the concept of dimension of an irreducible closed set. This is defined in terms of the field of rational functions.

Let $X \subset \mathbb{A}^n$ be an irreducible closed set. Then $\mathcal{I}(X)$ is a prime ideal, and hence the algebra $K[X] = K[x_1, \ldots, x_n]/\mathcal{I}(X)$ does not have zero divisors. So we can form its field of fractions

$$K(X) = \left\{ \tfrac{f}{g} \mid f, g \in K[X], g \neq 0 \right\},$$

which is called the *field of rational functions on* X.

Let $\frac{f}{g}, \frac{f_1}{g_1} \in K(X)$. By definition these are equal if and only if $fg_1 - f_1 g = 0$ (that is, equal to the function that takes the value 0 in all points of X). For example, $\frac{1-y}{x} = \frac{x}{1+y}$ on the circle defined by $x^2 + y^2 = 1$.

Let $h \in K(X)$, and $v \in X$. Then h is said to be defined in v if there are $f, g \in K[X]$ with $h = \frac{f}{g}$ and $g(v) \neq 0$. In that case $h(v) = \frac{f(v)}{g(v)}$. The example above shows that for different v we may have to choose different f, g.

Let $f_1, \ldots, f_r$ be generators of $\mathcal{I}(X)$. Let $v \in X$, then by Proposition 1.2.6, the tangent space $T_v(X)$ can be identified with the subspace of K^n consisting of all $(a_1, \ldots, a_n)$ such that $\sum_{i=1}^n \frac{\partial f_j}{\partial x_i}(v) a_i = 0$ for $1 \leq j \leq r$. Define $h_{ij} = \frac{\partial f_j}{\partial x_i}$, and let $\bar{h}_{ij} \in K[X]$ be its restriction to X. Let M_X be the $r \times n$ matrix with $M_X(i,j) = \bar{h}_{ji}$. Then M_X is a matrix with coefficients in the field $K(X)$. (It depends on the choice of generators of $\mathcal{I}(X)$; however what we will do in the sequel will be independent of that choice.) By ρ_X we denote its rank.

In order to formulate the next theorem, we need some concepts from the theory of fields. Let $E \supset F$ be fields. A *transcendence basis of* E *over* F is a maximal (with respect to inclusion) set of elements of E that is algebraically independent over F. It is known that every transcendence basis has the same cardinality ([Hun80], Chapter VI, Theorem 1.9). The size of a transcendence basis of E over F is called the *transcendence degree of* E *over* F, and denoted $\mathrm{trdeg}(E/F)$.

A *derivation* of E is a map $\delta : E \to E$ with $\delta(a+b) = \delta(a) + \delta(b)$ and $\delta(ab) = \delta(a)b + a\delta(b)$ for all $a, b \in E$. If $\delta(a) = 0$ for all $a \in F$ then we say that δ is an F-*derivation.* The set of all F-derivations of E is denoted $\mathrm{Der}_F(E)$. We note that this is a vector space over E.

Theorem 1.3.1 *Let the notation be as above. Set* $s = n - \rho_X$.

(i) *For* $v \in X$ *we have* $\dim T_v(X) \geq s$.

(ii) *Let* $U = \{v \in X \mid \dim T_v(X) = s\}$, *then* U *is non-empty and open in* X.

(iii) $s = \dim_{K(X)} \mathrm{Der}_K(K(X))$.

(iv) s *is equal to the transcendence degree of* $K(X)$.

Proof. For $v \in X$ let $M_X(v)$ denote the matrix with entries $\bar{h}_{ji}(v)$. Then $\dim T_v(X) = n - \mathrm{rank}(M_X(v))$. Since $\mathrm{rank}(M_X(v)) \leq \rho_X$, we have proved (i).

Let $d_1, \ldots, d_t$ be the determinants of all $\rho_X \times \rho_X$-submatrices of M_X. These are elements of $K[X]$, and $\dim T_v(X) > s$ if and only if $d_i(v) = 0$ for $1 \leq i \leq t$. So $C = \{v \in X \mid \dim T_v(X) > s\}$ is closed in X. We cannot have $C = X$ as otherwise all d_i are zero, contradicting $\mathrm{rank}(M_X) = \rho_X$. As $U = X \setminus C$, (ii) also holds.

Note that $K(X) = K(\bar{x}_1, \dots, \bar{x}_n)$. Let δ_0 be the derivation of K mapping everything to 0. Then by [Coh91], §5.4, Theorem 4.2, there is a unique derivation δ of $K(X)$, extending δ_0, with $\delta(\bar{x}_i) = u_i$, where $u_i \in K(X)$, if and only the u_i satisfy $\sum_{i=1}^n \frac{\partial f_j}{\partial x_i}(\bar{x}_1, \dots, \bar{x}_n)u_i = 0$ for $1 \leq j \leq r$. In other words, there is a bijection between $\mathrm{Der}_K K(X)$ and the space consisting of all $u = (u_1, \dots, u_n) \in K(X)^n$ such that $M_X u^T = 0$. The latter space has dimension s over $K(X)$; hence so has the former space.

For the final assertion we first show that $K(X)$ is separably generated over K. This means that there is a transcendence basis $t_1, \dots, t_r$ of $K(X)$ over K such that the algebraic extension $K(X) \supset K(t_1, \dots, t_r)$ is separable. This only presents a problem if the characteristic is $p > 0$. Let $u_1, \dots, u_s \in K(X)$ be linearly independent over K. Let $\alpha_1, \dots, \alpha_s \in K$ be such that $\sum_i \alpha_i u_i^p = 0$. Since K is algebraically closed there are $\beta_i \in K$ with $\beta_i^p = \alpha_i$. So $(\sum_i \beta_i u_i)^p = 0$. From this it follows that all β_i are zero, so the same holds for the α_i. [Lan02], Chapter VIII, Proposition 4.1 implies that $K(X)$ is separably generated over K. [Lan02], Chapter VIII, Proposition 5.5 says that the dimension of $\mathrm{Der}_K K(X)$ is equal to the transcendence degree of $K(X)$ over K. □

Definition 1.3.2 *Let $X \subset \mathbb{A}^n$ be an irreducible closed set. Then the* dimension *of X is defined as the transcendence degree of $K(X)$ (over K).*

Let s be as in Theorem 1.3.1. A point $v \in X$ is called *singular* if $\dim T_v(X) > s$, otherwise it is non-singular. By the same theorem, the set of singular points is closed in X.

Example 1.3.3 Let the notation be as in Examples 1.1.1, 1.1.4 and 1.2.7. The dimension of $R_1(K)$ is 4, and there is one singular point consisting of the matrix with all coefficients 0.

Theorem 1.3.4 *Let $X \subset Y \subset \mathbb{A}^n$ be two irreducible closed sets. Then $\dim X \leq \dim Y$. If $\dim X = \dim Y$ then $X = Y$.*

Proof. Write $d = \dim X$. Let $\rho : K[Y] \to K[X]$ be the restriction map. The ring $K[X]$ contains a transcendence basis $f_1, \dots, f_d$ over K of the field $K(X)$. Let $g_1, \dots, g_d \in K[Y]$ be such that $\rho(g_i) = f_i$, $1 \leq i \leq d$. Let $p \in K[X_1, \dots, X_d]$, then $\rho(p(g_1, \dots, g_d)) = p(f_1, \dots, f_d)$. This implies that $g_1, \dots, g_d$ are algebraically independent over K, whence $\dim Y \geq d$.

Now suppose that $\dim Y = d$ as well. Let $R = K[g_1, \dots, g_d]$, then by the above, ρ maps R injectively into $K[X]$. Let $h \in K[Y]$ be such that $\rho(h)$ is zero. Since $g_1, \dots, g_d$ are algebraically independent over K, they form a transcendence basis of $K(Y)$. Let $E \subset K(Y)$ be the field of fractions of R. It follows that h is algebraic over E. Let $q \in E[T]$ be its minimal polynomial. There is a non-zero $a \in R$ such that aq has coefficients in R. Let $\rho(aq)$ denote the polynomial in $K[X][T]$ obtained by applying ρ to the coefficients of aq. Then $0 = \rho(aq(h)) = \rho(aq)(\rho(h)) = \rho(aq)(0)$. Therefore, as ρ is injective on R, $aq(0) = 0$ as well. But $aq(0) = a(q(0))$; thus $q(0) = 0$. Since $q \in E[T]$ is irreducible, we infer that $q = T$ and consequently $h = 0$. We conclude that $\mathcal{I}(X) = \mathcal{I}(Y)$, whence $X = Y$. □

1.3.1 Specializations and generic points

Here we define the notion of generic point, which we occasionally use in connection with algebraic groups.

Let $X \subset \mathbb{A}^n$ be a closed set. Let $K' \supset K$ be a field extension. Let $f_1, \dots, f_r$ generate $\mathcal{I}(X)$ and $v = (v_1, \dots, v_n)$ be a point with coordinates in K'. If $f_i(v) = 0$ for $1 \leq i \leq r$, then we say that v is a *generalized point* of X, and we write $v \in X(K')$. For such a point $K[v]$ will denote the subring of K' consisting of all $p(v)$ for $p \in K[x_1, \dots, x_n]$. This ring has no zero divisors (as subring of a field), and hence we can form its field of fractions denoted $K(v)$.

Let $v' \in X(K')$ be a generalized point of X, and let $v \in X$. Then we say that v is a *specialization* of v' if for all $p \in K[x_1, \dots, x_n]$ we have that $p(v') = 0$ implies that $p(v) = 0$. In that case we have a surjective ring homomorphism $\phi : K[v'] \to K$ given by $\phi(p(v')) = p(v)$. (Note that all elements of $K[v']$ can be written $p(v')$ for a polynomial p.) This map is well defined. Indeed, suppose that $p_1(v') = p_2(v')$. Then $(p_1 - p_2)(v') = 0$, and hence $(p_1 - p_2)(v) = 0$, yielding $\phi(p_1(v')) = \phi(p_2(v'))$.

A generalized point $v' \in X(K')$ is called a *generic point* of X if all $v \in X$ are specializations of v', or equivalently, if for $p \in K[x_1, \dots, x_n]$ we have $p(v') = 0$ if and only if $p \in \mathcal{I}(X)$.

Proposition 1.3.5 *Let X be an irreducible closed set. Suppose that X has a generic point $v' \in X(K')$. Then $K(X) \cong K(v')$.*

Proof. Let $h \in K(X)$ and let $f, g \in K[X]$ be such that $h = \frac{f}{g}$. Then $g(v') \neq 0$ (otherwise $g = 0$). So we can define a homomorphism $\psi : K(X) \to K(v')$ by $\psi(h) = h(v')$. This is clearly surjective. Furthermore, $\psi(h) = 0$ forces $f(v') = 0$ implying $f = 0$ and therefore $h = 0$. So ψ is injective as well. □

In particular, the dimension of X equals the transcendence degree of $K(v')$.

Example 1.3.6 Let the notation be as in Examples 1.1.1 and 1.1.4. Let $L = K(u, v, w, \alpha)$ be the rational function field over K in four indeterminates. Set

$$a_0 = \begin{pmatrix} u & v & w \\ \alpha u & \alpha v & \alpha w \end{pmatrix}.$$

Then $a_0 \in R_1(L)$ is a generic point of $R_1(K)$. It follows that the dimension of $R_1(K)$ is 4, in accordance with Example 1.3.3. We can also conclude that $R_1(K)$ is irreducible, as only irreducible closed sets can have generic points.

1.4 Dominant maps

Let $X \subset \mathbb{A}^n$, $Y \subset \mathbb{A}^m$ be closed sets. A regular map $h : X \to Y$ is called *dominant* if $h(X)$ is dense in Y.

Lemma 1.4.1 *Let the notation be as above. The regular map h is dominant if and only if its comorphism $h^* : K[Y] \to K[X]$ is injective.*

Proof. Suppose that h is dominant and $f \in K[Y]$. If $h^*(f) = 0$ then $f(w) = 0$ for all $w \in h(X)$. Since $h(X)$ is dense in Y this implies that $f = 0$. Conversely, if $h(X)$ is not dense in Y then $\overline{h(X)} \neq Y$, so there is an $f \in K[Y]$, $f \neq 0$ such that $f(w) = 0$ for all $w \in h(X)$. This means that $h^*(f) = 0$ and h^* is not injective. □

Proposition 1.4.2 *Let $h : X \to Y$ be as above. Assume that X, Y are irreducible. Suppose that there is a non-singular $u \in X$ such that $h(u)$ is non-singular in Y, and the differential $\mathrm{d}_u h : T_u(X) \to T_{h(u)}(Y)$ is surjective. Then h is dominant.*

Proof. Here we prove this for the special case where $X = \mathbb{A}^n$, $Y = \mathbb{A}^m$. A treatment of the more general case is contained in [GW09], Theorem A.3.4 and [Spr98], Theorem 4.3.6.

Let $K[x_1, \dots, x_n]$, $K[y_1, \dots, y_m]$ be the coordinate rings of $\mathbb{A}^n$, $\mathbb{A}^m$ respectively. Let $h_1, \dots, h_m \in K[x_1, \dots, x_n]$ be such that $h(v) = (h_1(v), \dots, h_m(v))$, for $v \in \mathbb{A}^n$. By (1.2) the j-th coordinate of $\mathrm{d}_u h(v)$ is $\sum_{i=1}^n \frac{\partial h_j}{\partial x_i}(u) v_i$. So from our hypothesis it follows that the $n \times m$-matrix $(\frac{\partial h_j}{\partial x_i})$ (which has coefficients in the function field $K(x_1, \dots, x_n)$) has rank m.

Suppose that h^* is not injective. Then there is an $f \in K[y_1, \dots, y_m]$ such that $f \neq 0$ but $h^*(f) = 0$. We take such an f of smallest degree. Then $g = f(h_1, \dots, h_m)$ is the zero polynomial in $K[x_1, \dots, x_n]$. Hence

$$0 = \frac{\partial g}{\partial x_i} = \sum_{j=1}^{m} \frac{\partial f}{\partial y_j}(h_1, \dots, h_m) \frac{\partial h_j}{\partial x_i}.$$

Since the matrix $(\frac{\partial h_j}{\partial x_i})$ has rank m, this imples that $\frac{\partial f}{\partial y_j}(h_1, \dots, h_m) = 0$ for $1 \leq j \leq m$. By the hypothesis on the degree of f it follows that $\frac{\partial f}{\partial y_j} = 0$ for all j. If the characteristic is 0, it follows that f is a non-zero constant, which is not possible. If the characteristic is $p > 0$, the conclusion is that f is a polynomial in $y_1^p, \dots, y_m^p$. That means that $f = f_0^p$, for some $f_0 \in K[y_1, \dots, y_m]$. But then $h^*(f_0) = 0$, contradicting the assumption on the degree of f. We see that h^* is injective, and we conclude by Lemma 1.4.1. □

Theorem 1.4.3 *Suppose that K is of characteristic zero. Let $X \subset \mathbb{A}^n$, $Y \subset \mathbb{A}^m$ be irreducible closed sets, and $h : X \to Y$ a dominant regular map. Then there is a non-empty open subset $U \subset X$ such that for all $u \in U$ we have the following: u, $h(u)$ are non-singular points of X and Y respectively, and $\mathrm{d}_u h : T_u(X) \to T_{h(u)}(Y)$ is surjective.*

Proof. Write $K[\mathbb{A}^n] = K[x_1, \ldots, x_n]$ and let $\bar{x}_i$ be the restriction of x_i to X. So $K[X] = K[\bar{x}_1, \ldots, \bar{x}_n]$.

We start with a construction regarding derivations of $K[X]$. Let δ be a K-derivation of $K[X]$ (i.e., a K-linear map which is a derivation). Define a regular map $\tau(\delta) : X \to \mathbb{A}^n$ by $\tau(\delta)(v) = (\delta(\bar{x}_1)(v), \ldots, \delta(\bar{x}_n)(v))$. Let $f \in K[x_1, \ldots, x_n]$ and $\bar{f} \in K[X]$ its restriction to X. Then $f : \mathbb{A}^n \to \mathbb{A}^1$ is a regular map, so for a $v \in \mathbb{A}^n$ we can consider its differential $\mathrm{d}_v f : \mathbb{A}^n \to \mathbb{A}^1$. Let $v \in X$; then by (1.2) $\mathrm{d}_v f(\tau(\delta)(v)) = \sum_{i=1}^n \frac{\partial f}{\partial x_i}(v)\delta(\bar{x}_i)(v)$. Since $\delta(\bar{f}) = \sum_{i=1}^n \frac{\partial f}{\partial \bar{x}_i}\delta(\bar{x}_i)$, we infer

$$\mathrm{d}_v f(\tau(\delta)(v)) = \delta(\bar{f})(v). \tag{1.4}$$

In particular, if $f \in \mathcal{I}(X)$ then $\mathrm{d}_v f(\tau(\delta)(v)) = 0$. So by Proposition 1.2.6, $\tau(\delta)(v) \in T_v(X)$.

By Lemma 1.4.1, $h^* : K[Y] \to K[X]$ is injective. Let R be the image of h^*; then R is a subring of $K[X]$, isomorphic to $K[Y]$. Let E be the field of fractions of R, then $E \subset K(X)$. Write $d = \dim X$, $e = \dim Y$. By Theorem 1.3.1 we have $\mathrm{trdeg}(E/K) = e$, $\mathrm{trdeg}(K(X)/K) = d$ and hence $\mathrm{trdeg}(K(X)/E) = d - e$. Write $\mathcal{D} = \mathrm{Der}_K(K(X))$, $\mathcal{E} = \mathrm{Der}_E(K(X))$; then $\mathcal{E} \subset \mathcal{D}$. By Theorem 1.3.1 we have $\dim_{K(X)} \mathcal{D} = d$. Since the characteristic is assumed to be 0, $K(X)$ is automatically separably generated over E, and therefore, [Lan02], Chapter VIII, Proposition 5.5, says that $\dim_{K(X)} \mathcal{E} = d - e$.

Let $\delta_1', \ldots, \delta_d'$ be a basis of $\mathcal{D}$ containing a basis $\delta_{e+1}', \ldots, \delta_d'$ of $\mathcal{E}$. There is a non-zero $f_0 \in K[X]$ such that $f_0\delta_i'(\bar{x}_j) \in K[X]$ for all i, j. Set $\delta_i = f_0\delta_i'$ for $1 \le i \le d$. Then the δ_i are derivations of $K(X)$, mapping $K[X]$ into itself. Furthermore, $\delta_1, \ldots, \delta_d$, $\delta_{e+1}, \ldots, \delta_d$ are bases of $\mathcal{D}$ and $\mathcal{E}$ respectively. For $1 \le i \le d$ and $v \in X$ define $\tau_i(v) = \tau(\delta_i)(v)$. As seen above, $\tau_i(v) \in T_v(X)$. So $\mathrm{d}_v h(\tau_i(v)) \in T_{h(v)}(Y)$.

Let $h_1, \ldots, h_m \in K[x_1, \ldots, x_n]$ be such that $h(v) = (h_1(v), \ldots, h_m(v))$, for $v \in X$. Define $A_{ij} : X \to K$ by $A_{ij}(v) = \sum_{i=1}^n \frac{\partial h_j}{\partial x_i}(v)\delta_i(\bar{x}_i)(v)$. Then $A_{ij} \in K[X]$ and $\mathrm{d}_v h(\tau_i(v)) = (A_{i1}(v), \ldots, A_{im}(v))$. Define the $e \times m$ matrix A, with coefficients in $K(X)$, by $A(i,j) = A_{ij}$, for $1 \le i \le e$, $1 \le j \le m$. So, if $A(v)$ denotes the matrix $(A_{ij}(v))$, then the rows of $A(v)$ lie in $T_{h(v)}(Y)$, for $v \in X$.

We claim that the rank of A is e. If not, there are $u_1, \ldots, u_e \in K(X)$, not all zero, such that $\sum_{i=1}^e u_i A_{ij} = 0$ for $1 \le j \le m$. After possibly multiplying all u_i by the same non-zero element of $K[X]$, we have $u_i \in K[X]$ for all i. Then $\delta = \sum_{i=1}^e u_i\delta_i$ is a K-derivation of $K[X]$. Let $q \in K[\mathbb{A}^m]$, and write $\bar{q}$ for its restriction to Y. Then $\bar{q} \circ h : X \to \mathbb{A}^1$ is a regular map and $\mathrm{d}_v(\bar{q} \circ h) = \mathrm{d}_{h(v)}\bar{q} \circ \mathrm{d}_v h$ by Lemma 1.2.3. Consider the map $\tau(\delta) : X \to \mathbb{A}^n$. We have

$\tau(\delta)(v) = \sum_{i=1}^{e} u_i(v)\tau_i(v)$. Then $\mathrm{d}_v h(\tau(\delta)(v)) = \sum_{i=1}^{e} u_i(v)\mathrm{d}_v h(\tau_i(v))$. But the j-th coordinate of this is $\left(\sum_{i=1}^{e} u_i A_{ij}\right)(v)$, which is 0. We conclude that $\mathrm{d}_v(\bar{q}\circ h)(\tau(\delta)(v)) = 0$, for all $v \in X$. So by (1.4), $\delta(\bar{q}\circ h) = 0$. Now $\bar{q}\circ h = h^*(\bar{q})$. Therefore, δ maps R and E to zero. Hence δ is contained in $\mathcal{E}$. But that is not possible, as δ is a linear combination of $\delta_1, \dots, \delta_e$. So the claim is proved.

Let $U_1 \subset X$ consist of all $v \in X$ such that the rank of $A(v)$ is e. Then U_1 is open in X and non-empty (by a similar argument as in the proof of Theorem 1.3.1(ii)). Let $S \subset Y$ be the set of non-singular points. Then S is non-empty and open in Y by Theorem 1.3.1. So $U_2 = h^{-1}(S)$ is non-empty and open in X. Let U_3 be the set of non-singular points of X; then U_3 is non-empty and open in X. Set $U = U_1 \cap U_2 \cap U_3$; then U is non-empty and open in X. Let $u \in U$. Since $u \in U_1$ we have that $\dim \mathrm{d}_u h(T_u(X)) \geq e$. As $u \in U_2$, $\dim T_{h(u)}(Y) = e$. So $\mathrm{d}_u h(T_u(X)) = T_{h(u)}(Y)$. □

Next we show that for a dominant map $h : X \to Y$ the image $h(X)$ is not only dense in Y, but even contains an open set of Y.

Theorem 1.4.4 *Let $X \subset \mathbb{A}^n$, $Y \subset \mathbb{A}^m$ be irreducible closed sets, and $h : X \to Y$ a dominant regular map. Let $U \subset X$ be non-empty and open in X. Then $h(U)$ contains a non-empty open set of Y.*

Proof. First suppose that there is an $f \in K[X]$, $f \neq 0$, such that $U = \{v \in X \mid f(v) \neq 0\}$. Let $A = h^*(K[Y])$, which by Lemma 1.4.1 is isomorphic to $K[Y]$. Write $K[\mathbb{A}^n] = K[x_1, \dots, x_n]$, and let $\bar{x}_i$ be the restriction of x_i to X. Write $B = K[X]$, then $B = A[\bar{x}_1, \dots, \bar{x}_n]$. Set $L = K(X)$; then we can apply [Jac79], Chapter IX, Theorem 2, from which it follows that there is a non-zero $a \in A$ such that every homomorphism $A \to K$ that does not map a to zero, can be extended to a homomorphism $B \to K$, which does not map f to zero. Set $g = (h^*)^{-1}(a)$.

Set $W = \{w \in Y \mid g(w) \neq 0\}$. We show that $h(U)$ contains W. Let $w \in W$. Sending $c \in A$ to $(h^*)^{-1}(c)(w)$ is a homomorphism $\sigma : A \to K$, with $\sigma(a) \neq 0$. Let $\phi : K[X] \to K$ be an extension of σ, such that $\phi(f) \neq 0$. Define $v \in \mathbb{A}^n$ by $v_j = \phi(\bar{x}_j)$, $1 \leq j \leq n$. Let $p \in K[\mathbb{A}^n]$, and let $\bar{p}$ be its restriction to X. Then

$$p(v) = p(\phi(\bar{x}_1), \dots, \phi(\bar{x}_n)) = \phi(p(\bar{x}_1, \dots, \bar{x}_n)) = \phi(\bar{p}). \tag{1.5}$$

It follows that $p(v) = 0$ for all p such that $\bar{p} = 0$, and thus $v \in X$. Also, since $\phi(f) \neq 0$ we have $f(v) \neq 0$, whence $v \in U$. Let $q \in K[Y]$ then again by (1.5), $q(h(v)) = h^*(q)(v) = \phi(h^*(q)) = \sigma(h^*(q)) = q(w)$. This implies that $h(v) = w$ and we are done in this case.

In general an open set U is the union of a finite number of sets of the form $\{v \in X \mid f(v) \neq 0\}$. So the theorem follows directly from the special case treated above. □

1.5 Gröbner bases

In this section we give a very short introduction into the theory of Gröbner bases and the algorithm to compute them (much more extensive treatments can be found in, for example, [CLO15], [BW93]). Gröbner bases provide computational tools to solve various problems regarding ideals in polynomial rings. Their main selling point is that they can be used to solve an amazing range of problems for which no other general algorithms are known. Their main drawback is that in practice they can be difficult to compute.

Every discussion concerning Gröbner bases starts by defining an order on the monomials and keeping that fixed throughout. Set $\mathbb{N} = \mathbb{Z}_{\geq 0}$; we use the convention that for $\alpha \in \mathbb{N}^n$, $\alpha_1, \ldots, \alpha_n$ denote its coordinates, i.e., $\alpha = (\alpha_1, \ldots, \alpha_n)$. A total order $<$ on $\mathbb{N}^n$ is said to be a *monomial order* if

$$\alpha < \beta \text{ implies } \alpha + \gamma < \beta + \gamma \text{ for all } \gamma \in \mathbb{N}^n, \tag{1.6}$$

$$\text{every subset of } \mathbb{N}^n \text{ has a smallest element.} \tag{1.7}$$

Example 1.5.1 The *lexicographical order* is defined as follows: $\alpha <_{\text{lex}} \beta$ if $\alpha_i < \beta_i$ where i is the smallest index such that $\alpha_i \neq \beta_i$. By induction on n it is shown that $<_{\text{lex}}$ is a monomial order.

The *degree lexicographical order* is defined by looking first at the degree, and when the degrees are equal at the lexicographical order. The degree of $\alpha \in \mathbb{N}^n$ is the number $|\alpha| = \alpha_1 + \cdots + \alpha_n$. So the order $<_{\text{dlex}}$ is defined as follows: $\alpha <_{\text{dlex}} \beta$ if $|\alpha| < |\beta|$ or $|\alpha| = |\beta|$ and $\alpha <_{\text{lex}} \beta$. Since there is a finite number of monomials of a given degree, $<_{\text{dlex}}$ is obviously a monomial order.

A monomial order $<$ with the property that $|\alpha| < |\beta|$ implies $\alpha < \beta$ is *degree compatible.*

Let k be an arbitrary field and consider the polynomial ring $R = k[x_1, \ldots, x_n]$. For $\alpha = (\alpha_1, \ldots, \alpha_n) \in \mathbb{N}^n$ we set $x^\alpha = x_1^{\alpha_1} \cdots x_n^{\alpha_n}$; these are the monomials of R.

Let $<$ be a monomial order. We use it to order the monomials of R by $x^\alpha < x^\beta$ if and only if $\alpha < \beta$. Then (1.6) amounts to *multiplicativity*, i.e., $x^\alpha < x^\beta$ implies $x^\alpha x^\gamma < x^\beta x^\gamma$. Furthermore, (1.7) is equivalent to the *descending chain condition*: there is no infinite descending chain of monomials.

From now on we fix a monomial order $<$. Let $f \in R$. Then the largest monomial occurring in f, with respect to $<$, is called the *leading monomial* of f, and denoted $\mathrm{LM}(f)$. A monomial x^α is said to be a *factor* of a monomial x^β if $x^\beta = x^\alpha x^\gamma$ for some $\gamma \in \mathbb{N}^n$. Let $G \subset R$, and $f \in R$ be such that there is no $g \in G$ such that $\mathrm{LM}(g)$ is a factor of a monomial occurring in f. Then f is said to be in *normal form* modulo G. Now we describe the *reduction algorithm.*

Algorithm 1.5.2 Input: *a finite set $G \subset R$, and $f \in R$.*
Output: *an $f' \in R$ such that f' is in normal form modulo G, and $f - f'$ lies*

in the ideal generated by G.

1. *Set $f' := 0$, $h := f$.*
2. *If $h = 0$, return f', otherwise go to step 3.*
3. *Set $x^\alpha = \mathrm{LM}(h)$, and let λ be its coefficient in h. If there is a $g \in G$ such that $\mathrm{LM}(g)$ is a factor of x^α, let x^γ be such that $x^\gamma \mathrm{LM}(g) = x^\alpha$, and set $h := h - \frac{\lambda}{\mu} x^\gamma g$, where μ is the coefficient of $\mathrm{LM}(g)$ in g. Otherwise set $f' := f' + \lambda x^\alpha$, and $h := h - \lambda x^\alpha$. Go to step 2.*

It is clear that this algorithm terminates as $\mathrm{LM}(h)$ decreases every step, and $<$ satisfies the descending chain condition. Moreover, throughout the algorithm $f - (f' + h)$ lies in the ideal generated by G. Therefore the correct output is returned.

Unfortunately the output of the algorithm is not unique. It can happen that more than one $\mathrm{LM}(g)$ for $g \in G$ is a factor of x^α in step 3. Then a choice on which the output can depend has to be made. In the sequel we say that $f \in R$ *reduces to zero modulo a* $G \subset R$ if there is a way of making the choices in step 3 of Algorithm 1.5.2, so that, on inputs G and f, it returns 0.

Definition 1.5.3 *Let $I \subset R$ be an ideal. A set $G \subset I$ is called a* Gröbner basis *of I if for all $f \in I$ there is a $g \in G$ such that* $\mathrm{LM}(g)$ *is a factor of* $\mathrm{LM}(f)$*.*

An ideal I of R always has a finite Gröbner basis. In order to see this one uses Dickson's lemma, stating that every set of monomials has a finite subset generating the same ideal. So the ideal generated by $\mathrm{LM}(f)$ for $f \in I$, is also generated by a finite set $\{\mathrm{LM}(g_1), \dots, \mathrm{LM}(g_s)\}$, where $g_i \in I$. Then $\{g_1, \dots, g_s\}$ is a Gröbner basis of I.

It is straightforward to see that if G is a Gröbner basis, the output of Algorithm 1.5.2 is unique, and moreover, it is 0 if and only if f lies in the ideal generated by G.

Moreover, given a finite generating set of I, there is an algorithm to compute a Gröbner basis of I called the *Buchberger algorithm.* To describe it we need the notion of S-polynomial. Let $f \neq g$ be two polynomials, x^α, x^β be the unique elements of lowest degree such that $\mathrm{LM}(x^\alpha f) = \mathrm{LM}(x^\beta g)$ and λ, μ be the coefficients of respectively $\mathrm{LM}(g)$ in g and $\mathrm{LM}(f)$ in f. Then $S(f,g) = \lambda x^\alpha f - \mu x^\beta g$ is the S-polynomial of f, g. The next theorem is called Buchberger's criterion.

Theorem 1.5.4 *Let $G \subset R$ generate the ideal I. Then G is a Gröbner basis of I if and only if $S(g_1, g_2)$ reduces to 0 modulo G, for all $g_1, g_2 \in G$.*

Proof.(Sketch) Let $f \in I$. We must show that there is a $g \in G$ such that $\mathrm{LM}(g)$ divides $\mathrm{LM}(f)$. There are $g_i \in G$ such that $f = c_1 x^{\alpha_1} g_1 + \cdots + c_s x^{\alpha_s} g_s$, for certain $c_i \in k$, $\alpha_i \in \mathbb{N}^n$, $g_i \in G$. Write $M_i = \mathrm{LM}(x^{\alpha_i} g_i)$, and assume that $M_1 = \cdots = M_r > M_{r+1} \geq \cdots \geq M_s$. Among all expressions of this form take one where M_1 is minimal, and among all expressions with that M_1, an expression where r is minimal. Suppose that $r > 1$ and write

$$f = c_1(x^{\alpha_1} g_1 - x^{\alpha_2} g_2) + (c_1 + c_2)x^{\alpha_2} g_2 + \cdots + x^{\alpha_s} g_s.$$

Since the leading monomials of $x^{\alpha_1} g_1$ and $x^{\alpha_2} g_2$ are equal, the expression between brackets is divisible by an S-polynomial. By reducing this S-polynomial modulo G we can rewrite $x^{\alpha_1} g_1 - x^{\alpha_2} g_2$ so that we obtain a similar expression for f, but with r reduced, or in case $r = 2$ and $c_1 + c_2 = 0$, with M_1 reduced. Both are impossible, so we conclude that $r = 1$ in the original expression. Therefore, $\mathrm{LM}(g_1)$ divides $\mathrm{LM}(f)$. □

The algorithm for computing a Gröbner basis starts with a finite set G of generators of I. If there are $g_1, g_2 \in G$ such that a normal form of $S(g_1, g_2)$ modulo G (computed with Algorithm 1.5.2) is not zero, then we add this normal form to G. This process is repeated until all normal forms of $S(g_1, g_2)$ for $g_1, g_2 \in G$ reduce to zero modulo G. This algorithm terminates because the ideal generated by $\mathrm{LM}(g)$, for $g \in G$, increases every step (and as a consequence of Theorem 1.1.3, there are no infinite strictly increasing chains of ideals).

Definition 1.5.5 *Let G be a Gröbner basis of the ideal I. Then G is said to be* reduced *if the coefficient of* $\mathrm{LM}(g)$ *is 1 for all $g \in G$, and no monomial in a $g \in G$ is divisible by a monomial in* $\mathrm{LM}(G \setminus \{g\})$.

Theorem 1.5.6 *Every ideal has a unique reduced Gröbner basis.*

Proof.(Sketch.) First one shows that an ideal has a reduced Gröber basis by starting with any Gröbner basis G' and reducing $g \in G'$ modulo $G' \setminus \{g\}$, until no further reductions are possible. Next, if $G, \widetilde{G}$ are two reduced Gröbner bases of the same ideal, then one shows that $\mathrm{LM}(G) = \mathrm{LM}(\widetilde{G})$. Let $g \in G$, $\tilde{g} \in \widetilde{G}$ have the same leading monomial. In $h = g - \tilde{g}$ this leading monomial cancels. On the other hand, no monomial in h can be divisible by a monomial in $\mathrm{LM}(G) = \mathrm{LM}(\widetilde{G})$. Therefore, $h = 0$. □

Remark 1.5.7 We note that only arithmetical operations are needed to compute a Gröbner basis. So if the generators of an ideal all have coefficients in a field $k' \subset k$, the elements of the Gröbner basis computed from the given generating set have coefficients in k'.

Remark 1.5.8 There are some bounds on the complexity of the algorithms for computing a Gröbner basis, indicating that in the worst case the complexity can be as bad as doubly exponential (see for example, [BW93], Appendix).

In view of the wide range of applications of Gröbner bases, a lot of effort has gone into producing computer implementations that are as efficient as possible. Although tremendous progress has been made, everyone who works with Gröbner bases has the experience of relatively small sets of polynomials leading to huge Gröbner bases, which occasionally are difficult or even impossible to compute.

1.6 Elimination

Let $<$ be a monomial order with the following property: any monomial having at least one of $x_1, \dots, x_l$ with a positive exponent is larger than all monomials in $k[x_{l+1}, \dots, x_n]$. Then $<$ is said to be of *l-elimination type.* An example is the lexicographical order, which is of l-elimination type for all l between 1 and $n-1$. Gröbner bases with respect to such monomial orders have the following important property.

Lemma 1.6.1 *Let G be a Gröbner basis of the ideal $I \subset k[x_1, \dots, x_n]$, with respect to a monomial order of l-elimination type. Then $G \cap k[x_{l+1}, \dots, x_n]$ is a Gröbner basis of $I \cap k[x_{l+1}, \dots, x_n]$.*

Proof. Let $f \in I \cap k[x_{l+1}, \dots, x_n]$. Then $f \in I$ and hence there is $g \in G$ such that $\mathrm{LM}(g)$ is a factor of $\mathrm{LM}(f)$. This implies that $g \in k[x_{l+1}, \dots, x_n]$. □

This result has a few applications. The first that we discuss is an algorithm to compute the closure of the image of a projection. For this we let the ground field be denoted K, which, as usual, is assumed to be algebraically closed. Let $\pi_l : \mathbb{A}^n \to \mathbb{A}^{n-l}$ be given by $\pi_l(v_1, \dots, v_n) = (v_{l+1}, \dots, v_n)$. Let $X = \mathcal{V}(I) \subset \mathbb{A}^n$ be a closed set. Then the image $\pi_l(X) \subset \mathbb{A}^{n-l}$ is not necessarily closed. However, the following holds.

Lemma 1.6.2 *The closure of $\pi_l(X)$ is equal to $\mathcal{V}(I \cap K[x_{l+1}, \dots, x_n])$.*

Proof. Set $I_l = I \cap K[x_{l+1}, \dots, x_n]$. Then the statement of the lemma is equivalent to $\mathcal{V}(I_l) = \mathcal{V}(\mathcal{I}(\pi_l(X)))$. It is clear that $I_l \subset \mathcal{I}(\pi_l(X))$; hence $\mathcal{V}(I_l) \supset \mathcal{V}(\mathcal{I}(\pi_l(X)))$.

Let $f \in K[x_{l+1}, \dots, x_n]$ be zero on $\pi_l(X)$. Then f (seen as element of $K[x_1, \dots, x_n]$) is zero on X. Hence by the Nullstellensatz (Theorem 1.1.2) there is an $m > 0$ with $f^m \in I$. But $f^m \in K[x_{l+1}, \dots, x_n]$. So $f^m \in I_l$. We conclude that $\mathcal{I}(\pi_l(X)) \subset \sqrt{I_l}$. So $\mathcal{V}(I_l) = \mathcal{V}(\sqrt{I_l}) \subset \mathcal{V}(\mathcal{I}(\pi_l(X)))$. □

Lemmas 1.6.1 and 1.6.2 and the algorithm to compute a Gröbner basis of an ideal yield an immediate algorithm to compute polynomials defining the closure of $\pi_l(X)$ (given polynomials defining X). More generally we can use this technique to compute the closure of the image of any regular map. Let $X \subset \mathbb{A}^n$ be a closed set, defined by the polynomials $f_1, \ldots, f_r$. Let $h : X \to \mathbb{A}^m$ be a regular map, given by $h(v) = (h_1(v), \ldots, h_m(v))$, where $h_i \in K[x_1, \ldots, x_n]$. Let $K[y_1, \ldots, y_m]$, $K[x_1, \ldots, x_n, y_1, \ldots, y_m]$ be the coordinate rings of $\mathbb{A}^m$ and $\mathbb{A}^n \times \mathbb{A}^m$ respectively. We also write $K[x, y]$ instead of the latter ring. Then the graph of h,

$$\Gamma = \{(v, h(v)) \in \mathbb{A}^n \times \mathbb{A}^m \mid v \in X\}$$

is a closed set in $\mathbb{A}^n \times \mathbb{A}^m$ defined by the polynomials $\{f_1, \ldots, f_r, y_1 - h_1, \ldots, y_m - h_m\}$. The closure of the image, $\overline{h(X)}$, is the same as the closure of the image of $\pi_n(\Gamma)$ and we can compute it by using Gröbner bases with respect to an elimination order.

Corollary 1.6.3 *Let $X \subset \mathbb{A}^n$ be defined over k. Let $h : X \to \mathbb{A}^m$ be a regular map defined over k. Then $\overline{h(X)}$ is defined over k.*

Proof. Let $f_1, \ldots, f_r \in k[x_1, \ldots, x_n]$ generate $\mathcal{I}(X)$. Let h_j and Γ be as above. Let $J \subset K[x, y]$ be the ideal generated by the f_i and the $y_j - h_j$. Suppose that $g \in K[x, y]$ vanishes on Γ. We have that $g = g(x_1, \ldots, x_n, h_1, \ldots, h_m) + g'$, where $g' \in J$. Now $g(x_1, \ldots, x_n, h_1, \ldots, h_m)$ vanishes on X; hence it lies in the ideal generated by the f_i. Therefore $J = \mathcal{I}(\Gamma)$ and the algorithm for computing defining polynomials of $\overline{h(X)}$ computes a Gröbner basis G of $\mathcal{I}(\Gamma) \cap K[y_1, \ldots, y_m]$. This is a radical ideal because $\mathcal{I}(\Gamma)$ is one. Hence G generates $\mathcal{I}(\overline{h(X)})$. Moreover, by Remark 1.5.7, the elements in G have coefficients in k. □

1.7 Dimension of ideals

Let $I \subset k[x_1, \ldots, x_n]$ be an ideal. A set of indeterminates $\{x_{i_1}, \ldots, x_{i_r}\}$ is said to be *independent* modulo I if $k[x_{i_1}, \ldots, x_{i_r}] \cap I = 0$. The *dimension* of I is defined to be the maximal cardinality of an independent set of indeterminates modulo I.

Lemma 1.7.1 *Let $I \subset k[x_1, \ldots, x_n]$ be an ideal. Then the following are equivalent:*

(i) *I is zero-dimensional,*

(ii) *for $1 \le i \le n$ there is a univariate polynomial $f_i(x_i)$ contained in I,*

(iii) *$\dim(k[x_1, \ldots, x_n]/I)$ is finite.*

Proof. If I is zero-dimensional, then by definition $I \cap k[x_i] \neq 0$, so we get the polynomials $f_i(x_i)$. The reverse implication is immediate. Note that $k[x_1, \dots, x_n]/I$ is spanned by the monomials that are in normal form modulo I. If I contains the polynomials $f_i(x_i)$, the exponent of x_i in a monomial in normal form does not exceed $\deg(f_i)$ and (ii) implies (iii). Finally, if $k[x_1, \dots, x_n]/I$ is finite-dimensional then we let f_i be the minimal polynomial of the image of x_i in this ring. □

The proof also gives a straightforward way to compute the polynomials $f_i(x_i)$. We just calculate the minimal polynomial of the image of x_i in $k[x_1, \dots, x_n]/I$. For this we do not need a basis of this ring; a basis of the subspace consisting of all powers of x_i suffices. In order to compute this basis it is enough to have any Gröbner basis of I.

Lemma 1.7.2 *Let $I \subset K[x_1, \dots, x_n]$ be a prime ideal. Then* $\dim(I) = \dim(\mathcal{V}(I))$.

Proof. (See [BW93], Lemma 7.25 for more details.) Since I is prime, it is radical. Let $X = \mathcal{V}(I)$, then $K[X] = K[x_1, \dots, x_n]/I$. Now let $\{x_{i_1}, \dots, x_{i_r}\}$ be a maximal independent set. Write $\bar{x}_i$ for the image of x_i in $K[X]$. Then $K(X)$ is an algebraic extension of $K(\bar{x}_{i_1}, \dots, \bar{x}_{i_r})$. So the transcendence degree of $K(X)$ is the transcendence degree of this last field, which is r. □

In [BW93], §9.3 the following is proved.

Proposition 1.7.3 *Let $I \subset k[x_1, \dots, x_n]$ be an ideal. Let $<$ be a degree-compatible monomial order, and G a Gröbner basis of I with respect to $<$. Let $\mathcal{T} = \{x_{i_1}, \dots, x_{i_r}\}$ be a set of indeterminates of maximal cardinality such that the leading monomial of every element of G involves at least one indeterminate outside $\mathcal{T}$. Then* $\dim(I) = r$ *and $\mathcal{T}$ is independent modulo I.*

This gives an immediate algorithm to compute the dimension of an ideal, and a maximal independent set. So in view of Lemma 1.7.2, we also have an algorithm to compute the dimension of an irreducible closed set X if generators of $\mathcal{I}(X)$ are known.

Example 1.7.4 Let the notation be as in Example 1.1.4. A set $\mathcal{T}$, as in Proposition 1.7.3 is $\mathcal{T} = \{x_{11}, x_{23}, x_{12}, x_{22}\}$. Again we find that the dimension of $R_1(K)$ is 4 (as in Example 1.3.3).

1.8 Ideal quotients

Let $I \subset k[x_1, \ldots, x_n]$ be an ideal. Let f be a polynomial. Then the ideal $I : f = \{g \in k[x_1, \ldots, x_n] \mid fg \in I\}$ is called the *quotient* of I by f.

Note that $I : f^i \subset I : f^{i+1}$. Since, as a consequence of Theorem 1.1.3, there are no infinite increasing chains of ideals, there is an s with $I : f^s = I : f^t$ for all $t \geq s$. Then $I : f^s = \cup_{i \geq 0} I : f^i$. This last ideal is denoted $I : f^\infty$. In view of Lemma 1.6.1, the next lemma yields a straightforward algorithm to compute it.

Lemma 1.8.1 *Let $J \subset k[x_1, \ldots, x_n, y]$ be the ideal generated by I and $1 - yf$. Then $I : f^\infty = J \cap k[x_1, \ldots, x_n]$.*

Proof. Let $g \in J \cap k[x_1, \ldots, x_n]$. Then $g = q_1p_1 + \cdots + q_sp_s + q_{s+1}(1 - yf)$, where $q_i \in k[x_1, \ldots, x_n, y]$ and $p_i \in I$. In this expression we formally substitute $\frac{1}{f}$ for y and multiply by a high enough power of f to clear denominators. We get $f^e g = \sum_i \tilde{q}_i p_i$, where $\tilde{q}_i \in k[x_1, \ldots, x_n]$, $p_i \in I$. So $g \in I : f^\infty$.

Suppose that $f^e g \in I$. Since $(yf)^e = 1 \bmod J$ we get $g = y^e f^e g \bmod J = 0 \bmod J$. Hence $g \in J$. □

1.9 Radical of an ideal

Let X be an irreducible closed set defined by the polynomials $f_1, \ldots, f_s$. In order to compute a basis of the tangent space $T_v(X)$ at the point $v \in X$, we need a generating set of $\mathcal{I}(X)$ (Proposition 1.2.6). By Theorem 1.1.2, the latter ideal is equal to the radical of the ideal generated by the f_i. So in general, if we want to compute tangent spaces, we need an algorithm to compute the radical of an ideal. This is especially important when dealing with algebraic groups, as the tangent space at the identity is its Lie algebra. In this section we outline an algorithm to compute the radical of an ideal.

Let $f \in k[y]$ and $f = p_1^{k_1} \cdots p_s^{k_s}$ be its factorization into irreducibles. Then the polynomial $p_1 \cdots p_s$ is called the *square-free part* of f. Let $f(x_i) \in k[x_i] \cap I$. It is clear that its square-free part lies in $\sqrt{I}$. The following is Proposition 8.14 of [BW93].

Proposition 1.9.1 *Let k be a perfect field, and $I \subset k[x_1, \ldots, x_n]$ a zero-dimensional ideal. Then I is radical if and only if I contains a square-free polynomial $f_i(x_i) \in k[x_i]$.*

Based on these observations we have an algorithm for computing generators for the radical of a zero-dimensional ideal I, in case the base field is perfect. Initially, S is a generating set of I. For $1 \le i \le n$ compute a $f_i(x_i) \in I$ (see Section 1.7), and add the square-free part of f_i to S. (In characteristic zero this is equal to $f_i/\gcd(f_i, f_i')$.) After this procedure $S \subset \sqrt{I}$. Moreover, by Proposition 1.9.1 S generates a radical ideal and thus generates the radical of I.

In [Kem02], an extension of this algorithm is described for the case where the base field is of the form $L = k(t_1, \dots, t_m)$, where k is a perfect field of characteristic $p > 0$. We refer to the mentioned paper for the details. Note that in the algorithm for computing the radical, given below, it is necessary to compute radicals of zero-dimensional ideals over fields of the form $k(t_1, \dots, t_m)$.

The idea for computing the radical of an ideal of higher dimension is to reduce to the case of zero-dimensional ideals. Let $I \subset k[x_1, \dots, x_n]$ be an ideal of dimension > 0. Let $\mathcal{T} = \{x_{i_1}, \dots, x_{i_r}\}$ be an independent set modulo I of maximal size (see Section 1.7). Let $k' = k(x_{i_1}, \dots, x_{i_r})$ denote the rational function field in the indeterminates of $\mathcal{T}$. Let $\mathcal{T}^c = \{x_1, \dots, x_n\} \setminus \mathcal{T}$ be the complement of $\mathcal{T}$. Then $k'[\mathcal{T}^c]$ will denote the polynomial ring over k' in the indeterminates of $\mathcal{T}^c$. The next results deal with the relation between ideals in $k[x_1, \dots, x_n]$ and $k'[\mathcal{T}^c]$. If we fix a monomial order on $k'[\mathcal{T}^c]$ then the coefficient of the leading monomial of an $h \in k'[\mathcal{T}^c]$ lies in k'. We denote this coefficient by $\mathrm{LC}(h)$. If h happens to lie in $k[x_1, \dots, x_n]$ then $\mathrm{LC}(h) \in k[x_{i_1}, \dots, x_{i_r}]$.

Lemma 1.9.2 *Let $J \subset k'[\mathcal{T}^c]$ be an ideal with Gröbner basis G (with respect to any monomial order). Assume that $G \subset k[x_1, \dots, x_n]$. Let f be the least common multiple of $\mathrm{LC}(g)$ as g runs through G. Then $J \cap k[x_1, \dots, x_n] = I : f^\infty$, where I is the ideal of $k[x_1, \dots, x_n]$ generated by G.*

Proof. Let $g \in I : f^\infty$. Then $f^s g \in I \subset J$. But $f \in k'$; hence $g = \frac{1}{f^s} f^s g \in J \cap k[x_1, \dots, x_n]$.

Let $p \in J \cap k[x_1, \dots, x_n]$. We use induction on $\mathrm{LM}(p)$ to prove $p \in I : f^\infty$. For $p = 0$ this is clear. If $p \neq 0$, then there is a $g \in G$ such that $\mathrm{LM}(g)$ divides $\mathrm{LM}(p)$. Set $p_1 = p - \frac{\mathrm{LC}(p)}{\mathrm{LC}(g)} ug$, where u is a monomial such that $u\mathrm{LM}(g) = \mathrm{LM}(p)$. We multiply this expression by f. Since $\mathrm{LC}(g)$ divides f we get $fp_1 \in J \cap k[x_1, \dots, x_n]$. Also $\mathrm{LM}(fp_1) = \mathrm{LM}(p_1)$ as f lies in k'. Hence by induction $fp_1 \in I : f^\infty$. Furthermore, $f \frac{\mathrm{LC}(p)}{\mathrm{LC}(g)} ug \in I$. So $fp \in I : f^\infty$ and therefore $p \in I : f^\infty$. □

Now we let $<_1$, $<_2$ be any monomial orders on the monomials involving the indeterminates from $\mathcal{T}^c$ and $\mathcal{T}$ respectively. For a monomial x^α let α' be defined by $\alpha_i' = \alpha_i$ if $x_i \in \mathcal{T}^c$, otherwise $\alpha_i' = 0$. Furthermore, $\alpha'' = \alpha - \alpha'$. Then we define a monomial order $<$ on the monomials of $k[x_1, \dots, x_n]$ by $x^\alpha < x^\beta$ if $x^{\alpha'} <_1 x^{\beta'}$ or $\alpha' = \beta'$ and $x^{\alpha''} <_2 x^{\beta''}$. Such an order is called an *inverse block order* with respect to $\mathcal{T}$. The next lemma follows from the definition of inverse block order.

Lemma 1.9.3 *Let $<$ be an inverse block order with respect to $\mathcal{T}$. Let $G \subset k[x_1, \ldots, x_n]$ be a Gröbner basis with respect to $<$ (of the ideal it generates). Then G is also a Gröbner basis in $k'[\mathcal{T}^c]$ (of the ideal it generates), with respect to the restriction of $<$ to $k'[\mathcal{T}^c]$.*

The following proposition follows from this lemma and Lemma 1.9.2.

Proposition 1.9.4 *Let $<$ and G be as in Lemma 1.9.3. Let $I \subset k[x_1, \ldots, x_n]$ be the ideal generated by G. The monomials of $k'[\mathcal{T}^c]$ are ordered using the restriction of $<$. Let f be the least common multiple of $\mathrm{LC}(g)$ as g runs through G (where we view those g as elements of $k'[\mathcal{T}^c]$). Let J be the ideal of $k'[\mathcal{T}^c]$ generated by G. Then $J \cap k[x_1, \ldots, x_n] = I : f^\infty$.*

Lemma 1.9.5 *Let $I \subset k[x_1, \ldots, x_n]$ be an ideal, and f a polynomial in the same ring. Let s be such that $I : f^s = I : f^\infty$. Let I' be the ideal generated by I and f^s. Then $I = I' \cap (I : f^\infty)$.*

Proof. $\subset$ is trivial. Let $g \in I' \cap (I : f^\infty)$; then $f^s g \in I$ and $g = p + f^s h$ for certain $p \in I$, $h \in k[x_1, \ldots, x_n]$. This implies that $f^{2s} h \in I$ and hence $h \in I : f^{2s} = I : f^s$. So $f^s h \in I$, and therefore $g \in I$. □

Now we can formulate an algorithm for computing the radical of an ideal.

Algorithm 1.9.6 Input: *an ideal $I \subset k[x_1, \ldots, x_n]$, where k is a perfect field.* Output: *the radical of I.*

1. *Let $\mathcal{T} = \{x_{i_1}, \ldots, x_{i_r}\}$ be an independent set modulo I of maximal size. If $\mathcal{T} = \emptyset$, compute the radical of I using the algorithm for the zero-dimensional case. Otherwise go to step 2.*
2. *Let I' be the ideal of $k'[\mathcal{T}^c]$ generated by I. Compute $M' = \sqrt{I'}$ using the algorithm for the zero-dimensional case.*
3. *Compute $M = M' \cap k[x_1, \ldots, x_n]$ (Lemma 1.9.2).*
4. *Apply Proposition 1.9.4 to get an $f \in k[\mathcal{T}]$ with $I' \cap k[x_1, \ldots, x_n] = I : f^\infty$.*
5. *Recursively compute the radical J of the ideal generated by I and f.*
6. *The radical of I is the intersection of J and M.*

Proposition 1.9.7 *Algorithm 1.9.6 terminates and is correct.*

Proof. First we prove that the algorithm is correct. Let s be such that $I : f^s = I : f^\infty$. Let U be the ideal generated by I and f^s. Then by Lemma 1.9.5, $I = U \cap I : f^\infty$. This implies $\sqrt{I} = \sqrt{U} \cap \sqrt{I : f^\infty}$. Now $\sqrt{U} = J$. Moreover,

$$\sqrt{I : f^\infty} = \sqrt{I' \cap k[x_1, \dots, x_n]} = \sqrt{I'} \cap k[x_1, \dots, x_n] = M.$$

In order to show termination we note that $f \in k[x_{i_1}, \dots, x_{i_r}]$, and therefore $f \notin I$. Hence the ideal generated by I and f is larger than I. Since there are no infinite increasing chains of ideals, the algorithm has to terminate. □

Due to the number of Gröbner bases that have to be computed, it is not always easy to apply this algorithm to a given ideal; see Example 3.4.4 and Remark 3.6.5 for a brief discussion of some examples.

Remark 1.9.8 Along with the radical, the *primary decomposition* of an ideal I is of importance. This is a finite number of ideals, $Q_1, \dots, Q_r$, such that I is their intersection and the radical of each Q_i is a prime ideal. Algorithms to compute the primary decomposition are structured in a similar way to the radical algorithm: first a zero-dimensional ideal is computed in a polynomial ring over a function field, and the primary decomposition of that ideal is then used to obtain the primary decomposition of the original ideal. Here we do not go into the details, and refer readers to [BW93], Section 8.7.

1.10 Notes

One of the standard texts on algebraic geometry to which we referred in this chapter is [Sha94]. Our treatment of tangent spaces is based on [Che58]. Many results regarding dimension and dominant maps have been taken from [Che55b]. Standard references for the theory of Gröbner bases containing a wealth of details and applications are [CLO15] and [BW93].

Chapter 2

Lie Algebras

When studying algebraic groups, Lie algebras come up almost immediately. They are constructed as the tangent space at the identity of an algebraic group. They capture a lot of the structure of the corresponding algebraic group (especially when the base field is of characteristic 0), but at the same time they are linear spaces, and hence the powerful tools of linear algebra can be employed to investigate them. This chapter describes some of the main classical results on the structure of Lie algebras.

In the first sections (Sections 2.1 to 2.6) we are mainly concerned with nilpotency and solvability. The importance of Cartan subalgebras is stressed early on. In Sections 2.7 to 2.9 we look at the classification of the semisimple Lie algebras over fields of characteristic 0. Subsequently (Sections 2.10, 2.11) we outline the main points of the representation theory of the semisimple Lie algebras. The final two sections are devoted to reductive Lie algebras and the Jacobson-Morozov theorem. For some of the main theoretical constructions we also indicate algorithms for performing them.

In this chapter it is necessary to be brief, as otherwise its length would grow excessively. On the other hand, where possible, the main ideas behind the proofs and constructions are given. Long technical arguments have occasionally been replaced by references.

2.1 Basic constructions

This section is devoted to some basic concepts (such as algebra, representation, centre and derivation) that we will be using throughout this chapter.

2.1.1 Algebras

An *algebra* A over the field k is a vector space over k, equipped with a bilinear map $m : A \times A \to A$. The map m is called the *multiplication*. Very often the infix operator $\cdot$ is used instead of m: we write $a \cdot b$, or more simply ab, instead of $m(a, b)$. For Lie algebras the convention is to write $[a, b]$ instead of $m(a, b)$ (Example 2.1.1 demonstrates the reason for this).

An algebra A is said to be *associative* if $a\cdot(b\cdot c) = (a\cdot b)\cdot c$ for all $a, b, c \in A$.

An algebra A is called a *Lie algebra* if for all $a, b, c \in A$ we have $[a, a] = 0$ and $[a, [b, c]] + [c, [a, b]] + [b, [c, a]] = 0$. The second of these identities is called the *Jacobi identity.* The first identity implies $[a, b] + [b, a] = 0$ for all $a, b \in A$.

Throughout this book we will use the common convention to denote Lie algebras by fraktur letters, such as $\mathfrak{g}$, $\mathfrak{h}$. The reason is that most Lie algebras that we consider come from algebraic groups, and if the algebraic group is denoted G, then its Lie algebra is $\mathfrak{g}$. This simple convention makes it easier to track which Lie algebra belongs to which group.

Example 2.1.1 By $M_n(k)$ we denote the space of all $n \times n$ matrices with coefficients in k. With the usual matrix product this is an associative algebra. Define $[a, b] = ab - ba$. Then a straightforward calculation shows that $M_n(k)$ together with $[\ ,\ ]$ is a Lie algebra, which is denoted $\mathfrak{gl}(n, k)$.

Let V be an n-dimensional vector space over k. Then the set of all linear maps $V \to V$ is denoted by $\mathrm{End}(V)$. The composition of mappings makes $\mathrm{End}(V)$ into an associative algebra. With $[\ ,\ ]$ defined by $[S, T] = ST - TS$, the same space turns into a Lie algebra, denoted by $\mathfrak{gl}(V)$. By choosing a basis of V, and mapping each linear map to its matrix with respect to that basis, $\mathrm{End}(V)$ and $\mathfrak{gl}(V)$ are seen to be isomorphic to $M_n(k)$ and $\mathfrak{gl}(n, k)$ respectively.

Let A be an algebra; then a subspace B which is closed under the multiplication map is called a subalgebra. If B is also closed under multiplication by elements of A (from the left and the right) then B is called a (two-sided) ideal. In that case the quotient space A/B inherits the algebra structure from A; so when A is associative or Lie, the same holds for A/B.

Let A be an algebra, and suppose that there are ideals I, J of A such that $A = I \oplus J$. Then $I \cdot J = J \cdot I = 0$, so the multiplication of A is completely determined by I and J. Conversely, if B_1, B_2 are two algebras over the same field k, then we can construct the direct sum of vector spaces $A = B_1 \oplus B_2$. Defining $(b_1 + b_2)(c_1 + c_2) = b_1c_1 + b_2c_2$ (for $b_1, c_1 \in B_1$, $b_2, c_2 \in B_2$) yields an algebra structure on A. If both B_i are associative or Lie, then the same holds for A. The B_i are ideals in A, and A is their direct sum.

2.1.2 Homomorphisms and representations

Let $\mathfrak{g}, \mathfrak{h}$ be Lie algebras over the field k. A linear map $f : \mathfrak{g} \to \mathfrak{h}$ with $f([x, y]) = [f(x), f(y)]$ for all $x, y \in \mathfrak{g}$ is said to be a *homomorphism of Lie algebras.* If in addition it is bijective, it is called an *isomorphism of Lie algebras.* It is a very difficult problem to list all Lie algebras up to isomorphism, which has only been solved to a high degree of completeness in dimension up to 3, although for some higher dimensions a lot is known.

A *representation* of a Lie algebra $\mathfrak{g}$ is a homomorphism $\rho : \mathfrak{g} \to \mathfrak{gl}(V)$, where V is a vector space defined over the same field k as $\mathfrak{g}$. In this context, V is said to be a $\mathfrak{g}$-module. If there can be no confusion about which representation is used we write $x \cdot v$ ($x \in \mathfrak{g}$, $v \in V$) instead of $\rho(x)v$.

Let V be a $\mathfrak{g}$-module, corresponding to the representation ρ. A subspace $W \subset V$ such that $x \cdot w \in W$ for all $x \in \mathfrak{g}$ and $w \in W$ is said to be a *submodule.* If V has no submodules other than 0 and V itself, it is called *irreducible.* In that case also the representation ρ is called irreducible. If V is the direct sum of irreducible submodules, then the module V and the representation ρ are called *completely reducible.*

Let $\rho : \mathfrak{g} \to \mathfrak{gl}(V)$ be a representation of the Lie algebra $\mathfrak{g}$. Define a bilinear form $\tau : \mathfrak{g} \times \mathfrak{g} \to k$ by $\tau(x, y) = \mathrm{Tr}(\rho(x)\rho(y))$. This is called the *trace form* associated to ρ. By the next lemma we have $\tau([x, y], z) = \tau(x, [y, z])$ for all $x, y, z \in \mathfrak{g}$. This property is summarized by saying that τ is an *invariant* bilinear form.

Lemma 2.1.2 *Let V be a finite-dimensional vector space. Let $x, y, z \in \mathfrak{gl}(V)$. Then* $\mathrm{Tr}([x, y]z) = \mathrm{Tr}(x[y, z])$.

A *composition series* of the $\mathfrak{g}$-module V is a series of submodules, $0 = V_0 \subset V_1 \subset \cdots \subset V_s = V$ such that the quotients V_i/V_{i-1} are irreducible $\mathfrak{g}$-modules, for $1 \leq i \leq s$. It is routine to show that every finite-dimensional $\mathfrak{g}$-module has a composition series.

Define $\mathrm{ad}_{\mathfrak{g}} : \mathfrak{g} \to \mathfrak{gl}(\mathfrak{g})$ by $\mathrm{ad}_{\mathfrak{g}}(x)(y) = [x, y]$. By the Jacobi identity this is a representation of $\mathfrak{g}$. It is called the *adjoint representation* of $\mathfrak{g}$. The submodules of $\mathfrak{g}$ under the adjoint representation are exactly the ideals of $\mathfrak{g}$.

If no confusion can arise about which Lie algebra is meant, then we write ad instead of $\mathrm{ad}_{\mathfrak{g}}$.

2.1.3 Structure constants

Let $\mathfrak{g}$ be an n-dimensional Lie algebra over the field k. Let $x_1, \ldots, x_n$ be a basis of $\mathfrak{g}$; then there are $c_{ij}^l \in k$ such that

$$[x_i, x_j] = \sum_{l=1}^{n} c_{ij}^l x_l.$$

Obviously, the c_{ij}^l completely determine the multiplication in $\mathfrak{g}$. Therefore they are called *structure constants.*

When designing algorithms for computations with Lie algebras, one has to decide how to present a Lie algebra by a finite amount of data. One of the most convenient ways is by giving a table of structure constants. Therefore, in this chapter, when we describe an algorithm operating on Lie algebras, we will always assume that the Lie algebras are given in this way. The table of structure constants of $\mathfrak{g}$ of course depends on the choice of a basis in $\mathfrak{g}$. So elements of $\mathfrak{g}$ can be given by their coefficient vectors relative to this basis. A subalgebra is given by a set of elements of $\mathfrak{g}$ forming a basis of it. A homomorphism of Lie algebras is given by a list of images of the basis of its domain.

Example 2.1.3 Let $\mathfrak{g}$ be a 6-dimensional Lie algebra over the field k of characteristic 0, with basis $\{x_1, x_2, x_3, x_4, x_5, x_6\}$, and multiplication table:

	x_1	x_2	x_3	x_4	x_5	x_6
x_1	0	0	x_1	x_5	$-\frac{1}{2}x_6$	0
x_2	0	0	$2x_2$	$x_3 - \frac{1}{2}x_6$	x_1	0
x_3	$-x_1$	$-2x_2$	0	$2x_4$	x_5	0
x_4	$-x_5$	$-x_3 + \frac{1}{2}x_6$	$-2x_4$	0	0	0
x_5	$\frac{1}{2}x_6$	$-x_1$	$-x_5$	0	0	0
x_6	0	0	0	0	0	0

This Lie algebra arises in the theory of Lie groups in connection with differential equations. It is a Lie algebra admitted by the heat equation (see [BK89], Section 4.2.4).

Note that if $\mathfrak{g}$ is a subalgebra of $\mathfrak{gl}(V)$, it is straightforward to obtain a table of structure constants.

Example 2.1.4 Since the trace of a commutator is zero, the subspace $\mathfrak{sl}(V) = \{a \in \mathfrak{gl}(V) \mid \mathrm{Tr}(a) = 0\}$ is in fact a subalgebra. By choosing a basis of V and mapping each $a \in \mathfrak{gl}(V)$ to its matrix with respect to that basis we get an isomorphism $\mathfrak{sl}(V) \to \mathfrak{sl}(n, k)$, where $\mathfrak{sl}(n, k)$ is the Lie algebra of traceless $n \times n$ matrices with coefficients in k. It is a routine exercise to write a basis of $\mathfrak{sl}(n, k)$ and the structure constants with respect to that basis. The simplest case, $\mathfrak{sl}(2, k)$, has basis

$$h = \begin{pmatrix} 1 & 0 \\ 0 & -1 \end{pmatrix},\ e = \begin{pmatrix} 0 & 1 \\ 0 & 0 \end{pmatrix},\ f = \begin{pmatrix} 0 & 0 \\ 1 & 0 \end{pmatrix}.$$

The corresponding multiplication table is

$$[h, e] = 2e, \quad [h, f] = -2f, \quad [e, f] = h.$$

2.1.4 Centre, centralizer and normalizer

Let $\mathfrak{g}$ be a Lie algebra over the field k. Let $E \subset \mathfrak{g}$ be a set of elements of $\mathfrak{g}$, then

$$\mathfrak{c}_{\mathfrak{g}}(E) = \{x \in \mathfrak{g} \mid [x, y] = 0 \text{ for all } y \in E\}$$

is a subalgebra of $\mathfrak{g}$ (this follows from the Jacobi identity), called the *centralizer* of E. The subalgebra $\mathfrak{c}_{\mathfrak{g}}(\mathfrak{g})$ is called the *centre* of $\mathfrak{g}$, and denoted $\mathfrak{c}(\mathfrak{g})$. The Jacobi identity implies that $\mathfrak{c}(\mathfrak{g})$ is an ideal of $\mathfrak{g}$.

Let $\mathfrak{u}$ be a subspace of $\mathfrak{g}$. Using the Jacobi identity, we see that

$$\mathfrak{n}_{\mathfrak{g}}(\mathfrak{u}) = \{x \in \mathfrak{g} \mid [x, y] \in \mathfrak{u} \text{ for all } y \in \mathfrak{u}\}$$

is a subalgebra of $\mathfrak{g}$. It is called the *normalizer* of $\mathfrak{u}$.

It is straightforward to formulate algorithms to compute centralizers and normalizers. We briefly indicate how this is done for the normalizer. Let $x_1, \ldots, x_n$, and $u_1, \ldots, u_m$ be bases of $\mathfrak{g}$ and $\mathfrak{u}$ respectively (where the u_j are given as linear combinations of the x_i). Let $x = \sum_{i=1}^{n} \alpha_i x_i$. Then $x \in \mathfrak{n}_{\mathfrak{g}}(\mathfrak{u})$ if and only if there are $\beta_{jr} \in k$, for $1 \leq j, r \leq m$ such that $[x, u_j] = \sum_{r=1}^{m} \beta_{jr} u_r$, for $1 \leq j \leq r$. Expanding all elements as linear combinations of $x_1, \ldots, x_n$ and using the table of structure constants of $\mathfrak{g}$, we obtain a set of linear equations for the α_i and β_{jr}. We solve these equations using the standard Gaussian elimination algorithm, and strip off the part corresponding to the β_{jr}.

2.2 Jordan decomposition of a linear transformation

This section is an intermezzo containing some material from linear algebra. We look at a decomposition of a linear transformation into a sum of a semisimple and a nilpotent transformation, called the Jordan decomposition.

Here k will be a field, and V a finite-dimensional vector space over k. Let $a \in \mathrm{End}(V)$, then by $k[a]$ we denote the associative algebra with one over k generated by a. Then V is a $k[a]$-module. Analogously to modules over a Lie algebra, V is said to be *irreducible* if the only $k[a]$-stable subspaces are 0 and V itself. Furthermore, V is a *completely reducible* $k[a]$-module if it can be written as the direct sum of irreducible $k[a]$-modules. In that case a is said to be semisimple. On the other hand, a is said to be *nilpotent* if there is an $m > 0$ with $a^m = 0$.

Let $K \supset k$ be an extension field. Write $V_K = K \otimes V$. Define $a_K \in \mathrm{End}(V_K)$ by $a_K \cdot \mu \otimes v = \mu \otimes (a \cdot v)$, where $\mu \in K$, $v \in V$.

Theorem 2.2.1 *Let $a \in \mathrm{End}(V)$.*

1. *a is semisimple if and only if its minimal polynomial is square-free.*
2. *If a_K is semisimple, the same is true for a. If k is perfect, then the converse holds as well.*
3. *Assume that k is perfect and that K is its algebraic closure. Then a is semisimple if and only if there is a basis of V_K relative to which a_K is diagonal.*

Proof. For (i) we refer to [Jac75], Chapter IV, Theorem 5. For (ii), let $f_a \in k[x]$ be the minimal polynomial of a. Then f_a, seen as an element of $K[x]$, is the minimal polynomial of a_K as well. This immediately implies the statement.

Suppose that a is semisimple. By (ii) a_K is semisimple as well. Hence V_K splits as a direct sum of irreducible $K[a_K]$-modules. Since K is algebraically closed, the only irreducible $K[a_K]$-modules are of dimension 1. The required basis is the union of the bases of these irreducible submodules. For the converse we note that if a_K is diagonalizable, then it is semisimple. Again we use (ii). □

Corollary 2.2.2 *Suppose that k is perfect. Let A denote the associative algebra with one generated by $a_1, \dots, a_l \in \mathrm{End}(V)$. Suppose that the a_i pairwise commute and are semisimple. Then all elements of A are semisimple.*

Proof. Let K be the algebraic closure of k. For all $a \in A$, a_K leaves invariant the eigenspaces in V_K of $(a_i)_K$. This implies that we can find a basis of V_K with respect to which all $(a_i)_K$ are diagonal. We conclude by Theorem 2.2.1(iii). □

For a proof of the next proposition we refer to [Jac79], Chapter III, Theorem 16, or [Hum78], Section 4.2.

Proposition 2.2.3 *Suppose that k is perfect. Let $a \in \mathrm{End}(V)$. Then there are unique $s, n \in \mathrm{End}(V)$ such that s is semisimple, n is nilpotent, $[s, n] = 0$ and $a = s + n$. Moreover, there are polynomials $g_s, g_n \in k[x]$ without constant terms such that $s = g_s(a)$ and $n = g_n(a)$.*

The decomposition $a = s + n$ of the proposition is called the *Jordan decomposition* of a. The cited proof of the proposition in [Jac79] immediately yields an algorithm to compute s and n. Since we omit the proof, we also do not go into this algorithm, but refer to [Gra00], Section A.2.

Proposition 2.2.4 *Let $a \in \mathrm{End}(V)$, with Jordan decomposition $a = s + n$. Let $U \subset V$ be a subspace. Then a leaves U invariant if and only if both s, n do.*

Proof. This follows from the fact that s, n are polynomials in a. □

Proposition 2.2.5 *Let $a \in \mathrm{End}(V)$, and consider* $\mathrm{ad}a : \mathrm{End}(V) \to \mathrm{End}(V)$, $\mathrm{ad}a(b) = ab - ba$. *If a is nilpotent then* $\mathrm{ad}a$ *is nilpotent. If the base field is perfect, then the statement that a is semisimple implies that* $\mathrm{ad}a$ *is semisimple as well.*

Proof. The first statement follows from the formula $(\mathrm{ad}a)^n(b) = \sum_{i=0}^{n} \binom{n}{i} (-1)^{n-i} a^i b a^{n-i}$. For the second statement, we may assume that the base field is algebraically closed (Theorem 2.2.1(ii)). Then V has a basis consisting of eigenvectors v_i of a, corresponding to the eigenvalues λ_i, $1 \le i \le n$ (Theorem 2.2.1(iii)). Hence $\mathrm{End}(V)$ has a basis consisting of elements e_{ij}, $1 \le i, j \le n$, where $e_{ij}(v_l) = \delta_{jl} v_i$. But $\mathrm{ad}a(e_{ij}) = (\lambda_i - \lambda_j) e_{ij}$. Therefore, by Theorem 2.2.1(iii), $\mathrm{ad}a$ is semisimple. □

2.3 Derivations

Let A be an algebra. A *derivation* of A is a linear map $d : A \to A$ such that $d(a \cdot b) = d(a) \cdot b + a \cdot d(b)$ for all $a, b \in A$. The set of all derivations of A is denoted $\mathrm{Der}(A)$. It forms a vector space and even a Lie algebra with the product $[d_1, d_2] = d_1 \circ d_2 - d_2 \circ d_1$.

Let $\mathfrak{g}$ be a Lie algebra, and $x \in \mathfrak{g}$. The Jacobi identity implies that $\mathrm{ad}x$ is a derivation of $\mathfrak{g}$. Derivations of this form are said to be *inner*.

Let A be an algebra defined over a field of characteristic 0. Let d be a *nilpotent* derivation of A, i.e., there is an $n > 0$ with $d^n = 0$. Then we define its exponential as $\exp d = \sum_{i=0}^{n-1} d^i/i!$. Then $\exp d$ is an automorphism of A (see [Jac79], Section I.2).

For later use we record the following lemma.

Lemma 2.3.1 *Let V be a finite-dimensional vector space over a field of characteristic 0. Let $\mathfrak{g} \subset \mathfrak{gl}(V)$ be a Lie subalgebra. Let $x \in \mathfrak{g}$ be a nilpotent endomorphism of V. Then* $\exp(x)y\exp(-x) = \exp(\mathrm{ad}x)(y)$ *for all $y \in \mathfrak{g}$.*

Proof. Note that all exponentials involved are finite sums (Proposition 2.2.5). For $z \in \mathfrak{gl}(V)$ define $L_z, R_z : \mathfrak{gl}(V) \to \mathfrak{gl}(V)$, by $L_z(a) = za$, $R_z(a) = az$. Then L_z and R_z commute, so that $\exp(\mathrm{ad}x) = \exp(L_x - R_x) = \exp(L_x)\exp(-R_x) = L_{\exp(x)}R_{\exp(-x)}$, implying the statement of the lemma. □

2.4 Nilpotency

In this section we look at nilpotent Lie algebras. One of the main theorems in this context is Engel's, stating that the nilpotency of a Lie algebra $\mathfrak{g}$ is equivalent to all $\mathrm{ad}x$ being nilpotent endomorphisms for all $x \in \mathfrak{g}$. We also introduce the nilradical, which is the largest nilpotent ideal of a Lie algebra.

Let $\mathfrak{g}$ be a finite-dimensional Lie algebra over the field k. For two subspaces $\mathfrak{u}, \mathfrak{v} \subset \mathfrak{g}$ define $[\mathfrak{u}, \mathfrak{v}]$ to be the subspace spanned by all $[u, v]$ for $u \in \mathfrak{u}$, $v \in \mathfrak{v}$. It is called the *product space* of $\mathfrak{u}, \mathfrak{v}$. There is an obvious algorithm for computing a basis of $[\mathfrak{u}, \mathfrak{v}]$, given bases of $\mathfrak{u}, \mathfrak{v}$. Furthermore, if $\mathfrak{u}, \mathfrak{v}$ are ideals of $\mathfrak{g}$ then so is their product space.

Define $\mathfrak{g}^1 = \mathfrak{g}$ and for $j \geq 1$, $\mathfrak{g}^{j+1} = [\mathfrak{g}, \mathfrak{g}^j]$. Then $\mathfrak{g} = \mathfrak{g}^1 \supset \mathfrak{g}^2 \supset \mathfrak{g}^3 \cdots$ is a sequence of ideals of $\mathfrak{g}$, called the *lower central series* of $\mathfrak{g}$. By repeatedly using the algorithm for computing a product space, we can compute this series.

The Lie algebra $\mathfrak{g}$ is said to be *nilpotent* if there is a $c \geq 0$ with $\mathfrak{g}^{c+1} = 0$. The minimal such c is called the *nilpotency class* of $\mathfrak{g}$.

2.4.1 Engel's theorem

Proposition 2.4.1 *Let V be a finite-dimensional vector space over the field k, and $\mathfrak{g} \subset \mathfrak{gl}(V)$ a Lie subalgebra. Suppose that all $x \in \mathfrak{g}$ are nilpotent endomorphisms of V. Then there exists a non-zero $v \in V$ such that $x \cdot v = 0$ for all $x \in \mathfrak{g}$.*

Proof. The proof is by induction on $\dim \mathfrak{g}$, the case $\dim \mathfrak{g} = 1$ being trivial. If $\dim \mathfrak{g} > 1$ we let $\mathfrak{a}$ be a maximal proper subalgebra. By the adjoint representation $\mathfrak{a}$ acts on $\mathfrak{g}$, and $\mathfrak{a}$ is a submodule. So we get a representation $\rho : \mathfrak{a} \to \mathfrak{gl}(\mathfrak{g}/\mathfrak{a})$. Let $x \in \mathfrak{g}$; then x is nilpotent, implying the same for $\mathrm{ad}_{\mathfrak{g}} x$ (Proposition 2.2.5). So for $x \in \mathfrak{a}$ we have that $\rho(x)$ is nilpotent as well. By induction there is a non-zero $\bar{y} \in \mathfrak{g}/\mathfrak{a}$ such that $\rho(x)\bar{y} = 0$ for all $x \in \mathfrak{a}$. Let $y \in \mathfrak{g}$ be a preimage of $\bar{y}$. Then the span of $\mathfrak{a}$ and y is a subalgebra which therefore is equal to $\mathfrak{g}$. By induction the space W of all $v \in V$ such that $x \cdot v = 0$ for all $x \in \mathfrak{a}$ is non-zero. This space is stabilized by y. It follows that there is a non-zero $v \in W$ such that $y \cdot v = 0$. □

Lemma 2.4.2 *Let $\mathfrak{g}$ be a finite-dimensional Lie algebra over k, V a finite-dimensional vector space over k, and $\rho : \mathfrak{g} \to \mathfrak{gl}(V)$ an irreducible representation. Let $\mathfrak{a} \subset \mathfrak{g}$ be an ideal such that $\rho(x)$ is nilpotent for all $x \in \mathfrak{a}$. Then $\rho(x) = 0$ for all $x \in \mathfrak{a}$.*

Proof. Let W be the subspace of V consisting of all $v \in V$ such that $\rho(x)v = 0$ for all $x \in \mathfrak{a}$. This space is non-zero by the previous proposition. Moreover, it is stabilized by $\mathfrak{g}$. □

Theorem 2.4.3 (Engel) *Let $\mathfrak{g}$ be a finite-dimensional Lie algebra. Then $\mathfrak{g}$ is nilpotent if and only if $\mathrm{ad}x$ is a nilpotent endomorphism for all $x \in \mathfrak{g}$.*

Proof. Suppose that $\mathrm{ad}x$ is nilpotent for all $x \in \mathfrak{g}$. Let $0 = \mathfrak{g}_0 \subset \mathfrak{g}_1 \subset \cdots \subset \mathfrak{g}_s = \mathfrak{g}$ be a composition series of $\mathfrak{g}$ with respect to the adjoint representation. By Lemma 2.4.2, $[x, \mathfrak{g}_{i+1}] \subset \mathfrak{g}_i$ for $0 \leq i < s$, and all $x \in \mathfrak{g}$. Therefore, $\mathfrak{g}^i \subset \mathfrak{g}_{s-i+1}$. The other direction is trivial. □

Remark 2.4.4 Let $\mathfrak{g}$ be a nilpotent Lie algebra of nilpotency class c. Then $(\mathrm{ad}x)^c = 0$ for all $x \in \mathfrak{g}$. On the other hand, $(\mathrm{ad}x)^j = 0$ for a fixed integer j and all elements x does not imply that the nilpotency class is bounded by j. In other words, it may happen that $c > j$. See [CdG09] for methods to construct examples of this phenomenon.

Example 2.4.5 To be able to conclude that $\mathfrak{g}$ is nilpotent, it is not enough that $\mathrm{ad}x$ is nilpotent for all x in a basis of $\mathfrak{g}$. As an example, consider $\mathfrak{sl}(2, k)$

from Example 2.1.4. This Lie algebra is spanned by the elements e, f (given in the mentioned example) and $\begin{pmatrix} 1 & -1 \\ 1 & -1 \end{pmatrix}$, which are all nilpotent.

2.4.2 Nilradicals

Lemma 2.4.6 *Let $\mathfrak{g}$ be a Lie algebra and $\mathfrak{a}, \mathfrak{b}$ two nilpotent ideals of $\mathfrak{g}$. Then $\mathfrak{a} + \mathfrak{b}$ is a nilpotent ideal of $\mathfrak{g}$.*

Proof. Here we write $[u_1, \dots, u_n]$ instead of $[u_1, [u_2, [\cdots, [u_{n-1}, u_n] \cdots]]]$. For $1 \le i \le t$ let $x_i \in \mathfrak{a}$ and $y_i \in \mathfrak{b}$. Set $z_i = x_i + y_i$. Then $[z_1, \dots, z_t]$ expands as a sum of elements $[u_1, \dots, u_t]$ where each u_i either is x_i or y_i. Moreover, if there are r x_i's, the element lies in $\mathfrak{a}^r$. An analogous statement holds for the y_i's and $\mathfrak{b}$. It follows that by choosing t sufficiently large, all $[u_1, \dots, u_t]$ are zero. □

Let $\mathfrak{g}$ be a finite-dimensional Lie algebra. The lemma implies that $\mathfrak{g}$ has a unique maximal nilpotent ideal. This ideal is called the *nilradical* and denoted $\mathfrak{nr}(\mathfrak{g})$. In Section 2.6.5 we will indicate an algorithm for computing the nilradical.

2.5 Cartan subalgebras

Cartan subalgebras form important tools for the investigation of the structure of a Lie algebra. By the adjoint representation, a Cartan subalgebra of a Lie algebra $\mathfrak{g}$ acts on $\mathfrak{g}$. Relative to this action we can decompose $\mathfrak{g}$ as a sum of root spaces. This decomposition forms the basis of the structure theory of semisimple Lie algebras. Also good use can be made of it when dealing with non-semisimple Lie algebras. For example, in Section 2.6.2 a proof of Cartan's criterion for solvability will be given using the root space decomposition. A second example is the algorithm for computing the nilradical (Section 2.6.5), which uses Cartan subalgebras.

2.5.1 Primary decomposition

Here we consider a decomposition of a vector space relative to the action of a nilpotent Lie algebra. Cartan subalgebras arise as those nilpotent subalgebras for which this decomposition is the most interesting.

Let V be a finite-dimensional vector space over the field k. For a $b \in \mathrm{End}(V)$ we set

$$V_0(b) = \{v \in V \mid b^r v = 0 \text{ for some } r > 0\}.$$

Let $a \in \mathrm{End}(V)$ with minimal polynomial $f_a \in k[x]$. Let $f_a = p_1^{m_1} \cdots p_r^{m_r}$ be its factorization into a product of distinct monic irreducible polynomials. For $1 \leq i \leq r$ let h_i be the product of all $p_j^{m_j}$ except $p_i^{m_i}$ (i.e., $h_i = f_a/p_i^{m_i}$). These polynomials have greatest common divisor equal to 1, so there are $q_i \in k[x]$ with $q_1h_1 + \cdots + q_rh_r = 1$. Set $e_i = q_ih_i(a)$. Then the e_i are orthogonal idempotents with sum 1. So if we set $V_i = e_iV$ then $V = V_1 \oplus \cdots \oplus V_r$. Moreover, it can be shown that $V_i = V_0(p_i(a))$ and that the minimal polynomial of the restriction of a to V_i is $p_i^{m_i}$ ([Gra00], Lemma A.2.2).

Definition 2.5.1 *Let $\mathfrak{a} \subset \mathfrak{gl}(V)$ be a Lie subalgebra. A* primary decomposition *of V relative to the action of $\mathfrak{a}$ is a decomposition of V into a direct sum of $\mathfrak{a}$-invariant subspaces $V = V_1 \oplus \cdots \oplus V_r$ such that the minimal polynomial of the restriction of each $a \in \mathfrak{a}$ to V_i, $1 \leq i \leq r$, is a power of an irreducible polynomial. The primary decomposition is said to be* collected *if for $i \neq j$ there is an $a \in \mathfrak{a}$ such that the restrictions of a to V_i and V_j have coprime minimal polynomials.*

For a proof of the next lemma we refer to [Jac79], Section II.4.

Lemma 2.5.2 *Let $a, b \in \mathfrak{gl}(V)$, and suppose that $[a, [a, \cdots, [a, b] \cdots]] = 0$. Let $p \in k[x]$. Then $V_0(p(a))$ is invariant under b.*

Theorem 2.5.3 *Let $\mathfrak{a}$ be a nilpotent Lie subalgebra of $\mathfrak{gl}(V)$. Then V has a unique collected primary decomposition relative to $\mathfrak{a}$.*

Proof. If all $a \in \mathfrak{a}$ have irreducible minimal polynomials, there is nothing to prove. Otherwise take an $a \in \mathfrak{a}$ whose minimal polynomial factors into at least two coprime factors. Then we can form the primary decomposition relative to a as shown above. By the previous lemma, all components are invariant under $\mathfrak{a}$. So we can conclude by induction on $\dim V$ that a collected primary decomposition exists. We omit the proof of uniqueness; see [Jac79], Section II.4. □

2.5.2 Fitting decomposition

The Fitting decomposition of a Lie algebra relative to a nilpotent subalgebra is related to the primary decomposition, but it is coarser, having only two components. However, on many occasions these two components are all that matter, so it is useful to study the Fitting decomposition in its own right.

Let $\mathfrak{a} \subset \mathfrak{gl}(V)$ be a nilpotent Lie subalgebra, and $V = V_1 \oplus \cdots \oplus V_m$ the collected primary decomposition relative to $\mathfrak{a}$. There is at most one component V_{i_0} such that all $a \in \mathfrak{a}$ act nilpotently on it. We denote this component by $V_0(\mathfrak{a})$. (And $V_0(\mathfrak{a}) = 0$ if there is no such component V_{i_0}.) It is called the *Fitting zero component* of V relative to $\mathfrak{a}$. Furthermore, we let $V_1(\mathfrak{a})$ be the sum of all other components. This is called the *Fitting one component* of V relative to $\mathfrak{a}$.

Now let $\mathfrak{g}$ be a finite-dimensional Lie algebra, and $\mathfrak{a}$ a nilpotent subalgebra. We view $\mathfrak{g}$ as an $\mathfrak{a}$-module via the adjoint representation. By $\mathfrak{g}_0(\mathfrak{a})$ and $\mathfrak{g}_1(\mathfrak{a})$ we denote the corresponding Fitting zero and one components.

Lemma 2.5.4 *Let $x \in \mathfrak{g}_0(\mathfrak{a})$. Then the primary components in the collected primary decomposition of $\mathfrak{g}$ relative to $\mathfrak{a}$ are invariant under* $\mathrm{ad}x$.

Proof. Let $a \in \mathfrak{a}$, then $(\mathrm{ad}a)^s x = 0$ for some $s > 0$. The result now follows from Lemma 2.5.2, along with the construction of the primary decomposition in the proof of Theorem 2.5.3. □

Corollary 2.5.5 *We have that $\mathfrak{g}_0(\mathfrak{a})$ is a subalgebra and $[\mathfrak{g}_0(\mathfrak{a}), \mathfrak{g}_1(\mathfrak{a})] \subset \mathfrak{g}_1(\mathfrak{a})$.*

Proposition 2.5.6 *Write $\mathfrak{g}_0$ for $\mathfrak{g}_0(\mathfrak{a})$. We have $\mathfrak{n}_{\mathfrak{g}}(\mathfrak{g}_0) = \mathfrak{g}_0$.*

Proof. Let $x \in \mathfrak{n}_{\mathfrak{g}}(\mathfrak{g}_0)$, and decompose $x = x_0 + x_1$, $x_0 \in \mathfrak{g}_0$, $x_1 \in \mathfrak{g}_1(\mathfrak{a})$. Then for $y \in \mathfrak{g}_0$ we get $[x_1, y] \in \mathfrak{g}_0$. By Corollary 2.5.5, $[x_1, y] \in \mathfrak{g}_1(\mathfrak{a})$, so that $[x_1, y] = 0$. As $\mathfrak{a} \subset \mathfrak{g}_0$, it follows that $[x_1, \mathfrak{a}] = 0$, whence $x_1 \in \mathfrak{g}_0$. So $x_1 = 0$. □

It will be useful on various occasions to be able to compute the Fitting one component $\mathfrak{g}_1(\mathfrak{a})$. For that we write $[\mathfrak{a}^l, \mathfrak{g}] = [\mathfrak{a}, [\mathfrak{a}, \cdots, [\mathfrak{a}, \mathfrak{g}] \cdots]]$ (l factors $\mathfrak{a}$). Since $[\mathfrak{a}^l, \mathfrak{g}] = [\mathfrak{a}^l, \mathfrak{g}_0(\mathfrak{a})] + \mathfrak{g}_1(\mathfrak{a})$ we see that the series of subspaces $[\mathfrak{a}^l, \mathfrak{g}]$ is descending, and when $[\mathfrak{a}^l, \mathfrak{g}] = [\mathfrak{a}^{l+1}, \mathfrak{g}]$ we have $\mathfrak{g}_1(\mathfrak{a}) = [\mathfrak{a}^l, \mathfrak{g}]$. By employing the algorithm for computing a product space (Section 2.4) repeatedly, we can compute the Fitting one component.

2.5.3 Cartan subalgebras of Lie algebras

Let $\mathfrak{g}$ be a finite-dimensional Lie algebra over the field k. A subalgebra $\mathfrak{h}$ of $\mathfrak{g}$ is called a *Cartan subalgebra* if $\mathfrak{h}$ is nilpotent and $\mathfrak{n}_{\mathfrak{g}}(\mathfrak{h}) = \mathfrak{h}$.

The question whether every Lie algebra has a Cartan subalgebra is not completely settled yet. Below we will give an algorithm for computing a Cartan subalgebra for Lie algebras defined over a field of size at least $\dim \mathfrak{g} + 1$. So Lie algebras over adequate fields do have Cartan subalgebras.

The next proposition says that a nilpotent subalgebra is Cartan if and only if the corresponding Fitting zero component is as small as possible.

Proposition 2.5.7 *Let $\mathfrak{h} \subset \mathfrak{g}$ be a nilpotent subalgebra. Then $\mathfrak{h}$ is a Cartan subalgebra if and only if $\mathfrak{g}_0(\mathfrak{h}) = \mathfrak{h}$.*

Proof. Suppose that $\mathfrak{n}_{\mathfrak{g}}(\mathfrak{h}) = \mathfrak{h}$. Since $\mathfrak{h}$ is nilpotent, $\mathfrak{h} \subset \mathfrak{g}_0(\mathfrak{h})$. Both $\mathfrak{h}$ and $\mathfrak{g}_0(\mathfrak{h})$ are $\mathfrak{h}$-modules via the adjoint representation and the quotient $\mathfrak{g}_0(\mathfrak{h})/\mathfrak{h}$ is a $\mathfrak{h}$-module as well. Moreover, all elements of $\mathfrak{h}$ act nilpotently on all these modules. If $\mathfrak{g}_0(\mathfrak{h})$ is larger than $\mathfrak{h}$, the quotient contains a non-zero element annihilated by $\mathfrak{h}$ (Proposition 2.4.1). A preimage of such an element lies in $\mathfrak{n}_{\mathfrak{g}}(\mathfrak{h})$ but not in $\mathfrak{h}$, which is excluded. The converse implication follows from Proposition 2.5.6. □

For a proof of the next theorem we refer to [Jac79], Section IX.2, or [Hum78], Section 16. For the structure and representation theory of semisimple Lie algebras this fact is essential, because it allows us to develop the theory relative to one fixed Cartan subalgebra.

Theorem 2.5.8 *Suppose the ground field of $\mathfrak{g}$ is algebraically closed and of characteristic 0. Let G be the subgroup of the automorphism group of $\mathfrak{g}$ generated by all $\exp(\mathrm{ad}y)$ where $y \in \mathfrak{g}$ is such that $\mathrm{ad}y$ is nilpotent. Let $\mathfrak{h}$ and $\mathfrak{h}'$ be Cartan subalgebras of $\mathfrak{g}$. Then there is a $\sigma \in G$ such that $\sigma(\mathfrak{h}) = \mathfrak{h}'$.*

The next proposition underpins an algorithm to compute a Cartan subalgebra. For an $x \in \mathfrak{g}$ we write $\mathfrak{g}_0(x)$ instead of $\mathfrak{g}_0(\mathfrak{a})$, where $\mathfrak{a}$ is the subalgebra spanned by x.

Proposition 2.5.9 *Let $\Omega \subset k$ have at least $\dim \mathfrak{g} + 1$ elements. Let $x \in \mathfrak{g}$ and suppose $\mathfrak{b} = \mathfrak{g}_0(x)$ has an element y such that $\mathrm{ad}_{\mathfrak{b}} y$ is not nilpotent. Then there is a $c_0 \in \Omega$ such that $\mathfrak{g}_0(x + c_0(y - x))$ is properly contained in $\mathfrak{g}_0(x)$.*

Proof. Let $\mathfrak{g} = \mathfrak{b} \oplus \mathfrak{g}_1(x)$ be the Fitting decomposition of $\mathfrak{g}$ with respect to the subalgebra spanned by x. By Corollary 2.5.5 we infer that the linear transformations $A_c = \mathrm{ad}(x + c(y - x))$ stabilize $\mathfrak{b}$ and $\mathfrak{g}_1(x)$ for all $c \in k$. Let $f(X)$ be the characteristic polynomial of A_c. Then $f(X) = g(X)h(X)$, where $g(X)$ and $h(X)$ are the characteristic polynomials of the restriction of A_c to $\mathfrak{b}$ and $\mathfrak{g}_1(x)$ respectively. Write $g(X) = X^d + g_1(c)X^{d-1} + \cdots + g_d(c)$, and $h(X) = X^e + h_1(c)X^{e-1} + \cdots + h_e(c)$, where $g_i(c)$ and $h_i(c)$ are polynomials in c that, if non-zero, are of degree i. Since $\mathrm{ad}_{\mathfrak{b}} y$ is not nilpotent, there is an i with $g_i(1) \neq 0$. As $\mathfrak{b} = \mathfrak{g}_0(x)$, $h_e(0) \neq 0$. So g_i, h_e are non-zero. But $\deg g_i h_e \leq \dim \mathfrak{g}$, so there is a $c_0 \in \Omega$ such that $g_i(c_0)h_e(c_0) \neq 0$. Finally, $h_e(c_0) \neq 0$ implies that $\mathfrak{g}_0(x + c_0(y - x))$ is contained in $\mathfrak{b}$, and $g_i(c_0) \neq 0$ implies that the inclusion is strict. □

Corollary 2.5.10 *Let $\mathfrak{g}$ be a Lie algebra over a field of at least $\dim \mathfrak{g} + 1$ elements. Let $h \in \mathfrak{g}$ be such that $\mathrm{ad}_{\mathfrak{g}} h$ is semisimple. Then h lies in a Cartan subalgebra of $\mathfrak{g}$.*

Proof. Since h is semisimple, $\mathfrak{g}_0(h) = \mathfrak{c}_{\mathfrak{g}}(h)$, the centralizer of h. From the previous propositions it follows that there is an $x \in \mathfrak{g}_0(h)$ such that $\mathfrak{g}_0(x)$ is a Cartan subalgebra of $\mathfrak{g}$. Since $h \in \mathfrak{g}_0(x)$, the corollary is proved. □

The algorithm based on Proposition 2.5.9 is straightforward, assuming that the field k has at least $\dim \mathfrak{g} + 1$ elements. We start with $x = 0$ and $\mathfrak{b} = \mathfrak{g}_0(x)$ (which is $\mathfrak{g}$). If there is a $y \in \mathfrak{b}$ such that $\mathrm{ad}_{\mathfrak{b}} y$ is not nilpotent then, by trial and error, find a $c_0 \in k$ such that $\dim \mathfrak{g}_0(x + c_0(y - x)) < \dim \mathfrak{b}$, and replace $\mathfrak{b}$ by $\mathfrak{g}_0(x + c_0(y - x))$ and x by $x + c_0(y - x)$. This continues until there are no non-nilpotent elements in $\mathfrak{b}$. But then $\mathfrak{b}$ is a nilpotent Lie algebra by Engel's theorem (Theorem 2.4.3). Moreover, since $\mathfrak{b} = \mathfrak{g}_0(x)$ we have $\mathfrak{n}_{\mathfrak{g}}(\mathfrak{b}) = \mathfrak{b}$ (Proposition 2.5.6) so $\mathfrak{b}$ is a Cartan subalgebra.

This algorithm needs a method that finds a non-nilpotent element in a Lie algebra or decides that no such element exists. In Section 2.6.4 we will give a simple algorithm for this task that works over fields of characteristic 0.

Example 2.5.11 Let $\mathfrak{g}$ be the Lie algebra of Example 2.1.3. Then x_3 is a non-nilpotent element of $\mathfrak{g}$. Furthermore, $\mathfrak{g}_0(x_3)$ is spanned by x_3, x_6. This is a nilpotent subalgebra and hence a Cartan subalgebra. Denote it by $\mathfrak{h}$. Then the primary decomposition of $\mathfrak{g}$ with respect to $\mathfrak{h}$ is $\mathfrak{g} = \mathfrak{h} \oplus \mathfrak{g}_1 \oplus \mathfrak{g}_2 \oplus \mathfrak{g}_4 \oplus \mathfrak{g}_5$, where $\mathfrak{g}_i$ is spanned by x_i.

2.5.4 Weights and roots

This section is devoted to two concepts related to the primary decomposition of a $\mathfrak{g}$-module relative to a Cartan subalgebra of $\mathfrak{g}$. If the $\mathfrak{g}$-module under consideration is $\mathfrak{g}$ itself, the primary components are called root spaces. These have proved to be very useful for the investigation of the structure of $\mathfrak{g}$. Otherwise they are called weight spaces and play an important role in the representation theory of $\mathfrak{g}$.

Let $\mathfrak{g}$ be a finite-dimensional Lie algebra over the field k, and $\mathfrak{h} \subset \mathfrak{g}$ a Cartan subalgebra. Let $\rho : \mathfrak{g} \to \mathfrak{gl}(V)$ be a finite-dimensional representation of $\mathfrak{g}$. We suppose $\rho(\mathfrak{h})$ is *split*, which means that k contains the eigenvalues of $\rho(h)$ for all $h \in \mathfrak{h}$. Let $V = V_1 \oplus \cdots \oplus V_r$ be the collected primary decomposition of V with respect to $\rho(\mathfrak{h})$. Let $h \in \mathfrak{h}$; then the minimal polynomial of the restriction of $\rho(h)$ to V_i is a power of an irreducible polynomial $p_{h,i} \in k[X]$, and since $\rho(\mathfrak{h})$ is split, $p_{h,i} = X - \mu_i(h)$ where $\mu_i(h) \in k$. For each i, with $1 \leq i \leq r$, we get a function $\mu_i : \mathfrak{h} \to k$. Since the primary decomposition is collected, these functions are all different. They are called the *weights* of the $\mathfrak{g}$-module V with respect to $\mathfrak{h}$. The primary component corresponding to the weight μ is called the *weight space* of μ, and denoted V_μ.

If ρ is the adjoint representation, the zero function appears among the weights, and the corresponding weight space is $\mathfrak{h}$ itself by Proposition 2.5.7. In this situation the weights, *excluding the zero weight*, are called *roots*. The primary component corresponding to the root α is called the *root space* of α, and written $\mathfrak{g}_\alpha$, so

$$\mathfrak{g}_\alpha = \{x \in \mathfrak{g} \mid \text{ for all } h \in \mathfrak{h} \text{ there is an } l > 0 \text{ with } (\mathrm{ad} h - \alpha(h))^l(x) = 0\}$$

and the primary decomposition of $\mathfrak{g}$ with respect to $\mathrm{ad}_{\mathfrak{g}}\mathfrak{h}$ reads

$$\mathfrak{g} = \mathfrak{h} \oplus \mathfrak{g}_{\alpha_1} \oplus \cdots \oplus \mathfrak{g}_{\alpha_t},$$

where $\alpha_1, \dots, \alpha_t$ are the roots of $\mathfrak{g}$. This decomposition is called the *root space decomposition* of $\mathfrak{g}$ with respect to $\mathfrak{h}$.

Proposition 2.5.12 *Let $\rho : \mathfrak{g} \to \mathfrak{gl}(V)$ be a finite-dimensional representation of $\mathfrak{g}$. Let α be a root of $\mathfrak{g}$, and μ be a weight of V with respect to the same Cartan subalgebra $\mathfrak{h}$. Let $x \in \mathfrak{g}_\alpha$ and $v \in V_\mu$ then $\rho(x)v \in V_{\mu+\alpha}$. In particular, $\rho(x)v = 0$ if $\mu + \alpha$ is not a weight of V.*

Proof. By induction on m it follows that $(\rho(h) - (\alpha(h) + \mu(h)))^m(\rho(x)v)$ equals

$$\sum_{i=0}^{m} \binom{m}{i} \rho\left((\mathrm{ad}h - \alpha(h))^{m-i}(x)\right) (\rho(h) - \mu(h))^i(v).$$

But for sufficiently large m this is zero. □

Example 2.5.13 Let $\mathfrak{g}$ be the Lie algebra of Example 2.1.3, with root space decomposition as in Example 2.5.11. Let α_i denote the root corresponding to the root space $\mathfrak{g}_i$, for $i = 1, 2, 4, 5$. Then $\alpha_1 + \alpha_4 = \alpha_5$. Indeed, $[x_1, x_4] = x_5$.

2.6 Solvability

This section is devoted to the concept of solvability. One main result is Lie's theorem: a solvable matrix algebra (under an additional hypothesis) consists of upper triangular matrices. A second important theorem is a criterion due to Cartan: a Lie algebra fulfilling it is necessarily solvable. This is another cornerstone of the structure theory of semisimple Lie algebras. We also look at the maximal solvable ideal of a Lie algebra, called its solvable radical. We give algorithms for computing the solvable radical as well as for computing the nilradical.

Define $\mathfrak{g}^{(1)} = \mathfrak{g}$ and for $j \geq 1$, $\mathfrak{g}^{(j+1)} = [\mathfrak{g}^{(j)}, \mathfrak{g}^{(j)}]$. Then $\mathfrak{g} = \mathfrak{g}^{(1)} \supset \mathfrak{g}^{(2)} \supset \mathfrak{g}^{(3)} \cdots$ is a sequence of ideals of $\mathfrak{g}$, called the *derived series* of $\mathfrak{g}$. By repeatedly using the algorithm for computing a product space, we can compute this series. The second term, $\mathfrak{g}^{(2)} = [\mathfrak{g}, \mathfrak{g}]$ is called the *derived subalgebra* of $\mathfrak{g}$.

The Lie algebra $\mathfrak{g}$ is said to be *solvable* if there is an $s \geq 0$ with $\mathfrak{g}^{(s+1)} = 0$.

Lemma 2.6.1 *Let $\mathfrak{g}$ be a Lie algebra. Let $\mathfrak{a}$ be a solvable ideal of $\mathfrak{g}$ such that $\mathfrak{g}/\mathfrak{a}$ is solvable. Then $\mathfrak{g}$ is solvable.*

Proof. Suppose $\mathfrak{a}^{(s)} = 0$. We also have $\mathfrak{g}^{(t)} \subset \mathfrak{a}$ for some $t > 0$. This implies that $\mathfrak{g}^{(s+t)} = 0$. □

Lemma 2.6.2 *Let $\mathfrak{g}$ be a Lie algebra and $\mathfrak{a}$ and $\mathfrak{b}$ solvable ideals of $\mathfrak{g}$. Then $\mathfrak{a} + \mathfrak{b}$ is a solvable ideal of $\mathfrak{g}$.*

Proof. We have that $\mathfrak{a}/(\mathfrak{a} \cap \mathfrak{b})$ is solvable. Moreover, it is isomorphic to $(\mathfrak{a} + \mathfrak{b})/\mathfrak{b}$, which consequently is solvable. We now apply Lemma 2.6.1. □

Let $\mathfrak{g}$ be a finite-dimensional Lie algebra. Using the lemma we see that there is a unique maximal solvable ideal in $\mathfrak{g}$. It is called the *solvable radical* of $\mathfrak{g}$, and denoted $\mathfrak{sr}(\mathfrak{g})$.

2.6.1 Lie's theorem

In this section we assume the ground field k to be of characteristic 0. A subalgebra $\mathfrak{a}$ of $\mathfrak{gl}(V)$ where V is a vector space over k is called *split* if k contains the eigenvalues of all elements of $\mathfrak{a}$.

Lemma 2.6.3 *Let V be a finite-dimensional vector space over k. Let $c, a_i, b_i \in \mathrm{End}(V)$ be such that $c = \sum_{i=1}^{m}[a_i, b_i]$. Suppose $[c, b_i] = 0$ for all i. Then c is nilpotent.*

Proof. We have $c^r = c^{r-1}\sum_i[a_i, b_i] = \sum_i(c^{r-1}a_i)b_i - b_i(c^{r-1}a_i) = \sum_i[c^{r-1}a_i, b_i]$. Since the trace of a commutator is zero, it follows that $\mathrm{Tr}(c^r) = 0$ for all $r > 0$ and c is nilpotent. □

Lemma 2.6.4 *Let V be a finite-dimensional vector space over k and $\mathfrak{g} \subset \mathfrak{gl}(V)$ a Lie subalgebra. Suppose V is an irreducible $\mathfrak{g}$-module. Then the solvable radical of $\mathfrak{g}$ is equal to the centre of $\mathfrak{g}$.*

Proof. It is enough to show that $\mathfrak{sr}(\mathfrak{g}) \subset \mathfrak{c}(\mathfrak{g})$, which is equivalent to saying that $\mathfrak{a} = [\mathfrak{g}, \mathfrak{sr}(\mathfrak{g})]$ is zero. Suppose that it is non-zero. Since $\mathfrak{a}$ is solvable, there is an $s > 0$ with $\mathfrak{a}^{(s)} \neq 0$ and $\mathfrak{a}^{(s+1)} = 0$. Set $\mathfrak{b} = [\mathfrak{g}, \mathfrak{a}^{(s)}]$. For $c \in \mathfrak{b}$ there are $a_i \in \mathfrak{g}$, $b_i \in \mathfrak{a}^{(s)}$ such that $c = \sum_i[a_i, b_i]$. Lemma 2.6.3 implies that c is nilpotent. Now Lemma 2.4.2 yields $\mathfrak{b} = 0$, so that $\mathfrak{a}^{(s)} \subset \mathfrak{c}(\mathfrak{g})$. Also $\mathfrak{a}^{(s)} \subset \mathfrak{a}$, so for $c \in \mathfrak{a}^{(s)}$ there are $a_i \in \mathfrak{g}$, $b_i \in \mathfrak{sr}(\mathfrak{g})$ with $c = \sum_i[a_i, b_i]$. Since $c \in \mathfrak{c}(\mathfrak{g})$, by Lemma 2.6.3 we infer that c is nilpotent. This time Lemma 2.4.2 implies that $\mathfrak{a}^{(s)} = 0$, which is a contradiction. □

Theorem 2.6.5 (Lie) *Let $\mathfrak{g}$ be a finite-dimensional solvable Lie algebra over k, and $\rho : \mathfrak{g} \to \mathfrak{gl}(V)$ a finite-dimensional representation. Suppose $\rho(\mathfrak{g})$ is split. Then there is a basis of V with respect to which the matrix of $\rho(x)$ is upper triangular for all $x \in \mathfrak{g}$.*

Proof. First suppose that V is an irreducible $\mathfrak{g}$-module. Then we claim that $\dim V = 1$. By Lemma 2.6.4, $\rho(\mathfrak{g})$ is commutative. Let $x \in \mathfrak{g}$ and $W \subset V$ be a non-zero eigenspace of $\rho(x)$. Then $\rho(\mathfrak{g})$ stabilizes W, so $W = V$. Therefore every $\rho(x)$ acts as multiplication by a scalar. The irreducibility of V now proves the claim.

For the general case, let $0 = V_0 \subset V_1 \subset \cdots \subset V_m = V$ be a composition series of the $\mathfrak{g}$-module V. By our claim all quotients V_{i+1}/V_i are 1-dimensional. This immediately implies the theorem. □

2.6.2 Cartan's criterion for solvability

Theorem 2.6.6 (Cartan) *Let $\mathfrak{g}$ be a Lie algebra over a field k of characteristic 0. Let $\rho : \mathfrak{g} \to \mathfrak{gl}(V)$ be a finite-dimensional representation. Suppose the kernel of ρ is a solvable ideal of $\mathfrak{g}$ and* $\mathrm{Tr}(\rho(x)\rho(y)) = 0$ *for all $x, y \in [\mathfrak{g}, \mathfrak{g}]$. Then $\mathfrak{g}$ is solvable.*

Proof. First we assume that k is algebraically closed. Suppose that $[\mathfrak{g}, \mathfrak{g}] = \mathfrak{g}$. Let $\mathfrak{h}$ be a Cartan subalgebra of $\mathfrak{g}$ and $\alpha_1, \dots, \alpha_s$ be the roots of $\mathfrak{g}$ with respect to $\mathfrak{h}$. Set $\alpha_0 = 0$ so that $\mathfrak{g} = \mathfrak{g}_{\alpha_0} \oplus \cdots \oplus \mathfrak{g}_{\alpha_s}$. In view of Proposition 2.5.12, since $[\mathfrak{g}, \mathfrak{g}] = \mathfrak{g}$, $\mathfrak{h} = \mathfrak{g}_{\alpha_0}$ is spanned by the spaces $[\mathfrak{g}_\alpha, \mathfrak{g}_{-\alpha}]$ where α is a root such that $-\alpha$ is a root as well (so these include $\alpha = \alpha_0$). Let α be such a root and $x_\alpha \in \mathfrak{g}_\alpha$, $x_{-\alpha} \in \mathfrak{g}_{-\alpha}$, and $z_\alpha = [x_\alpha, x_{-\alpha}] \in \mathfrak{h}$. Let μ be a weight of V. Then there is a $q_\mu \in \mathbb{Q}$ such that $\mu(z_\alpha) = q_\mu \alpha(z_\alpha)$. (In order to see this, let $W \subset V$ be the subspace spanned by all $V_{\mu+i\alpha}$ for $i \in \mathbb{Z}$. For $x \in \mathfrak{g}$ such that $\rho(x)$ stabilizes W, we denote by $\mathrm{Tr}_W(\rho(x))$ the trace of the restriction of $\rho(x)$ to W. Then $\mathrm{Tr}_W(\rho(z_\alpha)) = 0$ as z_α is a commutator. But with $d_i = \dim V_{\mu+i\alpha}$ we also have $\mathrm{Tr}_W(\rho(z_\alpha)) = \sum_i d_i(\mu(z_\alpha) + i\alpha(z_\alpha))$.) Writing $d_\mu = \dim V_\mu$ we see that $0 = \mathrm{Tr}(\rho(z_\alpha)\rho(z_\alpha)) = \sum_\mu d_\mu \mu(z_\alpha)^2 = \alpha(z_\alpha)^2 \sum_\mu d_\mu q_\mu^2$. This implies that $\mu(z_\alpha) = 0$. Since $\mathfrak{h}$ is spanned by elements of the form z_α we see that $\mu = 0$ is the only weight of V. Hence all $\mathfrak{g}_\alpha$ for $\alpha \neq \alpha_0$ lie in the kernel of ρ (here we use Proposition 2.5.12). So $\mathfrak{g}/\ker\rho \cong \rho(\mathfrak{g})$ is a homomorphic image of $\mathfrak{h}$ and therefore nilpotent and hence solvable. So $\mathfrak{g}$ is solvable by Lemma 2.6.1. It follows that $\mathfrak{g} \neq [\mathfrak{g}, \mathfrak{g}]$ and we have a contradiction. So in fact, $\mathfrak{g} \neq [\mathfrak{g}, \mathfrak{g}]$. Now the restriction of ρ to $[\mathfrak{g}, \mathfrak{g}]$ has the same two properties as ρ. We conclude that $\mathfrak{g}$ is solvable.

If k is not algebraically closed, we tensor everything with the algebraic closure K of k. The two properties of ρ remain valid (note that $\mathrm{Tr}(\rho(x)\rho(y))$ is bilinear, so it is enough to have this for x, y in a basis of $[\mathfrak{g}, \mathfrak{g}]$). The conclusion is that $K \otimes \mathfrak{g}$ is solvable, implying the same for $\mathfrak{g}$. □

2.6.3 Computing the solvable radical

Throughout we let $\mathfrak{g}$ be a finite-dimensional Lie algebra over the field k of characteristic 0. Here we describe an algorithm for computing the solvable radical of $\mathfrak{g}$. It is based on a result which can be viewed as a converse to Cartan's criterion.

Lemma 2.6.7 *Let $\rho : \mathfrak{g} \to \mathfrak{gl}(V)$ be a finite-dimensional representation. Then all elements of $\rho([\mathfrak{g}, \mathfrak{g}] \cap \mathfrak{sr}(\mathfrak{g}))$ are nilpotent endomorphisms of V.*

Proof. First suppose V is an irreducible $\mathfrak{g}$-module. Then we claim that $\rho([\mathfrak{g}, \mathfrak{g}] \cap \mathfrak{sr}(\mathfrak{g})) = 0$. Let $x \in [\mathfrak{g}, \mathfrak{g}] \cap \mathfrak{sr}(\mathfrak{g})$. Then there are $y_i, z_i \in \mathfrak{g}$ such that $x = \sum_i [y_i, z_i]$, so that $\rho(x) = \sum_i [\rho(y_i), \rho(z_i)]$. Since $x \in \mathfrak{sr}(\mathfrak{g})$, we obtain $[\rho(x), \rho(\mathfrak{g})] = 0$ from Lemma 2.6.4. So by Lemma 2.6.3 it follows that $\rho(x)$ is nilpotent, and we can conclude using Lemma 2.4.2.

Now the lemma follows from the claim by considering a composition series of the $\mathfrak{g}$-module V. □

Theorem 2.6.8 $\mathfrak{sr}(\mathfrak{g}) = \{x \in \mathfrak{g} \mid \mathrm{Tr}(\mathrm{ad}x \cdot \mathrm{ad}y) = 0$ *for all* $y \in [\mathfrak{g}, \mathfrak{g}]\}$.

Proof. Let $\mathfrak{a}$ denote the right hand side. Let $0 = \mathfrak{g}_0 \subset \mathfrak{g}_1 \subset \cdots \subset \mathfrak{g}_s = \mathfrak{g}$ be a composition series of $\mathfrak{g}$ with respect to the adjoint representation. Let $u \in [\mathfrak{g}, \mathfrak{g}] \cap \mathfrak{sr}(\mathfrak{g})$. Then by Lemma 2.6.7, $\mathrm{ad}u$ is nilpotent. By Lemma 2.4.2 it follows that $\mathrm{ad}u(\mathfrak{g}_{i+1}) \subset \mathfrak{g}_i$, for $0 \le i < s$. Let $r \in \mathfrak{sr}(\mathfrak{g})$, and $x, y \in \mathfrak{g}$. Then $\mathrm{ad}[y, r](\mathfrak{g}_{i+1}) \subset \mathfrak{g}_i$ and $\mathrm{ad}x \cdot \mathrm{ad}[y, r](\mathfrak{g}_{i+1}) \subset \mathfrak{g}_i$. In particular, $\mathrm{ad}x \cdot \mathrm{ad}[y, r]$ is nilpotent. Using Lemma 2.1.2 we obtain $0 = \mathrm{Tr}(\mathrm{ad}x \cdot \mathrm{ad}[y, r]) = \mathrm{Tr}(\mathrm{ad}r \cdot \mathrm{ad}[x, y])$. We conclude that $r \in \mathfrak{a}$.

For the reverse inclusion, one first proves that $\mathfrak{a}$ is an ideal of $\mathfrak{g}$, using Lemma 2.1.2. Cartan's criterion (Theorem 2.6.6) implies that $\mathfrak{a}$ is solvable, so that $\mathfrak{a} \subset \mathfrak{sr}(\mathfrak{g})$. □

Let $\{x_1, \dots, x_n\}$ be a basis of $\mathfrak{g}$. In order to compute the solvable radical of $\mathfrak{g}$ we first compute a basis $\{u_1, \dots, u_m\}$ of $[\mathfrak{g}, \mathfrak{g}]$ (see Section 2.4). Then using the previous theorem, we have that $\sum_{i=1}^n \alpha_i x_i$ lies in $\mathfrak{sr}(\mathfrak{g})$ if and only if $\sum_{i=1}^n \mathrm{Tr}(\mathrm{ad}x_i \cdot \mathrm{ad}u_j)\alpha_i = 0$ for $1 \le j \le m$. Hence we can compute a basis of $\mathfrak{sr}(\mathfrak{g})$ by solving a system of m linear equations in n indeterminates.

Remark 2.6.9 There is an algorithm due to Rónyai for computing the solvable radical of a Lie algebra defined over a field of characteristic $p > 0$. The basic idea is to repeatedly compute the nilradical and construct quotients. For more details we refer to [Rón90] and [Gra00].

2.6.4 Finding a non-nilpotent element

In this section, $\mathfrak{g}$ is a finite-dimensional Lie algebra over a field k of characteristic 0. An element $x \in \mathfrak{g}$ is said to be *nilpotent* if $\mathrm{ad}_{\mathfrak{g}}x$ is a nilpotent linear map. We describe an algorithm for finding an element in $\mathfrak{g}$ that is not nilpotent (provided such elements exist). This is an essential step in the algorithm for computing a Cartan subalgebra.

Proposition 2.6.10 *Write* $\mathfrak{r} = \mathfrak{sr}(\mathfrak{g})$. *Let* A *be the set of* $x \in \mathfrak{r}$ *such that* $\mathrm{ad}_{\mathfrak{r}} x$ *is nilpotent. Let* B *be the set of* $x \in \mathfrak{r}$ *such that* $\mathrm{ad}_{\mathfrak{g}} x$ *is nilpotent. Then* $\mathfrak{nr}(\mathfrak{g}) = A = B$.

Proof. Obviously $\mathfrak{nr}(\mathfrak{g}) \subset A \subset B$. Let $x \in B$. Let $0 = \mathfrak{g}_0 \subset \mathfrak{g}_1 \subset \cdots \subset \mathfrak{g}_s$ be a composition series of $\mathfrak{g}$ with respect to the adjoint representation of $\mathfrak{g}$. Let $\sigma_i : \mathfrak{g} \to \mathfrak{gl}(\mathfrak{g}_i/\mathfrak{g}_{i-1})$ be the induced representation. By Lemma 2.6.4, $\sigma_i(x)$ lies in the centre of $\sigma_i(\mathfrak{g})$, so it spans an ideal of the latter algebra. Hence by Lemma 2.4.2, $\sigma_i(x) = 0$ and $\mathrm{ad}x(\mathfrak{g}_i) \subset \mathfrak{g}_{i-1}$. The set of all elements with that property forms a nilpotent ideal of $\mathfrak{g}$. So $x \in \mathfrak{nr}(\mathfrak{g})$. □

The next proposition yields an immediate algorithm to find a non-nilpotent element in $\mathfrak{g}$, provided one exists.

Proposition 2.6.11 *Let* $\{x_1, \dots, x_n\}$ *be a basis of* $\mathfrak{g}$. *Assume that* $\mathfrak{g}$ *is not nilpotent. Then* $\{x_1, \dots, x_n\} \cup \{x_i + x_j \mid 1 \le i < j \le n\}$ *contains a non-nilpotent element.*

Proof. If $\mathfrak{g}$ is solvable, then by Proposition 2.6.10, one of the x_i is not nilpotent. If $\mathfrak{g}$ is not solvable, then by Cartan's criterion (Theorem 2.6.6), there are x_i, x_j such that $\mathrm{Tr}(\mathrm{ad}x_i \cdot \mathrm{ad}x_j) \neq 0$. So

$$\mathrm{Tr}(\mathrm{ad}(x_i + x_j)^2) - \mathrm{Tr}((\mathrm{ad}x_i)^2) - \mathrm{Tr}((\mathrm{ad}x_j)^2) = 2\mathrm{Tr}(\mathrm{ad}x_i \cdot \mathrm{ad}x_j) \neq 0.$$

Thus at least one of x_i, x_j, $x_i + x_j$ is not nilpotent. □

2.6.5 Computing the nilradical

Here $\mathfrak{g}$ is a finite-dimensional Lie algebra over the field k of characteristic 0. We describe an algorithm to compute the nilradical of $\mathfrak{g}$. For that we make use of associative algebras. The *radical* of an associative algebra A is defined as the maximal nilpotent ideal of A, denoted $\mathrm{Rad}(A)$ (the existence of this is proved in the same way as for the nilradical). Over base fields of characteristic 0 there is a straightforward algorithm to compute the radical of A; see [Rón90].

Lemma 2.6.12 *Let* V *be a finite-dimensional vector space over* k, *and* $\mathfrak{a} \subset \mathfrak{gl}(V)$ *a solvable Lie subalgebra. Let* $A \subset \mathrm{End}(V)$ *be the associative algebra generated by* $\mathfrak{a}$, *along with the identity. Then* $A/\mathrm{Rad}(A)$ *is commutative.*

Proof. Let $0 = V_0 \subset V_1 \subset \cdots \subset V_s = V$ be a composition series of the $\mathfrak{a}$-module V. Let I be the set of $a \in A$ such that $a \cdot V_i \subset V_{i-1}$, $1 \le i < s$. Then I is a nilpotent ideal and $I \subset \mathrm{Rad}(A)$. For $x, y \in \mathfrak{a}$ we have $[x, y] \in I$

by Lemma 2.6.4. So the generators of A commute modulo $\mathrm{Rad}(A)$. Therefore $A/\mathrm{Rad}(A)$ is commutative. □

Proposition 2.6.13 *Write $\mathfrak{r} = \mathfrak{sr}(\mathfrak{g})$. Let $\mathfrak{h}$ be a Cartan subalgebra of $\mathfrak{r}$, and $\mathfrak{r} = \mathfrak{h} \oplus \mathfrak{r}_1$ the Fitting decomposition of $\mathfrak{r}$ with respect to $\mathfrak{h}$. Let A be the associative algebra generated by $\mathrm{ad}_{\mathfrak{r}_1} h$ for $h \in \mathfrak{h}$, along with the identity on $\mathfrak{r}_1$. Then $\mathfrak{nr}(\mathfrak{g}) = \mathfrak{r}_1 \oplus \{h \in \mathfrak{h} \mid \mathrm{ad}_{\mathfrak{r}_1} h \in \mathrm{Rad}(A)\}$.*

Proof. We remark that $\mathfrak{r}_1 = [\mathfrak{h}, \mathfrak{r}_1] \subset [\mathfrak{r}, \mathfrak{r}]$, and the latter is a nilpotent ideal of $\mathfrak{r}$ (Lemma 2.6.7) and lies in $\mathfrak{nr}(\mathfrak{r}) = \mathfrak{nr}(\mathfrak{g})$ (this equality follows from Proposition 2.6.10). If $\mathrm{ad}_{\mathfrak{r}_1} h \in \mathrm{Rad}(A)$, then $\mathrm{ad}_{\mathfrak{r}} h$ is nilpotent, so that $h \in \mathfrak{nr}(\mathfrak{g})$ by Proposition 2.6.10. We conclude that the right-hand side is contained in $\mathfrak{nr}(\mathfrak{g})$.

Let $x \in \mathfrak{nr}(\mathfrak{g})$; then $x \in \mathfrak{r}$ and write $x = h + r$, $h \in \mathfrak{h}$, $r \in \mathfrak{r}_1$. So $h \in \mathfrak{nr}(\mathfrak{g})$, and $a = \mathrm{ad}_{\mathfrak{r}_1} h$ is nilpotent. The image of a in the quotient $A/\mathrm{Rad}(A)$ is nilpotent as well. But the latter is a commutative semisimple associative algebra, and therefore has no nilpotent elements other than 0. It follows that $\mathrm{ad}_{\mathfrak{r}_1} h \in \mathrm{Rad}(A)$. □

This implies that the following procedure for computing the nilradical is correct. First compute the solvable radical $\mathfrak{r} = \mathfrak{sr}(\mathfrak{g})$. Second compute a Cartan subalgebra $\mathfrak{h} \subset \mathfrak{r}$, and the Fitting one component $\mathfrak{r}_1 = \mathfrak{r}_1(\mathfrak{h})$. Third compute a basis of the algebra A as in the proposition and compute its radical. Finally, compute the space $\{h \in \mathfrak{h} \mid \mathrm{ad}_{\mathfrak{r}_1} h \in \mathrm{Rad}(A)\}$, and return the sum of that with $\mathfrak{r}_1$.

Remark 2.6.14 For Lie algebras $\mathfrak{g}$ over finite fields of characteristic $p > 0$ the nilradical can be computed. One first computes the radical R of the associative algebra with one generated by $\mathrm{ad}\mathfrak{g}$, then the nilradical of $\mathfrak{g}$ is the set consisting of $x \in \mathfrak{g}$ such that $\mathrm{ad}x \in R$. (We refer to [Rón90], [Gra00], Section 2.2 for more details.)

2.7 Semisimple Lie algebras

The classification of the semisimple Lie algebras over algebraically closed fields of characteristic 0 is viewed by many as one of the main results in modern mathematics. It has many applications in such diverse fields as group theory, geometry and physics. Because of the connection between Lie algebras and algebraic groups, this classification is also of great importance for the theory of algebraic groups. Here and in subsequent sections we give an overview of the main constructions used to obtain this classification.

A finite-dimensional Lie algebra $\mathfrak{g}$ is called *simple* if $\dim \mathfrak{g} > 1$, and its only ideals are 0 and $\mathfrak{g}$ itself. Furthermore, $\mathfrak{g}$ is said to be *semisimple* if $\mathfrak{sr}(\mathfrak{g}) = 0$.

Example 2.7.1 Let $\mathfrak{g} = \mathfrak{sl}(2, k)$, where k is a field of characteristic 0 (see Example 2.1.4). By some elementary calculations it is seen that $\mathfrak{g}$ is simple.

A pivotal role in the theory of semisimple Lie algebras is played by the *Killing form.* This is the trace form corresponding to the adjoint representation, denoted $\kappa_{\mathfrak{g}}$, so $\kappa_{\mathfrak{g}}(x, y) = \mathrm{Tr}(\mathrm{ad}_{\mathfrak{g}} x \cdot \mathrm{ad}_{\mathfrak{g}} y)$. We also write κ instead of $\kappa_{\mathfrak{g}}$ if it is clear which Lie algebra we mean. From Lemma 2.1.2 we recall that $\kappa_{\mathfrak{g}}$ is invariant, i.e., $\kappa([x, y], z) = \kappa(x, [y, z])$.

Proposition 2.7.2 *Let $\mathfrak{g}$ be a finite-dimensional Lie algebra over a field of characteristic 0. Then $\mathfrak{g}$ is semisimple if and only if $\kappa_{\mathfrak{g}}$ is non-degenerate.*

Proof. Consider $R = \{x \in \mathfrak{g} \mid \kappa_{\mathfrak{g}}(x, y) = 0 \text{ for all } y \in \mathfrak{g}\}$, which is called the radical of $\kappa_{\mathfrak{g}}$. The invariance of $\kappa_{\mathfrak{g}}$ implies that R is an ideal of $\mathfrak{g}$. Moreover, by Cartan's criterion (Theorem 2.6.6) it is solvable. So if $\mathfrak{g}$ is semisimple, then $R = 0$ and $\kappa_{\mathfrak{g}}$ is non-degenerate.

Conversely, suppose that $\kappa_{\mathfrak{g}}$ is non-degenerate. Set $\mathfrak{a} = \mathfrak{sr}(\mathfrak{g})$, and suppose that it is non-zero. Then there is an $m > 0$ such that $\mathfrak{a}^{(m)} \neq 0$, and $\mathfrak{a}^{(m+1)} = 0$. But then $\mathfrak{a}^{(m)}$ must be contained in R, so we have a contradiction. □

Example 2.7.3 Let $\mathfrak{g}$ be as in Example 2.7.1. Let the basis elements be denoted x_i, $i = 1, 2, 3$. It is straightforward to write the matrix $(\kappa_{\mathfrak{g}}(x_i, x_j))$. This matrix is non-singular, and hence $\kappa_{\mathfrak{g}}$ is non-degenerate. So by the proposition it follows that $\mathfrak{g}$ is semisimple.

Proposition 2.7.4 *Let $\mathfrak{g}$ be a finite-dimensional semisimple Lie algebra over a field of characteristic 0. Then there are unique simple ideals $\mathfrak{a}_1, \dots, \mathfrak{a}_r$ of $\mathfrak{g}$ such that $\mathfrak{g} = \mathfrak{a}_1 \oplus \cdots \oplus \mathfrak{a}_r$.*

Proof. If $\mathfrak{g}$ is simple, there is nothing to prove. Otherwise, $\mathfrak{g}$ has a proper ideal $\mathfrak{b}$. Set $\mathfrak{b}^\perp = \{x \in \mathfrak{g} \mid \kappa(x, y) = 0 \text{ for all } y \in \mathfrak{b}\}$. By the invariance of κ this is an ideal of $\mathfrak{g}$. Set $\mathfrak{c} = \mathfrak{b} \cap \mathfrak{b}^\perp$. Again using the invariance of κ we have $\kappa([\mathfrak{c}, \mathfrak{c}], \mathfrak{g}) = 0$, so $[\mathfrak{c}, \mathfrak{c}] = 0$ (Proposition 2.7.2). Hence $\mathfrak{c}$ is a solvable ideal and therefore $\mathfrak{c} = 0$. As κ is non-degenerate, $\dim \mathfrak{b} + \dim \mathfrak{b}^\perp = \dim \mathfrak{g}$, whence $\mathfrak{g} = \mathfrak{b} \oplus \mathfrak{b}^\perp$. The Killing forms of $\mathfrak{b}$, $\mathfrak{b}^\perp$ are non-degenerate as well. So the proof of existence is finished by induction on the dimension.

In order to show uniqueness, let $\mathfrak{b}$ be a simple ideal of $\mathfrak{g}$. Set $\mathfrak{c}_i = \mathfrak{a}_i \cap \mathfrak{b}$. Note that $[\mathfrak{a}_i, \mathfrak{b}] \subset \mathfrak{c}_i$ so there must be an i with $\mathfrak{c}_i \neq 0$, but that implies $\mathfrak{a}_i = \mathfrak{b}$. □

Example 2.7.5 Let $\mathfrak{g}$ be as in Example 2.7.1. In Example 2.7.3 it is shown that $\mathfrak{g}$ is semisimple. If it is not simple, then it has a simple ideal of dimension 1, which is not possible. We conclude that $\mathfrak{g}$ is simple.

The next theorem is of fundamental importance for the representation theory of semisimple Lie algebras, and by considering the adjoint representation, for their structure theory as well. We do not go into the proof, but refer to [Jac79], Chapter III, Theorem 8 or [Hum78], Section 6.3.

Theorem 2.7.6 (Weyl) *Let $\mathfrak{g}$ be a finite-dimensional semisimple Lie algebra over a field k of characteristic 0. Let $\rho : \mathfrak{g} \to \mathfrak{gl}(V)$ be a finite-dimensional representation of $\mathfrak{g}$. Then there are irreducible $\mathfrak{g}$-modules, $V_1, \dots, V_r \subset V$ such that $V = V_1 \oplus \cdots \oplus V_r$.*

2.7.1 Derivations and the Jordan decomposition

In this section $\mathfrak{g}$ is a finite-dimensional semisimple Lie algebra over the field k of characteristic 0.

Proposition 2.7.7 *All derivations of $\mathfrak{g}$ are inner.*

Proof. Let $d \in \mathrm{Der}(\mathfrak{g})$ and $f : \mathfrak{g} \to k$ be the linear map defined by $f(x) = \mathrm{Tr}(d \cdot \mathrm{ad}x)$. Since the Killing form κ is non-degenerate (Proposition 2.7.2), there is a unique $x_f \in \mathfrak{g}$ with $f(y) = \kappa(x_f, y)$ for $y \in \mathfrak{g}$. Set $d' = d - \mathrm{ad}x_f$. A short calculation shows $[d', \mathrm{ad}x](y) = [d'(x), y]$ for $x, y \in \mathfrak{g}$, so that $[d', \mathrm{ad}x] = \mathrm{ad}d'(x)$ for $x \in \mathfrak{g}$. Hence for $y \in \mathfrak{g}$ we have $\kappa(d'(x), y) = \mathrm{Tr}(\mathrm{ad}d'(x) \cdot \mathrm{ad}y) = \mathrm{Tr}([d', \mathrm{ad}x] \cdot \mathrm{ad}y) = \mathrm{Tr}(d' \cdot \mathrm{ad}[x, y])$ (Lemma 2.1.2). But the latter is zero by the definition of d'. By the non-degeneracy of κ we now conclude that $d' = 0$, and $d = \mathrm{ad}x_f$. □

Lemma 2.7.8 *Let A be a finite-dimensional algebra over k. Let $d \in \mathrm{Der}(A)$ and $d = d_s + d_n$ be the Jordan decomposition of d (where d_s is semisimple and d_n is nilpotent). Then $d_s, d_n \in \mathrm{Der}(A)$.*

Proof. We may assume that k is algebraically closed. Let $\lambda \in k$ be an eigenvalue of d and set $A_\lambda = \{a \in A \mid (d - \lambda)^m(a) = 0 \text{ for an } m > 0\}$. By induction on m we have, for $a, b \in A$ and $\lambda, \mu \in k$:

$$(d - \lambda - \mu)^m(ab) = \sum_{i=0}^{m} \binom{m}{i} (d - \lambda)^{m-i}(a)(d - \mu)^i(b).$$

So $A_\lambda A_\mu \subset A_{\lambda+\mu}$. Furthermore, d_s acts as multiplication by λ on A_λ. Hence, for $a_\lambda \in A_\lambda$, $a_\mu \in A_\mu$ we infer that $d_s(a_\lambda a_\mu) = (\lambda + \mu)a_\lambda a_\mu$. Since $d_s(a_\lambda)a_\mu + a_\lambda d_s(a_\mu) = (\lambda + \mu)a_\lambda a_\mu$, and A is the direct sum of subspaces of the form A_λ, we conclude that $d_s \in \mathrm{Der}(A)$ and $d_n \in \mathrm{Der}(A)$. □

Theorem 2.7.9 *For $x \in \mathfrak{g}$ there are unique $x_n, x_s \in \mathfrak{g}$ with the following properties: $x = x_s + x_n$, $\mathrm{ad}x_s$ is semisimple, $\mathrm{ad}x_n$ is nilpotent, and $[x_s, x_n] = 0$.*

Proof. Let $d = \mathrm{ad}x \in \mathrm{Der}(\mathfrak{g})$ and $d = d_s + d_n$ be its Jordan decomposition. By Lemma 2.7.8, together with Proposition 2.7.7, there are $x_s, x_n \in \mathfrak{g}$ with $d_s = \mathrm{ad}x_s$, $d_n = \mathrm{ad}x_n$. Since the centre of $\mathfrak{g}$ is zero, we have $x = x_s + x_n$ and $[x_s, x_n] = 0$. The uniqueness of x_s, x_n follows from the uniqueness of the Jordan decomposition of d (Proposition 2.2.3). □

The elements x_s, x_n provided by this theorem are called the *semisimple* and *nilpotent parts* of x.

2.7.2 Levi's theorem

This section is a bit of an aside in the classification of the semisimple Lie algebras. We state Levi's theorem, which concerns Lie algebras that are not solvable and not semisimple. We omit the proof, for which we refer to [Jac79], Section III.9. Subsequently we indicate how to compute a subalgebra as provided by the theorem.

Theorem 2.7.10 (Levi) *Let $\mathfrak{g}$ be a finite-dimensional Lie algebra over a field k of characteristic 0. Suppose $\mathfrak{g}$ is not solvable. Then $\mathfrak{g}$ contains a semisimple subalgebra $\mathfrak{s}$ such that $\mathfrak{g} = \mathfrak{s} \oplus \mathfrak{sr}(\mathfrak{g})$ (direct sum of vector spaces).*

A decomposition as in the theorem is called a *Levi decomposition.* In order to compute a Levi decomposition we first compute a basis $\{x_1, \dots, x_m\}$ of a complement in $\mathfrak{g}$ to $\mathfrak{r} = \mathfrak{sr}(\mathfrak{g})$. Then there are $\gamma_{ij}^l \in k$ such that $[x_i, x_j] = \sum_{l=1}^m \gamma_{ij}^l x_l \bmod \mathfrak{r}$. The idea is to find $u_i \in \mathfrak{r}$ such that with $y_i = x_i + u_i$ we have $[y_i, y_j] = \sum_{l=1}^m \gamma_{ij}^l y_l$, for $1 \le i, j \le m$. The existence of such u_i is guaranteed by Levi's theorem.

Note that the derived series of $\mathfrak{r}$ yields a series of ideals $\mathfrak{r} = \mathfrak{r}_1 \supset \mathfrak{r}_2 \supset \cdots \supset \mathfrak{r}_r \supset \mathfrak{r}_{r+1} = 0$ of $\mathfrak{r}$, such that $[\mathfrak{r}_i, \mathfrak{r}_i] \subset \mathfrak{r}_{i+1}$.

We define a sequence of elements y_i^t of $\mathfrak{g}$ in the following way. Set $y_i^1 = x_i$, $1 \le i \le m$; then $[y_i^1, y_j^1] = \sum_{l=1}^m \gamma_{ij}^l y_l^1 \bmod \mathfrak{r}_1$. Now let $t \ge 1$ and suppose $y_i^t \in \mathfrak{g}$ have been determined such that $[y_i^t, y_j^t] = \sum_{l=1}^m \gamma_{ij}^l y_l^t \bmod \mathfrak{r}_t$. Let V_t be a complement in $\mathfrak{r}_t$ to $\mathfrak{r}_{t+1}$, i.e., $\mathfrak{r}_t = V_t \oplus \mathfrak{r}_{t+1}$. Set $y_i^{t+1} = y_i^t + v_i^t$, where $v_i^t \in V_t$. Then since $[v_i^t, v_j^t] \in \mathfrak{r}_{t+1}$, we infer that $[y_i^{t+1}, y_j^{t+1}] = \sum_{l=1}^m \gamma_{ij}^l y_l^{t+1} \bmod \mathfrak{r}_{t+1}$ is equivalent to

$$[y_i^t, v_j^t] + [v_i^t, y_j^t] - \sum_{l=1}^m \gamma_{ij}^l v_l^t = -[y_i^t, y_j^t] + \sum_{l=1}^m \gamma_{ij}^l y_l^t \bmod \mathfrak{r}_{t+1}.$$

But these amount to a set of linear equations for the v_i^t. Moreover, since the equations are modulo $\mathfrak{r}_{t+1}$ the left- and right-hand sides can be viewed as elements of V_t; so when solving the equations we can work in this space.

Furthermore, the equations have a solution by Levi's theorem applied to the Lie algebra $\mathfrak{g}/\mathfrak{r}_{t+1}$.

Example 2.7.11 Let $\mathfrak{g}$ be the Lie algebra of Example 2.1.3. The elements x_1, x_5, x_6 span the solvable radical of $\mathfrak{g}$. An example of a semisimple subalgebra complementing $\mathfrak{r}$ is the subalgebra spanned by $x_2, x_3 - \frac{1}{2}x_6, x_4$.

2.8 Root systems

A root system is a set of vectors in real vector space satisfying a few requirements. For the study of semisimple Lie algebras they have proved to be of enormous value, as the classification of the semisimple Lie algebras can be reduced to the classification of root systems. In this section we summarize some of their main properties.

First we briefly recall some constructions from linear algebra. Let V be a finite-dimensional vector space over $\mathbb{R}$. A bilinear form $(\ ,\ ) : V \times V \to \mathbb{R}$ that is symmetric ($(v, w) = (w, v)$ for all $v, w \in V$) and positive definite ($(v, v) > 0$ if $v \neq 0$) is called an *inner product* on V. Let $v_1, \dots, v_n$ be a basis of V, and define the $n \times n$ matrix B by $B(i, j) = (v_i, v_j)$. It is clear that B determines the inner product completely and therefore is called the matrix of the inner product $(\ ,\ )$ (relative to the given basis). Furthermore, B is symmetric and positive definite. Conversely, let B be a positive definite symmetric real $n \times n$ matrix. Then setting $(v_i, v_j) = B(i, j)$ defines an inner product on V.

By Sylvester's criterion, a real symmetric matrix B is positive definite if and only if all its principal minors are positive.

If we are given an inner product $(\ ,\ )$ on V, the *norm* of a $v \in V$ is $\|v\| = \sqrt{(v, v)}$. Furthermore, the *angle* between $v, w \in V$ is the real number $\theta \in [0, \pi]$ such that $\cos(\theta) = \frac{(v,w)}{\|v\|\|w\|}$.

Throughout this section we use the following notation

$$\langle v, w^\vee \rangle = \frac{2(v, w)}{(w, w)}.$$

Reflections will be of paramount importance. These are linear maps leaving a hyperplane pointwise fixed and mapping vectors orthogonal to this hyperplane to their negatives. If $v \in V$, $v \neq 0$, the reflection of V with respect to the hyperplane orthogonal to v is defined by

$$s_v(w) = w - \langle w, v^\vee \rangle v.$$

Lemma 2.8.1 *Let V be a real vector space with inner product $(\ ,\ )$. Let $v, w \neq 0$ be non-proportional elements of V with $\|v\| \leq \|w\|$ and such that $\langle v, w^\vee \rangle$*

and $\langle w, v^\vee \rangle$ are both integers. They have the same sign, and the following possibilities occur:

1. $|\langle v, w^\vee \rangle| = |\langle w, v^\vee \rangle| = 1$ *and* $\|w\|^2 = \|v\|^2$,
2. $|\langle v, w^\vee \rangle| = 1$, $|\langle w, v^\vee \rangle| = 2$ *and* $\|w\|^2 = 2\|v\|^2$,
3. $|\langle v, w^\vee \rangle| = 1$, $|\langle w, v^\vee \rangle| = 3$ *and* $\|w\|^2 = 3\|v\|^2$.

Proof. This follows immediately from the observation $\langle v, w^\vee \rangle \langle w, v^\vee \rangle = 4\cos^2(\theta)$ where θ is the angle between v and w. □

2.8.1 Cartan matrices

To a root system one associates an integral matrix by which the root system is completely determined. This is called the *Cartan matrix* of the root system. Here we give the classification of the Cartan matrices, which will later immediately yield the classification of the root systems.

Definition 2.8.2 *An $\ell \times \ell$ matrix C with entries in $\mathbb{Z}$ is called a* Cartan matrix *if*

1. $C(i,i) = 2$ *for* $1 \le i \le \ell$,
2. $C(i,j) \le 0$ *for* $i \ne j$,
3. *there is a diagonal matrix* $D = \mathrm{diag}(d_1, \dots, d_\ell)$ *with* $d_i \in \mathbb{Z}_{>0}$ *such that* $B = CD$ *is a positive definite symmetric matrix.*

Example 2.8.3 Set

$$C = \begin{pmatrix} 2 & -1 & 0 \\ -1 & 2 & -1 \\ 0 & -2 & 2 \end{pmatrix}.$$

Let $D = \mathrm{diag}(1,1,2)$. By Sylvester's criterion, CD is positive definite so C is a Cartan matrix.

Let C, C' be $\ell \times \ell$ Cartan matrices. Then C and C' are said to be *equivalent* if there is a permutation σ of $\{1, \dots, \ell\}$ with $C(i,j) = C'(\sigma(i), \sigma(j))$ for all i, j. The Cartan matrix C is called *indecomposable* if it is not equivalent to a block diagonal matrix.

Let C be an $\ell \times \ell$ Cartan matrix, and $B = CD$ the positive definite matrix from Definition 2.8.2. Let V be an ℓ-dimensional vector space over $\mathbb{R}$ with basis $\alpha_1, \dots, \alpha_\ell$. The matrix B defines an inner product on V by $(\alpha_i, \alpha_j) = B(i,j)$. Writing $D = \mathrm{diag}(d_1, \dots, d_\ell)$ we have $\langle \alpha_i, \alpha_j^\vee \rangle = 2B(i,j)/B(j,j) = 2C(i,j)d_j/C(j,j)d_j = C(i,j)$. This, together with Lemma 2.8.1 implies that $C(i,j)C(j,i) = 0, 1, 2, 3$. Furthermore, if $C(i,j)C(j,i) \ne 0$, at least one of $C(i,j)$, $C(j,i)$ is -1.

It is extremely useful to code the information present in a Cartan matrix C into a graph, called the *Dynkin diagram* of C. This graph has ℓ points with labels $1, \ldots, \ell$. Two points i and j are connected by $C(i,j)C(j,i) = 0, 1, 2, 3$ edges. If this number is > 1, we insert an arrow toward j if $|C(i,j)| > 1$, and we put an arrow toward i if $|C(j,i)| > 1$. So the number of edges is greater than 1 if and only if α_i and α_i have different lengths; and the arrow points toward the smaller of the two. This arrow can also be seen as a "smaller than" sign. It is straightforward to recover the Cartan matrix from its Dynkin diagram.

Example 2.8.4 The Dynkin diagram of the Cartan matrix of Example 2.8.3 is

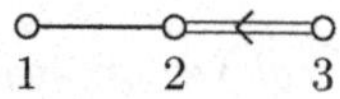

Theorem 2.8.5 *The Dynkin diagrams of the indecomposable Cartan matrices are given in the following list:*

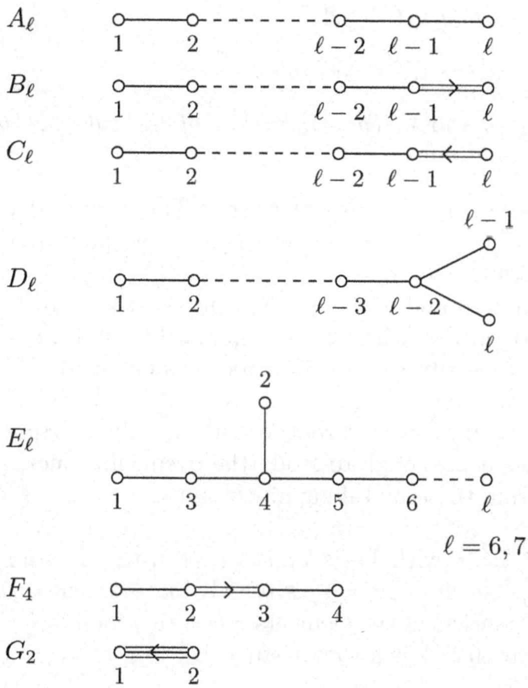

Proof. (Sketch.) Let C be an $\ell \times \ell$ Cartan matrix. Construct a real vector space V with basis $\{\alpha_1, \ldots, \alpha_\ell\}$, and inner product as above. Set $v_i = \lambda_i \alpha_i$, where $\lambda_i \in \mathbb{R}$ is a positive number such that v_i has length 1. Then

$$(v_i, v_i) = 1, \;\; 4(v_i, v_j)^2 = 0, 1, 2, 3, \quad \text{and} \quad (v_i, v_j) \leq 0,$$

for $1 \leq i \neq j \leq \ell$. A set of linearly independent vectors $A = \{v_1, \ldots, v_l\}$ in V having these properties is called an *allowable configuration.* By a long series of elementary arguments it is possible to narrow the list of Dynkin diagrams of an allowable configuration. This leads to the conclusion that the Dynkin diagram of C appears on the list (see [Jac79], Section IV.5). To show that every graph listed is the Dynkin diagram of a Cartan matrix it suffices to write the corresponding Cartan matrices C and the diagonal matrices D such that CD is symmetric, and prove that CD is positive definite using Sylvester's criterion. □

2.8.2 Root systems

Definition 2.8.6 *Let V be a real vector space with inner product $(\ ,\ )$. A finite subset Φ of non-zero elements of V is called a* root system *if*

1. *For $\alpha \in \Phi$, $\lambda \in \mathbb{R}$ we have $\lambda\alpha \in \Phi$ if and only if $\lambda = \pm 1$,*
2. *for all $\alpha, \beta \in \Phi$ we have that $s_\alpha(\beta) \in \Phi$,*
3. *for all $\alpha, \beta \in \Phi$ the number $\langle \alpha, \beta^\vee \rangle$ lies in $\mathbb{Z}$.*

The elements of Φ are called roots. *The dimension of the subspace of V spanned by Φ is called the* rank *of Φ.*

Let $\Phi_1 \subset V_1$, $\Phi_2 \subset V_2$ be root systems of rank ℓ. They are said to be *isomorphic* if there is a linear map $f : V_1 \to V_2$ with $f(\Phi_1) = \Phi_2$ and $\langle \alpha, \beta^\vee \rangle = \langle f(\alpha), f(\beta)^\vee \rangle$, for all $\alpha, \beta \in \Phi_1$.

Let $\Phi \subset V$ be a root system and $\Phi = \Phi_1 \cup \Phi_2$ with $(\alpha_1, \alpha_2) = 0$ for all $\alpha_1 \in \Phi_1$ and $\alpha_2 \in \Phi_2$. Let $W_i \subset V$ be the subspace spanned by Φ_i for $i = 1, 2$. Then it is straightforward to verify that Φ_i is a root system in W_i. If this happens Φ is called the *direct sum* of the root systems Φ_1 and Φ_2. A root system that is not the direct sum of other root systems is called *irreducible.* When studying root systems it is enough to study the irreducible ones, since the others can be formed from these by taking direct sums.

Example 2.8.7 Let $V = \mathbb{R}^{\ell+1}$ with basis $v_1, \ldots, v_{\ell+1}$ with the standard inner product $(v_i, v_j) = \delta_{ij}$. Set $\Phi = \{v = \sum_i k_i v_i \in V \mid k_i \in \mathbb{Z}$ and $(v, v) = 2$ and $\sum_i k_i = 0\}$. Then Φ consists of the elements $\pm(v_i - v_j)$ for $1 \leq i < j \leq \ell + 1$. It is immediately clear that Φ is a root system.

In the remainder of this section we let $\Phi \subset V$ be a root system.

Lemma 2.8.8 *Let $\alpha, \beta \in \Phi$ with $\alpha \neq \pm\beta$. If $(\alpha, \beta) > 0$ then $\alpha - \beta \in \Phi$, and if $(\alpha, \beta) < 0$ then $\alpha + \beta \in \Phi$.*

Proof. If $(\alpha,\beta) > 0$ then $\langle\alpha,\beta^\vee\rangle > 0$. Lemma 2.8.1 implies that $\langle\alpha,\beta^\vee\rangle = 1$ or $\langle\beta,\alpha^\vee\rangle = 1$. In the first case $s_\beta(\alpha) = \alpha - \beta$ lies in Φ. If $\langle\beta,\alpha^\vee\rangle = 1$ by the same argument $\beta - \alpha \in \Phi$. Hence $\alpha - \beta = -(\beta - \alpha)$ also lies in Φ. Finally, if $(\alpha,\beta) < 0$ we apply the above argument to $-\beta$. □

Lemma 2.8.9 *Let α, β be as in the previous lemma. Let r and q be the largest integers such that $\beta - r\alpha$ and $\beta + q\alpha$ are roots. Then $\beta + i\alpha$ are roots for $-r \leq i \leq q$, $r - q = \langle\beta,\alpha^\vee\rangle$, and $r + q \leq 3$.*

Proof. For the first statement, if $r = q = 0$ we have nothing to prove. So suppose that at least one of r and q is non-zero. Suppose further that there is an integer j with $-r < j < q$ such that $\beta + j\alpha$ is not a root. Set $\sigma = \beta - r\alpha$. Then there are integers $0 \leq s < t$ such that $\sigma + s\alpha$ and $\sigma + t\alpha$ are roots while $\sigma + (s+1)\alpha$ and $\sigma + (t-1)\alpha$ are not roots. By Lemma 2.8.8, $(\sigma + s\alpha, \alpha) \geq 0$ and $(\sigma + t\alpha, \alpha) \leq 0$. This means that $(\beta,\alpha) + (s-r)(\alpha,\alpha) \geq 0$ while $(\beta,\alpha) + (t-r)(\alpha,\alpha) \leq 0$. This entails $(t-s)(\alpha,\alpha) \leq 0$ which is a contradiction since $\alpha \neq 0$.

Set $S = \{\beta + i\alpha \mid -r \leq i \leq q\}$. The reflection s_α maps S bijectively onto itself. So the pre-image of $\beta - r\alpha$ must be $\beta + q\alpha$, implying that $r - q = \langle\beta,\alpha^\vee\rangle$.

We have just seen that $\langle\beta + q\alpha, \alpha^\vee\rangle = \langle\beta,\alpha^\vee\rangle + 2q = q + r$. By Lemma 2.8.1 we conclude that $q + r = 0, 1, 2, 3$. □

Now we use an order on the vector space V. We first choose a basis $\{v_1, \ldots, v_\ell\}$ of V. Let $v = \sum_i \lambda_i v_i$, then $v > 0$ if the first non-zero λ_i is positive. More generally, $v > w$ if $v - w > 0$ and $>$ is a total order on V. We call $<$ the *lexicographical order* relative to $v_1, \ldots, v_\ell$.

A root $\alpha \in \Phi$ is said to be *positive* if $\alpha > 0$, and *negative* if $\alpha < 0$. By Φ^+ we denote the set of positive roots and by Φ^- the set of negative roots. Since roots are either positive or negative, $\Phi = \Phi^+ \cup \Phi^-$. Note that $\Phi^- = -\Phi^+$. We stress that this partition depends on the order $<$, which in turn depends on the choice of a basis of V.

A root $\alpha \in \Phi$ is said to be *simple* (with respect to a fixed lexicographical order) if $\alpha > 0$ and α cannot be written as a sum $\alpha = \beta + \gamma$ where $\beta, \gamma \in \Phi$ are both positive.

Proposition 2.8.10 *Let $\Delta = \{\alpha_1, \ldots, \alpha_\ell\}$ be the set of all simple roots. Let $\alpha \in \Phi$ be a positive root.*

(i) *For $i \neq j$ we have that $\alpha_i - \alpha_j$ is not a root, and $(\alpha_i, \alpha_j) \leq 0$.*

(ii) *Δ is linearly independent.*

(iii) *There are unique non-negative integers k_i with $\alpha = \sum_{i=1}^{\ell} k_i \alpha_i$.*

(iv) *If $\alpha \notin \Delta$, there is an $\alpha_i \in \Delta$ such that $\alpha - \alpha_i \in \Phi$.*

Proof. Suppose $\alpha_i - \alpha_j$ is a root. If it is positive, then $\alpha_i = (\alpha_i - \alpha_j) + \alpha_j$ and α_i is not simple. If it is negative, then $\alpha_j = -(\alpha_i - \alpha_j) + \alpha_i$ is not simple and we have a contradiction. Lemma 2.8.8 implies that $(\alpha_i, \alpha_j) \le 0$.

It can be shown that if $v_1, \dots, v_n \in V$ are positive and $(v_i, v_j) \le 0$ for $i \ne j$, the v_i are linearly independent ([Jac79], Section IV.3, Lemma 1). In view of (i), this implies (ii).

For the third statement, if α is simple, we have nothing to prove. Otherwise $\alpha = \beta + \gamma$ where β and γ are positive roots, and both are $< \alpha$. By induction we may assume the statement for all positive roots $< \alpha$. Therefore it also holds for α. Uniqueness follows from (ii).

We cannot have $(\alpha, \alpha_i) \le 0$ for all i, as otherwise $\Delta \cup \{\alpha\}$ would be linearly independent, contrary to (iii). So there is an α_i with $(\alpha, \alpha_i) > 0$, implying $\alpha - \alpha_i \in \Phi$ by Lemma 2.8.8. □

This proposition allows us to define the *height* of a positive root α as $\mathrm{ht}(\alpha) = \sum_i k_i$, if $\alpha = \sum_i k_i \alpha_i$.

Lemma 2.8.11 *Suppose Φ is irreducible, and Δ is a set of simple roots of Φ. Then it is not possible to partition Δ as $\Delta = \Delta_1 \cup \Delta_2$ with Δ_i non-empty and $(\Delta_1, \Delta_2) = 0$.*

Proof. Using induction on $\mathrm{ht}(\alpha)$, Proposition 2.8.10 and Lemma 2.8.9, we prove that every positive root is a sum of elements of Δ_1 or a sum of elements of Δ_2. □

Proposition 2.8.12 *Let $\Delta = \{\alpha_1, \dots, \alpha_\ell\}$ be the set of simple roots and C be the $\ell \times \ell$ matrix with $C(i,j) = \langle \alpha_i, \alpha_j^\vee \rangle$. Then C is a Cartan matrix. Moreover, C does not depend on the choice of the lexicographical order (up to equivalence).*

Proof. In view of Proposition 2.8.10(i), we only need to show the existence of an integral diagonal matrix D such that CD is symmetric and positive definite. We may assume that Φ is irreducible. Let D' be the diagonal matrix with entries $2(\alpha_i, \alpha_i)$; then CD' is the matrix of the inner product $(\ , \ )$, so CD' is symmetric and positive definite. Set $D'' = (2(\alpha_1, \alpha_1))^{-1} D'$; then $D''(1,1) = 1$. Let $(\ , \ )''$ denote the inner product with matrix CD''. Then $2(\alpha_i, \alpha_j)''/(\alpha_j, \alpha_j)'' = C(i,j)$. Now consider the graph with ℓ nodes labeled $1, \dots, \ell$, where i and j are connected by an edge if $(\alpha_i, \alpha_j)'' \ne 0$. This graph is connected by Lemma 2.8.11. Suppose $(\alpha_i, \alpha_i)'' \in \mathbb{Q}$ and i, j are connected. Then it follows that $(\alpha_j, \alpha_j)'' \in \mathbb{Q}$ as well. Since the graph is connected, D'' has entries in $\mathbb{Q}$. Hence, there is a positive integer d such that $D = dD''$ has integral entries.

The second assertion is proved by considering the group W generated by the reflections s_α, for $\alpha \in \Phi$. Let C' be the Cartan matrix relative to a second set of simple roots Δ' (constructed using a second lexicographical order).

There is a $w \in W$ such that $w(\Delta) = \Delta'$. (In order to see that, let P and P' be the positive roots corresponding to Δ and Δ', respectively. Note that $|P| = |P'| = \frac{1}{2}|\Phi|$. Let $r = |P \cap P'|$, and suppose that $r < |P|$. Then there is an $\alpha \in \Delta \setminus P'$. By Lemma 2.8.18, for $\beta \in P \cap P'$ we have $s_\alpha(\beta) \in P$, so that $s_\alpha(\beta) \in P \cap s_\alpha(P')$. But also $\alpha = s_\alpha(-\alpha)$ lies in $P \cap s_\alpha(P')$. So $|P \cap s_\alpha(P')| \geq r + 1$. Continuing, we find a $w \in W$ such that $w(P') = P$, implying that $w(\Delta') = \Delta$.) Since all elements of W leave the inner product invariant, it follows that C' is equivalent to C. □

The matrix C of the previous proposition is called the Cartan matrix of Φ. The proposition states that it is uniquely determined by Φ (up to equivalence).

Example 2.8.13 Let Φ be as in Example 2.8.7 and $<$ be the lexicographical order with respect to the basis $v_1, \ldots, v_{\ell+1}$. Then the positive roots are $v_i - v_j$, for $j > i$. The simple roots are $v_i - v_{i+1}$, $1 \leq i \leq \ell$. The Dynkin diagram of the Cartan matrix is the one of type A_ℓ; see Theorem 2.8.5.

Suppose Φ is irreducible. Then by Lemma 2.8.11 the Dynkin diagram of the Cartan matrix of Φ is connected and appears in the list of Theorem 2.8.5. Now we show that for each diagram in that list there is a unique irreducible root system. We first formulate an algorithm for reconstructing Φ from its set of simple roots and the corresponding Cartan matrix.

Algorithm 2.8.14 *Input: the set of simple roots $\Delta = \{\alpha_1, \ldots, \alpha_\ell\}$ of the root system Φ and the Cartan matrix C of Φ relative to Δ.*
Output: the set of vectors Φ.

1. *Set $\Phi^+ := \Delta$ and $n := 1$.*
2. *For all $\gamma \in \Phi^+$, $\gamma = \sum_i k_i \alpha_i$ of height n and for all $\alpha_j \in \Delta$ we:*
   (a) *Determine the largest integer $r \geq 0$ such that $\gamma - r\alpha_j \in \Phi^+$.*
   (b) *Set $q := r - \sum_{i=1}^{\ell} k_i C(i,j)$.*
   (c) *If $q > 0$ then set $\Phi^+ := \Phi^+ \cup \{\gamma + \alpha_j\}$.*
3. *If in Step 2, Φ^+ has been enlarged, we set $n := n + 1$ and return to the beginning of Step 2. Otherwise we return $\Phi^+ \cup -\Phi^+$.*

Lemma 2.8.15 *Algorithm 2.8.14 returns the set Φ.*

Proof. Let β be a positive root. By induction on $\mathrm{ht}(\beta)$ and using Lemma 2.8.9 and Proposition 2.8.10(iv), it follows that after $\mathrm{ht}(\beta) - 1$ rounds of the iteration, the set Φ^+ contains β. □

Proposition 2.8.16 *Let C be an indecomposable Cartan matrix. Then there exists a unique (up to isomorphism) irreducible root system with Cartan matrix C.*

Proof. To prove existence, we can go through the list of Theorem 2.8.5, and for each type explicitly construct a root system. We refer to [Hum78], Section 12.1 for the details. For the types A_ℓ, D_ℓ and E_ℓ a more uniform construction is possible. This runs as follows. Let C be the corresponding Cartan matrix, and note that in these cases C is symmetric. Let $e_1, \dots, e_\ell$ be a basis of an ℓ-dimensional vector space V over $\mathbb{R}$. Define an inner product on V by $(e_i, e_j) = C(i, j)$, and set

$$\Phi = \{v = k_1 e_1 + \cdots + k_\ell e_\ell \mid k_i \in \mathbb{Z} \text{ and } (v, v) = 2\}.$$

Then $\Phi \subset V$ is a root system with basis of simple roots $\{e_1, \dots, e_\ell\}$ and Cartan matrix C.

Uniqueness can be proved by using Algorithm 2.8.14. Let Φ and Φ' be two root systems with bases of simple roots $\{\alpha_1, \dots, \alpha_\ell\}$ and $\{\alpha_1', \dots, \alpha_\ell'\}$. Suppose that both have Cartan matrix C. Then by Lemma 2.8.15 we infer that $\sum_i k_i \alpha_i \in \Phi$ if and only if $\sum_i k_i \alpha_i' \in \Phi'$. So mapping α_i to α_i' extends to a bijection $f : \Phi \to \Phi'$. Moreover, from Lemma 2.8.9 it follows that $\langle f(\alpha), f(\beta)^\vee \rangle = \langle \alpha, \beta^\vee \rangle$ for all $\alpha, \beta \in \Phi$. □

Remark 2.8.17 We can use Algorithm 2.8.14 to construct a root system Φ with a given Cartan matrix Φ. Let V be a real vector space with basis $\Delta = \{\alpha_1, \dots, \alpha_\ell\}$. Moreover, let D be an integral diagonal matrix such that $B = CD$ is symmetric and positive definite. Define an inner product on V by $(\alpha_i, \alpha_j) = B(i, j)$. Finally, apply Algorithm 2.8.14 with input Δ and C. The output is then a root system in V with Cartan matrix C.

2.8.3 The Weyl group

In this section $\Phi \subset V$ will be a root system, with a fixed basis of simple roots $\Delta = \{\alpha_1, \dots, \alpha_\ell\}$, and Cartan matrix $C = (\langle \alpha_i, \alpha_j^\vee \rangle)$. The *Weyl group*, denoted W, is the subgroup of $\mathrm{GL}(V)$ generated by the reflections s_α for $\alpha \in \Phi$. We have already seen this group in action in the proof of Proposition 2.8.12. Here we collect a number of useful results on this group.

We start by stating a few elementary facts about W. First we note that we can write $V = V' \oplus V''$, where V' is spanned by Φ and $(V', V'') = 0$. Then W acts trivially on V'', and the restriction of a $w \in W$ to V' is completely determined by its action on Φ. Hence W is finite as it is isomorphic to a subgroup of the group of all permutations of Φ. Second, the elements of W leave the inner product invariant, or $(w\lambda, w\mu) = (\lambda, \mu)$ for all $\lambda, \mu \in V$, $w \in W$. This also means that $\langle w\lambda, (w\mu)^\vee \rangle = \langle \lambda, \mu^\vee \rangle$, implying that for $\alpha \in \Phi$

and $w \in W$ we have $ws_\alpha w^{-1} = s_{w(\alpha)}$. The s_α for $\alpha \in \Delta$ are called *simple reflections.*

The next lemma follows from Proposition 2.8.10(iii), and the fact that $s_\alpha(\beta) = \beta - m_{\beta,\alpha}\alpha$, where $m_{\beta,\alpha} \in \mathbb{Z}$.

Lemma 2.8.18 *Let $\alpha \in \Delta$, then s_α permutes the set $\Phi^+ \setminus \{\alpha\}$.*

Lemma 2.8.19 *Let W' be the subgroup of W generated by all simple reflections. Let $\beta \in \Phi$ be a root; then there exist $\alpha \in \Delta$ and $w \in W'$ such that $w(\alpha) = \beta$.*

Proof. It suffices to prove this for the positive roots β. For that we use induction on $\mathrm{ht}(\beta)$. Suppose $\mathrm{ht}(\beta) > 1$. As seen in the proof of Proposition 2.8.10(iv), there is an $\alpha \in \Delta$ such that $(\alpha, \beta) > 0$. Set $\gamma = s_\alpha(\beta)$. By Lemma 2.8.18, γ is a positive root of smaller height. So the result holds for γ, and hence also for β. □

Theorem 2.8.20 *The Weyl group W is generated by the simple reflections.*

Proof. Let $\beta \in \Phi$, then by Lemma 2.8.19 there is a $w \in W'$ and $\alpha \in \Delta$ such that $w(\alpha) = \beta$. So $s_\beta = s_{w(\alpha)} = ws_\alpha w^{-1}$. Therefore, $s_\beta \in W'$ and hence $W = W'$. □

For brevity we will write $s_i = s_{\alpha_i}$. (Recall that $\Delta = \{\alpha_1, \dots, \alpha_\ell\}$.) By Theorem 2.8.20, every element of W can be written as a product $s_{i_1} \cdots s_{i_r}$ (note that $s_i^{-1} = s_i$). If w cannot be written as a product of fewer than r simple reflections, we say that this expression is *reduced.* In that case r is called the *length* of w, and we write $\mathcal{L}(w) = r$. We define the length of the identity as 0.

Lemma 2.8.21 *Let $w = s_{i_1} \cdots s_{i_r} \in W$. If $s_{i_1} \cdots s_{i_{r-1}}(\alpha_{i_r})$ is a negative root, then there is a $t \in \{1, \dots, r-1\}$ such that $w = s_{i_1} \cdots s_{i_{t-1}} s_{i_{t+1}} \cdots s_{i_{r-1}}$.*

Proof. Set $\gamma_a = s_{i_{a+1}} \cdots s_{i_{r-1}}(\alpha_{i_r})$ for $0 \le a \le r-2$ and $\gamma_{r-1} = \alpha_{i_r}$. Then $\gamma_0 < 0$ and $\gamma_{r-1} > 0$. Let γ_t for some $t > 0$ be the first positive root in this sequence. Then $s_{i_t}(\gamma_t) = \gamma_{t-1} < 0$. So by Lemma 2.8.18, $\gamma_t = \alpha_{i_t}$. Therefore $(s_{i_{t+1}} \cdots s_{i_{r-1}}) s_{i_r} (s_{i_{r-1}} \cdots s_{i_{t+1}}) = s_{s_{i_{t+1}} \cdots s_{i_{r-1}}(\alpha_{i_r})} = s_{\gamma_t} = s_{i_t}$. If we substitute this expression for s_{i_t} into $s_{i_1} \cdots s_{i_r}$ we arrive at the desired result. □

Corollary 2.8.22 *Let $w = s_{i_1} \cdots s_{i_r} \in W$, and suppose this expression is reduced. Then $w(\alpha_{i_r})$ is a negative root.*

Proof. If $w(\alpha_{i_r}) > 0$, then $s_{i_1} \cdots s_{i_{r-1}}(\alpha_{i_r}) = -w(\alpha_{i_r}) < 0$, and by Lemma 2.8.21, the given expression is not reduced. □

For $w \in W$ we define Φ_w as the set of $\alpha \in \Phi^+$ such that $w(\alpha) \in \Phi^-$, and set $n(w) = |\Phi_w|$.

Lemma 2.8.23 *Let $w \in W$ and $\alpha_i \in \Delta$. If $w(\alpha_i) > 0$, then $n(ws_i) = n(w) + 1$. On the other hand, if $w(\alpha_i) < 0$, then $n(ws_i) = n(w) - 1$.*

Proof. Use Lemma 2.8.18, along with the fact that s_i maps α_i to $-\alpha_i$. □

Lemma 2.8.24 *For $w \in W$ we have $\mathcal{L}(w) = n(w)$.*

Proof. Let $w = s_{i_1} \cdots s_{i_r}$ be reduced. Write $v = s_{i_1} \cdots s_{i_{r-1}}$. By Corollary 2.8.22, $w(\alpha_{i_r}) < 0$. From Lemma 2.8.18, we see that $\Phi_w = s_{i_r}(\Phi_v) \dot{\cup} \{\alpha_{i_r}\}$. The lemma now follows by induction on r. □

An immediate consequence of this lemma is that W has a unique longest element. Indeed, by interchanging Φ^+ and Φ^-, the set of simple roots changes to $-\Delta$. As seen in the proof of Proposition 2.8.12 there is a $w_0 \in W$ such that $w_0(\Delta) = -\Delta$. This w_0 maps all positive roots to negative ones, and has maximal length $|\Phi^+|$ by the lemma. Another consequence is the following.

Corollary 2.8.25 *Let $w \in W$ and write $w = s_{i_1} \cdots s_{i_r}$, with $\mathcal{L}(w) = r$. Let $\alpha_j \in \Delta$.*

(i) *If $w(\alpha_j) > 0$ then $\mathcal{L}(ws_j) = \mathcal{L}(w) + 1$ (in other words, $s_{i_1} \cdots s_{i_r} s_j$ is reduced).*

(ii) *If $w(\alpha_j) < 0$ then $\mathcal{L}(ws_j) = \mathcal{L}(w) - 1$, and there is an index $t \in \{1, \dots, r\}$ with $ws_j = s_{i_1} \cdots s_{i_{t-1}} s_{i_{t+1}} \cdots s_{i_r}$.*

Proof. The statements about $\mathcal{L}(ws_j)$ follow immediately from Lemmas 2.8.23 and 2.8.24. For the second statement of (ii), set $i_{r+1} = j$, $u = s_{i_1} \cdots s_{i_r} s_{i_{r+1}}$ and apply Lemma 2.8.21. □

The second statement of Corollary 2.8.25 is called the *exchange condition.* Note also that the proof of Lemma 2.8.21 gives an immediate algorithm to find the index t.

Let $\alpha_i, \alpha_j \in \Delta$, $i \neq j$. These roots span a root system of rank 2. By inspecting the root systems of rank 2, we see that if this root system is of type $A_1 + A_1$, A_2, B_2, G_2, the order of $s_i s_j$ is respectively 2, 3, 4 and 6. Furthermore, $s_i s_j \cdots = s_j s_i \cdots$, where the number of factors on both sides is equal to the order of $s_i s_j$. Let us call this an *elementary relation.*

Proposition 2.8.26 *Let $\hat{u} = s_{i_1} \cdots s_{i_t}$, $\hat{v} = s_{j_1} \cdots s_{j_t}$ be two reduced expressions of the same element $w \in W$. Then $\hat{u}$ can be rewritten to $\hat{v}$ by using elementary relations.*

Proof. We use induction on t. Write $i = i_t$, $j = j_t$. If $i = j$ then the proof is finished by induction. So suppose $i \neq j$. The proof is split into four cases, depending on the root system spanned by α_i and α_j. Here we consider the case where this root system is of type A_2, leaving the other cases to the reader. We repeatedly use the exchange condition. By leaving out one s_{i_l} from $\hat{u}$ we find a reduced expression $\hat{u}_1$ for ws_j, and set $\hat{u}_2 = \hat{u}_1 s_j$ (so that $\hat{u}_2$ is a reduced expression for w). By removing one element from $\hat{u}_2$, we then find a reduced expression $\hat{u}_3$ for ws_i, and set $\hat{u}_4 = \hat{u}_3 s_i$. Then $\hat{u}_4$ is a reduced expression for w, ending on $s_i s_j s_i$. As $\hat{u}$, $\hat{u}_4$ end with the same element, by induction we can find elementary relations to rewrite $\hat{u}$ to $\hat{u}_4$. Next we rewrite $s_i s_j s_i \mapsto s_j s_i s_j$ in $\hat{u}_4$, to obtain $\hat{u}_5$. Again by induction we find elementary relations to rewrite $\hat{u}_5$ to $\hat{v}$. □

We remark that this proof, along with the algorithm for performing the exchange condition, gives an algorithm for finding the elementary relations moving one reduced expression to the other.

An element $\lambda \in V$ is called an *integral weight* if $\langle \lambda, \alpha_i^\vee \rangle \in \mathbb{Z}$ for $1 \leq i \leq \ell$. The set of all integral weights is denoted by P. We have $\Phi \subset P$. For $1 \leq i \leq \ell$ let $\lambda_i \in P$ satisfy $\langle \lambda_i, \alpha_j^\vee \rangle = \delta_{ij}$ for $1 \leq j \leq \ell$. As C is non-singular this requirement determines λ_i uniquely. The λ_i are called the *fundamental weights* because of the following result.

Lemma 2.8.27 *Let $\lambda \in P$, and set $m_j = \langle \lambda, \alpha_j^\vee \rangle$ for $1 \leq j \leq \ell$. Then $\lambda = m_1\lambda_1 + \cdots + m_\ell\lambda_\ell$.*

Proof. By definition of the λ_i, $\langle \lambda - \sum_{i=1}^{\ell} m_i\lambda_i, \alpha_j^\vee \rangle = 0$. □

It follows that $P = \{m_1\lambda_1 + \cdots + m_\ell\lambda_\ell \mid m_i \in \mathbb{Z}\}$. For this reason P is also called the *weight lattice*. Furthermore, $\alpha_i = m_{i,1}\lambda_1 + \cdots + m_{i,\ell}\lambda_\ell$, and according to the lemma, $m_{i,j} = \langle \alpha_i, \alpha_j^\vee \rangle = C(i,j)$. In particular we see that the λ_i form a basis of V.

Let s_i be a simple reflection, then $s_i(\lambda_j) = \lambda_j - \delta_{ij}\alpha_i$. So $w(P) = P$ for all $w \in W$. Furthermore, we can easily compute $s_i(\lambda)$ for $\lambda \in P$.

We say that an integral weight $\lambda = m_1\lambda_1 + \cdots + m_\ell\lambda_\ell$ is *dominant* if $m_i \geq 0$ for all i. We denote the set of all dominant integral weights by P_+.

Lemma 2.8.28 *Let $\lambda, \mu \in P_+$ and $w \in W$ be such that $w(\lambda) = \mu$. Then $\lambda = \mu$.*

Proof. Let $w = s_{i_1} \cdots s_{i_t}$ be a reduced expression for w. Suppose $t > 0$. By Corollary 2.8.22, $w(\alpha_{i_t}) < 0$ so that $0 \leq (\lambda, \alpha_{i_t}) = (w(\lambda), w(\alpha_{i_t})) = (\mu, w(\alpha_{i_t})) \leq 0$. As a consequence $(\lambda, \alpha_{i_t}) = 0$, whence $s_{i_t}(\lambda) = \lambda$ and $ws_{i_t}(\lambda) = \mu$. But $\mathcal{L}(ws_{i_t}) = t - 1$ and we conclude by induction. □

Theorem 2.8.29 *Each element $\mu \in P$ is conjugate under W to exactly one element of P_+.*

Proof. Let $\mu \in P$. Define a sequence $\mu_l \in P$ as follows. First, $\mu_0 = \mu$. If, for $l \geq 0$, $\mu_l = m_1\lambda_1 + \cdots m_\ell\lambda_\ell$, and there is an i with $m_i < 0$, we set $\mu_{l+1} = s_i(\mu_l) = \mu_l - m_i\alpha_i$; if there is no such i the sequence stops. Let $r > l$ and write $\mu_l = \mu + \sum_{i=1}^{\ell} a_i\alpha_i$, $\mu_r = \mu + \sum_{i=1}^{\ell} b_i\alpha_i$; then $\sum_i b_i > \sum_i a_i$ and in particular $\mu_l \neq \mu_r$. Since the orbit $W \cdot \mu$ is finite, the constructed sequence is finite. But the final element of the sequence lies in P_+. Uniqueness follows from Lemma 2.8.28. □

Remark 2.8.30 The proof of Theorem 2.8.20 yields an immediate algorithm for finding the dominant weight conjugate under the W to a given $\mu \in P$.

Proposition 2.8.31 *Let $\lambda \in P_+$, and let $W_\lambda = \{w \in W \mid w(\lambda) = \lambda\}$ be its stabilizer. Then W_λ is generated by the simple reflections s_i where i is such that $\langle \lambda, \alpha_i^\vee \rangle = 0$.*

Proof. Let $1 \neq w \in W$ and $w = s_{i_1} \cdots s_{i_t}$ be a reduced expression. For $1 \leq j \leq t$ consider the element $\mu_j = s_{i_j} \cdots s_{i_t}(\lambda)$. We calculate $(\mu_j, \alpha_{i_{j-1}}) = (s_{i_j} \cdots s_{i_t}\lambda, \alpha_{i_{j-1}}) = (\lambda, s_{i_t} \cdots s_{i_j}(\alpha_{i_{j-1}}))$. But by Corollary 2.8.22, $s_{i_t} \cdots s_{i_{j-1}}(\alpha_{i_{j-1}})$ is a negative root. So $s_{i_t} \cdots s_{i_j}(\alpha_{i_{j-1}}) = -s_{i_t} \cdots s_{i_{j-1}}(\alpha_{i_{j-1}})$ is a positive root and $(\mu_j, \alpha_{i_{j-1}}) \geq 0$. Therefore $\mu_{j-1} = \mu_j - m\alpha_{i_{j-1}}$ where $m \geq 0$. We infer that $w(\lambda) = \lambda$ implies that $\lambda = \mu_j$ for all j and $\langle \lambda, \alpha_{i_j}^\vee \rangle = 0$. □

A subset $\Phi_0 \subset \Phi$ is said to be a *root subsystem* of Φ if for $\alpha \in \Phi_0$ we have $-\alpha \in \Phi_0$ and $s_\beta(\alpha) \in \Phi_0$ for all $\beta \in \Phi_0$. A root subsystem is a root system in its own right. Let $\beta_1, \ldots, \beta_s \in \Phi$ form a basis of simple roots of a root subsystem Φ_0 of Φ. Then the Weyl group W_0 of Φ_0 is the subgroup of W generated by s_{β_i}, $1 \leq i \leq s$. We say that W_0 is a *reflection subgroup* of W. We will assume that the β_i lie in Φ^+.

Proposition 2.8.32 *Let $w \in W$. The coset $W_0 w$ contains a unique element of shortest length.*

Proof. We claim the following. Let $v \in W$ and β be one of the β_i. Suppose $v^{-1}(\beta) > 0$. Then $\mathcal{L}(s_\beta v) > \mathcal{L}(v)$.

First we note that this is equivalent to $\mathcal{L}(v^{-1}s_\beta) > \mathcal{L}(v^{-1})$. For $u \in W$ we set

$$T_\pm(u) = \{\alpha \in \Phi^+ \mid s_\beta(\alpha) \in \Phi^\pm,\ u(\alpha) \in \Phi^-\}.$$

Then s_β maps $T_+(v^{-1})$ bijectively to $T_+(v^{-1}s_\beta)$, so both sets have the same size. Let $S_v = \{\alpha \in \Phi^+ \mid s_\beta(\alpha) \in \Phi^-,\ v^{-1}(\alpha) \in \Phi^+\}$. Then $-s_\beta$ maps $T_-(v^{-1}s_\beta)$ bijectively to S_v.

Now we show that $|S_v| > |T_-(v^{-1})|$. Let $\alpha \in T_-(v^{-1})$. From $s_\beta(\alpha) < 0$

we see that $\langle \alpha, \beta^\vee \rangle > 0$. But $v^{-1}(s_\beta(\alpha)) = v^{-1}(\alpha) - \langle \alpha, \beta^\vee \rangle v^{-1}(\beta)$, and as $v^{-1}(\alpha) < 0$ and $v^{-1}(\beta) > 0$ we see that $v^{-1}s_\beta(\alpha) < 0$. It follows that $-s_\beta$ maps $T_-(v^{-1})$ into S_v. But $\beta \in S_v$ does not lie in the image and we are done.

Now, using Lemma 2.8.24, $\mathcal{L}(v^{-1}s_\beta) = |T_+(v^{-1}s_\beta)| + |T_-(v^{-1}s_\beta)| = |T_+(v^{-1})| + |S_v| > |T_+(v^{-1})| + |T_-(v^{-1})| = \mathcal{L}(v^{-1})$.

Let $v \in W_0 w$ be of minimal length and β be one of the β_i. If $v^{-1}(\beta) < 0$, by arguments similar to the ones used earlier, we prove that $\mathcal{L}(s_\beta v) < \mathcal{L}(v)$ (this time $-s_\beta$ maps S_v into $T_-(v^{-1})$), contrary to the choice of v. It follows that $v^{-1}(\beta_i) > 0$ for $1 \leq i \leq s$.

Let $u \in W_0$ be such that $\mathcal{L}(uv) > \mathcal{L}(v)$. By $\mathcal{L}_0$ we denote the length function of W_0. Again let β be one of the β_i with $\mathcal{L}_0(s_\beta u) > \mathcal{L}_0(u)$. Then $u^{-1}(\beta)$ is a positive root in Φ_0 (Corollary 2.8.25). Hence $v^{-1}u^{-1}(\beta) > 0$. By the claim above, $\mathcal{L}(s_\beta uv) > \mathcal{L}(uv)$. So by induction on $\mathcal{L}_0(z)$, $\mathcal{L}(zv) > \mathcal{L}(v)$ for all $z \in W_0$, $z \neq 1$. This finishes the proof. □

Remark 2.8.33 Suppose further that all β_i lie in Δ. Let v be the unique shortest element of $W_0 w$. Then the proof of the previous proposition also shows that $\mathcal{L}(uv) = \mathcal{L}(u) + \mathcal{L}(v)$, for $u \in W_0$, as in this case $\mathcal{L}_0$ is the restriction of $\mathcal{L}$ to W_0.

We say that a $w \in W$ is a *shortest representative* if it is the unique element of shortest length of the coset $W_0 w$.

Lemma 2.8.34 *Let w be a shortest representative. Write $w = w' s_{\alpha_i}$ where $\mathcal{L}(w') = \mathcal{L}(w) - 1$. Then w' is a shortest representative.*

Proof. If it is not, we can write $w' = w_1 w''$, with $w_1 \in W_0$ and $\mathcal{L}(w'') < \mathcal{L}(w')$. Hence w and $w'' s_{\alpha_i}$ lie in the same right W_0-coset and $w'' s_{\alpha_i}$ is not a shortest representative. Therefore we can write $w'' s_{\alpha_i} = w_2 w'''$, with $w_2 \in W_0$ and w''' a shortest representative, $\mathcal{L}(w''') < \mathcal{L}(w'' s_{\alpha_i})$. But then $w = w_1 w_2 w'''$, and this implies $w = w'''$. We also have $\mathcal{L}(w''') \leq \mathcal{L}(w'') < \mathcal{L}(w') < \mathcal{L}(w)$, which is a contradiction. □

Now let R_m denote the set of shortest representatives of length m. Lemma 2.8.34 yields the following algorithm for computing R_{m+1} from R_m. Initially we set $R_{m+1} = \emptyset$. Then for $1 \leq i \leq \ell$ and $w \in R_m$ we perform the following step: if $\mathcal{L}(w s_{\alpha_i}) > \mathcal{L}(w)$ and $s_{\alpha_i} w^{-1}(\beta_j) > 0$ for $1 \leq j \leq s$ (the second condition is justified by the proof of Proposition 2.8.32) we add $w s_{\alpha_i}$ to R_{m+1}. Continuing until we obtain R_n such that $R_{n+1} = \emptyset$, we find all shortest representatives.

2.9 Classification of simple Lie algebras

Here we describe the classification of the semisimple Lie algebras over algebraically closed fields of characteristic 0. derived from the classification of the root systems. We will show how a root system is obtained from a semisimple Lie algebra, by which the Lie algebra is determined up to isomorphism. Finally we indicate how a semisimple Lie algebra can be constructed starting from a root system.

Many of the statements in this section hold under the assumptions that the base field is of characteristic 0 (but not necessarily algebraically closed) and that the semisimple Lie algebras have a split Cartan subalgebra (i.e., k contains the eigenvalues of each $\mathrm{ad}_{\mathfrak{g}}h$ for h in the Cartan subalgebra; see Section 2.5.4). We often just work with these hypotheses.

We start with some material on representations of $\mathfrak{sl}(2,k)$, that is surprisingly useful.

2.9.1 Representations of $\mathfrak{sl}(2,k)$

Let $\mathfrak{s} = \mathfrak{sl}(2,k)$ (see Example 2.1.4) be the 3-dimensional Lie algebra over k with basis h, e, f and

$$[h,e] = 2e,\ [h,f] = -2f,\ [e,f] = h.$$

The subspace spanned by h is a Cartan subalgebra of $\mathfrak{s}$. Here we completely describe the representations ρ of $\mathfrak{s}$ such that $\rho(h)$ is split.

Let $\rho : \mathfrak{s} \to \mathfrak{gl}(V)$ be a finite-dimensional representation of $\mathfrak{s}$. Since $\mathfrak{s}$ is simple (see Example 2.7.5), by Theorem 2.7.6 V is a direct sum of irreducible submodules. Therefore it suffices to describe the irreducible representations of $\mathfrak{s}$.

Theorem 2.9.1 *Let $\rho : \mathfrak{s} \to \mathfrak{gl}(V)$ be an irreducible representation of $\mathfrak{s}$. Assume that $\rho(h)$ is split. Then there is a basis $v_0, \dots, v_n$ of V such that $h \cdot v_i = (n-2i)v_i$, $f \cdot v_i = v_{i+1}$, $e \cdot v_i = i(n-i+1)v_{i-1}$ (where $v_{-1} = v_{n+1} = 0$).*

Proof. Since $\rho(h)$ is split, it has an eigenvalue μ and corresponding eigenvector w. Then $e \cdot w$ is an eigenvector of $\rho(h)$ with eigenvalue $\mu + 2$. So there is an eigenvector v_0 of $\rho(h)$, with eigenvalue λ, such that $e \cdot v_0 = 0$. For $i \geq 1$ set $v_i = f \cdot v_{i-1}$. Then v_i is an eigenvector of $\rho(h)$ with eigenvalue $\lambda - 2i$, so the v_i are linearly independent. By induction we obtain $e \cdot v_i = i(\lambda - i + 1)v_{i-1}$. Let n be such that $v_n \neq 0$, $v_{n+1} = 0$. Then $0 = e \cdot v_{n+1} = (n+1)(\lambda - n)v_n$, so $\lambda = n$. □

2.9.2 From Lie algebra to root system

Let $\mathfrak{g}$ be a semisimple Lie algebra over a field k of characteristic 0. We suppose that $\mathfrak{g}$ has a split Cartan subalgebra $\mathfrak{h}$. We let $\mathfrak{g} = \mathfrak{h} \oplus \mathfrak{g}_{\alpha_1} \oplus \cdots \oplus \mathfrak{g}_{\alpha_t}$ be the root space decomposition of $\mathfrak{g}$ with respect to $\mathfrak{h}$ (Section 2.5.4). Let $\Phi = \{\alpha_1, \ldots, \alpha_t\}$ be the set of roots. By Lie's theorem (Theorem 2.6.5) there exists a basis of $\mathfrak{g}_{\alpha_i}$ with respect to which the matrix of the restriction of $\mathrm{ad}_{\mathfrak{g}}(h)$ $(h \in \mathfrak{h})$ is upper triangular, with $\alpha_i(h)$ on the diagonal. This implies that the α_i are linear functions on $\mathfrak{h}$; in other words, they lie in the dual space $\mathfrak{h}^*$. It is the objective of this section to show that they form a root system in (a real form of) $\mathfrak{h}^*$. Throughout κ denotes the Killing form of $\mathfrak{g}$; by Proposition 2.7.2 it is non-degenerate. It will be convenient to define $\Phi^0 = \Phi \cup \{0\}$; note that $\mathfrak{g}_0 = \mathfrak{h}$.

From Proposition 2.5.12 it follows that $[\mathfrak{g}_\alpha, \mathfrak{g}_\beta] \subset \mathfrak{g}_{\alpha+\beta}$ for $\alpha, \beta \in \Phi^0$. Let $x \in \mathfrak{g}_\alpha$, $y \in \mathfrak{g}_\beta$ be non-zero, then $\mathrm{ad}x \cdot \mathrm{ad}y$ maps $\mathfrak{g}_\gamma$ into $\mathfrak{g}_{\gamma+\beta+\alpha}$. So if $\beta \neq -\alpha$ this map is nilpotent, whence $\kappa(x, y) = 0$. Now the non-degeneracy of κ immediately implies the following proposition.

Proposition 2.9.2 (i) *If $\alpha \in \Phi^0$, then also $-\alpha \in \Phi^0$.*

(ii) *For every $x \in \mathfrak{g}_\alpha$, $x \neq 0$, there is a $y \in \mathfrak{g}_{-\alpha}$ such that $\kappa(x, y) \neq 0$.*

(iii) *The restriction of κ to $\mathfrak{h}$ is non-degenerate.*

Proposition 2.9.2 enables us to define a bijective linear map $\mathfrak{h}^* \to \mathfrak{h}$, $\sigma \mapsto \hat{h}_\sigma$, where $\hat{h}_\sigma$ is defined by $\sigma(h) = \kappa(h, \hat{h}_\sigma)$ for all $h \in \mathfrak{h}$. Using this we get a non-degenerate bilinear form on $\mathfrak{h}^*$ by $(\sigma, \rho) = \kappa(\hat{h}_\sigma, \hat{h}_\rho)$. Note that this implies that $(\sigma, \sigma) = \sigma(\hat{h}_\sigma)$.

Proposition 2.9.3 (i) *Let $h \in \mathfrak{h}$. If $\alpha(h) = 0$ for all $\alpha \in \Phi$, then $h = 0$.*

(ii) *For $h \in \mathfrak{h}$ the linear transformation $\mathrm{ad}_{\mathfrak{g}}h$ is semisimple.*

(iii) *For $x \in \mathfrak{g}_\alpha$, $h \in \mathfrak{h}$ we have $[h, x] = \alpha(h)x$.*

(iv) *There are* $\dim \mathfrak{h}$ *linearly independent roots in Φ.*

Proof. As noted above, by Lie's theorem there is a basis of $\mathfrak{g}_\alpha$ with respect to which $\mathrm{ad}h$ acts by an upper triangular matrix with $\alpha(h)$ on the diagonal. This implies that, for $h_1, h_2 \in \mathfrak{h}$, we have $\kappa(h_1, h_2) = \sum_{\alpha\in\Phi}(\dim \mathfrak{g}_\alpha)\alpha(h_1)\alpha(h_2)$. So (i) follows by Proposition 2.9.2(iii).

Let $h_s, h_n \in \mathfrak{g}$ be the semisimple and nilpotent parts of h (Theorem 2.7.9). From Proposition 2.2.4 it follows that $h_n \in \mathfrak{n}_{\mathfrak{g}}(\mathfrak{h}) = \mathfrak{h}$. But as $\mathrm{ad}h_n$ is nilpotent, $\alpha(h_n) = 0$ for all $\alpha \in \Phi$, and by (i), $h_n = 0$.

The restriction of $\mathrm{ad}h$ to $\mathfrak{g}_\alpha$ is semisimple, with sole eigenvalue $\alpha(h)$, whence (iii).

Finally, (iv) follows from (i). □

Proposition 2.9.4 *Let $\alpha \in \Phi$.*

(i) *For $x_\alpha \in \mathfrak{g}_\alpha$, $x_{-\alpha} \in \mathfrak{g}_{-\alpha}$ we have $[x_\alpha, x_{-\alpha}] = \kappa(x_\alpha, x_{-\alpha})\hat{h}_\alpha$.*

(ii) $(\alpha, \alpha) \neq 0$.

(iii) *Let $e \in \mathfrak{g}_\alpha$, $e \neq 0$, and set $h = \frac{2\hat{h}_\alpha}{(\alpha,\alpha)}$. Then there is an $f \in \mathfrak{g}_{-\alpha}$ such that $[h, e] = 2e$, $[h, f] = -2f$ and $[e, f] = h$ (i.e., h, e and f span a subalgebra isomorphic to $\mathfrak{sl}(2, k)$).*

Proof. Using the invariance of κ, and Proposition 2.9.3(iii) we obtain for $h \in \mathfrak{h}$ that $\kappa([x_\alpha, x_{-\alpha}], h) = \alpha(h)\kappa(x_\alpha, x_{-\alpha})$. But this is also equal to $\kappa(\kappa(x_\alpha, x_{-\alpha})\hat{h}_\alpha, h)$, so (i) follows from Proposition 2.9.2(iii).

Now choose $x_\alpha, x_{-\alpha}$ such that $\kappa(x_\alpha, x_{-\alpha}) \neq 0$ (we can do that by Proposition 2.9.2(ii)). If $0 = (\alpha, \alpha) = \alpha(\hat{h}_\alpha)$, then $[\hat{h}_\alpha, x_\alpha] = [\hat{h}_\alpha, x_{-\alpha}] = 0$. So $\hat{h}_\alpha$, x_α and $x_{-\alpha}$ span a 3-dimensional solvable subalgebra $\mathfrak{r}$, where $[\mathfrak{r}, \mathfrak{r}]$ is spanned by $\hat{h}_\alpha$. By Lemma 2.6.7, $\mathrm{ad}\hat{h}_\alpha$ is nilpotent, contradicting Proposition 2.9.3(ii).

For (iii) we let $x_{-\alpha} \in \mathfrak{g}_{-\alpha}$ be such that $\kappa(e, x_{-\alpha}) \neq 0$, and we let f be a scalar multiple of $x_{-\alpha}$ such that $\kappa(e, f) = \frac{2}{(\alpha,\alpha)}$. □

Lemma 2.9.5 *Let $\alpha \in \Phi$. Then $\dim \mathfrak{g}_\alpha = 1$ and the integral multiples of α that are roots are $\pm\alpha$.*

Proof. Let $x_\alpha \in \mathfrak{g}_\alpha$ be non-zero. Let $\mathfrak{a}$ be the subspace of $\mathfrak{g}$ spanned by x_α, $\hat{h}_\alpha$, and all $\mathfrak{g}_{-m\alpha}$ for $m \geq 1$, $m \in \mathbb{Z}$. Using Proposition 2.9.4(i) we see that $\mathfrak{a}$ is a subalgebra. Setting $d_m = \dim \mathfrak{g}_{-m\alpha}$ we have $\mathrm{Tr}(\mathrm{ad}_\mathfrak{a}\hat{h}_\alpha) = \alpha(\hat{h}_\alpha)(1 - d_1 - 2d_2 - \cdots)$. However up to a scalar multiple, $\mathrm{ad}_\mathfrak{a}\hat{h}_\alpha$ is the commutator of $\mathrm{ad}_\mathfrak{a}x_\alpha$ and $\mathrm{ad}_\mathfrak{a}x_{-\alpha}$ (for a certain $x_{-\alpha} \in \mathfrak{g}_{-\alpha}$, see Proposition 2.9.4(i)). Hence $\mathrm{Tr}(\mathrm{ad}_\mathfrak{a}\hat{h}_\alpha) = 0$. Since $\alpha(\hat{h}_\alpha) = (\alpha, \alpha) \neq 0$ (Proposition 2.9.4(ii)), we infer that $d_m = 0$ for all $m \geq 2$, and $d_1 = 1$. □

Proposition 2.9.6 *Let $\alpha, \beta \in \Phi$, $\beta \neq \pm\alpha$ and r and q be the maximal integers such that $\beta - r\alpha$ and $\beta + q\alpha$ are in Φ. Then $\beta + i\alpha \in \Phi$ for $-r \leq i \leq q$ and $r - q = \frac{2(\beta,\alpha)}{(\alpha,\alpha)}$.*

Proof. Set $h = \frac{2\hat{h}_\alpha}{(\alpha,\alpha)}$. According to Proposition 2.9.4 there are $e \in \mathfrak{g}_\alpha$ and $f \in \mathfrak{g}_{-\alpha}$ such that h, e, f span a subalgebra $\mathfrak{s}$ isomorphic to $\mathfrak{sl}(2, k)$. Let V be the subspace spanned by all $\mathfrak{g}_{\beta+i\alpha}$, $-r \leq i \leq q$. Then V is an $\mathfrak{s}$-module, and the eigenvalues of h on V are $\beta(h) + 2i$. These are all distinct, and both 0 and 1 do not occur, so by Theorem 2.9.1, V must be irreducible, and the eigenvalues of h on V are $-m, -m+2, \ldots, m-2, m$. So $\beta(h) + 2q = m$ and $\beta(h) - 2r = -m$, whence $\beta(h) = r - q$. Furthermore, all eigenvalues $m - 2j$, $0 \leq j \leq m$, occur, so all $\beta + i\alpha$ must be roots. □

Now set $\ell = \dim \mathfrak{h}$. By Proposition 2.9.3(iv) there are linearly independent roots $\alpha_1, \ldots, \alpha_\ell \in \Phi$. Let $\mathfrak{h}^*_{\mathbb{Q}}$ be the span over $\mathbb{Q}$ of these roots.

Proposition 2.9.7 (i) *For $\sigma, \rho \in \mathfrak{h}^*$ we have $(\sigma, \rho) = \sum_{\alpha\in\Phi}(\sigma,\alpha)(\rho,\alpha)$.*

(ii) *For $\alpha, \beta \in \Phi$ we have $(\alpha, \beta) \in \mathbb{Q}$.*

(iii) *$\mathfrak{h}^*_{\mathbb{Q}}$ contains Φ.*

(iv) *The form $(\ ,\ )$ is positive definite on $\mathfrak{h}^*_{\mathbb{Q}}$.*

Proof. As $\dim \mathfrak{g}_\alpha = 1$ for all $\alpha \in \Phi$ (Lemma 2.9.5) we have $(\sigma, \rho) = \kappa(\hat{h}_\sigma, \hat{h}_\rho) = \sum_\alpha \alpha(\hat{h}_\sigma)\alpha(\hat{h}_\rho)$, which yields (i).

Using (i) we obtain $\frac{4}{(\alpha,\alpha)} = (\frac{2\alpha}{(\alpha,\alpha)}, \frac{2\alpha}{(\alpha,\alpha)}) = \sum_{\gamma\in\Phi} \frac{2(\gamma,\alpha)}{(\alpha,\alpha)} \frac{2(\gamma,\alpha)}{(\alpha,\alpha)}$, which is an integer by Proposition 2.9.6. Hence $(\alpha, \beta) = \frac{2(\alpha,\beta)}{(\alpha,\alpha)} \frac{(\alpha,\alpha)}{2}$ is rational.

Let $\alpha \in \Phi$. Then there are $a_1, \ldots, a_\ell \in k$ such that $\alpha = \sum_i a_i\alpha_i$. So $(\alpha, \alpha_j) = \sum_i(\alpha_i, \alpha_j)a_i$. Letting j run from 1 to ℓ we obtain a system of equations for the a_i, having rational coefficients by (ii). Since the form $(\ ,\ )$ is non-degenerate, it has a unique solution, which necessarily has coefficients in $\mathbb{Q}$.

For $\sigma \in \mathfrak{h}^*_{\mathbb{Q}}$ we have, by (i), $(\sigma, \sigma) = \sum_\alpha(\sigma, \alpha)^2$, which is non-negative by (ii). Furthermore, $(\sigma, \sigma) = 0$ implies $(\sigma, \alpha) = 0$ for all $\alpha \in \Phi$, entailing $\sigma = 0$ by the non-degeneracy of the form. □

Theorem 2.9.8 *Set $\mathfrak{h}^*_{\mathbb{R}} = \mathbb{R} \otimes \mathfrak{h}^*_{\mathbb{Q}}$, and extend the form $(\ ,\ )$ to this space. Then Φ is a root system in $\mathfrak{h}^*_{\mathbb{R}}$. Moreover, this root system does not depend (up to isomorphism) on the choice of the Cartan subalgebra $\mathfrak{h}$.*

Proof. Note that by Proposition 2.9.7, the given form is positive definite. For $\alpha, \beta \in \Phi$ we have $\langle \beta, \alpha^\vee \rangle \in \mathbb{Z}$ by Proposition 2.9.6. That same proposition implies that $s_\alpha(\beta) \in \Phi$. Moreover, Φ does not contain 0 and spans $\mathfrak{h}^*_{\mathbb{R}}$. Using Lemma 2.9.5 and Proposition 2.9.6 we infer that $a\alpha \notin \Phi$ for $\alpha \in \Phi$ and $a \in \mathbb{R}$, $a \neq \pm 1$.

Let $\tilde{\mathfrak{h}}$ be a second split Cartan subalgebra yielding the root system $\tilde{\Phi}$. In order to show that Φ and $\tilde{\Phi}$ are isomorphic we may assume that k is algebraically closed. Indeed, we can tensor $\mathfrak{g}$ with the algebraic closure of k, then the root systems that we find are the same. Then $\mathfrak{h}$, $\tilde{\mathfrak{h}}$ are conjugate by an inner automorphism of $\mathfrak{g}$ (Theorem 2.5.8). This implies that Φ and $\tilde{\Phi}$ are isomorphic. □

Remark 2.9.9 Note that the Weyl group of Φ acts on $\mathfrak{h}^*$ (not just on $\mathfrak{h}^*_{\mathbb{R}}$). We use the bijection $\mathfrak{h}^* \to \mathfrak{h}$ to define an action of W on $\mathfrak{h}$ as well. More precisely, for $\alpha \in \Phi$, $h \in \mathfrak{h}$ we set $s_\alpha(h) = h - \alpha(h)h_\alpha$, where $h_\alpha = \frac{2\hat{h}_\alpha}{(\alpha,\alpha)}$. This makes the following diagram commute:

$$\begin{array}{ccc} \mathfrak{h}^* & \xrightarrow{s_\alpha} & \mathfrak{h}^* \\ \downarrow & & \downarrow \\ \mathfrak{h} & \xrightarrow{s_\alpha} & \mathfrak{h}. \end{array}$$

2.9.3 Canonical generators and isomorphisms

In the previous section we showed that a semisimple Lie algebra determines a unique root system. Here we show that two semisimple Lie algebras having the same root systems are isomorphic. So a semisimple Lie algebra is determined (up to isomorphism) by its root system.

We let $\mathfrak{g}$, $\mathfrak{h}$ and Φ be as in the previous section. We note that $\mathfrak{h}$ is commutative ($[\mathfrak{h},\mathfrak{h}]=0$) by Proposition 2.9.3(ii). By Theorem 2.9.8, Φ is a root system and hence has a basis of simple roots $\Delta=\{\alpha_1,\dots,\alpha_\ell\}$. Using results shown in the previous section (most importantly Proposition 2.9.4), we see that there are non-zero $h_i\in\mathfrak{h}$, $e_i\in\mathfrak{g}_{\alpha_i}$, $f_i\in\mathfrak{g}_{-\alpha_i}$ such that

$$\begin{aligned} &[h_i,h_j]=0\\ &[e_i,f_j]=\delta_{ij}h_i\\ &[h_j,e_i]=\langle\alpha_i,\alpha_j^\vee\rangle e_i\\ &[h_j,f_i]=\langle-\alpha_i,\alpha_j^\vee\rangle f_i \quad \text{for } 1\le i,j\le\ell. \end{aligned} \tag{2.1}$$

Definition 2.9.10 *A set of non-zero elements* $\{h_i,e_i,f_i \mid 1\le i\le\ell\}$ *of* $\mathfrak{g}$ *satisfying* (2.1) *is called a* canonical generating set *of* $\mathfrak{g}$.

There is a straightforward algorithm for computing a canonical generating set. Indeed, for i from 1 to ℓ:

1. Let e_i be a non-zero element of $\mathfrak{g}_{\alpha_i}$.
2. Find an $f_i\in\mathfrak{g}_{-\alpha_i}$ such that $[[e_i,f_i],e_i]=2e_i$.
3. Set $h_i=[e_i,f_i]$.

Since a canonical generating set exists, this algorithm is correct. Indeed, let $\bar{h}_i$, $\bar{e}_i$, $\bar{f}_i$ be the elements of a canonical generating set. As the root spaces are 1-dimensional, there are non-zero scalars ν_i and μ_i with $\bar{e}_i=\nu_i e_i$, $\bar{f}_i=\mu_i f_i$. This immediately gives $\bar{h}_i=\nu_i\mu_i h_i$. From $[\bar{h}_i,\bar{e}_i]=2\bar{e}_i$ it follows that $\nu_i\mu_i=1$. But that implies that h_i, e_i, f_i, $1\le i\le\ell$, form a canonical generating set as well.

Proposition 2.9.11 *A canonical generating set generates* $\mathfrak{g}$.

Proof. Let $\mathfrak{a}$ be the subalgebra generated by a canonical generating set $\{h_i, e_i, f_i\}$. Let $\beta \in \Phi$ be a positive root, $\beta \notin \Delta$. By Proposition 2.8.10(iv) there is a simple root α_i such that $\gamma = \beta - \alpha_i \in \Phi$. By induction we may assume that $\mathfrak{g}_\gamma \subset \mathfrak{a}$. Let $\mathfrak{s}$ be the subalgebra spanned by h_i, e_i, f_i. Let V be the subspace of $\mathfrak{g}$ spanned by all $\mathfrak{g}_{\gamma+l\alpha_i}$, for $l \in \mathbb{Z}$. As seen in the proof of Proposition 2.9.6, V is an irreducible $\mathfrak{s}$-module. Let q and r be the maximal integers such that $\gamma + q\alpha_i$, $\gamma - r\alpha_i$ lie in Φ. Let v_0 span $\mathfrak{g}_{\gamma+q\alpha_i}$ and set $v_j = (\mathrm{ad} f_i)^j v_0$. Then $[h, v_0] = (\gamma + q\alpha_i)(h)v_0 = (q + r)v_0$ (see the proof of Proposition 2.9.6). Moreover, v_q spans $\mathfrak{g}_\gamma$, and by Theorem 2.9.1 we see that $[e_i, v_q] = q(r + 1)v_{q-1}$, which is non-zero as $q > 0$. So $\mathfrak{g}_\beta \subset \mathfrak{a}$. We can do the same with the negative roots. It follows that $\mathfrak{a} = \mathfrak{g}$. □

Theorem 2.9.12 *Let $\mathfrak{g}$, $\tilde{\mathfrak{g}}$ be semisimple Lie algebras over k having split Cartan subalgebras and corresponding root systems Φ and $\tilde{\Phi}$. Let Δ and $\tilde{\Delta}$ be respective bases of simple roots, and assume that the Cartan matrices of Φ and $\tilde{\Phi}$ with respect to Δ and $\tilde{\Delta}$ are identical. Let h_i, e_i, f_i and $\tilde{h}_i, \tilde{e}_i, \tilde{f}_i$ form canonical generating sets of $\mathfrak{g}$ and $\tilde{\mathfrak{g}}$, chosen relative to Δ and $\tilde{\Delta}$. Then there is a unique isomorphism $\phi : \mathfrak{g} \to \tilde{\mathfrak{g}}$ with $\phi(h_i) = \tilde{h}_i$, $\phi(e_i) = \tilde{e}_i$, $\phi(f_i) = \tilde{f}_i$.*

Proof. Uniqueness follows from Proposition 2.9.11. To prove existence we first note that there is an isomorphism of root systems $\Phi \to \tilde{\Phi}$, $\beta \mapsto \tilde{\beta}$. Write $\Delta = \{\alpha_1, \dots, \alpha_\ell\}$, $\tilde{\Delta} = \{\tilde{\alpha}_1, \dots, \tilde{\alpha}_\ell\}$. Let $\beta \in \Phi$ be positive. Then by Proposition 2.8.10(iv) there is a sequence $i_1, \dots, i_r$ such that $\alpha_{i_1} + \cdots + \alpha_{i_l} \in \Phi$ for $1 \leq l \leq r$ and $\alpha_{i_1} + \cdots + \alpha_{i_r} = \beta$. For each β we fix such a sequence and define $e_\beta = [e_{i_r}, [e_{i_{r-1}}, [\cdots [e_{i_2}, e_{i_1}] \cdots]]]$, $f_\beta = [f_{i_r}, [f_{i_{r-1}}, [\cdots [f_{i_2}, f_{i_1}] \cdots]]]$. Correspondingly we define the elements $e_{\tilde{\beta}}$ and $f_{\tilde{\beta}}$ in $\tilde{\mathfrak{g}}$. It can be shown that the structure constants of $\mathfrak{g}$ with respect to the basis $h_1, \dots, h_\ell$, e_β, f_β are rational numbers that can be determined using the Cartan matrix only (see [Jac79], Section IV.3, Theorem 2). A similar statement holds for $\tilde{\mathfrak{g}}$. So the structure constants are the same for $\mathfrak{g}$ and $\tilde{\mathfrak{g}}$. It follows that ϕ exists. □

2.9.4 Structure constants of semisimple Lie algebras

Here we comment on the missing piece of the classification of the semisimple Lie algebras (over fields of characteristic 0, and having a split Cartan subalgebra): for each root system there exists a corresponding semisimple Lie algebra. We start with a simple observation: $\mathfrak{g}$ is the direct sum of two ideals $\mathfrak{g}_1, \mathfrak{g}_2$ if and only if the root system of $\mathfrak{g}$ is the direct sum of the root systems of $\mathfrak{g}_1$ and $\mathfrak{g}_2$. This implies that it suffices to construct Lie algebras corresponding to the irreducible root systems. Those Lie algebras are then necessarily simple. One way to approach this is to explicitly construct a Lie algebra for each type appearing in Theorem 2.8.5. We refer to [Jac79], Section IV.6 for a detailed account of this. A second method, invented by Serre ([Ser66]), is to construct the Lie algebra by generators and relations starting from the Cartan matrix of the root system. A third method starts from the following theorem.

Theorem 2.9.13 (Chevalley) *Let $\mathfrak{g}$ be a semisimple Lie algebra over the field k of characteristic 0. Suppose $\mathfrak{g}$ has a split Cartan subalgebra $\mathfrak{h}$. Let Φ be the root system of $\mathfrak{g}$ relative to $\mathfrak{h}$, with basis of simple roots $\Delta = \{\alpha_1, \dots, \alpha_\ell\}$. Then there is a basis $h_1, \dots, h_\ell$ of $\mathfrak{h}$ and elements $x_\alpha \in \mathfrak{g}_\alpha$, for $\alpha \in \Phi$ such that*

$$\begin{aligned}
&[h_i, h_j] = 0 \quad \text{for } 1 \leq i, j \leq l,\\
&[h_i, x_\alpha] = \langle \alpha, \alpha_i^\vee \rangle x_\alpha \quad \text{for } 1 \leq i \leq l \text{ and } \alpha \in \Phi,\\
&[x_\alpha, x_{-\alpha}] = \sum_{i=1}^{l} n_i^\alpha h_i \quad \text{for } \alpha \in \Phi;\ \text{all } n_i^\alpha \in \mathbb{Z},\\
&[x_\alpha, x_\beta] = 0 \quad \text{for } \alpha, \beta \in \Phi \text{ such that } \alpha + \beta \notin \Phi \text{ and } \beta \neq -\alpha,\\
&[x_\alpha, x_\beta] = \varepsilon(\alpha, \beta)(r+1) x_{\alpha+\beta} \quad \text{for } \alpha, \beta \in \Phi \text{ such that } \alpha + \beta \in \Phi,\\
&\qquad \text{where } \varepsilon(\alpha, \beta) = \pm 1, \text{ and } r \text{ is the largest integer}\\
&\qquad \text{such that } \alpha - r\beta \in \Phi.
\end{aligned} \tag{2.2}$$

Moreover, the n_i^α are determined in the following way. For $\alpha \in \Phi$ set $\bar{\alpha} = \frac{2\alpha}{(\alpha,\alpha)}$. Then $\overline{\Phi} = \{\bar{\alpha} \mid \alpha \in \Phi\}$ is a root system with basis of simple roots $\{\bar{\alpha}_1, \dots, \bar{\alpha}_\ell\}$ and for $\alpha \in \Phi$ we have $\bar{\alpha} = \sum_i n_i^\alpha \bar{\alpha}_i$.

Proof. For this we first suppose that k is algebraically closed. Let h_i, e_i, f_i, $1 \leq i \leq \ell$ form a canonical generating set. Set $\bar{h}_i = -h_i$, $\bar{e}_i = -f_i$, $\bar{f}_i = -e_i$; these elements also satisfy (2.1), and hence form a canonical generating set. So by Theorem 2.9.12 there is a unique automorphism $\theta : \mathfrak{g} \to \mathfrak{g}$ with $\theta(h_i) = -h_i$, $\theta(e_i) = -f_i$, $\theta(f_i) = -e_i$. This implies that $\theta(\mathfrak{g}_\alpha) = \mathfrak{g}_{-\alpha}$ for all $\alpha \in \Phi$.

Let $\alpha \in \Phi$ be positive and $z \in \mathfrak{g}_\alpha$ be non-zero. Set $x_\alpha = \xi z$, where $\xi \in k$ is to be determined. Set $x_{-\alpha} = -\theta(x_\alpha)$. Then by Proposition 2.9.4, $[x_\alpha, x_{-\alpha}] = \kappa(x_\alpha, x_{-\alpha})\hat{h}_\alpha = \xi^2 \kappa(z, -\theta(z))\hat{h}_\alpha$. Now we choose ξ such that $\xi^2 \kappa(z, -\theta(z)) = \frac{2}{(\alpha,\alpha)}$, so that $[x_\alpha, x_{-\alpha}] = \frac{2\hat{h}_\alpha}{(\alpha,\alpha)}$.

The first two commutation relations are clear. The third commutation relation follows from the statements on $\overline{\Phi}$ (whose proof we leave to the reader). The fourth is also clear. Finally, the fifth is proved by a series of technical arguments involving root systems and the fact that $x_{-\alpha} = -\theta(x_\alpha)$, for which we refer to [Hum78], Section 25.

If k is not algebraically closed, we tensor with its algebraic closure $\bar{k}$, and conclude that $\bar{k} \otimes \mathfrak{g}$ has a basis satisfying (2.2). In particular, (2.2) is the multiplication table of a Lie algebra $\tilde{\mathfrak{g}}$, and since the structure constants are integers, we may take the ground field of $\tilde{\mathfrak{g}}$ to be k. Now $h_i, x_{\alpha_i}, x_{-\alpha_i}$ form a canonical generating set of $\tilde{\mathfrak{g}}$. Mapping this set to any canonical generating set of $\mathfrak{g}$ (chosen with respect to Δ) extends to an isomorphism $\tilde{\mathfrak{g}} \to \mathfrak{g}$ (Theorem 2.9.12). Hence $\mathfrak{g}$ also has the required basis. □

A basis of $\mathfrak{g}$ consisting of $h_1, \ldots, h_\ell \in \mathfrak{h}$, along with x_α for $\alpha \in \Phi$, satisfying the commutation relations (2.2), is called a *Chevalley basis* of $\mathfrak{g}$.

We remark that the only unknown constants in (2.2) are the $\varepsilon(\alpha, \beta)$. The strategy for proving the existence of a semisimple Lie algebra corresponding to the root system Φ now is to specify the $\varepsilon(\alpha, \beta)$ (for all $\alpha, \beta \in \Phi$ such that $\alpha + \beta \in \Phi$), and check that the Jacobi identity is satisfied. It then follows that (2.2) defines a semisimple Lie algebra with root system Φ. This approach was proposed by Tits ([Tit66]). Tits' method has been worked into an algorithm for obtaining the $\varepsilon(\alpha, \beta)$ by Casselman ([Cas14]). An alternative way to specify the $\varepsilon(\alpha, \beta)$, which is significantly simpler, can be found in [Kac90]. The construction is split into two parts. The first is for root systems of types A_ℓ, D_ℓ and E_ℓ. The Lie algebras corresponding to the root systems of the remaining types are constructed as subalgebras of the Lie algebras of the previous types; this then also yields a method for specifying the $\varepsilon(\alpha, \beta)$.

Here we briefly indicate, without proof, how this works for the simply laced types A_ℓ, D_ℓ and E_ℓ. Let Q denote the root lattice consisting of all integral linear combinations of the simple roots $\alpha_1, \ldots, \alpha_\ell$. We define a map $\eta : Q \times Q \to \{1, -1\}$ in the following way. Firstly, $\eta(\alpha_i, \alpha_j) = -1$ if $i = j$ or $i < j$ and $\langle \alpha_i, \alpha_j \rangle = -1$. Otherwise $\eta(\alpha_i, \alpha_j) = 1$. Secondly, we extend η to all of $Q \times Q$ by

$$\eta(\sum_i s_i\alpha_i, \sum_j t_j\alpha_j) = \prod_i \prod_j \eta(\alpha_i, \alpha_j)^{s_i t_j}.$$

Set $\epsilon(\alpha) = 1$ for $\alpha \in \Phi^+$, $\epsilon(\alpha) = -1$ for $\alpha \in \Phi^-$. Finally, we define $\varepsilon(\alpha, \beta) = \epsilon(\alpha)\epsilon(\beta)\epsilon(\alpha + \beta)\eta(\alpha, \beta)$.

Example 2.9.14 Let Φ be the root system of type A_2 with positive roots α_1, α_2, $\alpha_3 = \alpha_1 + \alpha_2$. Using the above procedure we arrive at the following multiplication table of $\mathfrak{g}$:

	h_1	h_2	x_{α_1}	x_{α_2}	x_{α_3}	$x_{-\alpha_1}$	$x_{-\alpha_2}$	$x_{-\alpha_3}$
h_1	0	0	$2x_{\alpha_1}$	$-x_{\alpha_2}$	x_{α_3}	$-2x_{-\alpha_1}$	$x_{-\alpha_2}$	$-x_{-\alpha_3}$
h_2	·	0	$-x_{\alpha_1}$	$2x_{\alpha_2}$	x_{α_3}	$x_{-\alpha_1}$	$-2x_{-\alpha_2}$	$-x_{-\alpha_3}$
x_{α_1}	·	·	0	$-x_{\alpha_3}$	0	h_1	0	$x_{-\alpha_2}$
x_{α_2}	·	·	·	0	0	0	h_2	$-x_{-\alpha_1}$
x_{α_3}	·	·	·	·	0	x_{α_2}	$-x_{\alpha_1}$	$h_1 + h_2$
$x_{-\alpha_1}$	·	·	·	·	·	0	$x_{-\alpha_3}$	0
$x_{-\alpha_2}$	·	·	·	·	·	·	0	0

Remark 2.9.15 Let $\alpha \in \Phi$, and x_α, h_i as in Theorem 2.9.13. Set $h = [x_\alpha, x_{-\alpha}]$. Then

$$\alpha(h) = \sum_{i=1}^{\ell} n_i^\alpha \alpha(h_i) = \sum_{i=1}^{\ell} n_i^\alpha \langle \alpha, \alpha_i^\vee \rangle = (\alpha, \bar{\alpha}) = 2.$$

This implies that $h = \frac{2\hat{h}_\alpha}{(\alpha,\alpha)}$ and that h, $e = x_\alpha$, $f = x_{-\alpha}$ satisfy the commutation relations of the basis of $\mathfrak{sl}(2, k)$ given in Example 2.1.4.

2.10 Universal enveloping algebras

Each Lie algebra $\mathfrak{g}$ has a universal enveloping algebra, which is an infinite-dimensional associative algebra, playing an important role in the representation theory of $\mathfrak{g}$. Here we mainly study universal enveloping algebras of semisimple Lie algebras. They are vital for the construction of the irreducible representations of those Lie algebras, as well as for the construction of the semisimple algebraic groups.

2.10.1 Poincaré-Birkhoff-Witt theorem

Let X be a set of symbols whose elements are called *letters*. A finite sequence of elements of X is called a *word*. This includes the empty word. The set of all words is denoted X^*. It is endowed with a binary operation $\cdot : X^* \times X^* \to X^*$, $u \cdot v = uv$ (this is just concatenation).

Let k be a field and let $k\langle X\rangle$ denote the k-span of X^*. By extending the operation $\cdot$ bilinearly to $k\langle X\rangle$, this space is made into an associative algebra, called the *free associative algebra* over k generated by X.

Let $\mathfrak{g}$ be a Lie algebra over k. For convenience we assume that $\mathfrak{g}$ is finite-dimensional, and let $B = \{x_1, \dots, x_n\}$ be a basis of $\mathfrak{g}$. Let $X = \{\bar{x}_1, \dots, \bar{x}_n\}$ be a set of symbols, and let $\phi : B \to X$, $\phi(x_i) = \bar{x}_i$ denote the corresponding bijection. We extend ϕ linearly to a map $\phi : \mathfrak{g} \to k\langle X\rangle$. Let I be the (two-sided) ideal of $k\langle X\rangle$ generated by the elements

$$\bar{x}_i\bar{x}_j - \bar{x}_j\bar{x}_i - \phi([x_i, x_j]), \text{ for } 1 \le j < i \le n.$$

The quotient algebra $\mathcal{U}(\mathfrak{g}) = k\langle X\rangle/I$ is called the *universal enveloping algebra* of $\mathfrak{g}$. We also denote it $\mathcal{U}$ if no confusion can arise concerning the Lie algebra used. By $\bar{x}_i$ we also denote the image of $\bar{x}_i$ in $\mathcal{U}(\mathfrak{g})$.

Theorem 2.10.1 (Poincaré-Birkhoff-Witt) *The monomials $\bar{x}_i^{m_1} \cdots \bar{x}_n^{m_n}$ form a basis of $\mathcal{U}(\mathfrak{g})$.*

Proof. Perhaps the most elegant way to prove this is to set up a Gröbner basis theory for the free associative algebra $k\langle X\rangle$. Here instead of S-polynomials we have the notion of *compositions*. The Jacobi identity entails that the composition of two generators of I reduces to 0 modulo these generators. This means that the given set of generators already forms a Gröbner basis of I. Therefore the set of monomials not divisible by a leading monomial of a generator forms a basis of the quotient. The leading monomials of the generators of I are exactly $\bar{x}_j\bar{x}_i$, where $j > i$ (relative to a carefully chosen ordering on the words in X^*). So the monomials given in the theorem form a basis of the quotient $\mathcal{U}(\mathfrak{g})$. For more details we refer to [Ber78] and [Gra00]. □

Corollary 2.10.2 *Compose ϕ with the projection map to obtain a linear map $\phi : \mathfrak{g} \to \mathcal{U}(\mathfrak{g})$. We have that ϕ is injective.*

From this corollary it follows that we can identify $\mathfrak{g}$ with the subspace of $\mathcal{U}(\mathfrak{g})$ spanned by (the cosets of) $\bar{x}_1, \dots, \bar{x}_n$. For this reason we denote these cosets by the same symbols as the basis elements of $\mathfrak{g}$, i.e., $x_1, \dots, x_n$. From the context it will always be clear whether we mean an element of $\mathfrak{g}$ or of $\mathcal{U}(\mathfrak{g})$. The basis elements $x_1^{m_1} \cdots x_n^{m_n}$ of $\mathcal{U}(\mathfrak{g})$ are called *monomials.*

Remark 2.10.3 The construction of $\mathcal{U}(\mathfrak{g})$ depends on the choice of a basis of $\mathfrak{g}$. However, it is easily seen that a different choice of basis leads to an isomorphic algebra. Also it can be shown that $\mathcal{U}(\mathfrak{g})$ satisfies a universal property, namely, if A is an algebra with one, and $\psi : \mathfrak{g} \to A$ is a linear map satisfying $\psi([x,y]) = \psi(x)\psi(y) - \psi(y)\psi(x)$, there is a unique algebra homomorphism $\pi : \mathcal{U}(\mathfrak{g}) \to A$ such that $\phi \circ \pi = \psi$.

Example 2.10.4 Let $\mathfrak{g} = \mathfrak{sl}(2,k)$ with basis f, h, e as in Example 2.1.4. Then $\mathcal{U}(\mathfrak{g})$ is generated by h, e, f, which satisfy

$$hf = fh - 2f, \quad eh = he - 2e, \quad ef = fe + h.$$

And $\mathcal{U}(\mathfrak{g})$ is spanned by the monomials $f^k h^m e^n$, for $k, m, n \in \mathbb{Z}_{\geq 0}$. Using the above relations we can rewrite a product of monomials as a linear combination of monomials. For example, by induction one proves that $hf^i = f^i h - 2if^i$ for all $i \geq 0$.

Now let $\rho : \mathfrak{g} \to \mathfrak{gl}(V)$ be a representation of $\mathfrak{g}$. In the obvious way we extend ρ to a homomorphism of associative algebras $\rho : k\langle X \rangle \to \mathrm{End}(V)$. Then $\rho(I) = 0$, so ρ induces a map $\rho : \mathcal{U}(\mathfrak{g}) \to \mathrm{End}(V)$. We see that a representation of $\mathfrak{g}$ induces a representation of $\mathcal{U}(\mathfrak{g})$.

Example 2.10.5 Let the notation be as in Example 2.10.4. Let ρ be the representation considered in Theorem 2.9.1. Using the notation of that theorem, $v_i = f^i \cdot v_0$. So $h \cdot v_i = (hf^i) \cdot v_0 = (f^i h - 2if^i) \cdot v_0 = (n-2i)f^i \cdot v_0 = (n-2i)v_i$ in accordance with Theorem 2.9.1.

2.10.2 Left ideals of universal enveloping algebras

It is the objective of this section to sketch a Gröbner basis theory for left ideals in universal enveloping algebras. This will become useful when we study representations of semisimple Lie algebras.

As before let $\mathfrak{g}$ be a Lie algebra over the field k and $\mathcal{U}(\mathfrak{g})$ or $\mathcal{U}$ denotes its universal enveloping algebra. A subspace I of $\mathcal{U}(\mathfrak{g})$ is a *left ideal* if $fg \in I$ for all $f \in \mathcal{U}(\mathfrak{g})$ and $g \in I$.

As in the previous section we let $x_1, \dots, x_n$ denote a fixed basis of $\mathfrak{g}$. By Theorem 2.10.1, a basis of $\mathcal{U}(\mathfrak{g})$ is formed by the monomials $x_1^{k_1} \cdots x_n^{k_n}$. Let m be such a monomial, then the degree of m, denoted $\deg(m)$, is the number $\sum_i k_i$. We define the order $<_{\mathsf{dlex}}$ on the set of monomials as follows: if $\deg(m_1) < \deg(m_2)$ then $m_1 <_{\mathsf{dlex}} m_2$, otherwise, write $m_1 = x_1^{k_1} \cdots x_n^{k_n}$, $m_2 = x_1^{l_1} \cdots x_n^{l_n}$, suppose the degrees are equal, and let i be minimal such that $k_i \neq l_i$; if $k_i < l_i$ then $m_1 <_{\mathsf{dlex}} m_2$.

Using the ordering $<_{\mathsf{dlex}}$ every $f \in \mathcal{U}$ has a largest monomial, called the *leading monomial* and denoted $\mathrm{LM}(f)$. Let m_1, m_2 be two monomials, written as above. In general the product $m_1 m_2$ is not equal to a monomial, but rather to a linear combination of monomials. However, because $<_{\mathsf{dlex}}$ is compatible with the degree,

$$\mathrm{LM}(m_1 m_2) = x_1^{k_1+l_1} \cdots x_n^{k_n+l_n}. \tag{2.3}$$

We say that the monomial m_1 is a *factor* of m_2 if $k_i \leq l_i$ for all i. This is justified by (2.3): if m_1 is a factor of m_2 then there is a monomial m_3 such that $\mathrm{LM}(m_3 m_1) = m_2$. Subsequently we define the notion of normal form, and formulate a normal form algorithm exactly as for polynomial rings; see Section 1.5 where we replace "ideal" by "left ideal". Also in this case the normal form algorithm is guaranteed to terminate, as by (2.3), the leading monomial of the element h (as in Algorithm 1.5.2) decreases every step. Furthermore, the notion of Gröbner basis of a left ideal is defined exactly as in Definition 1.5.3.

Also we define S-elements analogously to S-polynomials. Let $f_1, f_2 \in \mathcal{U}$, and $m_i = \mathrm{LM}(f_i)$, $i = 1, 2$. Write m_1, m_2 as above. Define $s_i = \max(k_i, l_i)$ for $1 \leq i \leq n$. Let c_i for $i = 1, 2$ be the coefficient of m_i in f_i. Then

$$S(f_1, f_2) = c_2 x_1^{s_1-k_1} \cdots x^{s_n-k_n} f_1 - c_1 x_1^{s_1-l_1} \cdots x_n^{s_n-l_n} f_2$$

is called the S-element of f_1, f_2. Note that by (2.3), the leading monomials of the two summands in the definition of S-element are the same, and thus cancel.

Now it can be shown that a set $G \subset \mathcal{U}$ is a Gröbner basis of the left ideal it generates if and only if the normal form of $S(g_1, g_2)$ modulo G is 0 for all $g_1, g_2 \in G$. Indeed, the proof of Theorem 1.5.4 can be copied almost verbatim when necessary using (2.3). This yields an algorithm for computing a Gröbner basis of a left ideal that is completely analogous to the algorithm in Section 1.5. In this case termination is ensured by considering a polynomial ring $R = k[y_1, \dots, y_n]$ and mapping the leading monomial $x_1^{k_1} \cdots x_n^{k_n}$ of an element of the Gröbner basis under construction to its corresponding monomial $y_1^{k_1} \cdots y_n^{k_n}$ in R. Then each time G increases, the ideal in R generated by the monomials corresponding to the leading monomials of G also increases.

Example 2.10.6 Let $\mathfrak{g}$ be the 3-dimensional Lie algebra with basis y_1, y_2, y_3, satisfying $[y_1, y_2] = y_3$, $[y_1, y_3] = [y_2, y_3] = 0$. Let I be the left ideal of $\mathcal{U}(\mathfrak{g})$ generated by $g_1 = y_1^2$, $g_2 = y_2^2$. Then $S(g_1, g_2) = y_2^2 y_1^2 - y_1^2 y_2^2 = -4y_1y_2y_3 + 2y_3^2$, which, after multiplying by a scalar, yields the element $g_3 = y_1y_2y_3 - \frac{1}{2}y_3^2$. Now $S(g_1, g_3) = -\frac{3}{2}y_1y_3^2$, yielding $g_4 = y_1y_3^2$. Furthermore, $S(g_2, g_3) = \frac{3}{2}y_2y_3^2$, $S(g_3, g_4) = \frac{1}{2}y_3^3$, and we get $g_5 = y_2y_3^2$, $g_6 = y_3^3$. The normal form of the remaining S-elements modulo the g_i is always 0. So $\{g_1, \dots, g_6\}$ form a Gröbner basis of I.

The quotient $\mathcal{U}(\mathfrak{g})/I$ is spanned by the (cosets of the) monomials that do not have a leading monomial of a g_i as a factor. These are 1, y_1, y_2, y_3, y_1y_2, y_2y_3, y_2y_3, y_3^2. So $\dim \mathcal{U}(\mathfrak{g})/I = 8$.

2.10.3 Integral forms

Here we let $\mathfrak{g}$ be a semisimple Lie algebra over the field k of characteristic 0, with a fixed split Cartan subalgebra $\mathfrak{h}$. We describe a basis of $\mathcal{U} = \mathcal{U}(\mathfrak{g})$ such that the $\mathbb{Z}$-span of this basis forms a subring of $\mathcal{U}$. This $\mathbb{Z}$-span is said to be an integral form of $\mathcal{U}$, and denoted $\mathcal{U}_\mathbb{Z}$. We also derive commutation relations in $\mathcal{U}_\mathbb{Z}$, making it possible to compute products of basis elements rather efficiently.

We let $h_1, \dots, h_\ell$, along with x_α for $\alpha \in \Phi$, be the elements of a fixed Chevalley basis of $\mathfrak{g}$ (Section 2.9.4). For $\alpha \in \Phi$ we consider the divided powers

$$x_\alpha^{(n)} = \frac{x_\alpha^n}{n!} \in \mathcal{U}.$$

Let $\mathcal{U}_\mathbb{Z}$ denote the subring of $\mathcal{U}$ generated by all $x_\alpha^{(n)}$, for $\alpha \in \Phi$, $n \geq 0$.

Let $\alpha, \beta \in \Phi$ be such that $\alpha \neq \pm\beta$. Then $x_\beta^{(m)} x_\alpha^{(n)}$ is equal to $x_\alpha^{(n)} x_\beta^{(m)}$ plus a sum of products of $x_\gamma^{(k)}$. We derive precise formulas for this.

Set $\Psi = \{i\alpha + j\beta \mid i, j \in \mathbb{Z}\} \cap \Phi$. Then Ψ is a root system of rank 2. So it is of type $A_1 + A_1$, A_2, B_2, or G_2. By inspecting these root systems we see that the set $R_{\alpha,\beta} = \{i\alpha + j\beta \mid i, j \in \mathbb{Z}_{\geq 0}\} \cap \Phi$ can be one of the following:

I $R_{\alpha,\beta} = \{\alpha, \beta\}$,

II $R_{\alpha,\beta} = \{\alpha, \beta, \alpha + \beta\}$,

III $R_{\alpha,\beta} = \{\alpha, \beta, \alpha + \beta, \alpha + 2\beta\}$,

IV $R_{\alpha,\beta} = \{\alpha, \beta, \alpha + \beta, 2\alpha + \beta\}$,

V $R_{\alpha,\beta} = \{\alpha, \beta, \alpha + \beta, 2\alpha + \beta, 3\alpha + \beta, 3\alpha + 2\beta\}$,

VI $R_{\alpha,\beta} = \{\alpha, \beta, \alpha + \beta, \alpha + 2\beta, \alpha + 3\beta, 2\alpha + 3\beta\}$,

VII $R_{\alpha,\beta} = \{\alpha, \beta, \alpha + \beta, 2\alpha + \beta, \alpha + 2\beta\}$.

We remark that the sets V, VI and VII can only occur if Ψ is of type G_2.

In the following proposition we define $x_\gamma^{(r)} = 0$ if $r < 0$. Because of this convention, the summations are finite, although their indices run over $\mathbb{Z}$. Furthermore, for roots γ, δ, with $\gamma \neq \pm\delta$ we define $N_{\gamma,\delta} = \varepsilon(\gamma,\delta)(r+1)$, where r is the maximal integer with $\gamma - r\delta \in \Phi$, and ε as in Theorem 2.9.13, so that $[x_\gamma, x_\delta] = N_{\gamma,\delta}x_{\gamma+\delta}$. Also we set $N_{\gamma,\delta} = 0$ if at least one of $\gamma, \delta, \gamma+\delta$ does not lie in Φ.

Proposition 2.10.7 *Let α, β be as above. Set $c_{\alpha,\beta} = N_{\alpha,\beta}$, $c_{\alpha,\alpha+\beta} = \frac{1}{2}N_{\alpha,\alpha+\beta}$, $c_{\beta,\alpha+\beta} = \frac{1}{2}N_{\beta,\alpha+\beta}$, $c_{\alpha,2\alpha+\beta} = \frac{1}{3}N_{\alpha,2\alpha+\beta}$, $c_{\beta,\alpha+2\beta} = \frac{1}{3}N_{\beta,\alpha+2\beta}$ and $c_{\alpha+\beta,\alpha+2\beta} = \frac{1}{3}N_{\alpha+\beta,\alpha+2\beta}$. Then $x_\beta^{(m)}x_\alpha^{(n)}$ is given by*

$$\text{I } x_\beta^{(m)}x_\alpha^{(n)} = x_\alpha^{(n)}x_\beta^{(m)},$$

$$\text{II } x_\beta^{(m)}x_\alpha^{(n)} = \sum_{p\in\mathbb{Z}}(-1)^p c_{\alpha,\beta}^p x_\alpha^{(n-p)}x_\beta^{(m-p)}x_{\alpha+\beta}^{(p)},$$

$$\text{III } x_\beta^{(m)}x_\alpha^{(n)} = \sum_{p,q\in\mathbb{Z}}(-1)^p c_{\alpha,\beta}^{p+q} c_{\beta,\alpha+\beta}^q x_\alpha^{(n-p-q)}x_\beta^{(m-p-2q)}x_{\alpha+\beta}^{(p)}x_{\alpha+2\beta}^{(q)},$$

$$\text{IV } x_\beta^{(m)}x_\alpha^{(n)} = \sum_{p,q\in\mathbb{Z}}(-1)^p c_{\alpha,\beta}^{p+q} c_{\alpha,\alpha+\beta}^q x_\alpha^{(n-p-2q)}x_\beta^{(m-p-q)}x_{\alpha+\beta}^{(p)}x_{2\alpha+\beta}^{(q)},$$

$$\text{V } x_\beta^{(m)}x_\alpha^{(n)} = \sum_{p,q,r,s\in\mathbb{Z}}(-1)^{p+r} c_{\alpha,\beta}^{p+q+r+s} c_{\alpha,\alpha+\beta}^{q+r+s} c_{\alpha,2\alpha+\beta}^r c_0^s \cdot x_\alpha^{(n-p-2q-3r-3s)}x_\beta^{(m-p-q-r-2s)}x_{\alpha+\beta}^{(p)}x_{2\alpha+\beta}^{(q)}x_{3\alpha+\beta}^{(r)}x_{3\alpha+2\beta}^{(s)},$$

$$\text{where } c_0 = \tfrac{1}{2}(N_{\alpha,\beta}N_{\alpha+\beta,2\alpha+\beta} + c_{\alpha,2\alpha+\beta}N_{\beta,3\alpha+\beta})$$

$$\text{VI } x_\beta^{(m)}x_\alpha^{(n)} = \sum_{p,q,r,s\in\mathbb{Z}}(-1)^{p+r}2^s c_{\alpha,\beta}^{p+q+r+2s} c_{\beta,\alpha+\beta}^{q+r+s} c_{\beta,\alpha+2\beta}^r c_{\alpha+\beta,\alpha+2\beta}^s \cdot x_\alpha^{(n-p-q-r-2s)}x_\beta^{(m-p-2q-3r-3s)}x_{\alpha+\beta}^{(p)}x_{\alpha+2\beta}^{(q)}x_{\alpha+3\beta}^{(r)}x_{2\alpha+3\beta}^{(s)},$$

$$\text{VII } x_\beta^{(m)}x_\alpha^{(n)} = \sum_{p,q,r\in\mathbb{Z}}(-1)^p c_{\alpha,\beta}^{p+q+r} c_{\alpha,\alpha+\beta}^q c_{\beta,\alpha+\beta}^r \cdot x_\alpha^{(n-p-2q-r)}x_\beta^{(m-p-q-2r)}x_{\alpha+\beta}^{(p)}x_{2\alpha+\beta}^{(q)}x_{\alpha+2\beta}^{(r)}.$$

Proof. The formula in case I is obvious. By induction on n it is shown that

$$x_\beta x_\alpha^{(n)} = \sum_{p=0}^{n}\frac{(-1)^p}{p!}\left(\prod_{i=1}^{p}N_{\alpha,(i-1)\alpha+\beta}\right)x_\alpha^{(n-p)}x_{p\alpha+\beta}. \tag{2.4}$$

This yields the formulas for $m = 1$. The induction step is done separately for each case. Here we do it for case III, in which case the formula for $m = 1$ is

$$x_\beta x_\alpha^{(n)} = x_\alpha^{(n)}x_\beta - c_{\alpha,\beta}x_\alpha^{(n-1)}x_{\alpha+\beta}.$$

Using the induction hypothesis we have

$$x_\beta^{(m+1)}x_\alpha^{(n)} = \frac{1}{m+1}x_\beta x_\beta^{(m)}x_\alpha^{(n)}$$
$$= \frac{1}{m+1}\sum_{p,q\in\mathbb{Z}}(-1)^p c_{\alpha,\beta}^{p+q}c_{\beta,\alpha+\beta}^{q}x_\beta x_\alpha^{(n-p-q)}x_\beta^{(m-p-2q)}x_{\alpha+\beta}^{(p)}x_{\alpha+2\beta}^{(q)}.$$

By the formula for $m = 1$ we have

$$x_\beta x_\alpha^{(n-p-q)}x_\beta^{(m-p-2q)}x_{\alpha+\beta}^{(p)}x_{\alpha+2\beta}^{(q)} =$$
$$(x_\alpha^{(n-p-q)}x_\beta - c_{\alpha,\beta}x_\alpha^{(n-p-q-1)}x_{\alpha+\beta})x_\beta^{(m-p-2q)}x_{\alpha+\beta}^{(p)}x_{\alpha+2\beta}^{(q)}.$$

In view of $x_{\alpha+\beta}x_\beta^{(m-p-2q)} = x_\beta^{(m-p-2q)}x_{\alpha+\beta} - N_{\beta,\alpha+\beta}x_\beta^{(m-p-2q-1)}x_{\alpha+2\beta}$, which follows from (2.4), the right-hand side equals

$$(m-p-2q+1)x_\alpha^{(n-p-q)}x_\beta^{(m-p-2q+1)}x_{\alpha+\beta}^{(p)}x_{\alpha+2\beta}^{(q)} \quad (2.5)$$
$$-(p+1)c_{\alpha,\beta}x_\alpha^{(n-p-q-1)}x_\beta^{(m-p-2q)}x_{\alpha+\beta}^{(p+1)}x_{\alpha+2\beta}^{(q)} \quad (2.6)$$
$$+2(q+1)c_{\alpha,\beta}c_{\beta,\alpha+\beta}x_\alpha^{(n-p-q-1)}x_\beta^{(m-p-2q-1)}x_{\alpha+\beta}^{(p)}x_{\alpha+2\beta}^{(q+1)}. \quad (2.7)$$

So $x_\beta^{(m+1)}x_\alpha^{(n)}$ is equal to a sum of three summations divided by $m+1$. In the summation arising from (2.6) we substitute p for $p+1$, and in the one arising from (2.7) we substitute q for $q+1$. Then the three summations are seen to merge into one, multiplied by $m+1$. Hence the latter factor cancels, and the formula is proved by induction. □

For $n \geq 0$ and $h \in \mathfrak{h}$ we set

$$\binom{h}{n} = \frac{h(h-1)\cdots(h-n+1)}{n!},$$

which is a well defined element of $\mathcal{U}$.

Lemma 2.10.8 *Let $\alpha \in \Phi$, $h \in \mathfrak{h}$ and $p \in k[x]$. Then*

$$x_\alpha^{(n)}p(h) = p(h - n\alpha(h))x_\alpha^{(n)}.$$

Proof. Suppose $p = x^m$. Then by induction on m we see that $x_\alpha p(h) = p(h - \alpha(h))x_\alpha$. This formula follows for general p by linearity. Using this, the lemma is proved by induction on n. □

Lemma 2.10.9 *For $h \in \mathfrak{h}$, $\alpha \in \Phi$ and $r \in \mathbb{Z}$ we have*

$$\binom{h+r}{m} x_{-\alpha}^{(n)} = x_{-\alpha}^{(n)} \binom{h+r-n\alpha(h)}{m},$$

$$x_{\alpha}^{(n)} \binom{h+r}{m} = \binom{h+r-n\alpha(h)}{m} x_{\alpha}^{(n)}.$$

Proof. Both formulas immediately follow from Lemma 2.10.8. □

Lemma 2.10.10 *Let $\alpha \in \Phi$, and write $e = x_\alpha$, $f = x_{-\alpha}$, $h = [x_\alpha, x_{-\alpha}]$. Then*

$$e^{(m)} f^{(n)} = \sum_{j=0}^{\min(m,n)} f^{(n-j)} \binom{h-m-n+2j}{j} e^{(m-j)}.$$

Proof. As seen in Remark 2.9.15, h, e, f satisfy $[h,e] = 2e$, $[h,f] = -2f$, $[e,f] = h$. By induction on n, also using Lemma 2.10.9, it follows that $ef^{(n)} = f^{(n)}e + f^{(n-1)}(h-n+1)$. The proof is finished by induction on m. Here again Lemma 2.10.9 is of use. One crucial step is to show that

$$(m-j+1)\binom{h-m-n+2j-2}{j} + (h-n+j)\binom{h-m-n+2j-2}{j-2} = (m+1)\binom{h-m-n+2j-1}{j}.$$

This is achieved by using the well-known formula $\binom{n+1}{k} = \binom{n}{k} + \binom{n}{k-1}$, which here yields

$$\binom{h-m-n+2j-2}{j-2} = \binom{h-m-n+2j-1}{j} - \binom{h-m-n+2j-2}{j}.$$

By substituting this and also using $A\binom{A-1}{j} = (A-j)\binom{A}{j}$, the induction step is completed. □

Now we consider elements of the form

$$\left(\prod_{\alpha\in\Phi^+} x_{-\alpha}^{(n_\alpha)}\right) \binom{h_1}{k_1} \cdots \binom{h_\ell}{k_\ell} \left(\prod_{\alpha\in\Phi^+} x_{\alpha}^{(m_\alpha)}\right),$$

where the products on the left and the right are taken in any fixed order on Φ^+. We call such an element an *integral monomial.*

Theorem 2.10.11 (Kostant) *$\mathcal{U}_{\mathbb{Z}}$ is a free $\mathbb{Z}$-module spanned by the integral monomials.*

Proof. By Lemma 2.10.10 we have

$$x_{\alpha_i}^{(n)} x_{-\alpha_i}^{(n)} = \binom{h_i}{n} + \sum_{j=0}^{n-1} x_{-\alpha_i}^{(n-j)} \binom{h_i - 2n + 2j}{j} x_{\alpha_i}^{(n-j)}.$$

Now with induction on n we see that $\binom{h_i}{n} \in \mathcal{U}_\mathbb{Z}$. So all integral monomials lie in $\mathcal{U}_\mathbb{Z}$. Moreover, they span $\mathcal{U}$ over k, so their $\mathbb{Z}$-span is a free $\mathbb{Z}$-module. We need to show that this span is closed under multiplication. Using Proposition 2.10.7, and Lemmas 2.10.9, 2.10.10 we can almost rewrite a product of integral monomials as a linear combination of integral monomials. In order to complete the process we also use the obvious relation $x_\alpha^{(n)} x_\alpha^{(n)} = \binom{m+n}{m} x_\alpha^{(n+m)}$. From applications of Lemmas 2.10.9, 2.10.10 we get factors of the form

$$\binom{\sum_{i=1}^{\ell} n_i h_i + r}{j}$$

which we rewrite using the formula $\binom{a+b}{j} = \sum_{i=0}^{j} \binom{a}{i}\binom{b}{j-i}$. Finally, for rewriting a product of the form $\binom{h_i}{m}\binom{h_i}{n}$ we use the formula

$$\binom{h}{m}\binom{h}{n} = \sum_{i=0}^{m} \binom{m}{i}\binom{n+i}{m}\binom{h}{n+i}, \text{ where } m \leq n,$$

which can be proved by induction on m (for $m = 0, \dots, n$). □

Remark 2.10.12 Note that the previous proof gives an immediate method to rewrite the product of two integral monomials as a linear combination of integral monomials. So we have an algorithm to perform the multiplication in $\mathcal{U}_\mathbb{Z}$. We can also use the basis consisting of the monomials as a basis of $\mathcal{U}(\mathfrak{g})$. This yields an algorithm for performing the multiplication in $\mathcal{U}(\mathfrak{g})$, which is more efficient than the straightforward approach.

2.11 Representations of semisimple Lie algebras

In this section $\mathfrak{g}$ is a semisimple Lie algebra over the algebraically closed field k of characteristic 0. We let $\mathfrak{h}$ be a fixed Cartan subalgebra of $\mathfrak{g}$, and Φ the root system of $\mathfrak{g}$ with respect to $\mathfrak{h}$. Also we fix a basis of simple roots $\Delta = \{\alpha_1, \dots, \alpha_\ell\} \subset \Phi$ corresponding to the set of positive roots Φ^+. By W we denote the Weyl group of Φ.

For $\alpha \in \Phi$ we fix a non-zero $x_\alpha \in \mathfrak{g}_\alpha$ throughout. For example, we can take the elements of a fixed Chevalley basis for this. Furthermore, we fix a canonical generating set h_i, e_i, f_i, for $1 \leq i \leq \ell$, where $e_i \in \mathfrak{g}_{\alpha_i}$, $f_i \in \mathfrak{g}_{-\alpha_i}$. The h_i are uniquely determined by $h_i = \frac{2\hat{h}_{\alpha_i}}{(\alpha_i,\alpha_i)}$. An important role is played by elements μ of the dual space $\mathfrak{h}^*$, which are called *weights*. For such a μ we have $\mu(\hat{h}_{\alpha_i}) = \kappa(\hat{h}_\mu, \hat{h}_{\alpha_i}) = (\mu, \alpha_i)$, so that $\mu(h_i) = \langle \mu, \alpha_i^\vee \rangle$. The weight μ is said to be *integral* if $\langle \mu, \alpha_i^\vee \rangle \in \mathbb{Z}$ for $1 \leq i \leq \ell$. We see that the set of integral weights is exactly the set P considered in Section 2.8.3.

Let $\mu \in \mathfrak{h}^*$. A v in a $\mathfrak{g}$-module V is called a *weight vector* of *weight* μ if $h \cdot v = \mu(h)v$ for all $h \in \mathfrak{h}$. The space $V_\mu \subset V$ consisting of all weight vectors of weight μ is called the *weight space* of weight μ.

We define a partial order on $\mathfrak{h}^*$ by $\mu \preceq \lambda$ if $\mu = \lambda - \sum_{i=1}^{\ell} k_i \alpha_i$, where the k_i are non-negative integers.

2.11.1 Highest weight modules

Let V be a (possibly infinite-dimensional) $\mathfrak{g}$-module. As seen in Section 2.10.1, V is also a $\mathcal{U}(\mathfrak{g})$-module in a natural way. Let $\lambda \in \mathfrak{h}^*$. Then V is called a *highest weight module* over $\mathfrak{g}$ of highest weight λ if there is a non-zero weight vector $v_\lambda \in V$ (called *highest weight vector*) of weight λ such that

$$x_\alpha \cdot v_\lambda = 0 \text{ for all positive roots } \alpha \text{ and } x_\alpha \in \mathfrak{g}_\alpha, \tag{2.8}$$

$$\mathcal{U}(\mathfrak{g}) \cdot v_\lambda = V. \tag{2.9}$$

Proposition 2.11.1 *Let V be a highest weight module over $\mathfrak{g}$ of highest weight λ. Then $\dim V_\lambda = 1$ and V is the direct sum of weight spaces V_μ, where $\mu \preceq \lambda$. All such weight spaces are finite-dimensional.*

Proof. Let $\mathfrak{n}_-$, $\mathfrak{n}_+$ be the subalgebras of $\mathfrak{g}$ spanned by the negative and positive root vectors respectively. By Theorem 2.10.1, every element of $\mathcal{U}$ is a sum of elements of the form $u^- u^0 u^+$, where $u^- \in \mathcal{U}(\mathfrak{n}_-)$, $u^0 \in \mathcal{U}(\mathfrak{h})$, $u^+ \in \mathcal{U}(\mathfrak{n}_+)$. By (2.8), $u^+ \cdot v_\lambda = 0$, unless $u^+ = 1$. Furthermore, $u^0 \cdot v_\lambda$ is proportional to v_λ. So by (2.9), V is spanned by $\mathcal{U}(\mathfrak{n}_-) \cdot v_\lambda$.

Write $u^- = \prod_{\alpha>0} x_{-\alpha}^{k_\alpha}$. Then $u^- \cdot v_\lambda$ is a weight vector of weight $\lambda - \sum_{\alpha>0} k_\alpha \alpha$ and V is a sum of weight spaces V_μ, where $\mu \preceq \lambda$. This sum is direct as generally eigenvectors corresponding to different eigenvalues are linearly independent. Also it follows that $u^- \cdot v_\lambda$ is not of weight λ unless $u^- = 1$. Therefore $\dim V_\lambda = 1$. Finally, for a given μ with $\mu \preceq \lambda$ the set of sequences $(k_\alpha)_{\alpha>0}$, with $\lambda - \sum_\alpha k_\alpha \alpha = \mu$, is finite. It follows that $\dim V_\mu$ is finite. □

Let $B(\lambda)$ be the left ideal of $\mathcal{U}(\mathfrak{g})$ generated by $h - \lambda(h)$, for $h \in \mathfrak{h}$, along with x_α, for $x_\alpha \in \mathfrak{g}_\alpha$, where $\alpha > 0$. It is straightforward to see that the normal forms of the S-elements of the generators of $B(\lambda)$ are all zero. Hence the given generators form a Gröbner basis of $B(\lambda)$ (Section 2.10.2). So, if we set $M(\lambda) = \mathcal{U}(\mathfrak{g})/B(\lambda)$, and let v_λ denote the image of 1 in $M(\lambda)$, then $M(\lambda)$ is spanned by the elements

$$\prod_{\alpha>0} x_{-\alpha}^{k_\alpha} \cdot v_\lambda.$$

Furthermore, $M(\lambda)$ is a $\mathfrak{g}$-module, as $B(\lambda)$ is a left ideal. Also v_λ is a weight vector of weight λ. It follows that $M(\lambda)$ is a highest weight module over $\mathfrak{g}$ of weight λ. It is called a *Verma module.*

Let $U \subset M(\lambda)$ be a proper $\mathfrak{g}$-submodule then U does not contain v_λ. So the sum $W(\lambda)$ of all proper $\mathfrak{g}$-submodules of $M(\lambda)$ is again a proper $\mathfrak{g}$-submodule. Set $V(\lambda) = M(\lambda)/W(\lambda)$; then $V(\lambda)$ is an irreducible highest weight module over $\mathfrak{g}$ of highest weight λ.

Proposition 2.11.2 *Up to isomorphism, $V(\lambda)$ is the unique irreducible highest weight module over $\mathfrak{g}$ of highest weight λ.*

Proof. Let V' be a second such module and v'_λ be a fixed weight vector of weight λ. Mapping $v_\lambda \mapsto v'_\lambda$ extends to a surjective homomorphism of $\mathfrak{g}$-modules, $M(\lambda) \to V'$. So V' is isomorphic to $M(\lambda)/U$, where U is a $\mathfrak{g}$-submodule of $M(\lambda)$. Therefore, $U \subset W(\lambda)$. If the inclusion is strict then $M(\lambda)/U$ contains the image of $W(\lambda)$ as a non-trivial submodule, which is excluded as V' is irreducible. □

Lemma 2.11.3 *Let $\lambda \in \mathfrak{h}^*$ be a dominant integral weight, so that $m_i = \lambda(h_i)$ is a non-negative integer. Let J be a $\mathfrak{g}$-submodule of $M(\lambda)$ containing $f_i^{m_i+1} \cdot v_\lambda$. Set $V = M(\lambda)/J$. Then V is finite-dimensional. Moreover, let μ be a weight of V, and $w \in W$; then $w(\mu)$ also is a weight of V and* $\dim V_\mu = \dim V_{w(\mu)}$.

Proof. First we show that the e_i and f_i act locally nilpotently on V. This means that for all $v \in V$ there is an integer m_v such that $e_i^{m_v} \cdot v = 0$, $f_i^{m_v} \cdot v = 0$. For e_i this is obvious, as e_i maps weight vectors of weight μ to weight vectors of weight $\succ \mu$. By induction on n the following formula is proved

$$x_\beta^n x_\alpha = \sum_{j=0}^{n} \binom{n}{j} \left(\prod_{i=1}^{j} N_{\beta,\alpha+(i-1)\beta} \right) x_{\alpha+j\beta} x_\beta^{n-j}.$$

Now the maximal j such that $\alpha + j\beta$ can be a root is 3. So it follows that $f_i^n x_\alpha \in \mathcal{U} f_i^{n-3}$. Let

$$u = \prod_{\alpha>0} x_{-\alpha}^{k_\alpha},$$

then by using the above repeatedly, we conclude that there is an $m > 0$ such that $f_i^m u \in \mathcal{U} f_i^{m_i+1}$. As V is spanned by various $u \cdot v_\lambda$, it follows that f_i acts locally nilpotently on V.

Now we show the last statement. It is enough to do that for $w =$

s_{α_i}. Let $\phi : \mathfrak{g} \to \mathfrak{gl}(V)$ denote the representation afforded by V. Then $\exp(\phi(e_i))$, $\exp(\phi(f_i))$ are well-defined endomorphisms of V. Define $\tau_i = \exp(\phi(e_i))\exp(-\phi(f_i))\exp(\phi(e_i))$. Then it can be shown that τ_i maps V_μ to $V_{s_{\alpha_i}(\mu)}$ (see Lemma 5.2.14).

To show that V is finite-dimensional we use that the number of dominant weights μ with $\mu \preceq \lambda$ is finite ([Hum78], Lemma 13.2B). This, together with Theorem 2.8.29, Proposition 2.11.1, proves that V is finite-dimensional. □

Theorem 2.11.4 *Let $\lambda \in \mathfrak{h}^*$ be a dominant integral weight. Then $V(\lambda)$ is finite-dimensional. Moreover, let μ be a weight of $V(\lambda)$, and $w \in W$, then $w(\mu)$ also is a weight of $V(\lambda)$ and $\dim V(\lambda)_\mu = \dim V(\lambda)_{w(\mu)}$.*

Proof. By v_λ we also denote its image in $V(\lambda)$. Set $m_i = \langle \lambda, \alpha_i^\vee \rangle$, then $m_i \in \mathbb{Z}_{\geq 0}$ and $\lambda(h_i) = m_i$. Set $u_i = f_i^{m_i+1} \cdot v_\lambda$; according to Lemma 2.11.3, it suffices to show that $u_i = 0$.

First, we have $e_j \cdot u_i = 0$, when $j \neq i$, as in that case $[e_i, f_j] = 0$. From Lemma 2.10.10 we infer that $e_i f_i^{m_i+1} = f_i^{m_i+1} e_i + f_i^{m_i}(h_i - m_i)$, implying that $e_i \cdot u_i = 0$. As the e_i generate the subalgebra spanned by the positive root vectors, we see that $x_\alpha \cdot u_i = 0$ for all $\alpha > 0$. Hence u_i generates a highest weight module of highest weight $\lambda - (m_i + 1)\alpha_i$. The irreducibility of $V(\lambda)$ now forces $u_i = 0$. □

2.11.2 Classification of irreducible modules

Here we use the results of the previous subsection to obtain the classification of the irreducible $\mathfrak{g}$-modules.

Proposition 2.11.5 *Let $\rho : \mathfrak{g} \to \mathfrak{gl}(V)$ be a finite-dimensional representation. Let $\mathfrak{h}' \subset \mathfrak{g}$ be a Cartan subalgebra. Then the elements of $\rho(\mathfrak{h}')$ are semisimple.*

Proof. Let h_i', e_i', f_i', $1 \leq i \leq \ell$ be a canonical set of generators of $\mathfrak{g}$ with $h_i' \in \mathfrak{h}'$. Then the h_i' form a basis of $\mathfrak{h}'$. Let $\mathfrak{a}_i$ denote the subalgebra with basis h_i', e_i', f_i'. Then $\mathfrak{a}_i$ is isomorphic to $\mathfrak{sl}(2,k)$. So by Theorem 2.9.1, together with Weyl's theorem, $\rho(h_i')$ is diagonalizable. As the h_i' commute, the $\rho(h_i')$ are simultaneously diagonalizable. So there is a basis of V with respect to which every element of $\rho(\mathfrak{h})$ is represented by a diagonal matrix. □

Corollary 2.11.6 *Let ρ be as in the previous proposition. Let $h \in \mathfrak{g}$ be such that $\mathrm{ad}_{\mathfrak{g}} h$ is semisimple. Then $\rho(h)$ is semisimple.*

Proof. This follows from Proposition 2.11.5 and the fact that h lies in a Cartan subalgebra (Corollary 2.5.10). □

Theorem 2.11.7 *Let V be an irreducible $\mathfrak{g}$-module. Then there is a unique dominant integral weight $\lambda \in \mathfrak{h}^*$ such that V is isomorphic to $V(\lambda)$.*

Proof. By Proposition 2.11.5, V is spanned by weight vectors relative to $\mathfrak{h}$. Let μ be a weight of V. By the representation theory of $\mathfrak{sl}(2,k)$ it follows that $\mu(h_i) = \langle \mu, \alpha_i^\vee \rangle$ is an integer for all i; so μ is integral. Let v_μ be a weight vector of weight μ. Then $e_i \cdot v_\mu$ is a weight vector of weight $\mu + \alpha_i \succ \mu$. So there is a weight vector v_λ of weight λ such that $e_i \cdot v_\lambda = 0$ for all i. The submodule of V generated by v_λ is a highest weight module of highest weight λ. As V is irreducible, it is equal to V. By Proposition 2.11.2, V is isomorphic to $V(\lambda)$. Finally, the representation theory of $\mathfrak{sl}(2,k)$ shows that λ is dominant. □

Together with Theorem 2.11.4, this yields a bijection between the set of finite-dimensional irreducible $\mathfrak{g}$-modules and the set of dominant integral weights.

2.11.3 Path model

In view of the results of the previous section, the following question immediately comes to mind: given $\mathfrak{g}$ and a dominant integral weight λ, what can we say about $V(\lambda)$? Here we will describe a method to determine the weights of $V(\lambda)$, along with their multiplicities (by definition, the multiplicity of a weight is the dimension of the corresponding weight space).

Using this method, we can also determine $\dim V(\lambda)$. However, we also remark that for $\dim V(\lambda)$ there is an elegant formula called Weyl's dimension formula. It says that the dimension of $V(\lambda)$ is equal to a product of easily computable terms indexed by the positive roots. Here we will not go into it, but refer to [Jac79].

For the weights of $V(\lambda)$ and their multiplicities, the most efficient known algorithm is based on Freudenthal's formula (see [Hum78] and [MP82], [Gra00] for the computational aspects of it). We briefly describe a different method, based on the path model, which was invented by Littelmann ([Lit94], [Lit97]). It is less efficient than the algorithms based on Freudenthal's formula, but it has other features which will be useful when dealing with semisimple algebraic groups.

Let P denote the weight lattice, as in Section 2.8.3. It is spanned by the fundamental weights $\lambda_1, \dots, \lambda_\ell$. Set $P_\mathbb{R} = \mathbb{R} \otimes P$, and identify $1 \otimes \lambda_i$ with λ_i, so that we write an element in $P_\mathbb{R}$ uniquely as $\sum_i a_i \lambda_i$, $a_i \in \mathbb{R}$. We consider piecewise linear paths, $\pi : [0,1] \to P_\mathbb{R}$, with $\pi(0) = 0$. Such a path is given by two sequences $\bar{\mu} = (\mu_1, \dots, \mu_r)$, where $\mu_i \in P_\mathbb{R}$ and $\bar{a} = (a_0 = 0, a_1, \dots, a_r = 1)$, with $a_i \in \mathbb{R}$ and $a_i < a_{i+1}$. The path π corresponding to these sequences is given by

$$\pi(t) = \sum_{i=1}^{s-1} (a_i - a_{i-1})\mu_i + (t - a_{s-1})\mu_s \quad \text{for } a_{s-1} \leq t \leq a_s.$$

This means that

$$\begin{aligned}&\pi(t)=\mu_1 t \quad \text{for } 0\le t\le a_1,\\ &\pi(t)=a_1\mu_1+(t-a_1)\mu_2 \quad \text{for } a_1\le t\le a_2,\\ &\pi(t)=a_1\mu_1+(a_2-a_1)\mu_2+(t-a_2)\mu_3 \quad \text{for } a_2\le t\le a_3,\\ &\text{etc.}\end{aligned}$$

In other words, the path π is obtained by joining the points 0, $a_1\mu$, $a_1\mu+(a_2-a_1)\mu_2$ until $\sum_{i=1}^{r}(a_i-a_{i-1})\mu_i$ by straight lines.

Example 2.11.8 Let the root system Φ be of type B_2. Let the simple roots be α_1, α_2, with α_1 long, and α_2 short. Then the Cartan matrix C, and the matrix B of the inner product are

$$C=\begin{pmatrix}2 & -2\\ -1 & 2\end{pmatrix},\quad B=\begin{pmatrix}4 & -2\\ -2 & 2\end{pmatrix}.$$

The positive roots are $\alpha_1,\alpha_2,\alpha_1+\alpha_2,\alpha_1+2\alpha_2$. Let $\bar{\mu}=(\lambda_1-4\lambda_2, 3\lambda_1-4\lambda_2,-\lambda_1+4\lambda_2)$, and $\bar{a}=(0,\frac{1}{2},\frac{3}{4},1)$. The root system, the fundamental weights and the path π corresponding to these sequences are drawn in the following picture. The construction of the path $f_{\alpha_1}\pi$ will be described in Example 2.11.9. It is included in the picture to avoid the need for two pictures.

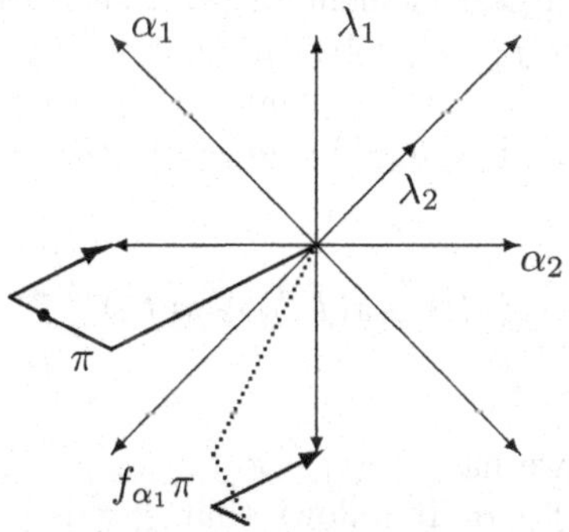

Let the path π be defined by the sequences $\bar{\mu}$, $\bar{a}$. Then we say that $(\bar{\mu},\bar{a})$ is a *presentation* of π, and we write $\pi=(\bar{\mu},\bar{a})$. Note that we can always make a presentation of a path longer in a trivial way. For example, (μ_1,μ_2,μ_3) and $(0,\frac{1}{2},\frac{3}{4},1)$ define the same path as $(\mu_1,\mu_1,\mu_2,\mu_3)$ and $(0,\frac{1}{4},\frac{1}{2},\frac{3}{4},1)$. *By doing this we can always assume that a given $a\in[0,1]$ occurs among the a_i in the second sequence.*

Write $\bar{\mu}=(\mu_1,\dots,\mu_r)$, $\bar{a}=(0=a_0,a_1,\dots,a_r=1)$. Then the presentation $(\bar{\mu},\bar{a})$ is called *short* if $\mu_i\neq\mu_{i+1}$ for $1\le i<r$. Suppose the presentation $(\bar{\mu},\bar{a})$ is not short, and let i be such that $\mu_i=\mu_{i+1}$. Then set $\bar{\mu}'=(\mu_1,\dots,\mu_{i-1},\mu_{i+1},\dots,\mu_r)$ and $\bar{a}'=(0=a_0,\dots,a_{i-1},a_{i+1},\dots,a_r)$. Then also $\pi=(\bar{\mu}',\bar{a}')$. Repeating this we find a short presentation of π.

Now let $\pi=(\bar{\mu},\bar{a})$, $\sigma=(\bar{\nu},\bar{b})$ be two paths given by short presentations.

We write $\pi = \sigma$ if and only if $\bar{\mu} = \bar{\nu}$ and $\bar{a} = \bar{b}$. It is straightforward to see that this is equivalent to $\pi(t) = \sigma(t)$ for all $t \in [0,1]$.

Let Π be the set of all piecewise linear paths $\pi : [0,1] \to P_{\mathbb{R}}$. Then for a simple root α we define operators $e_\alpha, f_\alpha : \Pi \to \Pi \cup \{0\}$. Here 0 is just a symbol, and $f_\alpha \pi = 0$ means something like "$f_\alpha \pi$ is not defined".

Let $\pi \in \Pi$, and define $h_\alpha^\pi : [0,1] \to \mathbb{R}$ by $h_\alpha^\pi(t) = \langle \pi(t), \alpha^\vee \rangle$. Set

$$m_\alpha^\pi = \min\{n \in \mathbb{Z} \mid \text{ there is a } t \in [0,1] \text{ with } h_\alpha^\pi(t) = n\},$$

i.e., the smallest integer attained by h_α^π.

Let e_+ be the smallest element of $[0,1]$ with $h_\alpha^\pi(e_+) = m_\alpha^\pi$. If $m_\alpha^\pi \leq -1$, let $e_- < e_+$ be maximal such that $h_\alpha^\pi(e_-) = m_\alpha^\pi + 1$ (i.e., h_α^π descends from $m_\alpha^\pi + 1$ to m_α^π when t runs between e_- and e_+). Note that e_- exists, because $h_\alpha^\pi(0) = 0$. If $m_\alpha^\pi > -1$ then we set $e_\alpha \pi = 0$. Otherwise

$$(e_\alpha\pi)(t) = \begin{cases} \pi(t) & 0 \leq t \leq e_- \\ s_\alpha(\pi(t) - \pi(e_-)) + \pi(e_-) & e_- \leq t \leq e_+ \\ \pi(t) + \alpha & e_+ \leq t \leq 1. \end{cases}$$

Note that for $t = e_+$ we have $\langle \pi(t) - \pi(e_-), \alpha^\vee \rangle = -1$, so that $s_\alpha(\pi(e_+) - \pi(e_-)) + \pi(e_-) = \pi(e_+) + \alpha$. Hence $e_\alpha \pi$ is continuous at $t = e_+$ and therefore belongs to Π.

Now let f_- be the largest element of $[0,1]$ such that $h_\alpha^\pi(f_-) = m_\alpha^\pi$. If $h_\alpha^\pi(1) - m_\alpha^\pi \geq 1$ then let $f_+ > f_-$ be minimal such that $h_\alpha^\pi(f_+) = m_\alpha^\pi + 1$ (i.e., h_α^π increases from m_α^π to $m_\alpha^\pi + 1$ when t runs from f_- to f_+). Note that f_+ exists as $h_\alpha^\pi(1) \geq m_\alpha^\pi + 1$. If $h_\alpha^\pi(1) - m_\alpha^\pi < 1$, we set $f_\alpha \pi = 0$. Otherwise

$$(f_\alpha\pi)(t) = \begin{cases} \pi(t) & 0 \leq t \leq f_- \\ s_\alpha(\pi(t) - \pi(f_-)) + \pi(f_-) & f_- \leq t \leq f_+ \\ \pi(t) - \alpha & f_+ \leq t \leq 1. \end{cases}$$

In this case, for $t = f_+$ we have $\langle \pi(t) - \pi(f_-), \alpha^\vee \rangle = 1$, so that $s_\alpha(\pi(f_+) - \pi(f_-)) + \pi(f_-) = \pi(f_+) - \alpha$. It follows that $f_\alpha \pi$ is continuous at f_+, and hence $f_\alpha \pi \in \Pi$.

Example 2.11.9 We use the notation of Example 2.11.8. We have

$$h_{\alpha_1}^\pi(t) = \begin{cases} t & \text{for } 0 \leq t \leq \frac{1}{2} \\ 3t - 1 & \text{for } \frac{1}{2} \leq t \leq \frac{3}{4} \\ -t + 2 & \text{for } \frac{3}{4} \leq t \leq 1. \end{cases}$$

(Draw the graph of this function.) We see that $m_{\alpha_1}^\pi = 0$, and $f_- = 0$. Also $h_{\alpha_1}^\pi(1) - m_{\alpha_1}^\pi = 1$, and we see that $f_+ = \frac{2}{3}$. Hence $(f_{\alpha_1}\pi)(t) = s_{\alpha_1}(\pi(t))$ for $0 \leq t \leq \frac{2}{3}$, and $(f_{\alpha_1}\pi)(t) = \pi(t) - \alpha_1$ for $\frac{2}{3} \leq t \leq 1$. Write $\mu_1 = \lambda_1 - 4\lambda_2$, $\mu_2 = 3\lambda_1 - 4\lambda_2$, $\mu_3 = -\lambda_1 + 4\lambda_2$. Set $\nu_1 = s_{\alpha_1}(\mu_1) = -\lambda_1 - 2\lambda_2$, and

$\nu_2 = s_{\alpha_1}(\mu_2) = -3\lambda_1 + 2\lambda_2$. Then the sequences $(\nu_1, \nu_2, \mu_2, \mu_3)$, $(0, \frac{1}{2}, \frac{2}{3}, \frac{3}{4}, 1)$ yield a presentation of $f_{\alpha_1}\pi$. It is shown in the picture of Example 2.11.8. The part of π until the dot (i.e., until $t = \frac{2}{3}$) is reflected with s_{α_1}, and drawn with a dotted line. The remainder of π is added at the end of the dotted line.

Note that it is possible to compute $f_{\alpha_1}\pi$ by just looking at the sequences in a presentation of π. We do not need to draw pictures. We formalize this procedure in the next two propositions.

Proposition 2.11.10 *Let $\pi \in \Pi$ be defined by $(\mu_1, \dots, \mu_r)$ and $(a_0 = 0, a_1, \dots, a_r = 1)$. Assume that $m_\alpha^\pi \leq -1$, and that $e_- = a_s$, $e_+ = a_q$ for certain indices $s < q$. Then $e_\alpha \pi$ is defined by $(\mu_1, \dots, \mu_s, s_\alpha(\mu_{s+1}), \dots, s_\alpha(\mu_q), \mu_{q+1}, \dots, \mu_r)$ and $(a_0, \dots, a_r)$.*

Proof. The occurrence of $\mu_1, \dots, \mu_s$ in the sequence for $e_\alpha\pi$ is clear as $(e_\alpha\pi)(t) = \pi(t)$ for $0 \leq t \leq a_s$. Now let $a_i \leq t \leq a_{i+1}$, where $s \leq i \leq q-1$. Then

$$\pi(t) = \pi(e_-) + (a_{s+1} - a_s)\mu_{s+1} + \dots + (a_i - a_{i-1})\mu_i + (t - a_i)\mu_{i+1}.$$

Hence $s_\alpha(\pi(t) - \pi(e_-)) + \pi(e_-)$ equals

$$\pi(e_-) + (a_{s+1} - a_s)s_\alpha(\mu_{s+1}) + \dots + (a_i - a_{i-1})s_\alpha(\mu_i) + (t - a_i)s_\alpha(\mu_{i+1}).$$

From this we see that the $\mu_1, \dots, \mu_s$ are followed by $s_\alpha(\mu_{s+1}), \dots, s_\alpha(\mu_q)$. For $a_q \leq t \leq a_{q+1}$ we have $\pi(t) = \pi(a_q) + (t - a_q)\mu_{q+1}$. Furthermore, since $a_q = e_+$, $(e_\alpha\pi)(a_q) = \pi(a_q) + \alpha$. So $(e_\alpha\pi)(t) = \pi(t) + \alpha$ (by definition) $= (e_\alpha\pi)(a_q) + (t - a_q)\mu_{q+1}$. We see that the next element in the sequence for $e_\alpha\pi$ is μ_{q+1}. We can continue the argument like this, and arrive at the statement of the proposition. □

Proposition 2.11.11 *Let $\pi \in \Pi$ be defined by $(\mu_1, \dots, \mu_r)$ and $(a_0 = 0, a_1, \dots, a_r = 1)$. Assume that $h_\alpha^\pi(1) - m_\alpha^\pi \geq 1$, and that $f_- = a_s$, $f_+ = a_q$ for certain indices $s < q$. Then $f_\alpha\pi$ is defined by $(\mu_1, \dots, \mu_s, s_\alpha(\mu_{s+1}), \dots, s_\alpha(\mu_q), \mu_{q+1}, \dots, \mu_r)$ and $(a_0, \dots, a_r)$.*

Proof. This proceeds in exactly the same way as the previous proposition. □

Example 2.11.12 Let π be the path from Example 2.11.8. Then $\pi = (\bar{\mu}, \bar{a})$, where $\bar{\mu} = (\mu_1, \mu_2, \mu_3)$, and $\bar{a} = (0, \frac{1}{2}, \frac{3}{4}, 1)$. As seen in Example 2.11.9, $f_- = 0$ and $f_+ = \frac{2}{3}$. Now f_+ does not occur in $\bar{a}$. Therefore we make the presentation of π longer. Set $\bar{\mu}' = (\mu_1, \mu_2, \mu_2, \mu_3)$, and $\bar{a}' = (0, \frac{1}{2}, \frac{2}{3}, \frac{3}{4}, 1)$. Then also $\pi = (\bar{\mu}', \bar{a}')$. Using this presentation we can easily compute a presentation for $f_{\alpha_1}\pi$. Set $\bar{\nu} = (s_{\alpha_1}(\mu_1), s_{\alpha_1}(\mu_2), \mu_2, \mu_3)$. Then $f_{\alpha_1}\pi = (\bar{\nu}, \bar{a}')$.

Remark 2.11.13 Let $\pi = (\bar{\mu}, \bar{a})$, where $\bar{\mu} = (\mu_1, \dots, \mu_r)$ and $\bar{a} = (a_0 = 0, a_1, \dots, a_r = 1)$. The graph of h_α^π is obtained by drawing straight lines between the values attained at the a_i. Moreover, for $a_i \leq t \leq a_{i+1}$ we have $h_\alpha^\pi(t) = h_\alpha^\pi(a_i) + (t - a_i)\langle \mu_{i+1}, \alpha^\vee \rangle$. Using this and Propositions 2.11.10 and 2.11.11 it is straightforward to compute $f_\alpha \pi$ and $e_\alpha \pi$.

Let $\lambda \in P$ be dominant, and let π_λ be the path defined by (λ) and $(0, 1)$ (i.e., a straight line from the origin to λ). Let $\Pi(\lambda)$ to be the set of paths that can be obtained from π_λ by applying the path operators $f_{\alpha_i}, e_{\alpha_i}, 1 \leq i \leq \ell$. For a proof of the following theorem we refer to [Lit94] and [Lit97].

Theorem 2.11.14 (Littelmann) *Let $V(\lambda)$ be the irreducible $\mathfrak{g}$-module with highest weight λ. For every path $\pi \in \Pi(\lambda)$ we have $\pi(1) \in P$. Moreover, the number of paths ending in a $\mu \in P$ is equal to the multiplicity of μ in $V(\lambda)$.*

Remark 2.11.15 It can be shown ([Lit94], [Lit97]) that for two paths π, π' we have $e_\alpha \pi = \pi'$ if and only if $f_\alpha \pi' = \pi$. This implies that all paths in $\Pi(\lambda)$ can be obtained by applying only the f_α, starting with π_λ. Secondly, the action of the e_α can be read once we know that action of the f_α.

Let Γ_λ be the labeled directed graph that has $\Pi(\lambda)$ as vertices and an edge from π_1 to π_2, labeled i, if $f_{\alpha_i}\pi_1 = \pi_2$. The graph Γ_λ is called the *crystal graph* corresponding to the irreducible $\mathfrak{g}$-module $V(\lambda)$ (more precisely, it is the crystal graph of the irreducible representation of the quantum group $\mathcal{U}_q(\mathfrak{g})$ with highest weight λ; here we do not go into that, but refer to [Jan96] for an introduction into quantum groups and crystal graphs and to [Kas96] for a proof that using the path model we obtain the crystal graph). So the crystal graph summarizes the action of the path operators on $\Pi(\lambda)$.

Example 2.11.16 Let the root system be of type A_2 with fundamental weights λ_1, λ_2 and simple roots $\alpha_1 = 2\lambda_2 - \lambda_2$, $\alpha_2 = -\lambda_1 + 2\lambda_2$. Set $\lambda = 2\lambda_1$ and $\pi_1 = ((\lambda), (0, 1))$ (i.e., $\pi_1 = \pi_\lambda$). Also define the following paths

$$\begin{aligned}
\pi_2 &= ((-2\lambda_1 + 2\lambda_2, 2\lambda_1), (0, \tfrac{1}{2}, 1)) \\
\pi_3 &= ((-2\lambda_1 + 2\lambda_2), (0, 1)) \\
\pi_4 &= ((-2\lambda_2, 2\lambda_1)), (0, \tfrac{1}{2}, 1)) \\
\pi_5 &= ((-2\lambda_2, -2\lambda_1 + 2\lambda_2), (0, \tfrac{1}{2}, 1)) \\
\pi_6 &= ((-2\lambda_2), (0, 1)).
\end{aligned}$$

Some computations show that the crystal graph Γ_λ is

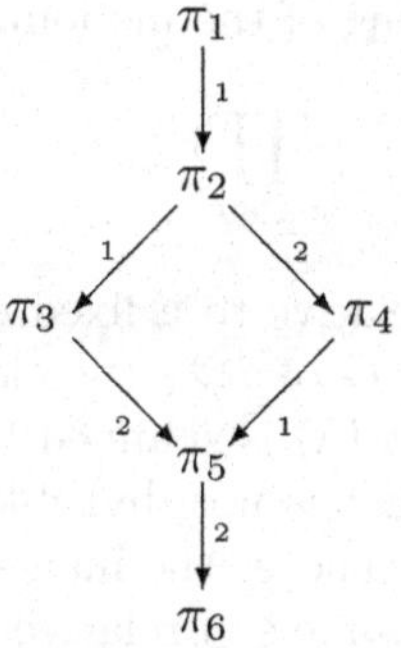

2.11.4 Constructing irreducible modules

In this section we sketch an algorithm that, given a semisimple Lie algebra $\mathfrak{g}$, and a dominant integral weight λ, constructs the irreducible highest weight module $V(\lambda)$. By this we mean that it computes a basis of $V(\lambda)$, and that an algorithm computes the matrix of an arbitrary element of $\mathfrak{g}$ with respect to that basis.

Let $\mathfrak{n}_-$ and $\mathfrak{n}_+$ be the subalgebras of $\mathfrak{g}$ spanned by the negative and positive root vectors respectively. We consider the Verma module $M(\lambda)$ (Section 2.11.1). As a vector space $M(\lambda)$ is isomorphic to $\mathcal{U}(\mathfrak{n}_-)$, by $u^- \mapsto u^- \cdot v_\lambda$. So using this isomorphism we can also view $M(\lambda)$ as an algebra, isomorphic to $\mathcal{U}(\mathfrak{n}_-)$. Throughout we use two notations for elements of $M(\lambda)$. If we write an element as $u^- \cdot v_\lambda$ (with $u^- \in \mathcal{U}(\mathfrak{n}_-)$), we are mainly concerned with the structure of $M(\lambda)$ as a $\mathfrak{g}$-module. If we simply write u for an element of $M(\lambda)$, then we view it as an algebra.

Recall that e_i, f_i, h_i denote the elements of the fixed canonical generating set of $\mathfrak{g}$. Let $I(\lambda)$ be the left ideal of $M(\lambda)$ generated by $f_i^{m_i+1}$, where $m_i = \langle \lambda, \alpha_i^\vee \rangle$, $1 \leq i \leq \ell$.

Proposition 2.11.17 *Let $W(\lambda)$ be the unique maximal proper $\mathfrak{g}$-submodule of $M(\lambda)$. Then $W(\lambda) = I(\lambda)$.*

Proof. Set $W_i = \mathcal{U}(\mathfrak{g}) \cdot f_i^{m_i+1} v_\lambda$. As in the proof of Theorem 2.11.4, W_i is a highest weight module with highest weight vector $w_i = f_i^{m_i+1} \cdot v_\lambda$. In particular, it is a proper $\mathfrak{g}$-submodule, whence $W_i \subset W(\lambda)$.

Since $I(\lambda)$, as $\mathcal{U}(\mathfrak{n}_-)$-module is generated by the $f_i^{m_i+1} \cdot v_\lambda$, we have $I(\lambda) \subset W(\lambda)$.

It also follows that $W_i = \{u^- \cdot w_i \mid u^- \in \mathcal{U}(\mathfrak{n}_-)\}$, and hence $W_i \subset I(\lambda)$. As $W_i \subset W(\lambda)$, also $R = W_1 + \cdots + W_\ell$ is contained in $W(\lambda)$. By Lemma 2.11.3, $M(\lambda)/R$ is finite-dimensional. Using Weyl's theorem (Theorem 2.7.6) and the fact that $M(\lambda)/R$ is generated as a $\mathfrak{g}$-module by the image of v_λ, it then also follows that $M(\lambda)/R$ is irreducible. Hence $R = W(\lambda)$ by Proposition 2.11.2. Since $W_i \subset I(\lambda)$, we conclude that $W(\lambda) \subset I(\lambda)$. □

The algorithm based on the preceding proposition is straightforward. We use the basis of $M(\lambda)$ consisting of the monomials

$$\prod_{\alpha>0} x_{-\alpha}^{k_\alpha},$$

where the product is taken relative to a fixed ordering of the positive roots. We compute a Gröbner basis G of $I(\lambda)$ (Section 2.10.2). Then the quotient $M(\lambda)/I(\lambda) = M(\lambda)/W(\lambda) = V(\lambda)$ is spanned by the elements $u^- \cdot v_\lambda$, where u^- is a monomial as above that is not divisible by any leading monomial of an element of G, and v_λ also denotes the image of this vector in $V(\lambda)$.

Computing the matrix of an $x \in \mathfrak{g}$ relative to this basis is easy. Let $w = u^- \cdot v_\lambda$ be a basis element of $V(\lambda)$. Then we rewrite xu^- as $v^{-1}v^0v^+$, where $v^- \in \mathcal{U}(\mathfrak{n}_-)$, $v^0 \in \mathcal{U}(\mathfrak{h})$, $v^+ \in \mathcal{U}(\mathfrak{n}_+)$. If $v^+ \neq 1$ then $x \cdot w = 0$. Since v_λ is a weight vector, we can compute a scalar ξ such that $v^0 \cdot v_\lambda = \xi v_\lambda$. Furthermore, using the Gröbner basis G we can compute the normal form $\sum_i c_i u_i^-$ of v^- modulo $I(\lambda)$ where $c_i \in k$ and the u_i^- are monomials in $\mathcal{U}(\mathfrak{n}_-)$ not divisible by a leading monomial of an element of G. Then $x \cdot w = \sum_i \xi c_i u_i^- \cdot v_\lambda$.

Example 2.11.18 Let $\mathfrak{g}$ be the simple Lie algebra of type A_2. The positive roots are $\alpha_1, \alpha_2, \alpha_1 + \alpha_2$. Hence $\mathfrak{n}_-$ is spanned by $f_1 = x_{-\alpha_1}$, $f_2 = x_{-\alpha_2}$, $f_3 = x_{-\alpha_1-\alpha_2}$. The signs of the structure constants can be chosen such that $[f_1, f_2] = f_3$ (see Example 2.9.14). Let $\lambda = \lambda_1 + \lambda_2$ (where the λ_i are the fundamental weights; see Section 2.8.3). Then $I(\lambda)$ is generated by f_1^2, f_2^2. In Example 2.10.6 a Gröbner basis of $I(\lambda)$ is computed, showing that $\dim V(\lambda) = 8$. In fact, the representation afforded by $V(\lambda)$ is isomorphic to the adjoint representation of $\mathfrak{g}$.

This algorithm can be made much more efficient on the basis of some observations. First, we define the weight of a monomial $\prod_{\alpha>0} x_{-\alpha}^{k_\alpha}$ as $\sum_\alpha k_\alpha \alpha$. Since the commutation relations of the generators of $\mathcal{U}(\mathfrak{n}_-)$ are homogeneous with respect to this weight, the weight function is multiplicative, that is, the product of two monomials of weight μ and ν is a linear combination of monomials of weight $\mu + \nu$. Also note that, if $u^- \in \mathcal{U}(\mathfrak{n}_-)$ is homogeneous of weight μ, then $u^- \cdot v_\lambda$ is a weight vector in $M(\lambda)$, or $V(\lambda)$, of weight $\lambda - \mu$. Furthermore, since $I(\lambda)$ is generated by monomials, it has a basis consisting of homogeneous elements.

By the methods outlined in Section 2.11.3 we can compute the set of weights $P(\lambda)$ of $V(\lambda)$, along with their multiplicities. We set

$$D(\lambda) = P(\lambda) \cup \{\mu - \alpha \mid \mu \in P(\lambda), \alpha \in \Phi^+\}.$$

For $\mu \in D(\lambda)$ let B_μ be a basis of the subspace of $I(\lambda)$ consisting of the elements that are homogeneous of weight $\lambda - \mu$. Let $\widetilde{G}$ be the union of the sets B_μ as μ runs through $D(\lambda)$. Then $\widetilde{G}$ is a Gröbner basis of $I(\lambda)$.

By the same arguments used for polynomial rings, it can be shown that a left ideal of $\mathcal{U}(\mathfrak{n}_-)$ has a unique reduced Gröbner basis (see Theorem 1.5.6). By computing normal forms of $g \in \widetilde{G}$ modulo $\widetilde{G} \setminus \{g\}$, we can construct the reduced Gröbner basis G of $I(\lambda)$. Let G_μ be the set of elements of G that are homogeneous of weight μ. It follows that G is the union of G_μ, where μ runs through $D(\lambda)$.

Let $\mu = \lambda - \sum_{i=1}^{\ell} k_i\alpha_i$ be an element of $D(\lambda)$. We define the height of μ as $\sum_i k_i$. We order the elements of $D(\lambda)$ in a way compatible with the height (so elements of smaller height are smaller). In order to construct G we run through $D(\lambda)$ according to this ordering (starting with the smallest element, i.e., λ). For each μ that we encounter we calculate the following.

- If $\mu \notin P(\lambda)$, we add to G the set of monomials of weight $\lambda-\mu$, not divisible by the leading monomials of elements of G previously constructed.
- If $\mu \in P(\lambda)$, let m_μ be its multiplicity. By computing S-elements corresponding to elements of G constructed previously, enlarge G with elements of weight $\lambda - \mu$, until the number of monomials of weight $\lambda - \mu$ not divisible by the leading monomials of the elements of G is equal to m_μ.

This terminates because the general algorithm for computing a Gröbner basis, as outlined in Section 2.10.2, terminates.

We remark that in order to perform the multiplication in $\mathcal{U}(\mathfrak{g})$ it is a good idea to use the basis and commutation formulas given in Section 2.10.3.

Example 2.11.19 Let the notation be as in Example 2.11.18. Here $P(\lambda) = \{\lambda, \lambda-\alpha_1, \lambda-\alpha_2, \lambda-\alpha_1-\alpha_2, \lambda-2\alpha_1-\alpha_2, \lambda-\alpha_1-2\alpha_2, \lambda-2\alpha_1-2\alpha_2\}$, and all weights have multiplicity 1, except $\lambda-\alpha_1-\alpha_2$ whose multiplicity is 2. To obtain $D(\lambda)$ we add the weights $\lambda-2\alpha_1, \lambda-2\alpha_2, \lambda-3\alpha_1-\alpha_2, \lambda-\alpha_1-3\alpha_2, \lambda-3\alpha_1-2\alpha_2, \lambda-2\alpha_1-3\alpha_2, \lambda-3\alpha_1-3\alpha_2$. Now we run through $D(\lambda)$, taking the weights of smaller height first. In the process we build the Gröbner basis G. The first time we add something to G is when we consider the weights of height 2, $\lambda-2\alpha_1$, $\lambda-2\alpha_2$, and we add the elements f_1^2, f_2^2. Next consider the weight $\mu = \lambda-2\alpha_1-\alpha_2$, of multiplicity 1. The monomials of weight $2\alpha_1+\alpha_2$ are $f_1^2f_2$, f_1f_3. However, the first one is divisible by a leading monomial of an element of G, so we can discard it. There remains one monomial, and since this equals the multiplicity we add nothing to G. Now take $\mu = \lambda-2\alpha_1-2\alpha_2$. This weight has multiplicity 1. But there are two monomials of weight $2\alpha_1+2\alpha_2$, not divisible by a leading monomial of an element of G, namely $f_1f_2f_3$, f_3^2. So here we need to compute an S-element, i.e., $S(f_1^2, f_2^2)$, which leads to adding $f_1f_2f_3 - \frac{1}{2}f_3^2$ to G. Finally, considering the weights of heights 5 and 6 leads to the addition of $f_1f_3^2$, $f_2f_3^2$, f_3^3. We see that this way of proceeding requires just one S-element to be constructed; the rest we get (almost) for free.

Example 2.11.20 Let $\mathfrak{g}$ be the simple Lie algebra of type E_6, and $\lambda = \lambda_4 + \lambda_6$ (here we use the ordering of simple roots given in Theorem 2.8.5). Then $\dim V(\lambda) = 51975$. A Gröbner basis of $I(\lambda)$, using the implementation of this algorithm in GAP4, takes 255 seconds.

Now let $\mathfrak{g}$ be of type E_7, and $\lambda = \lambda_5$. Then $\dim V(\lambda) = 27664$ and the algorithm takes 406 seconds to compute the Gröbner basis. This illustrates that the running time does not depend only on the dimension of the module constructed. It is also influenced by the number of positive roots, as the complexity of the multiplication in $\mathcal{U}(\mathfrak{n}_-)$ depends exponentially on that number (see [Gra01]).

2.12 Reductive Lie algebras

By Weyl's theorem all representations of a semisimple Lie algebra over a field of characteristic 0 are completely reducible. Here we study a wider class, the reductive Lie algebras, with the property that every representation satisfying an extra hypothesis (see Corollary 2.12.4), is completely reducible. In this section all algebras are defined over a field k of characteristic 0.

Definition 2.12.1 *A Lie algebra $\mathfrak{g}$ is called* reductive *if its adjoint representation is completely reducible. A subalgebra $\mathfrak{a}$ of the Lie algebra $\mathfrak{g}$ is said to be* reductive in $\mathfrak{g}$ *if the $\mathfrak{a}$-module $\mathfrak{g}$ is completely reducible.*

By Weyl's theorem all semisimple Lie algebras over k are reductive. However, there are reductive Lie algebras that are not semisimple.

Proposition 2.12.2 *Let $\mathfrak{g}$ be a finite-dimensional Lie algebra over k. Then the following statements are equivalent.*

(i) *$\mathfrak{g}$ is reductive,*

(ii) *$\mathfrak{g} = \mathfrak{s} \oplus \mathfrak{d}$ (direct sum of ideals), where $\mathfrak{s}$ is semisimple and $\mathfrak{d}$ is abelian,*

(iii) *$\mathfrak{g}$ has a finite-dimensional representation such that the associated trace form is non-degenerate,*

(iv) *$\mathfrak{g}$ has a faithful completely reducible representation.*

Proof. We will prove that (i) implies (ii), and so on, until (iv) implies (i).

For the first implication we note that $\mathfrak{g}$ is reductive and can be written as the sum of ideals that are irreducible $\mathfrak{g}$-modules. If the dimension of such an ideal is > 1, then it is a simple Lie algebra. Otherwise it is abelian (being 1-dimensional). So (ii) follows.

Let $\mathrm{ad}_{\mathfrak{s}}$ denote the adjoint representation of $\mathfrak{s}$. The corresponding trace form is the Killing form, and is non-degenerate (Proposition 2.7.2). Let $a_1, \dots, a_n$ be a basis of $\mathfrak{d}$. Set $V = k^n$ with basis $v_1, \dots, v_n$, and let $\rho : \mathfrak{d} \to \mathfrak{gl}(V)$ be defined by $\rho(a_i)v_j = \delta_{ij}$. Then the trace form associated to ρ is non-degenerate. Now we let $\mathfrak{g}$ act on $\mathfrak{s} \oplus V$ by $(x+a)\cdot(y+v) = [x,y] + \rho(a)v$. Then the trace form corresponding to this module is non-degenerate.

Assume (iii). Let $\rho : \mathfrak{g} \to \mathfrak{gl}(V)$ be a representation such that the corresponding trace form, τ, is non-degenerate. Let $0 = V_0 \subset V_1 \subset \cdots \subset V_s = V$ be a composition series of the $\mathfrak{g}$-module V. For $1 \leq i \leq s$ let ρ_i denote the induced representation of $\mathfrak{g}$ on V_i/V_{i-1}. Let $x \in \mathfrak{g}$ be such that $\rho_i(x) = 0$ for $1 \leq i \leq s$. For all $y \in \mathfrak{g}$ we have $\rho_i(x)\rho_i(y) = 0$, whence $\rho(x)\rho(y)V_i \subset V_{i-1}$, for all i. So $\rho(x)\rho(y)$ is nilpotent, and $\tau(x,y) = 0$. Since τ is non-degenerate it follows that $x = 0$. Let W be the direct sum of the V_i/V_{i-1}. Then W is a $\mathfrak{g}$-module. We have just seen that it is faithful. Moreover, it is completely reducible as the V_i/V_{i-1} are irreducible.

Let ρ be a faithful completely reducible representation of $\mathfrak{g}$. Let $x \in \mathfrak{sr}(\mathfrak{g})$. By Lemma 2.6.4, $\rho(x)$ commutes with $\rho(\mathfrak{g})$. Since ρ is faithful we conclude that $\mathfrak{sr}(\mathfrak{g})$ is the centre of $\mathfrak{g}$ and $\mathrm{ad}(\mathfrak{g})$ is a semisimple subalgebra of $\mathfrak{gl}(\mathfrak{g})$. By Weyl's theorem (Theorem 2.7.6), the adjoint representation of $\mathfrak{g}$ is completely reducible. □

Now we state a theorem whose proof we omit; for that we refer to [Jac79], Chapter III, Theorem 10. The corollary is an immediate consequence, in view of Proposition 2.12.2.

Theorem 2.12.3 *Let $\mathfrak{g} \subset \mathfrak{gl}(V)$, where V is a finite-dimensional vector space over k. Then the $\mathfrak{g}$-module V is completely reducible if and only if $\mathfrak{g} = \mathfrak{s} \oplus \mathfrak{d}$ (direct sum of ideals), where $\mathfrak{s}$ is semisimple, $\mathfrak{d}$ is the centre of $\mathfrak{g}$, and the elements of $\mathfrak{d}$ are semisimple.*

Corollary 2.12.4 *Let $\mathfrak{g}$ be a reductive Lie algebra and write $\mathfrak{g} = \mathfrak{s} \oplus \mathfrak{d}$ (notation as in Proposition 2.12.2). Let $\rho : \mathfrak{g} \to \mathfrak{gl}(V)$ be a finite-dimensional representation. Then ρ is completely reducible if and only if $\rho(x)$ is semisimple for all $x \in \mathfrak{d}$.*

Proposition 2.12.5 *Let $\mathfrak{g}$ be a semisimple Lie algebra over k. By κ we denote its Killing form. Let $\mathfrak{a}$ be a subalgebra of $\mathfrak{g}$ such that the restriction of κ to $\mathfrak{a}$ is non-degenerate, and such that $\mathfrak{a}$ contains the semisimple and nilpotent parts of its elements (note that by Theorem 2.7.9 these are well-defined elements of $\mathfrak{g}$). Then $\mathfrak{a}$ is reductive in $\mathfrak{g}$.*

Proof. By Proposition 2.12.2(iii) $\mathfrak{a}$ is reductive. So by (ii) of that same proposition, $\mathfrak{a} = \mathfrak{s} \oplus \mathfrak{d}$ (direct sum of ideals) with $\mathfrak{s}$ semisimple, and $\mathfrak{d}$ abelian. Let $x \in \mathfrak{d}$ and let $s, n \in \mathfrak{g}$ be its semisimple and nilpotent parts. Then $n \in \mathfrak{a}$. Because $\mathrm{ad}n$ can be written as a polynomial in $\mathrm{ad}x$ without constant term (Proposition 2.2.4), we have $[n, \mathfrak{a}] = 0$, whence $n \in \mathfrak{d}$. Let $y \in \mathfrak{a}$; then y commutes with n, so that $(\mathrm{ad}_{\mathfrak{g}}y)(\mathrm{ad}_{\mathfrak{g}}n)$ is nilpotent and $\kappa(y,n) = 0$. As the restriction of κ to $\mathfrak{a}$ is non-degenerate we infer that $n = 0$, and the conclusion follows from Theorem 2.12.3 □

2.13 The Jacobson-Morozov theorem

Let $\mathfrak{g}$ be a Lie algebra, and let $h, e, f \in \mathfrak{g}$ be non-zero elements satisfying the commutation relations of $\mathfrak{sl}(2, k)$ (Example 2.1.4). Then (h, e, f) is said to be an *$\mathfrak{sl}_2$-triple.*

Here we prove the Jacobson-Morozov theorem, stating that a nilpotent element in a semisimple Lie algebra can be embedded into an $\mathfrak{sl}_2$-triple.

Lemma 2.13.1 *Let $\mathfrak{g}$ be a Lie algebra over a field of characteristic 0. Let $e, h \in \mathfrak{g}$ be such that $[h, e] = 2e$ and $h \in [e, \mathfrak{g}]$. Then there is an $f \in \mathfrak{g}$ such that (h, e, f) is an $\mathfrak{sl}_2$-triple.*

Proof. There is a $z \in \mathfrak{g}$ with $h = [e, z]$. Set $E = \mathrm{ad}e$, $H = \mathrm{ad}h$, $Z = \mathrm{ad}z$. Let $U = \mathfrak{c}_{\mathfrak{g}}(e)$. Then $[h, U] \subset U$. Using $[Z, E^k] = \sum_{i=0}^{k-1} E^i[Z, E]E^{k-1-i}$ and $E^l H = HE^l - 2lE^l$ we see that $[Z, E^k] = -k(H - k + 1)E^{k-1}$. Let $x \in U \cap E^{k-1}\mathfrak{g}$ for some $k \geq 1$. Then $x = E^{k-1}y$ for some $y \in \mathfrak{g}$. Since $x \in U$ we have $0 = Ex = E^k y$. Moreover

$$-k(H - k + 1)x = -k(H - k + 1)E^{k-1}y = ZE^k y - E^k Zy = -E^k(Zy).$$

Hence $(H - k + 1)x \in E^k\mathfrak{g}$. From $[h, e] = 2e$ it follows that E is nilpotent. So there exists an $m > 0$ such that $(H - m)(H - m + 1) \cdots (H - 1)Hu = 0$ for all $u \in U$. Therefore the eigenvalues of H on U are non-negative integers. In particular, $H + 2$ is non-singular on U. Moreover, $[h, z] + 2z \in U$. So there is a $y \in U$ such that $(H + 2)y = [h, z] + 2z$. Set $f = z - y$. Then (h, e, f) is an $\mathfrak{sl}_2$-triple. □

Theorem 2.13.2 (Jacobson-Morozov) *Let $\mathfrak{g}$ be a semisimple Lie algebra over a field of characteristic 0. Let $e \in \mathfrak{g}$ be nilpotent. Then there are $h, f \in \mathfrak{g}$ such that (h, e, f) is an $\mathfrak{sl}_2$-triple.*

Proof. Consider the space $\mathrm{End}(\mathfrak{g})$ of all linear maps $\mathfrak{g} \to \mathfrak{g}$. Let $E \in \mathrm{End}(\mathfrak{g})$ be nilpotent. Then we can write $\mathfrak{g} = U_1 \oplus \cdots \oplus U_r$ (direct sum of subspaces), such that U_i has a basis $u_0^i, \ldots, u_{m_i}^i$ with $Eu_j^i = u_{j-1}^i$ for $j \geq 1$ and $Eu_0^i = 0$. Now define

$$v_j^i = j!(m_i - j + 1)(m_i - j + 2) \cdots m_i u_j^i.$$

Then $Ev_j^i = j(m_i - j + 1)v_{j-1}^i$. Define $H, F \in \mathrm{End}(\mathfrak{g})$ by $Hv_j^i = (m_i - 2j)v_j^i$,

$Fv_j^i = v_{j+1}^i$. Then $[E,F] = H$, $[H,E] = 2E$, $[H,F] = -2F$. So the Lie algebra $\mathfrak{a}$ spanned by H,E,F is isomorphic to $\mathfrak{sl}(2,k)$ and U_i is an irreducible $\mathfrak{a}$-module (compare Theorem 2.9.1).

We make $\mathrm{End}(\mathfrak{g})$ into a $\mathfrak{g}$-module by $x \cdot T = [\mathrm{ad}x, T]$. Then $\mathrm{ad}\mathfrak{g}$ is a submodule, and therefore by Weyl's theorem (Theorem 2.7.6) there exists a complementary subspace $M \subset \mathrm{End}(\mathfrak{g})$, with $\mathrm{End}(\mathfrak{g}) = \mathrm{ad}\mathfrak{g} \oplus M$. Set $E = \mathrm{ad}e$, and let $H, F \in \mathrm{End}(\mathfrak{g})$ be as above. Write $H = H_1 + H_2$ ($H_1 \in \mathrm{ad}\mathfrak{g}$, $H_2 \in M$) and $F = F_1 + F_2$ ($F_1 \in \mathrm{ad}\mathfrak{g}$, $F_2 \in M$). Then from $2E = [H,E]$ and $E \in \mathrm{ad}\mathfrak{g}$ it follows that $[H_1, E] = 2E$. Also, $H = [E,F]$ implies that $[E, F_1] = H_1$.

Let $h, f_1 \in \mathfrak{g}$ be such that $H_1 = \mathrm{ad}h$ and $F_1 = \mathrm{ad}f_1$. Then $[h,e] = 2e$ and $[e, f_1] = h$. So by Lemma 2.13.1 we infer that there exists an $f \in \mathfrak{g}$ such that (h,e,f) is an $\mathfrak{sl}_2$-triple. □

These results also yield a straightforward algorithm to compute an $\mathfrak{sl}_2$-triple containing a given nilpotent $e \in \mathfrak{g}$. We assume, as usual, that $\mathfrak{g}$ is given by a multiplication table. We compute a basis of the space $[e,\mathfrak{g}]$ and by solving a set of linear equations we find an $h \in [e,\mathfrak{g}]$ such that $[h,e] = 2e$. This set of linear equations has solutions by Theorem 2.13.2. Finally, completing the triple with an f is again done by solving a set of linear equations (we find the space V consisting of all $y \in \mathfrak{g}$ with $[h,y] = -2y$, and in there find an f with $[e,f] = h$). These linear equations have a solution by Lemma 2.13.1.

Example 2.13.3 Let $\mathfrak{g}$ be the simple Lie algebra of type A_2 with basis and multiplication table as in Example 2.9.14. Let $e = x_{\alpha_1} + x_{\alpha_2} + x_{\alpha_3}$. Then $[e,\mathfrak{g}]$ is spanned by $x_{\alpha_1}, x_{\alpha_2}, x_{\alpha_3}, h_1, h_2, x_{-\alpha_2} - x_{-\alpha_1}$. Set $h = a_1 x_{\alpha_1} + \cdots + a_6(x_{-\alpha_2} - x_{-\alpha_1})$. Then

$$[h,e] = -a_1 x_{\alpha_3} + a_2 x_{\alpha_3} + a_4(2x_{\alpha_1} - x_{\alpha_2} + x_{\alpha_3}) \\ + a_5(-x_{\alpha_1} + 2x_{\alpha_2} + x_{\alpha_3}) + a_6(h_1 - h_2 + x_{\alpha_1} + x_{\alpha_2}).$$

This is equal to $2e$ if and only if $a_4 = a_5 = 2$, $a_6 = 0$, $-a_1 + a_2 = -2$. So we can take $h = 2x_{\alpha_1} + 2h_1 + 2h_2$. The space consisting of $y \in \mathfrak{g}$ with $[h,y] = -2y$ is spanned by $x_{-\alpha_2}, x_{\alpha_1} - x_{-\alpha_1} + h_1$. By solving a system of linear equations we get $f = 2x_{-\alpha_2} - 2(x_{\alpha_1} - x_{-\alpha_1} + h_1)$.

2.14 Notes

As can be seen from the various references in the text, our main sources for the material of this chapter are [Hum78] and [Jac79]. One diversion from these books is that we introduce Cartan matrices before root systems, and treat the concept in its own right. This resembles the way in which Kac-Moody algebras are defined ([Kac90]). The main reference for the algorithms is [Gra00].

Gröbner bases of ideals in universal enveloping algebras (Section 2.10.2) were first studied in [AL88] and [KRW90].

Kostant's theorem (Theorem 2.10.11) appeared in [Kos66]. Also [Hum78] and [Ste67] present proofs of it. The proof outlined here is, as far as we know, not given elsewhere.

The algorithm for constructing irreducible highest weight modules is taken from [Gra01].

Chapter 3

Linear Algebraic Groups: Basic Constructions

This chapter mainly serves as an introduction to algebraic groups. We define algebraic groups "naively" as closed sets in affine space, such that the group operations are regular maps. Subsequently we describe some basic properties and constructions regarding algebraic groups. Among these, one of the most prominent certainly is the construction of the Lie algebra of an algebraic group. This will play an important role throughout the book, for the theory as well as for the algorithms.

We will also look at some basic algorithmic problems for algebraic groups, among which is the construction of the Lie algebra. Also we will briefly comment on ways to specify an algebraic group suitable for computation.

We use the same convention as in Chapter 1: K denotes an algebraically closed field, and $k, k', \ldots$ are fields that are not necessarily algebraically closed.

3.1 Definition and first properties

Definition 3.1.1 *Let $G \subset K^n$ be a closed set equipped with a group structure. Then G is called a* linear algebraic group *if the group operations $\cdot : G \times G \to G$ and $\iota : G \to G$ are regular maps. (Here ι denotes inversion.)*

In the following we simply say "algebraic group" instead of "linear algebraic group". The identity element of G will often be denoted e.

Example 3.1.2 Let $G = \{x \in \mathbb{C} \mid x^n - 1 = 0\}$. Then the product and the inverse are respectively given by $(x, y) \mapsto xy$ and $x \mapsto x^{n-1}$. Both are clearly regular maps. Hence G is an algebraic group.

Example 3.1.3 Let V be an n-dimensional vector space over K. Let $\mathrm{GL}(V)$ be the group of invertible linear transformations of V. In order to see that this is an algebraic group, the following trick is commonly used. We denote an element of K^{n^2+1} by $((a_{ij}), b)$, where $1 \le i, j \le n$. Correspondingly we consider the polynomial ring $K[x_{ij}, d]$, which is short for $K[x_{11}, x_{12}, \ldots, x_{nn}, d]$. We embed $\mathrm{GL}(V)$ into K^{n^2+1} by $g \mapsto ((g_{ij}), \frac{1}{\det(g)})$, where (g_{ij}) is the matrix of

g with respect to a fixed basis of V. The image of this map is clearly a group. It consists of the zeros of the polynomial $d\det((x_{ij})) - 1$. Hence it is a closed set. We have $K[\mathrm{GL}(V)] = K[x_{ij}, d]/(d\det(x_{ij}) - 1)$. The multiplication map is clearly regular. Inversion is regular as well because $d = \frac{1}{\det(x_{ij})}$ on $\mathrm{GL}(V)$. In the sequel we will write $K[\mathrm{GL}(V)] = K[x_{ij}, \frac{1}{\det(x_{ij})}]$. There is an implicit choice of basis involved here: when we want to avoid possible confusion about that we will talk about $\mathrm{GL}(n, K)$ instead, which is the group of invertible $n \times n$ matrices over K.

Example 3.1.4 Let $d \in \mathbb{Z}$, $d \neq 0$. Let G be the subset of $\mathrm{GL}(2, K)$ consisting of all

$$\begin{pmatrix} \alpha & \beta \\ d\beta & \alpha \end{pmatrix}.$$

It is straightforward to see that this is a group. As a subgroup of $\mathrm{GL}(2, K)$ it is defined by polynomial equations in the matrix entries, so it is a linear algebraic group.

Example 3.1.5 Let K have characteristic not 2. Let G be the subset of $\mathrm{GL}(3, K)$ consisting of all 3×3 matrices (a_{ij}) such that $p(a_{ij}) = 0$ with p running through the following list:

$$\begin{aligned}
&-\tfrac{1}{4}x_{32}^2 + x_{31}x_{33},\\
&x_{23}x_{31} - \tfrac{1}{2}x_{22}x_{32} + x_{21}x_{33},\\
&-x_{23}^2 + x_{13}x_{33},\\
&-2x_{22}x_{23} + x_{13}x_{32} + x_{12}x_{33},\\
&-4x_{21}x_{23} + x_{12}x_{32},\\
&-\tfrac{1}{4}x_{12}^2 + x_{11}x_{13},\\
&-x_{22}^2 + 2x_{21}x_{23} + x_{13}x_{31} + x_{11}x_{33},\\
&-2x_{21}x_{22} + x_{12}x_{31} + x_{11}x_{32},\\
&-x_{21}^2 + x_{11}x_{31},\\
&x_{13}x_{21} - \tfrac{1}{2}x_{12}x_{22} + x_{11}x_{23},\\
&-2x_{21}x_{23}^2 - 2x_{13}x_{23}x_{31} + x_{13}x_{22}x_{32}.
\end{aligned}$$

Then G is a subgroup of $\mathrm{GL}(3, K)$. But note that this is far from obvious! So it is a linear algebraic group.

Theorem 3.1.6 *An algebraic group G has no singular points.*

Proof. There is a $g \in G$ which is non-singular (Theorem 1.3.1(ii)). Let $g' \in G$. Then left multiplication by $g'g^{-1}$ is an isomorphism $G \to G$ mapping g to g'. Hence g' is non-singular as well. □

Lemma 3.1.7 *Let G be an algebraic group, and $A, B \subset G$ with A non-empty and open and B dense. Then $G = AB$.*

Proof. From Lemma 1.1.5 it follows that A^{-1} is open. Let $g \in G$. By the same lemma $A^{-1}g$ is open in G. But a non-empty open set and a dense set must have a point in common. Hence there are $a \in A$ and $b \in B$ with $a^{-1}g = b$ or $g = ab$. □

Lemma 3.1.8 *Let G be an algebraic group. Let H be a subgroup (not necessarily closed). Then the closure $\overline{H}$ is an algebraic subgroup of G.*

Proof. We have to show that $\overline{H}$ is a group. Of course, H is dense in $\overline{H}$. But $g \mapsto g^{-1}$ is an isomorphism of closed sets. So by Lemma 1.1.5(ii), $\overline{H^{-1}} = \overline{H}^{-1}$. Since $\overline{H^{-1}} = \overline{H}$, we conclude that $\overline{H}$ is closed under inversion.

In the same way for $h \in H$ we have $h\overline{H} = \overline{H}$. Therefore, $H\overline{H} = \overline{H}$. Hence for $h' \in \overline{H}$ we get $Hh' \subset \overline{H}$. Also $\overline{H}h'$ is closed, and contains Hh' as a dense subset (by Lemma 1.1.5). In other words, $\overline{Hh'} = \overline{H}h'$. Putting it together: $\overline{H}h' = \overline{Hh'} \subset \overline{H}$. Hence $\overline{H}$ is a group. □

Let G and H be algebraic groups, and $\alpha : G \to H$ a map. Then α is called a *morphism* of algebraic groups if α is a regular map and a group homomorphism.

Proposition 3.1.9 *Let $\alpha : G \to H$ be a morphism of algebraic groups. Then $\alpha(G)$ is an algebraic subgroup of H. Moreover, if G and α are defined over k, then so is $\alpha(G)$.*

Proof. Write $M = \alpha(G)$, and let $\overline{M}$ be its closure. By Lemma 3.1.8 $\overline{M}$ is an algebraic group. By Theorem 1.4.4, M contains an open set U in $\overline{M}$. But M is dense in $\overline{M}$, hence by Lemma 3.1.7, $\overline{M} = UM \subset M$. It follows that M is an algebraic subgroup of H. The last statement follows from Corollary 1.6.3. □

We note that the elimination techniques based on Gröbner bases (see Section 1.6) give an immediate algorithm to find polynomials defining $\alpha(G)$ if both G and H are explicitly given by sets of defining polynomials.

3.2 Connected components

Theorem 3.2.1 *Let G be an algebraic group, and let $X_1, \ldots, X_m$ its irreducible components. There is exactly one component containing the identity element e of G. This component is a normal subgroup of finite index in G. The X_i are pairwise disjoint.*

Proof. Suppose X_1 contains e and X_i contains e as well, for an $i \neq 1$. By Theorem 1.1.9, $X_1 \times X_i$ is irreducible. The multiplication map $X_1 \times X_i \to G$ is a regular map. Hence $\overline{X_1 X_i}$ is irreducible (Lemma 1.1.6). So it is contained in a component X_j. But both X_1 and X_i are contained in $\overline{X_1 X_i}$. Therefore $X_1 = X_i = X_j$.

Now X_1^{-1} is closed and irreducible, and it contains e. Hence it is equal to X_1. Let $g \in X_1$, then $g^{-1} \in X_1^{-1} = X_1$. Using that, an analogous argument shows that $gX_1 = X_1$. Hence X_1 is a subgroup. In the same way we see that $g^{-1}X_1 g = X_1$ for $g \in G$. Hence X_1 is normal.

Note that for $g \in G$, gX_1 is irreducible and closed. Let $A \subset G$ be such that gX_1 for $g \in A$ are exactly the cosets of X_1 in G. They are all irreducible, closed and disjoint. Hence they are the irreducible components of G. Therefore they are finite in number. □

So the irreducible components of G coincide with its connected components. The component X_1 is called the *connected component of the identity* (or, more briefly, the *identity component*) of G, and denoted G°.

This immediately leads to the following algorithmic problem: given an algebraic group G, find its connected components. If G is explicitly given by a set of polynomial equations, algorithms for the primary decomposition of an ideal (Remark 1.9.8) can be used. However, there are some difficulties. If G is defined over k, then it is possible that not all components are defined over that field. (An example is the algebraic group defined by $\det(g)^3 = 1$. This group has three components, defined by $\det(g) = \zeta^i$, $i = 0, 1, 2$, where ζ is a primitive third root of unity. So if k does not contain ζ, then only the identity component, G° is defined over k.) To obtain all components of G the algorithm for the primary decomposition has to be modified to also produce a possibly small extension $k' \supset k$ over which all components are defined. If we are only interested in the identity component, this problem does not arise, as G° is defined over k, if G is ([Bor91], Proposition 1.2). The second difficulty is practical: the Gröbner basis computations needed for the primary decomposition tend to get out of hand rather quickly. Finally we remark that in characteristic 0 a different approach is possible to find G°: first compute the Lie algebra $\mathfrak{g}$ of G, and then defining polynomials of the connected algebraic group with Lie algebra $\mathfrak{g}$ (Section 4.5).

The following lemma and proposition can often be used to show that a given algebraic group is connected. The proposition is a direct consequence of the lemma.

Lemma 3.2.2 *Let G be an algebraic group. Let I be a set, and for each $i \in I$ let X_i be an irreducible closed set and $\phi_i : X_i \to G$ a regular map. Set $Y_i = \phi_i(X_i)$, and assume that each Y_i contains the identity element $e \in G$. Let H denote the smallest algebraic subgroup of G containing all Y_i, $i \in I$. Then H is connected, and there exist $s > 0$ and $(i_1, \dots, i_s) \in I^s$ and signs $\epsilon_j = \pm 1$, $1 \leq j \leq s$ such that $H = Y_{i_1}^{\epsilon_1} \cdots Y_{i_s}^{\epsilon_{i_s}}$.*

Proof. Without loss of generality we may assume that the sets Y_i^{-1} occur among the Y_j. For $\lambda = (i_1, \ldots, i_m) \in I^m$ set $Y_\lambda = Y_{i_1} \cdots Y_{i_m}$. Then Y_λ is the image of the product morphism $X_{i_1} \times \cdots \times X_{i_m} \to G$. So from Lemma 1.1.6, Theorem 1.1.9 it follows that the closure $\overline{Y}_\lambda$ is irreducible.

Let $\mu = (j_1, \ldots, j_n) \in I^n$; we claim that $\overline{Y}_\lambda \overline{Y}_\mu \subset \overline{Y_\lambda Y_\mu}$. Indeed, using Lemma 1.1.5, we infer that for $g' \in \overline{Y}_\mu$, $\overline{Y}_\lambda g'$ is closed and contains $Y_\lambda g'$ as a dense subset. So $\overline{Y}_\lambda g' = \overline{Y_\lambda g'} \subset \overline{Y_\lambda Y_\mu}$.

Let $s > 0, \lambda \in I^s$ be such that $\overline{Y}_\lambda$ is of maximal dimension among all Y_μ, $\mu \in I^m$, for $m > 0$. Let $\mu \in I^n$. Since $e \in Y_\mu$ we see, using the claim above, and Theorem 1.3.4:

$$\overline{Y}_\lambda \subset \overline{Y}_\lambda \overline{Y}_\mu \subset \overline{Y_\lambda Y_\mu} = \overline{Y}_\lambda.$$

Also $\overline{Y}_\mu \subset \overline{Y}_\lambda \overline{Y}_\mu \subset \overline{Y}_\lambda$. By taking $\mu = \lambda$ we see that $\overline{Y}_\lambda$ is closed under taking products. Furthermore, there is a $\mu \in I^m$ such that $Y_\mu = Y_\lambda^{-1}$, so $Y_\lambda^{-1} = Y_\mu \subset \overline{Y}_\lambda$. Since $g \mapsto g^{-1}$ is a regular map, Y_λ^{-1} is also dense in $\overline{Y}_\lambda$. On the other hand, $\overline{Y}_\lambda^{-1}$ is closed and contains Y_λ^{-1} as a dense subset. So $\overline{Y}_\lambda^{-1} = \overline{Y_\lambda^{-1}} \subset \overline{Y}_\lambda$. We conclude that $\overline{Y}_\lambda$ is a group containing all Y_i. It follows that $H = \overline{Y}_\lambda$.

Now $\overline{Y}_\lambda$ is the closure of the image of a regular map. Hence Y_λ contains a non-empty open set U of $\overline{Y}_\lambda$ (Theorem 1.4.4). Since $\overline{Y}_\lambda$ is irreducible, U is also dense in it (Lemma 1.1.7). Hence by Lemma 3.1.7 we infer that $\overline{Y}_\lambda = UU \subset Y_\lambda \cdot Y_\lambda$. Hence $H = Y_\lambda \cdot Y_\lambda$, completing the proof. □

Proposition 3.2.3 *Let G_i for $i \in I$ be connected algebraic subgroups of the algebraic group G. Let $H \subset G$ be the subgroup generated by all the G_i. Then H is a connected algebraic group. Moreover, there are $j_1, \ldots, j_m \in I$ such that $H = G_{j_1} \cdots G_{j_m}$.*

As an application we prove a result on subgroups generated by commutators. Let $\mathcal{G}$ be a group, and $g, h \in \mathcal{G}$. Then $[g, h] = ghg^{-1}h^{-1}$ denotes the commutator of g and h. Let $\mathcal{H}_1, \mathcal{H}_2$ be subgroups of $\mathcal{G}$. Then $[\mathcal{H}_1, \mathcal{H}_2]$ denotes the smallest subgroup of $\mathcal{G}$ containing all $[h_1, h_2]$ for $h_i \in \mathcal{H}_i$.

Proposition 3.2.4 *Let G be an algebraic group with algebraic subgroups H_1 and H_2. Suppose H_1 is connected. Then $[H_1, H_2]$ is a connected algebraic subgroup of G.*

Proof. For $i \in H_2$ let $\phi_i : H_1 \to G$ be defined by $\phi_i(h) = [i, h]$. Let H denote the smallest algebraic subgroup of G containing all $\phi_i(H_1)$, $i \in H_2$. Lemma 3.2.2 shows that H is connected and generated by a finite number of the $\phi_i(H_1)$. So $H = [H_1, H_2]$. □

It follows that, for a connected algebraic group G, its commutator subgroup, $[G, G]$ is a connected algebraic subgroup. If G is not connected, it can still be shown that $[G, G]$ is an algebraic subgroup, but it is not necessarily connected.

Let $\mathcal{G}$ be a group and $\mathcal{N}_1$ and $\mathcal{N}_2$ normal subgroups. Then the product $\mathcal{N}_1\mathcal{N}_2$ is a normal subgroup. If moreover the $\mathcal{N}_i$ are solvable, the same holds for $\mathcal{N}_1\mathcal{N}_2$. So by Lemma 3.2.2 it follows that an algebraic group G has a unique maximal connected solvable normal algebraic subgroup. This is called the *radical* of G, and denoted $R(G)$. If $R(G)$ consists of the identity only, and G is connected, G is said to be *semisimple*.

This immediately leads to the problem to devise an algorithm for finding the radical of an algebraic group. In general we do not know of a solution for this. In characteristic 0 it is possible to compute the Lie algebra $\mathfrak{r}$ of $R(G)$ (Theorem 4.3.22), after which we can compute defining polynomials of the connected algebraic group R with Lie algebra $\mathfrak{r}$ (Section 4.5). Since $R(G)$ is connected it is equal to R.

3.3 Semidirect products

Let $\mathcal{G}$ and $\mathcal{N}$ be groups. Suppose $\mathcal{G}$ acts on $\mathcal{N}$ by group automorphisms. We will write this action on the right, i.e., we have a map $\mathcal{G}\times\mathcal{N} \to \mathcal{N}$, $(g, n) \mapsto n^g$, such that $n^{g_1g_2} = (n^{g_1})^{g_2}$, and $(n_1n_2)^g = n_1^g n_2^g$. Then the product $\mathcal{G}\times\mathcal{N}$ can be made into a group by stipulating $(g_1, n_1)(g_2, n_2) = (g_1g_2, n_1^{g_2}n_2)$. This group is called the *semidirect product* of $\mathcal{G}$ and $\mathcal{N}$ and is denoted $\mathcal{G} \ltimes \mathcal{N}$. Identifying $\mathcal{G}, \mathcal{N}$ with $\mathcal{G}\times 1$, $1\times\mathcal{N}$ respectively, we see that the semidirect product is equal to $\mathcal{G}\mathcal{N}$, $\mathcal{G}\cap\mathcal{N}$ consists of the identity, and $\mathcal{N}$ is a normal subgroup. Now let $\mathcal{H}$ be a group with subgroups $\mathcal{G}$ and $\mathcal{N}$ having these properties, i.e., $\mathcal{H} = \mathcal{G}\mathcal{N}$, $\mathcal{G}\cap\mathcal{N}$ consists of the identity, and $\mathcal{N}$ is normal. Then $\mathcal{H}$ is isomorphic to the semidirect product $\mathcal{G} \ltimes \mathcal{N}$, where $\mathcal{G}$ acts on $\mathcal{N}$ by $n^g = g^{-1}ng$.

We can perform the same constructions for algebraic groups G and N. Now we require that the maps $n \mapsto n^g$ are automorphisms of algebraic groups. Furthermore, an algebraic group H is said to be the semidirect product of subgroups G and N if $H = GN$, $G \cap N$ consists of the identity, N is normal in H and the map $G \times N \to H$, $(g, n) \mapsto gn$ is an isomorphism of closed sets (i.e., it is regular with the inverse being regular as well). Because of this last condition, it can happen that as an abstract group H is the semidirect product of G, N, while not being the semidirect product as an algebraic group.

The direct product of algebraic groups $G \times N$ is a special case of this construction, where the action of G on N is trivial.

3.4 The Lie algebra of an algebraic group

The Lie algebra of an algebraic group G reveals a lot of information about G. If G is connected and the base field is of characteristic 0, the Lie algebra even determines G completely, as we will see in Chapter 4. It is no surprise, therefore, that Lie algebras are useful tools when doing computations with algebraic groups.

Let G be a linear algebraic group. Its Lie algebra is the tangent space at the identity e, which we equip with a Lie algebra structure. Following Section 1.2 we define the Lie algebra of G as the space

$$\mathrm{Lie}(G) = \{\delta \in K[G]^* \mid \delta(fh) = f(e)\delta(h) + h(e)\delta(f) \text{ for all } f, h \in K[G]\}$$

(where $K[G]^*$ is the dual space of $K[G]$, i.e., the space of all linear maps $K[G] \to K$).

Now we want to equip $\mathrm{Lie}(G)$ with a Lie algebra structure. For that we first define a multiplication on $K[G]^*$ that makes it into an associative algebra. Then we show that $\mathrm{Lie}(G)$ is closed under taking commutators by identifying it with an algebra of derivations. This immediately implies that $\mathrm{Lie}(G)$ is a Lie algebra.

We identify $K[G \times G]$ with $K[G] \otimes K[G]$, see Lemma 1.1.8. We define a map $\Delta : K[G] \to K[G] \otimes K[G]$ by $\Delta(f)(a, b) = f(ab)$, for $a, b \in G$. This is the cohomomorphism of the multiplication map $G \times G \to G$. So if $\Delta(f) = \sum_i f_i \otimes f_i'$ then $f(ab) = \sum_i f_i(a) f_i'(b)$ for all $a, b \in G$.

We also define a map $\Delta \otimes 1 : K[G] \otimes K[G] \to K[G] \otimes K[G] \otimes K[G]$ by $(\Delta \otimes 1)(f \otimes g) = \Delta(f) \otimes g$. Similarly we have a map $1 \otimes \Delta$.

Lemma 3.4.1 $(\Delta \otimes 1) \circ \Delta = (1 \otimes \Delta) \circ \Delta$.

Proof. Let $f \in K[G]$ and write $\Delta(f) = \sum_i f_i \otimes f_i'$. Let $a, b, c \in G$; then

$$((\Delta \otimes 1) \circ \Delta)(f)(a, b, c) = \sum_i f_i(ab) f_i'(c) = \Delta(f)(ab, c) = f((ab)c).$$

This is equal to $f(a(bc))$, which in turn is equal to $((1 \otimes \Delta) \circ \Delta)(f)(a, b, c)$. □

The map Δ is called the comultiplication, while the property expressed by the previous lemma is called coassociativity.

Now let $\gamma, \delta \in K[G]^*$, and define their product by $\gamma\delta = (\gamma \otimes \delta) \circ \Delta$. In other words, if $\Delta(f) = \sum_i f_i \otimes f_i'$ then $(\gamma\delta)(f) = \sum_i \gamma(f_i)\delta(f_i')$. The fact that this multiplication is associative follows from a short calculation using Lemma 3.4.1:

$$\begin{aligned}(\gamma\delta)\epsilon &= (((\gamma \otimes \delta) \circ \Delta) \otimes \epsilon) \circ \Delta = (\gamma \otimes \delta \otimes \epsilon) \circ (\Delta \otimes 1) \circ \Delta \\ &= (\gamma \otimes \delta \otimes \epsilon) \circ (1 \otimes \Delta) \circ \Delta = \gamma(\delta\epsilon).\end{aligned}$$

Let $\varepsilon : K[G] \to K$ be defined by $\varepsilon(f) = f(e)$. Then ε is the identity element for the multiplication on $K[G]^*$. In order to see that, define $\rho_{g'} : K[G] \to K[G]$ by $\rho_{g'}(f)(g) = f(gg')$. For $f \in K[G]$ write $\Delta(f) = \sum_i f_i \otimes f_i'$; then $\rho_{g'}(f) = \sum_i f_i'(g')f_i$, so that $\delta\varepsilon(f) = \sum_i \delta(f_i)f_i'(e) = \delta(\rho_e(f)) = \delta(f)$. Similarly, $\varepsilon\delta(f) = \delta(f)$.

We now consider derivations of $K[G]$. These are linear maps $D : K[G] \mapsto K[G]$ with $D(f_1f_2) = D(f_1)f_2 + f_1D(f_2)$. The set of all derivations of $K[G]$ is denoted $\mathrm{Der}(K[G])$. As seen in Section 2.3, $\mathrm{Der}(K[G])$ forms a Lie algebra with the product $[D_1, D_2] = D_1 \circ D_2 - D_2 \circ D_1$.

For $g \in G$ we define the linear map $\lambda_g : K[G] \to K[G]$ by $\lambda_g(f)(g') = f(gg')$. A derivation D of $K[G]$ is said to be *left invariant* if $\lambda_g \circ D = D \circ \lambda_g$ for all $g \in G$. We denote the set of all left invariant derivations of $K[G]$ by $\mathrm{Der}_\ell(K[G])$. It is a Lie subalgebra of $\mathrm{Der}(K[G])$.

Let $\delta \in \mathrm{Lie}(G)$ and $f \in K[G]$, then the map $f * \delta : G \to K$ defined by $f * \delta(g) = \delta(\lambda_g(f))$ is called the *right convolution* of f by δ. Write $\Delta(f) = \sum_i f_i \otimes f_i'$, then

$$f * \delta = \sum_i \delta(f_i')f_i. \tag{3.1}$$

Indeed, $\lambda_g(f)(g') = f(gg') = \sum_i f_i(g)f_i'(g')$, so that $\lambda_g(f) = \sum_i f_i(g)f_i'$. Hence $f * \delta(g) = \sum_i \delta(f_i')f_i(g)$.

Now define $D_\delta : K[G] \to K[G]$ by $D_\delta(f) = f * \delta$. Let $f_1, f_2 \in K[G]$ and $g \in G$, then

$$\begin{aligned} D_\delta(f_1f_2)(g) &= \delta(\lambda_g(f_1)\lambda_g(f_2)) = \lambda_g(f_1)(e)\delta(\lambda_g(f_2)) + \lambda_g(f_2)(e)\delta(\lambda_g(f_1)) \\ &= f_1(g)(f_2 * \delta) + f_2(g)(f_1 * \delta). \end{aligned}$$

Hence D_δ is a derivation of $K[G]$. Moreover, $(\lambda_g \circ D_\delta)(f)(g') = \delta(\lambda_{gg'}(f)) = \delta(\lambda_{g'}(\lambda_g(f))) = (D_\delta \circ \lambda_g)(f)(g')$. Hence D_δ lies in $\mathrm{Der}_\ell(K[G])$. So we get a linear map $\eta : \mathrm{Lie}(G) \to \mathrm{Der}_\ell(K[G])$, by $\eta(\delta) = D_\delta$.

Let $\mathrm{End}(K[G])$ denote the space of all K-linear maps $K[G] \to K[G]$. We define a map $\theta : \mathrm{End}(K[G]) \to K[G]^*$ by $\theta(T)(f) = T(f)(e)$. A short calculation shows that for $D \in \mathrm{Der}(K[G])$ we have $\theta(D) \in \mathrm{Lie}(G)$. So by restriction we get a map $\theta : \mathrm{Der}_\ell(K[G]) \to \mathrm{Lie}(G)$.

Lemma 3.4.2 *θ is the inverse of η. In particular, η is bijective.*

Proof. On the one hand, $\theta(D_\delta)(f) = (f * \delta)(e) = \delta(f)$. On the other hand, if we set $\delta = \theta(D)$ then $D_\delta(f)(g) = \delta(\lambda_g(f)) = D(\lambda_g(f))(e) = \lambda_g(D(f))(e) = D(f)(g)$. □

Proposition 3.4.3 *For $\gamma, \delta \in \mathrm{Lie}(G)$, also $[\gamma, \delta] = \gamma\delta - \delta\gamma$ lies in $\mathrm{Lie}(G)$. In particular, $\mathrm{Lie}(G)$ is a Lie algebra.*

Proof. Let $\gamma, \delta \in \mathrm{Lie}(G)$. Let $f \in K[G]$ and write $\Delta(f) = \sum_i f_i \otimes f_i'$. Then $\theta(D_\gamma \circ D_\delta)(f) = ((f * \delta) * \gamma)(e) = \sum_i \delta(f_i')(f_i * \gamma)(e)$ by (3.1). But since $(f_i * \gamma)(e) = \gamma(f_i)$, this is equal to $\gamma\delta(f)$. So $\theta([D_\gamma, D_\delta]) = \gamma\delta - \delta\gamma$. □

Now we study the Lie algebra of $G = \mathrm{GL}(V)$, where V is an n-dimensional vector space over K. This group is viewed as a closed subset of K^{n^2+1} (see Example 3.1.3). We have $K[G] = K[x_{ij}, \frac{1}{\det(x_{ij})}]$, and set $d = \frac{1}{\det(x_{ij})}$. Let $\delta \in \mathrm{Lie}(G)$; then δ has $\mathfrak{m}_e^2$ in its kernel (notation as in Section 1.2). Moreover, $\det((x_{ij})) = x_{11} \cdots x_{nn} \bmod \mathfrak{m}_e^2$ (as $x_{ij} \in \mathfrak{m}_e$ if $i \neq j$) and $\delta(x_{11} \cdots x_{nn}) = \sum_{i=1}^n \delta(x_{ii})$. So $0 = \delta(\det((x_{ij}))d - 1) = \delta(x_{11} \cdots x_{nn} d) = \sum_{i=1}^n \delta(x_{ii}) + \delta(d)$. Hence $\delta(d)$ is determined once we know the $\delta(x_{ij})$. There are no algebraic relations between the x_{ij}; hence we can choose $a_{ij} \in K$ and define $\delta \in \mathrm{Lie}(G)$ by $\delta(x_{ij}) = a_{ij}$. So there is a bijective linear map from the set of all $n \times n$ matrices over K to $\mathrm{Lie}(G)$.

Let $\gamma, \delta \in \mathrm{Lie}(G)$ be defined by $\gamma(x_{ij}) = a_{ij}$ and $\delta(x_{ij}) = b_{ij}$. We want to compute c_{ij} such that $\gamma\delta(x_{ij}) = c_{ij}$. For this we need $\Delta(x_{ij})$. We have $\Delta(x_{ij})(g, g') = x_{ij}(gg') = \sum_{k=1}^n g_{ik} g'_{kj}$, so that $\Delta(x_{ij}) = \sum_{k=1}^n x_{ik} \otimes x_{kj}$. Therefore $c_{ij} = \sum_{k=1}^n a_{ik} b_{kj}$. We conclude that if we identify γ and δ with their matrices (a_{ij}) and (b_{ij}), the product $\gamma\delta$ corresponds to the product of the matrices. Therefore, the Lie algebra of G is isomorphic to the Lie algebra $\mathfrak{gl}(n, K)$ (Example 2.1.1).

Now let $H \subset \mathrm{GL}(V)$ be an algebraic subgroup. As a subgroup of $\mathrm{GL}(V)$, H is defined by a set of polynomial equations in the x_{ij}. Let $\mathcal{I}(H) \subset K[G]$ be the vanishing ideal of H. Let $F_1, \dots, F_r$ generate $\mathcal{I}(H)$. From Section 1.2 we recall that $\mathcal{D} = \{a + b\varepsilon \mid a, b \in K \text{ and } \varepsilon^2 = 0\}$ denotes the algebra of dual numbers. Then from Section 1.2 we get that a matrix $a = (a_{ij})$ lies in $\mathrm{Lie}(H)$ if and only if $F_j(I_n + \varepsilon a) = 0$ for $1 \leq j \leq r$ (where I_n is the $n \times n$ identity matrix).

Also from Sections 1.2, and 1.9 we get a straightforward algorithm to compute the Lie algebra of H. First we compute generators of the radical of the ideal generated by the polynomials that define H. By the Nullstellensatz this is the vanishing ideal of H. Let, as above, $F_1, \dots, F_r$ be its generators. Then $a = (a_{ij})$ lies in $\mathrm{Lie}(H)$ if and only if

$$\sum_{1 \leq i,j \leq n} \frac{\partial F_k}{\partial x_{ij}}(I_n) a_{ij} = 0 \text{ for } 1 \leq k \leq r. \tag{3.2}$$

However, in practice, due to the needed Gröbner basis computations, calculating the radical of an ideal is a difficult task. If we are unable to compute the radical, the matrix space defined by the equations (3.2) (or equivalently by equations of the form $F_j(I_n + \varepsilon a) = 0$) *contains* the Lie algebra of H, but may not be equal to it.

For that reason, before computing the radical, it usually is a good idea to compute the dimension of the group H using Proposition 1.7.3. By differentiating the polynomials defining H and solving the system of linear equations

(3.2), we obtain a linear space $\mathfrak{h}$ of matrices. If the dimension of $\mathfrak{h}$ equals $\dim H$, then $\mathfrak{h} = \mathrm{Lie}(H)$. Remark 3.6.5 illustrates the advantages that this approach can have.

Example 3.4.4 Let $G \subset \mathrm{GL}(3, K)$ be the algebraic group of Example 3.1.5. Let I be the ideal of $K[x_{ij}, d]$ generated by the polynomials given there, along with $d\det(x_{ij}) - 1$. Then it takes 0.04 seconds to compute the radical of I, and show that it is equal to I. So the Lie algebra of G can be computed by differentiating the polynomials given in Example 3.1.5. This leads to the following equations for a matrix $(a_{ij}) \in \mathrm{Lie}(G)$:

$$a_{31} = -\tfrac{1}{2}a_{32} + a_{21} = a_{13} = -2a_{23} + a_{12} = -2a_{22} + a_{11} + a_{33} = 0.$$

So $\mathrm{Lie}(G)$ is 4-dimensional and spanned by

$$\begin{pmatrix} 0 & 0 & 0 \\ 1 & 0 & 0 \\ 0 & 2 & 0 \end{pmatrix}, \begin{pmatrix} 0 & 2 & 0 \\ 0 & 0 & 1 \\ 0 & 0 & 0 \end{pmatrix}, \begin{pmatrix} 2 & 0 & 0 \\ 0 & 1 & 0 \\ 0 & 0 & 0 \end{pmatrix}, \begin{pmatrix} 0 & 0 & 0 \\ 0 & 1 & 0 \\ 0 & 0 & 2 \end{pmatrix}.$$

3.5 Subgroups and subalgebras

This section has two main results. The next proposition has a characterization of the Lie algebra of an algebraic subgroup of an algebraic group. Theorem 3.5.4 states that any algebraic group is isomorphic to an algebraic matrix group. So we do not lose any examples by just considering algebraic matrix groups.

Proposition 3.5.1 *Let G be an algebraic group and $H \subset G$ an algebraic subgroup. Set $J = \mathcal{I}(H) \subset K[G]$. Then* $\mathrm{Lie}(H) = \{\delta \in \mathrm{Lie}(G) \mid \delta(J) = 0\} = \{\delta \in \mathrm{Lie}(G) \mid J * \delta \subset J\}$.

Proof. The first equality follows from Proposition 1.2.5. For the second equality let $\delta \in \mathrm{Lie}(G)$ be such that $\delta(J) = 0$. Then for $f \in J$ and $h \in H$ we get $(f * \delta)(h) = \delta(\lambda_h(f)) = 0$ as $\lambda_h(f) \in J$. So $f * \delta$ vanishes on H and therefore $f * \delta \in J$. Conversely, suppose $J * \delta \subset J$. Then for $f \in J$: $\delta(f) = \delta(\lambda_e(f)) = (\delta * f)(e) = 0$. □

Lemma 3.5.2 *Let G be an algebraic group and $f \in K[G]$. Then the set $\{\lambda_g(f) \mid g \in G\}$ spans a finite-dimensional subspace of $K[G]$.*

Proof. Write $\Delta(f) = \sum_i f_i \otimes f_i'$. Then $\lambda_g(f)(g') = \sum_i f_i(g)f_i'(g')$. Hence $\lambda_g(f) = \sum_i f_i(g)f_i'$. So the space in question is contained in the space spanned by the f_i'. □

Corollary 3.5.3 *Let G be an algebraic group and $V \subset K[G]$ a subspace. Let $g \in G$ and suppose $\lambda_g(V) \subset V$. Then $\lambda_g(V) = V$.*

Proof. For $f \in V$ let $V(f) \subset K[G]$ be the smallest subspace of V stable under λ_g. Now $\lambda_g : V(f) \to V(f)$ is injective. Furthermore, λ_g has no kernel. Since $V(f)$ is finite-dimensional (previous lemma), λ_g is surjective on $V(f)$. In particular, there is an $f' \in K[G]$ with $\lambda_g(f') = f$. □

Theorem 3.5.4 *Let G be an algebraic group defined over K. Then there exists a finite-dimensional vector space V over K, an algebraic subgroup $H \subset \mathrm{GL}(V)$, and an isomorphism of algebraic groups $\rho : G \to H$.*

Proof. By Lemma 3.5.2 there exists a finite-dimensional subspace V of $K[G]$ so that $K[G]$ is generated (as algebra) by V and V is stable under all λ_g for $g \in G$. Define $\rho : G \to \mathrm{GL}(V)$ by $\rho(g)(f) = \lambda_g(f)$. Note that Corollary 3.5.3 assures that $\rho(g) \in \mathrm{GL}(V)$. Moreover, ρ is injective as V generates $K[G]$.

Let $v_1, \dots, v_s$ be a basis of V, which we extend to a basis $(v_j)_{j \geq 1}$ of $K[G]$. Let $f \in K[G]$ and write $\Delta(f) = \sum_i f_i \otimes f_i'$. Then for $g \in G$ we see that $\lambda_g(f) = \sum_i f_i(g) f_i' = \sum_{j \geq 1} h_j(g) v_j$, where $h_j \in K[G]$ (almost all of which are 0). If in addition $f \in V$, then $h_j = 0$ for $j > s$. We conclude that there are $h_{ij} \in K[G]$ such that $\lambda_g(v_i) = \sum_j h_{ij}(g) v_j$ and hence ρ is a regular map.

By Proposition 3.1.9, $H = \rho(G)$ is an algebraic subgroup of $\mathrm{GL}(V)$. For $g \in G$ we have $v_i(g) = \sum_j h_{ij}(g) v_j(e)$, where e denotes the identity element of G. Therefore v_i lies in the image of ρ^*. As the v_i generate $K[G]$ we infer that ρ^* is surjective. Finally, that implies that the inverse of ρ is a regular map, so that ρ is an isomorphism of algebraic groups. □

3.6 Examples

Example 3.6.1 The group $\mathrm{GL}(1, K)$ is also denoted $\mathbb{G}_{\mathrm{m}}$ and called the *multiplicative group*. It consists of the non-zero elements of K. We have $\mathrm{Lie}(\mathbb{G}_{\mathrm{m}}) = \mathfrak{gl}(1, K)$, which is 1-dimensional.

Example 3.6.2 Consider the group $\mathbb{G}_{\mathrm{a}} = K$, where the group operation is addition. This is called the *additive group*. We have $K[\mathbb{G}_{\mathrm{a}}] = K[t]$. Hence a $\delta \in \mathrm{Lie}(\mathbb{G}_{\mathrm{a}})$ is determined by the value $\delta(t)$. So $\mathrm{Lie}(\mathbb{G}_{\mathrm{a}})$ is the unique 1-dimensional Lie algebra.

Example 3.6.3 Let the field K be of characteristic $p > 0$. Let

$$G = \left\{ \begin{pmatrix} a & 0 & 0 \\ 0 & a^p & b \\ 0 & 0 & 1 \end{pmatrix} \mid a, b \in K, a \neq 0 \right\}.$$

Then G is an algebraic subgroup of $\mathrm{GL}(3, K)$. Writing $K[\mathrm{GL}(3, K)] = K[x_{11}, \dots, x_{33}, d]/(d\det(x_{ij}) - 1)$, G is defined by the polynomials

$$\{x_{12}, x_{13}, x_{21}, x_{31}, x_{32}, x_{22} - x_{11}^p, x_{33} - 1\}.$$

The ideal generated by these polynomials is radical. So we find the Lie algebra by differentiating. It is spanned by e_{11}, e_{23} (where e_{ij} denotes the 3×3 matrix with a 1 on position (i, j) and zeros elsewhere).

Example 3.6.4 Let K be of characteristic not 2. Let V be an n-dimensional vector space over K and $B : V \times V \to K$ be a non-degenerate bilinear form. We set $G = \{g \in \mathrm{GL}(V) \mid B(gv, gw) = B(v, w) \text{ for all } v, w \in V\}$. Let $v_1, \dots, v_n$ be a basis of V and write B for the matrix $(B(v_i, v_j))$. Then $g \in G$ if and only if $g^T Bg = B$. It follows that G is an algebraic group. We want to compute its Lie algebra. For this we first determine a generic point of G (see Section 1.3.1).

Let $L = \{X \in \mathfrak{gl}(n, K) \mid X^T B + BX = 0\}$. Then L is a linear space. Let $X_1, \dots, X_d$ be a basis of L. Let $K' = K(T_1, \dots, T_d)$ be the rational function field over K in d indeterminates. Set $X^* = \sum_{i=1}^d T_i X_i$. Write I_n for the $n \times n$ identity matrix. Then $\det(I_n - X^*)$ lies in $K[T_1, \dots, T_n]$. Moreover, it is not 0 because it takes the value 1 if we substitute 0 for each of the T_i. Set $g^* = (I_n + X^*)(I_n - X^*)^{-1}$ (note that the coefficients of g^* are well defined elements of K'). Then $g^*(I_n - X^*) = (I_n - X^*)g^* = I_n + X^*$. So

$$(I_n - (X^*)^T)(g^*)^T Bg^*(I_n - X^*) = (I_n + (X^*)^T)B(I_n + X^*) = B + (X^*)^T BX^*$$

(as $(X^*)^T B + BX^* = 0$). But the latter is also equal to $(I_n - (X^*)^T)B(I_n - X^*)$. Now since $I_n - X^*$ and $I_n - (X^*)^T$ are invertible, we get $(g^*)^T Bg^* = B$.

Let $g \in G$ be such that $\det(I_n + g) \neq 0$ and set $X = -(I_n - g)(I_n + g)^{-1}$. Then $(I_n - X)g = I_n + X$. Now we multiply the relation $g^T Bg = B$ from the left by $I_n - X^T$ and from the right by $I_n - X$. Then after some manipulation we get $X^T B + BX = 0$, i.e., $X \in L$. It is straightforward to show that X has no eigenvectors with eigenvalue 1. So $I_n - X$ is invertible, and $g = (I_n + X)(I_n - X)^{-1}$. Hence g is a specialization of g^*. Let p be a polynomial with $p(g^*) = 0$. Then the polynomial $g \mapsto \det(I_n + g)p(g)$ is zero on G and hence zero on G°. But G° is irreducible, and $g \mapsto \det(I_n + g)$ is not zero on G° (take $g = I_n$). It follows that p is zero on G°. The conclusion is that g^* is a generic point of G°.

Now $(I_n - g^*)(I_n - X^*) = -2X^*$ and $(I_n + g^*)(I_n - X^*) = 2I_n$. Hence $\det(I_n + g^*) \neq 0$ and $X^* = -(I_n - g^*)(I_n + g^*)^{-1}$. From this it follows that $K(g^*) = K'$, and therefore $\dim(G^\circ) = \dim(L)$ (see Proposition 1.3.5).

Among the generators of $\mathcal{I}(G)$ we can take the polynomial equations that follow from $g^T Bg = B$. If we take $I_n + \varepsilon X$ in place of g, we get $X^T B + BX = 0$. This implies that $\mathrm{Lie}(G) \subset L$. However, $\dim(\mathrm{Lie}(G)) = \dim(G) = \dim(L)$. We infer that $\mathrm{Lie}(G) = L$.

Remark 3.6.5 If it were known that the polynomial equations coming from $g^T Bg = B$ define a radical ideal, most of the argument of the previous example would become unnecessary. However, it does not seem to be straightforward to show that. We have checked by computer for $n = 2, 3, 4$ and $B = I_n$, that the ideal is indeed radical. The run times were 0.0 (s), 0.03 (s) and 26726 (s) respectively. The limits of the algorithm for computing the radical become clear. The computation time used to compute the dimension of G for the same n, were 0.0 (s), 0.0 (s), 0.3 (s). We see that, in order to compute the Lie algebra of an algebraic group, it is a good idea to compute the dimension before attempting a radical computation.

Example 3.6.6 Let $G = \mathrm{SL}(n, K) = \{g \in \mathrm{GL}(n, K) \mid \det(g) = 1\}$. Since G is defined by a single polynomial, it has codimension 1 (see [Sha94], Section I.6, Theorem 2), in other words, $\dim(\mathrm{SL}(n, K)) = n^2 - 1$. Now $\det(I_n + \varepsilon x) = 1 + \varepsilon \mathrm{Tr}(x)$. It follows that $\{x \in \mathfrak{gl}(n, K) \mid \mathrm{Tr}(x) = 0\} \supset \mathrm{Lie}(G)$. But since the dimension is $n^2 - 1$ in both cases, we have equality. This Lie algebra is denoted by $\mathfrak{sl}(n, K)$ (see Example 2.1.4).

Example 3.6.7 Here we restrict to the case where the field K has characteristic 0. Let a be a nilpotent linear transformation of a finite-dimensional vector space V. We set $\exp(a) = \sum_{k\geq 0} \frac{a^k}{k!}$ (note that this is a finite sum, so $\exp(a)$ is well-defined). Suppose now that $a \neq 0$ and set $G = \{\exp(ta) \mid t \in K\}$; then G is a group. Let $\mathbb{G}_a$ be the additive group (Example 3.6.2). Define $\sigma : \mathbb{G}_a \to \mathrm{GL}(V)$ by $\sigma(a) = \exp(ta)$. This is clearly a morphism of algebraic groups. As G is the image of σ it is a connected algebraic subgroup of $\mathrm{GL}(V)$ (Proposition 3.1.9).

We claim that σ has a regular inverse. Let T be an indeterminate over K and consider $\exp(Ta)$. There is a $v \in V$ with $av \neq 0$ but $a^2 v = 0$. So $\exp(Ta)v = v + Tav$. Let λ be a linear function on V such that $\lambda(av) = 1$, then $T = \lambda(\exp(Ta)v) - \lambda(v)$. It follows that there is a polynomial function p on $\mathrm{GL}(V)$ such that $T = p(\exp(Ta))$. The inverse of σ is given by $g \mapsto p(g)$, so our claim is proved.

The conclusion is that G is isomorphic to $\mathbb{G}_a$ and hence has dimension 1. We use dual numbers to compute the differential of σ. We have $\sigma(0 + \varepsilon \cdot 1) = I_V + \varepsilon a$, where I_V is the identity on V. It follows that $\mathrm{d}_0\sigma(1) = a$, so $a \in \mathrm{Lie}(G)$. Since the latter is 1-dimensional we conclude that a spans $\mathrm{Lie}(G)$.

Lemma 3.6.8 *Write $K[x_{ij}]$ for the subring of $K[\mathrm{GL}(n, K)]$ generated by all x_{ij}, $1 \leq i, j \leq n$. Let $G \subset \mathrm{GL}(n, K)$ be an algebraic subgroup defined by polynomials $f_1, \dots, f_s \in K[x_{ij}]$ of degree 1. Then $\mathcal{I}(G) \subset K[\mathrm{GL}(n, K)]$ is generated by $f_1, \dots, f_s$.*

Proof. We may assume that the f_i are triangularized, i.e., every polynomial has a unique leading term x_{ij}, with coefficient 1, that does not occur in the other polynomials. Set $y_{kl} = x_{kl}$ if x_{kl} does not occur as a leading term (and we say that y_{kl} is non-leading), and $y_{kl} = f_i$ if x_{kl} is the leading term of f_i (in this case we say that y_{kl} is leading). Let $p \in K[x_{ij}]$ vanish on G. Write p as a linear combination of monomials in the y_{kl}. Then $p = p_1 + p_2$, where p_1 is a linear combination of monomials that do not involve any leading y_{kl}, and p_2 consists of the monomials that have at least one leading y_{kl}.

Let V be the space of all $a \in M_n(K)$ such that $f_i(a) = 0$ for $1 \leq i \leq s$. Note that $G \subset V$ is open, defined by the condition $\det \neq 0$. Furthermore, an $a \in V$ is uniquely determined by the a_{kl}, where k, l are such that y_{kl} is non-leading. Hence $V \cong K^r$, where r is the number of non-leading y_{kl}. Also $K[V] = K[y_{kl} \mid \text{non-leading}\,]$ and $p_1 \in K[V]$. But p and p_2 vanish on G, and hence so does p_1. It follows that $p_1 = 0$ (Lemma 1.1.7) and p lies in the ideal generated by the f_i.

Finally, let $q \in K[\mathrm{GL}(n, K)]$ vanish on G. Then $q = \frac{1}{\det(x_{ij})^m} p$ with $p \in K[x_{ij}]$ that also vanishes on G. It follows that q lies in the ideal generated by the f_i. □

Example 3.6.9 Let $A \subset M_n(K)$ be an associative matrix algebra containing the identity I_n. Let G be the set of $a \in A$ such that $\det(a) \neq 0$. Let $a \in G$, then left multiplication by a is an injective hence also surjective map $A \to A$. Since $I_n \in A$, there is $b \in A$ with $ab = I_n$. It follows that $a^{-1} \in G$, and G is an algebraic subgroup of $\mathrm{GL}(n, K)$.

There is a set of linear polynomials in $K[x_{ij}]$ that define G (the polynomials that define A as a linear subspace of $M_n(K)$). So by the previous lemma, $\mathcal{I}(G)$ is generated by these linear polynomials. Since they are homogeneous of degree 1, $\mathrm{Lie}(G)$ is defined by the same polynomials. It follows that $\mathrm{Lie}(G) = A$ with the commutator as Lie bracket.

Example 3.6.10 Let V be an n-dimensional vector space, and $U \subset W$ two subspaces. Set $G = \{g \in \mathrm{GL}(V) \mid g \cdot w \equiv w \bmod U \text{ for all } w \in W\}$. Then G is a group. We claim that it is algebraic. In order to see this we extend a basis of U to a basis of W, which we then extend to a basis of V. If we write elements of $\mathrm{GL}(V)$ with respect to this basis we see that $g \in \mathrm{GL}(V)$ lies in G if and only if g satisfies a set of polynomial equations of degree 1 (they are of the form $x_{ij} = 0$ for certain i, j or $x_{ii} - 1 = 0$ for certain i). So by Lemma 3.6.8, $\mathcal{I}(G)$ is generated by the same polynomials. Therefore, $\mathrm{Lie}(G)$ is defined by the homogeneous parts of degree 1 of these polynomials. Hence $\mathrm{Lie}(G) = \{a \in \mathfrak{gl}(V) \mid a \cdot W \subset U\}$.

3.7 Morphisms and representations

Let G, H be algebraic groups and $\alpha : G \to H$ a morphism. We write $\mathrm{d}\alpha$ for the differential $\mathrm{d}_e\alpha : \mathrm{Lie}(G) \to \mathrm{Lie}(H)$. We claim that it is a homomorphism of Lie algebras. We already know that it is a linear map, so we need to show that $\mathrm{d}\alpha([\gamma, \delta]) = [\mathrm{d}\alpha(\gamma), \mathrm{d}\alpha(\delta)]$. We make use of the definitions of Sections 1.2 and 3.4. The map $\mathrm{d}\alpha$ is defined by $\mathrm{d}\alpha(\delta) = \delta \circ \alpha^*$. By Δ_G, Δ_H we denote the comultiplications relative to G and H respectively. Let $f \in K[H]$ and write $\Delta_H(f) = \sum_i f_i \otimes f_i'$. Then for $g, h \in G$ we have $\alpha^*(f)(gh) = f(\alpha(g)\alpha(h)) = \sum_i f_i(\alpha(g))f_i'(\alpha(h))$. Hence $\Delta_G(\alpha^*(f)) = \sum_i \alpha^*(f_i) \otimes \alpha^*(f_i')$. From this it follows that $\mathrm{d}\alpha(\gamma\delta)(f) = (\mathrm{d}\alpha(\gamma)\mathrm{d}\alpha(\delta))(f)$. This implies that $\mathrm{d}\alpha$ is a homomorphism of Lie algebras.

Example 3.7.1 Let G be an algebraic group with Lie algebra $\mathfrak{g} = \mathrm{Lie}(G)$. Let $\mu : G \times G \to G$ denote the multiplication map. We want to determine $\mathrm{d}_{(e,e)}\mu : T_{(e,e)}(G \times G) \to \mathfrak{g}$. We know that $T_{(e,e)}(G \times G)$ is isomorphic to $\mathfrak{g} \oplus \mathfrak{g}$ (Proposition 1.2.8). We first make this isomorphism a bit more explicit, using the construction in the proof of Proposition 1.2.8. Let $\sigma_1, \sigma_2 : G \to G \times G$ be defined by $\sigma_1(g) = (g, e)$, $\sigma_2(g) = (e, g)$ respectively. Then a $(\delta, \delta') \in \mathfrak{g} \oplus \mathfrak{g}$ corresponds to $\mathrm{d}\sigma_1(\delta) + \mathrm{d}\sigma_2(\delta') \in T_{(e,e)}(G \times G)$.

We let $\Delta = \mu^*$ be the comultiplication. Let $f \in K[G]$ and write $\psi = \Delta(f) = \sum_i f_i \otimes f_i' \in K[G] \otimes K[G]$ (which we identify with $K[G \times G]$). Then $\sigma_1^*(\psi) = \sum_i f_i'(e)f_i = f$. Similarly, $\sigma_2^*(\psi) = \sum_i f_i(e)f_i' = f$. Therefore, $\mathrm{d}\sigma_1(\delta)(\psi) = \sum_i \delta(f_i)f_i'(e)$ and $\mathrm{d}\sigma_2(\delta')(\psi) = \sum_i f_i(e)\delta'(f_i')$. So we get $\mathrm{d}\mu(\delta, \delta')(f) = (\delta, \delta')(\mu^*(f)) = (\delta, \delta')(\psi) = \sum_i \delta(f_i)f_i'(e) + \sum_i f_i(e)\delta'(f_i') = \delta(f) + \delta'(f)$. Hence $\mathrm{d}\mu(\delta, \delta') = \delta + \delta'$.

Let $\iota : G \to G$ be defined by $\iota(g) = g^{-1}$ and $\sigma : G \to G \times G$ be defined by $\sigma(g) = (g, \iota(g))$. Then $\mu \circ \sigma(g) = e$. Hence $0 = \mathrm{d}(\mu \circ \sigma) = \mathrm{d}\mu \circ \mathrm{d}\sigma$. But $\mathrm{d}_e(\sigma)(\delta) = (\delta, \mathrm{d}\iota(\delta))$. So $\mathrm{d}\mu(\delta, \mathrm{d}\iota(\delta)) = 0$ and $\mathrm{d}\iota(\delta) = -\delta$.

A morphism $\rho : G \to \mathrm{GL}(V)$ is called a *rational representation* of G. An example is the map σ of Example 3.6.7. In this context we also say that V is a *rational G-module* and write $g \cdot v$ (or gv) instead of $\rho(g)v$ (for $g \in G$, $v \in V$).

The differential $\mathrm{d}\rho : \mathrm{Lie}(G) \to \mathfrak{gl}(V)$ is a representation of $\mathrm{Lie}(G)$. Occasionally it is convenient to compute the differential as in Section 1.2 Let G be a closed set in $\mathbb{A}^m$; write $K[\mathbb{A}^m] = K[x_1, \dots, x_m]$. Let $\delta \in \mathrm{Lie}(G)$ be given by $\delta(x_i) = a_i$ for $1 \le i \le m$. Write $a = (a_1, \dots, a_m)$. Then $\mathrm{d}\rho(\delta)$ is given by the relation $\rho(e + a\varepsilon) = \rho(e) + \mathrm{d}\rho(\delta)\varepsilon$.

Example 3.7.2 Let $\rho_1 : G \to \mathrm{GL}(V)$ and $\rho_2 : G \to \mathrm{GL}(W)$ be two rational representations of G. Then we form the rational representation $\rho : G \to \mathrm{GL}(V \otimes W)$ by $\rho(g)(v \otimes w) = \rho_1(g)v \otimes \rho_2(g)w$. Let $\delta \in \mathrm{Lie}(G)$ be given by $\delta(x_i) = a_i$, $1 \le i \le m$. Then $\rho(e + a\varepsilon)(v \otimes w) = (\rho_1(e) + \mathrm{d}\rho_1(\delta)\varepsilon)v \otimes (\rho_2(e) +$

$\mathrm{d}\rho_2(\delta)\varepsilon)w = v \otimes w + (\mathrm{d}\rho_1(\delta)v \otimes w + v \otimes \mathrm{d}\rho_2(\delta)w)\,\varepsilon$. Hence $\mathrm{d}\rho(\delta)(v \otimes w) = \mathrm{d}\rho_1(\delta)v \otimes w + v \otimes \mathrm{d}\rho_2(\delta)w$.

Example 3.7.3 Let $\rho : G \to \mathrm{GL}(V)$ be a rational representation, where V has dimension n. For $k \geq 0$ we form the vector space $\mathrm{Sym}^k(V) \subset K[y_1, \ldots, y_n]$ which is spanned by all monomials of degree k. Let $e_1, \ldots, e_n$ be a basis of V, and for $g \in G$ write $\rho(g)e_i = \sum_{j=1}^n g_{ji}e_j$. We define $gy_i = \sum_{j=1}^n g_{ji}y_j$. So G acts on the space spanned by the y_i in the same way it acts on V. Now we construct the rational representation $\psi : G \to \mathrm{GL}(\mathrm{Sym}^k(V))$, by $\psi(g)(y_1^{k_1} \cdots y_n^{k_n}) = (gy_1)^{k_1} \cdots (gy_n)^{k_n}$. The differential of ψ is computed as in Example 3.7.2. It can be described in the following way. Let $\delta \in \mathrm{Lie}(G)$, and let $\mathrm{d}\rho(\delta)$ have matrix (a_{ij}) with respect to the basis $e_1, \ldots, e_n$. Then we define a map $d_\delta : K[y_1, \ldots, y_n] \to K[y_1, \ldots, y_n]$ by $d_\delta(y_i) = \sum_{j=1}^n a_{ji}y_j$, and $d_\delta(f_1f_2) = d_\delta(f_1)f_2 + f_1d_\delta(f_2)$ (i.e., d_δ is a derivation of $K[y_1, \ldots, y_n]$). Then $\mathrm{d}\psi(\delta)$ is the restriction of d_δ to $\mathrm{Sym}^k(V)$.

3.8 Adjoint representation

Let G be an algebraic group, and write $\mathfrak{g} = \mathrm{Lie}(G)$. For $g \in G$ we define the map $\mathrm{Int}(g) : G \to G$ by $\mathrm{Int}(g)(g') = gg'g^{-1}$. Then $\mathrm{Int}(g)$ is an isomorphism of algebraic groups. We denote its differential at e by $\mathrm{Ad}(g)$. So $\mathrm{Ad}(g) : \mathfrak{g} \to \mathfrak{g}$ is an automorphism of $\mathfrak{g}$. We get a map $\mathrm{Ad} : G \to \mathrm{GL}(\mathfrak{g})$, which is called the *adjoint representation* of G.

Let G be an algebraic subgroup of $\mathrm{GL}(n, K)$. Then $\mathrm{Int} : G \to G$ is the restriction of the linear map $M_n(K) \to M_n(K)$, given by $a \mapsto gag^{-1}$. It follows that the differential of $\mathrm{Int}(g)$ is the same map, that is, $\mathrm{Ad}(g)(a) = gag^{-1}$, for $a \in \mathfrak{g}$.

We continue to study the case where $G \subset \mathrm{GL}(n, K)$. We consider the differential of Ad. For that we use the ring $\mathcal{D}$ of dual numbers (Section 1.2). Let $\mathrm{GL}(n, \mathcal{D})$ be the group of of $n \times n$ matrices with coeffients in $\mathcal{D}$ whose determinant is invertible in $\mathcal{D}$. Let I_n denote the $n \times n$ identity matrix. Then for $a \in \mathfrak{gl}(n, K)$, we have $(I_n + a\varepsilon)(I_n - a\varepsilon) = I_n$, from which it follows that $I_n + a\varepsilon$ is invertible and that its inverse is $I_n - a\varepsilon$. As in Section 2.1.2, we denote the adjoint map of $\mathfrak{g}$ by ad. Now let $a, b \in \mathfrak{g}$, then

$$\mathrm{Ad}(I_n + a\varepsilon)(b) = (I_n + a\varepsilon)b(I_n - a\varepsilon) = b - [a, b]\varepsilon = \mathrm{Ad}(I_n)(b) - \mathrm{ad}a(b).$$

In view of (1.3) of Section 1.2.2, we conclude that the differential of Ad is the map $\mathrm{ad} : \mathfrak{g} \to \mathfrak{gl}(\mathfrak{g})$.

Theorem 3.8.1 *Let G be an algebraic group, $\mathfrak{g} = \text{Lie}(G)$. Then the adjoint representation* Ad *is a rational representation of G. If G is given as a subgroup of* $\text{GL}(n, K)$ *then* $\text{Ad}(g)(a) = gag^{-1}$*, for $g \in G$, $a \in \mathfrak{g}$. Furthermore,* $\text{dAd}(\delta)(\gamma) = [\delta, \gamma]$ *for $\gamma, \delta \in \text{Lie}(G)$.*

Proof. We show this above for algebraic subgroups of $\text{GL}(n, K)$. This implies the general case, because G is isomorphic to a closed subgroup of $\text{GL}(n, K)$ for some n by Theorem 3.5.4. □

Now fix a $\gamma \in \mathfrak{g}$ and let $u_\gamma : G \to \mathfrak{g}$ be the map defined by $u_\gamma(g) = \text{Ad}(g)(\gamma)$. This is a regular map, so it has a differential at $e \in G$, $\text{d}u_\gamma : \mathfrak{g} \to \mathfrak{g}$ (note that the tangent space to $\mathfrak{g}$ at any point is just $\mathfrak{g}$ itself).

Proposition 3.8.2 *We have* $\text{d}u_\gamma(\delta) = [\delta, \gamma]$.

Proof. Again, by Theorem 3.5.4 we may assume that $G \subset \text{GL}(n, K)$. Let $a, b \in \mathfrak{g}$, then using the same notation as above, $u_b(I_n + a\varepsilon) = (I_n + a\varepsilon)b(I_n - b\varepsilon) = u_b(I_n) - \text{ad}b(a)\varepsilon$. It follows that the differential of u_b is $-\text{ad}b$, proving the proposition. □

As an application we show that the Lie algebra of a connected solvable algebraic group is solvable as well.

Lemma 3.8.3 *Let G be an algebraic group, and let H_1 and H_2 be algebraic subgroups. Let H be the closure of $[H_1, H_2]$. Then* $\text{Lie}(H)$ *contains all elements $[\delta_1, \delta_2]$ for $\delta_1 \in \text{Lie}(H_1)$, $\delta_2 \in \text{Lie}(H_2)$.*

Proof. Note that by Lemma 3.1.8, H is an algebraic subgroup of G.

Let $h \in H_1$ and define $\alpha_h : G \to G$ by $\alpha_h(g) = hgh^{-1}g^{-1}$. Then $\alpha_h = \text{Int}(h)\iota$, where $\iota(g) = g^{-1}$. Hence by Example 3.7.1, $\text{d}\alpha_h = \text{dInt}(h) + \text{d}\iota$, so that $\text{d}\alpha_h(\delta) = \text{Ad}(h)(\delta) - \delta$. Now observe that α_h maps H_2 into H. So $\text{d}\alpha_h$ maps $\text{Lie}(H_2)$ into $\text{Lie}(H)$. It follows that $\text{Lie}(H)$ contains all elements of the form $\text{Ad}(h)(\delta) - \delta$, where $h \in H_1$ and $\delta \in \text{Lie}(H_2)$. The analogous argument implies that $\text{Lie}(H)$ also contains all those elements with $h \in H_2$ and $\delta \in \text{Lie}(H_1)$.

Now let $\delta_2 \in \text{Lie}(H_2)$, and define $\beta : H_1 \to \text{Lie}(H)$ by $\beta(h) = \text{Ad}(h)(\delta_2) - \delta_2$. By Proposition 3.8.2, the differential satisfies $\text{d}\beta(\delta_1) = [\delta_1, \delta_2]$. The lemma follows. □

Proposition 3.8.4 *Let G be a connected algebraic group. If G is a solvable group,* $\text{Lie}(G)$ *is a solvable Lie algebra.*

Proof. Let $G^{(1)} = G$, $G^{(l+1)} = [G^l, G^l]$ be the derived series of G. By an induction based on Proposition 3.2.4 we see that $G^{(l)}$ are algebraic subgroups. By another induction using Lemma 3.8.3 it follows that the l-th term of the derived series of $\text{Lie}(G)$ is contained in $\text{Lie}(G^{(l)})$. The proposition follows. □

Remark 3.8.5 Also for non-connected G it can be shown that the terms of the derived series are algebraic subgroups (see [Bor91], Section 2.3). So the previous proposition holds for non-connected groups as well.

3.9 Characters and diagonalizable groups

Let G be an algebraic group. A morphism of algebraic groups $\chi : G \to \mathbb{G}_{\mathrm{m}}$ is called a *character* of G. This means that χ is an element of $K[G]$ such that $\chi(gg') = \chi(g)\chi(g')$, and $\chi(g) \neq 0$ for all $g, g' \in G$. An example is $\det : G \to \mathbb{G}_{\mathrm{m}}$, for algebraic matrix groups $G \subset \mathrm{GL}(n, K)$.

Let χ and ψ be two characters of G. Then $\chi\psi$ (with $(\chi\psi)(g) = \chi(g)\psi(g)$) is again a character of G. Hence the set of all characters of G, denoted $X^*(G)$, has the structure of an abelian group. The following lemma holds for groups in general; a proof is contained in [Lan02], Section VI.4.

Lemma 3.9.1 (Dedekind) *Let G be a group and X the set of group homomorphisms $\lambda : G \to K^*$. Then X is a linearly independent set. (This means that if $\lambda_1, \dots, \lambda_s \in X$ are distinct and for some $a_i \in K$ we have $a_1\lambda_1(g) + \dots + a_s\lambda_s(g) = 0$ for all $g \in G$, all a_i are zero.)*

Example 3.9.2 Let $G = \mathrm{D}(n, K)$, the subgroup of $\mathrm{GL}(n, K)$ consisting of the diagonal matrices $\mathrm{diag}(a_1, \dots, a_n)$ with $a_1 \cdots a_n \neq 0$. For $1 \leq i \leq n$ define $\chi_i : G \to \mathbb{G}_{\mathrm{m}}$ by $\chi_i(\mathrm{diag}(a_1, \dots, a_n)) = a_i$, which are characters of G. Furthermore, $K[G] \cong K[\chi_1, \dots, \chi_n, \chi_1^{-1}, \dots, \chi_n^{-1}]$. Every monomial $\chi_1^{k_1} \cdots \chi_n^{k_n}$ where $k_i \in \mathbb{Z}$ is a character of G. The set of these monomials forms a basis of $K[G]$. Therefore Lemma 3.9.1 implies that $X^*(G)$ is precisely the set of these monomials. So as abelian groups, $X^*(G) \cong \mathbb{Z}^n$.

Let $\chi \in X^*(G)$ be defined by $\chi(\mathrm{diag}(a_1, \dots, a_n)) = a_1^{k_1} \cdots a_n^{k_n}$, for certain $k_i \in \mathbb{Z}$. We observe that the Lie algebra $\mathfrak{g}$ of G consists of all diagonal matrices. As in Section 1.2.2 we use dual numbers to compute the differential of χ, $\mathrm{d}\chi : \mathfrak{g} \to K$. We have $\chi(I_n + \varepsilon\mathrm{diag}(\alpha_1, \dots, \alpha_n)) = 1 + \sum_{i=1}^n k_i\alpha_i\varepsilon$. Hence $\mathrm{d}\chi(\mathrm{diag}(\alpha_1, \dots, \alpha_n)) = \sum_i k_i\alpha_i$.

An algebraic group G is said to be a *diagonalizable* if G is isomorphic to a subgroup of $\mathrm{D}(n, K)$ for some n. A connected diagonalizable algebraic group is called a *torus*. So if G is diagonalizable, then G° is a torus.

Proposition 3.9.3 *Let G be a linear algebraic group. The following are equivalent:*

(i) *G is diagonalizable,*

(ii) *$X^*(G)$ is a (vector space-) basis of $K[G]$,*

(iii) *every rational representation of G is a direct sum of 1-dimensional rational representations.*

Proof. To see that (i) implies (ii) we may assume that G is a subgroup of $\mathrm{D}(n,K)$. The restriction homomorphism $K[\mathrm{D}(n,K)] \to K[G]$ is surjective. So $K[G]$ is spanned by the restrictions of the characters of $\mathrm{D}(n,K)$. But these are characters of G. So $X^*(G)$ spans $K[G]$ and by Lemma 3.9.1 is a linearly independent set. The first implication is proved.

Suppose that (ii) holds. Let $\rho : G \to \mathrm{GL}(V)$ be a rational representation of G. By writing elements of $\mathrm{GL}(V)$ as matrices with respect to a fixed basis of V, we see that there are linear maps $L_i \in \mathrm{End}(V)$ and $f_i \in K[G]$ such that $\rho(g) = \sum_i f_i(g)L_i$, for $g \in G$. Since $X^*(G)$ is a basis of $K[G]$ we can rewrite this as $\rho(g) = \sum_{i=1}^r \chi_i(g)A_i$, where $\chi_i \in X^*(G)$ and $A_i \in \mathrm{End}(V)$. Observe that, for $\chi, \psi \in X^*(G)$, the map $(g,h) \mapsto \chi(g)\psi(h)$ is a character of the algebraic group $G \times G$. Expanding the relation $\rho(gh) = \rho(g)\rho(h)$, and using Lemma 3.9.1 for the characters of $G \times G$, we obtain that $A_i^2 = A_i$ for all i and $A_iA_j = 0$ for $i \neq j$. Also, by evaluating ρ at the identity element of G we see that $\sum_{i=1}^r A_i = I_V$ (the latter is the identity map on V). Set $V_i = A_iV$, then by what has just been shown, V is the direct sum of the V_i, and $g \in G$ acts as multiplication by $\chi_i(g)$ on V_i. This implies (iii).

For the final implication we consider an isomorphism of algebraic groups $\rho : G \to H \subset \mathrm{GL}(V)$ (see Theorem 3.5.4). By (iii) there is a basis of V such that all $\rho(g)$, for $g \in G$, are represented by diagonal matrices. So H is isomorphic to an algebraic subgroup of $\mathrm{D}(n,K)$ (where $n = \dim V$). Therefore G is diagonalizable. □

Example 3.9.4 Consider

$$G = \left\{ \begin{pmatrix} x & y \\ -y & x \end{pmatrix} \mid x^2 + y^2 \neq 0 \right\} \subset \mathrm{GL}(2,\mathbb{C}).$$

Let $\imath \in \mathbb{C}$ denote the imaginary unit, and set

$$B = \begin{pmatrix} -\imath & -1 \\ -\imath & 1 \end{pmatrix} \text{ and } g = \begin{pmatrix} x & y \\ -y & x \end{pmatrix} \in G.$$

We have $BgB^{-1} = \mathrm{diag}(x + \imath y, x - \imath y)$. We see that G is diagonalizable, and, since this yields an isomorphism with $\mathrm{D}(2,\mathbb{C})$, even a torus.

Proposition 3.9.5 *Let G be a diagonalizable algebraic group and $H \subset G$ an algebraic subgroup. Then H is also diagonalizable. Moreover, H is the intersection of the kernels of a finite number of characters of G.*

Proof. The first statement is obvious. Consider the restriction map ϕ : $K[G] \to K[H]$, which is surjective. Let $J \subset K[G]$ be the ideal generated by all $\chi - 1$ where $\chi \in X^*(G)$ is such that $\chi(H) = 1$. Then J is contained in the kernel of ϕ. Let $\chi_1, \dots, \chi_r \in X^*(G)$ be linearly independent modulo J, and suppose $\phi(a_1\chi_1 + \cdots + a_r\chi_r) = 0$, where not all a_i are zero. Choose the χ_i such that r is minimal. Write $\psi_i = \phi(\chi_i)$. If $\psi_1 = \psi_r$ then $\phi(a_2\chi_2 + \cdots + a_{r-1}\chi_{r-1} + (a_1 + a_r)\chi_r) = 0$, contradicting the minimality of r. Therefore there exists a $h_0 \in H$ with $\psi_1(h_0) \neq \psi_r(h_0)$. For $h \in H$ we evaluate $\sum_i a_i\psi_i$ in h_0h and multiply $\sum_i a_i\psi_i(h)$ by $\psi_r(h_0)$. Then we subtract and find that $\phi(\sum_{i=1}^{r-1} a_i(\psi_i(h_0) - \psi_r(h_0))\chi_i) = 0$, again contradicting the minimality of r. We conclude that J is the kernel of ϕ. So using Theorem 1.1.3 we see that H is defined as a subgroup of G by a finite number of equations $\chi(h) = 1$, where $\chi \in X^*(G)$. □

Example 3.9.6 We use the notation of Example 3.9.2. As abelian group $X^*(G)$ is isomorphic to $\mathbb{Z}^n$ by $e = (e_1, \dots, e_n) \mapsto \chi_e = \chi_1^{e_1} \cdots \chi_n^{e_n}$. For a $C \subset \mathbb{Z}^n$ we set

$$\mathrm{D}(C) = \{g \in G \mid \chi_e(g) = 1 \text{ for all } e \in C\}.$$

We observe that $\mathrm{D}(C) = \mathrm{D}(\Lambda)$, where $\Lambda \subset \mathbb{Z}^n$ is the lattice generated by C. Let $H \subset G$ be an algebraic subgroup. Then by Proposition 3.9.5 there is a lattice $\Lambda \subset \mathbb{Z}^n$ with $H = \mathrm{D}(\Lambda)$.

Proposition 3.9.7 *Let $\Lambda \subset \mathbb{Z}^n$ be a lattice of rank r, and $\mathrm{D}(\Lambda)$ as in the previous example. Then $\mathrm{D}(\Lambda) = F \times H$, where F is a finite group, and H is an r-dimensional torus. If Λ is pure (see Section 6.2), F is trivial and $\mathrm{D}(\Lambda)$ is a torus. If the base field is of characteristic 0, and $\mathrm{D}(\Lambda)$ is a torus, Λ is pure and F is trivial.*

Proof. For an $n \times n$ integral matrix A we define the map $\varphi_A : \mathrm{D}(\Lambda) \to \mathrm{D}(\Lambda)$ by $\varphi_A(\mathrm{diag}(a_1, \dots, a_n)) = \mathrm{diag}(b_1, \dots, b_n)$ with $b_i = \prod_{j=1}^n a_j^{A(j,i)}$. Then φ_A is a morphism of algebraic groups. A short calculation shows that $\varphi_A \circ \varphi_B = \varphi_{BA}$. Therefore, if $\det(A) = \pm 1$, φ_A is invertible and $\varphi_A^{-1} = \varphi_{A^{-1}}$. Furthermore, if $e \in \mathbb{Z}^n$, and χ_e is as in Example 3.9.6, then $\chi_e \circ \varphi_A = \chi_{Ae}$.

By Corollary 6.2.2 there are positive integers $d_1, \dots, d_r$ and a basis $v^1, \dots, v^n$ of $\mathbb{Z}^n$ such that $d_1v^1, \dots, d_rv^r$ is a basis of Λ. In particular, $\varphi_A(\mathrm{D}(\Lambda))$ consists of all $g = \mathrm{diag}(a_1, \dots, a_n)$ such that $1 = \chi_{d_iv^i} \circ \varphi_A^{-1}(g) = \chi_{d_iA^{-1}v^i}$ for $1 \leq i \leq r$. Now define A to be the matrix with columns $v^1, \dots, v^n$. Let $e^1, \dots, e^n$ denote the standard basis of $\mathbb{Z}^n$. Then $A^{-1}v^i = e^i$, $1 \leq i \leq n$. Furthermore, $\varphi_A(\mathrm{D}(\Lambda))$ consists of all $g = \mathrm{diag}(a_1, \dots, a_n)$ such that $1 = \chi_{d_ie^i}(g)$ for $1 \leq i \leq r$. It follows that $\varphi_A(\mathrm{D}(\Lambda))$ consists of all $\mathrm{diag}(a_1, \dots, a_n)$ such that $a_1 \cdots a_n \neq 0$ and $a_i^{d_i} = 1$. As φ_A is an isomorphism, the proof is finished. □

Remark 3.9.8 If the characteristic of the base field is 0, the above proof also shows that F is isomorphic to the torsion subgroup of $\mathbb{Z}^n/\Lambda$.

Remark 3.9.9 If the characteristic is $p > 0$, it is possible that $D(\Lambda)$ is a torus while Λ is not pure. For an example, take the lattice spanned by $(p, 0)$. In that case the purification Λ' of Λ defines the same group, i.e., $D(\Lambda) = D(\Lambda')$. So we can always assume that a torus is defined by a pure lattice.

Now let $T \subset D(n, K)$ be a torus, and let $\Lambda \subset \mathbb{Z}^n$ be the pure lattice with $T = D(\Lambda)$. We call Λ the *defining lattice* of T. As Λ is pure, there is a complementary lattice $\Lambda_c \subset \mathbb{Z}^n$ such that $\mathbb{Z}^n = \Lambda \oplus \Lambda_c$. From Proposition 3.9.3 it follows that a character of T is the restriction of a character of $D(n, K)$ to T. Therefore, $X^*(T)$ consists of the restrictions of χ_e for $e \in \Lambda_c$ (and these are all distinct characters of T). So we have an isomorphism $\Lambda_c \to X^*(T)$. We also note that, given a basis of Λ, it is possible to compute a basis of Λ_c by a Smith normal form computation; see Corollary 6.2.2.

A *cocharacter* of T is a morphism of algebraic groups $\mathbb{G}_{\mathrm{m}} \to T$. The set of cocharacters of T is denoted $X_*(T)$. It is an abelian group with group operation $(\lambda\mu)(t) = \lambda(t)\mu(t)$. Let $\lambda \in X_*(T)$, then $\lambda(t) = \mathrm{diag}(a_1(t), \dots, a_n(t))$, and each a_i is a character of $\mathbb{G}_{\mathrm{m}}$. Therefore there are integers m_i such that $a_i(t) = t^{m_i}$, $1 \leq i \leq n$. Furthermore, $\lambda(t) \in T$ if and only if $\sum_i e_i m_i = 0$ for all $e = (e_1, \dots, e_n) \in \Lambda$. Defining

$$\Lambda^{\perp} = \{(m_1, \dots, m_n) \in \mathbb{Z}^n \mid \sum_{i=1}^{n} e_i m_i = 0 \text{ for all } e = (e_1, \dots, e_n) \in \Lambda\},$$

we get an isomorphism $\Lambda^{\perp} \to X_*(T)$. In particular, $X^*(T)$ and $X_*(T)$ are abelian groups of the same rank, equal to the dimension of T. We also note that by the purification algorithm of Section 6.2 we can compute a basis of $\Lambda^{\perp}$.

The differential of the above cocharacter λ is given by $d\lambda(\alpha) = \alpha\mathrm{diag}(m_1, \dots, m_n)$. Let $m^1, \dots, m^s$ be a basis of $\Lambda^{\perp}$, corresponding to the cocharacters $\lambda_1, \dots, \lambda_s$. Define the morphism $\varphi : \mathbb{G}_{\mathrm{m}}^s \to T$ by $\psi(t_1, \dots, t_s) = \lambda_1(t_1) \cdots \lambda_s(t_s)$. By Example 3.7.1, $d\psi = d\lambda_1 + \cdots + d\lambda_s$. It follows that if the m^i are linearly independent over K, then $\varphi(\mathbb{G}_{\mathrm{m}}^s)$ is an algebraic subgroup of T of the same dimension as T. Since both are connected we must have $T = \varphi(\mathbb{G}_{\mathrm{m}}^s)$ (Theorem 1.3.4). In particular this holds if the characteristic of K is 0.

3.10 Jordan decomposition

In this section we show that the Lie algebra of an algebraic subgroup of $\mathrm{GL}(n, K)$ is closed under Jordan decompositions. Secondly, in $\mathrm{GL}(n, K)$ we define a multiplicative Jordan decomposition where an element is decomposed as the product of a semisimple and unipotent endomorphism. We prove that an algebraic subgroup of $\mathrm{GL}(n, K)$ is closed under this decomposition.

Throughout we set $G = \mathrm{GL}(n, K)$ and write $R_n = K[x_{11}, \ldots, x_{nn}]$. Recall that $K[G] = K[x_{ij}, \frac{1}{\det(x_{ij})}]$ and write $d = \frac{1}{\det(x_{ij})}$ (Example 3.1.3). Let $W_{l,t}$ be the space consisting of all fd^t, where $f \in R_n$ is homogeneous of degree l. Then $K[G]$ is the sum of the spaces $W_{l,t}$ (note that they may have non-zero intersection, due to the relation $d\det(x_{ij}) = 1$). We use the notation introduced in Section 3.4.

3.10.1 In the Lie algebra of an algebraic group

For $a \in \mathfrak{gl}(n, K)$ we denote the corresponding element of $\mathrm{Lie}(G)$ by δ_a, so $\delta_a(x_{ij}) = a_{ij}$. In Section 3.4 we showed that $\delta_a(d) = -\mathrm{Tr}(a)$. For $g \in G$ we have $\lambda_g(d) = \frac{1}{\det(g)}d$. Hence $(d * \delta_a)(g) = \delta_a(\lambda_g(d)) = -\mathrm{Tr}(a)/\det(g) = -\mathrm{Tr}(a)d(g)$. We conclude that $d*\delta_a = -\mathrm{Tr}(a)d$. Since $f \mapsto f*\delta_a$ is a derivation of $K[G]$ we obtain

$$fd^t * \delta_a = (f * \delta_a)d^t - t\mathrm{Tr}(a)fd^t. \tag{3.3}$$

As seen in Section 3.4, $\Delta(x_{ij}) = \sum_s x_{is} \otimes x_{sj}$, so by (3.1) we have $x_{ij} * \delta_a = \sum_s x_{is}a_{sj}$. Hence the entry on position (i, j) of the product of matrices $(x_{ij})a$ is $x_{ij} * \delta_a$. Since $f \mapsto f * \delta_a$ is a derivation of $K[G]$ this determines $f * \delta_a$ for all $f \in R_n$. In particular we see that if f is homogeneous of degree l, the same holds for $f * \delta_a$. So in view of (3.3), $*\delta_a$ leaves the spaces $W_{l,t}$ invariant.

Lemma 3.10.1 *If a is semisimple (nilpotent), the restriction of $*\delta_a$ to $W_{l,t}$ is semisimple (nilpotent).*

Proof. First let a be semisimple. Suppose $t = 0$. There is a basis of $W_{1,0}$ (i.e., the space of linear polynomials in the x_{ij}) consisting of eigenvectors, $\psi_1, \ldots, \psi_{n^2}$, of $*\delta_a$. Indeed, if $a = b^{-1}\mathrm{diag}(\lambda_1, \ldots, \lambda_n)b$, we can take the entries of $(x_{ij}) \cdot b^{-1}$. So the restriction of δ_a to $W_{1,0}$ is semsisimple. The monomials in the ψ_{ij} of degree l form a basis of $W_{l,0}$. But those monomials are also eigenvectors of $*\delta_a$. So $*\delta_a|_{W_{l,0}}$ is semisimple. The general case now follows from (3.3).

The case where a is nilpotent is analogous. Now $W_{1,0}$ has a basis $\psi_1, \ldots, \psi_{n^2}$ such that $\psi_i * \delta_a = \sum_{j>i} \alpha_{ij}\psi_j$. From this it follows that the restriction of $*\delta_a$ to $W_{l,0}$ is nilpotent. The general case follows again from (3.3) because $\mathrm{Tr}(a) = 0$ as a is nilpotent. □

Theorem 3.10.2 *Let $a \in \mathfrak{gl}(n, K)$ and $a = s+n$ be its Jordan decomposition. Let $H \subset G$ be an algebraic subgroup with $a \in \mathrm{Lie}(H)$. Then $s, n \in \mathrm{Lie}(H)$. (Here we identify an $x \in \mathfrak{gl}(n, K)$ with δ_x.)*

Proof. In view of Proposition 2.2.3, Lemma 3.10.1 implies that $*\delta_a|_{W_{l,t}} = *\delta_s|_{W_{l,t}} + *\delta_n|_{W_{l,t}}$ is the Jordan decomposition of $*\delta_a|_{W_{l,t}}$. Let $f \in \mathcal{I}(H)$. Then f lies in a sum of a finite number of spaces $W_{l,t}$; denote this sum by V.

Set $U = V \cap \mathcal{I}(H)$. Then by Proposition 3.5.1, $U * \delta_a \subset U$. Since the semisimple and nilpotent parts (in the Jordan decomposition) of a linear map leave invariant the same subspaces as the linear map itself (Proposition 2.2.4), we have $U * \delta_s \subset U$, and similarly for δ_n. So $f * \delta_s, f * \delta_n \in \mathcal{I}(H)$. Therefore, by Proposition 3.5.1, $s, n \in \mathrm{Lie}(H)$. □

3.10.2 In an algebraic group

Let k be a field. A $u \in \mathrm{GL}(n,k)$ is said to be *unipotent* if $u - I_n$ is nilpotent (where I_n denotes the $n \times n$ identity matrix).

Lemma 3.10.3 *Suppose that k is perfect. For $g \in \mathrm{GL}(n,k)$ there are unique $s, u \in \mathrm{GL}(n,k)$ such that s is semisimple, u is unipotent, $su = us$ and $g = su$. Moreover, s and u leave invariant the same subspaces of $V = k^n$ as g does.*

Proof. Let $g = s+n$ be the Jordan decomposition of g (Proposition 2.2.3). As g is invertible, 0 is not an eigenvalue of g. From the proof of Proposition 2.2.3 we infer that the square-free part of the minimal polynomial of g annihilates s. Therefore the minimal polynomial of s divides this square-free part. So 0 is not an eigenvalue of s either, and hence s is invertible as well. So s and $u = I_n + s^{-1}n$ have the required properties. Uniqueness follows from Proposition 2.2.3, because if s and u are as in the lemma, then $g = s + s(u - I_n)$ is the Jordan decomposition of g. □

The decomposition of the lemma is called the *multiplicative Jordan decomposition* of g. The elements s and u are respectively called the *semisimple* and *unipotent* parts of g. There is an obvious algorithm to compute the multiplicative Jordan decomposition: first compute the additive Jordan decomposition $g = s + n$ (see Section 2.2), then return s and $I_n + s^{-1}n$.

Theorem 3.10.4 *Let $H \subset G$ be an algebraic subgroup with $h \in H$. Let $h = su$ be its multiplicative Jordan decomposition. Then $s, u \in H$.*

Proof. The proof is very similar to the one of Theorem 3.10.2, we only indicate the main steps.

Here, for $g \in \mathrm{GL}(n,K)$, we use the map $\rho_g : K[G] \to K[G]$ defined by $\rho_g(f)(g') = f(g'g)$. These ρ_g are algebra automorphisms, with the property that $g \in H$ if and only if $\rho_g(\mathcal{I}(H)) = \mathcal{I}(H)$.

The first step is to show that if $g \in \mathrm{GL}(n,K)$ is semisimple or unipotent, the same holds for the restriction of ρ_g to $W_{l,t}$. This is done in the same way as in the proof of Lemma 3.10.1. Indeed, suppose that g is semisimple. Using the basis used in the mentioned proof shows that $\rho_g|_{W_{1,0}}$ is semisimple, and we continue in the same way.

As in the proof of Theorem 3.10.2, it follows that $\rho_g|_{W_{l,t}} = (\rho_s|_{W_{l,t}})(\rho_u|_{W_{l,t}})$ is the multiplicative Jordan decomposition of $\rho_g|_{W_{l,t}}$. In the same way, $\rho_s(U) \subset U$, implying $s \in H$. Similarly we infer that $u \in H$. □

3.11 The unipotent radical

Here we briefly indicate how the unipotent radical of an algebraic group is defined. However, we omit most proofs of the results that are necessary to underpin this definition.

Let G be an algebraic group. Let $R(G)$ be its radical (Section 3.2). By Theorem 3.5.4, there is a map $\rho : G \to \mathrm{GL}(n, K)$ providing an isomorphism of G onto an algebraic subgroup of $\mathrm{GL}(n, K)$. Let $S(n, K)$ denote the algebraic subgroup of $\mathrm{GL}(n, K)$ consisting of all upper triangular matrices. By the Lie-Kolchin theorem ([Bor91], Corollary 10.5), $\rho(R(G))$ can be conjugated into $S(n, K)$. So we may assume that $\rho(R(G)) \subset S(n, K)$. Denote by $R_u(G)$ the set of elements $g \in R(G)$ such that $\rho(g)$ is unipotent. Then $R_u(G)$ does not depend on the choice of ρ ([Bor91], Theorem 4.4). It is called the *unipotent radical* of G. By [Bor91], Theorem 10.6, $R_u(G)$ is a connected algebraic subgroup of G. It is obviously a normal subgroup.

The algebraic group G is said to be *reductive* if $R_u(G)$ is trivial.

Regarding the computation of the unipotent radical of an algebraic group, the same remarks hold as for the radical $R(G)$ (Section 3.2): in general this seems to be a difficult problem, but in characteristic 0 there is an algorithm using the Lie algebra, based on Theorem 4.3.22.

3.12 Algebraic groups acting on closed sets

Let G be an algebraic group and $X \subset \mathbb{A}^n$ be a closed set. We say that G acts on X if there is a regular map $\sigma : G \times X \to X$ such that

1. $\sigma(e, v) = v$ for all $v \in X$,
2. $\sigma(g_1, \sigma(g_2, v)) = \sigma(g_1 g_2, v)$ for all $v \in X$, and $g_1, g_2 \in G$.

If the map σ is understood, we also write $g \cdot v$ (or even gv) instead of $\sigma(g, v)$.

The orbit of a $v \in X$ is the set $G \cdot v = \{g \cdot v \mid g \in G\}$.

Examples are easily constructed. We can set $X = G$ and $g \cdot g' = gg'g^{-1}$ (Section 3.8). Here the orbits are the conjugacy classes of G. Another class of examples is constructed by using a rational representation $\rho : G \to \mathrm{GL}(V)$, and setting $X = V$.

For a $v \in X$ we consider the closure $\overline{Gv}$ of the orbit of v. A first observation is that $\overline{Gv}$ is stable under the action of G. Indeed, let $p_1, \dots, p_r \in K[X]$ define $\overline{Gv}$. For $g \in G$ define $p_i^g \in K[X]$ by $p_i^g(w) = p_i(gw)$. Then p_i^g vanishes on Gv, hence also on $\overline{Gv}$. This implies that $gw \in \overline{Gv}$ for all $w \in \overline{Gv}$.

Lemma 3.12.1 *Let G be a connected algebraic group, acting on the closed set X. Let $v \in X$. Then the orbit Gv is open in its closure $\overline{Gv}$.*

Proof. The map $G \to Gv$, $g \mapsto gv$ is regular. So the closure of its image, $\overline{Gv}$, is irreducible (Lemma 1.1.6). Hence Gv contains a non-empty open set U of $\overline{Gv}$ by Theorem 1.4.4. Hence Gv is the union of the open sets gU, $g \in G$. It follows that Gv is open in $\overline{Gv}$ as well. □

Example 3.12.2 Let $G = \mathrm{SL}(2, K)$ act by conjugation on the set $X = M_2(K)$. For $g = \begin{pmatrix} a & b \\ c & d \end{pmatrix}$, and $v = \begin{pmatrix} 0 & 1 \\ 0 & 0 \end{pmatrix}$ we have $gvg^{-1} = \begin{pmatrix} -ac & a^2 \\ -c^2 & ac \end{pmatrix}$. The closure of the orbit of v is defined by the polynomials $x_{11} + x_{22}$, $x_{11}x_{22} - x_{12}x_{21}$. We see that the zero matrix lies in the closure, but not in the orbit. In fact, Gv consists of the non-zero elements of $\overline{Gv}$ so it is open in its closure.

There are some immediate algorithms for computing polynomials defining the closure of a given orbit. However, we defer a discussion of those to Section 4.8.

3.13 Specification of an algebraic group

When computing with algebraic objects we must decide how the input and output of the algorithms are defined, that is, how the various objects are represented by a finite amount of data.

For finite-dimensional Lie algebras this is most conveniently done by a basis and a multiplication table (see Section 2.1.3). If the Lie algebra in question comes from an algebraic subgroup of $\mathrm{GL}(n, K)$, its basis will consist of elements of $\mathfrak{gl}(n, K)$. Moreover, if the algebraic group is defined over a subfield k, the basis elements can be chosen to lie in $\mathfrak{gl}(n, k)$.

For algebraic groups the situation is not so clear. The most obvious idea is to consider subgroups of $\mathrm{GL}(n, K)$, and represent them by a finite number of defining polynomials as in Example 3.1.5. By Theorem 3.5.4 any algebraic group can be presented in that way. Furthermore, when using Gröbner basis techniques to tackle computational questions, the defining polynomials are invariably needed. However, there are also algorithms for which the input is most conveniently given in a different way. For example, in characteristic 0, a connected algebraic group is completely determined by its Lie algebra (Theorem 4.2.2). This makes it possible to represent a connected algebraic group in characteristic 0 by a basis of its Lie algebra. On many occasions this is much more concise than the specification by a list of polynomials; compare Examples 3.1.5 and 3.4.4.

Yet another way to specify an algebraic group is by a finite set Γ of elements of $\mathrm{GL}(n, K)$: the algebraic group given by Γ is defined to be the closure of the group generated by Γ. (Note that by Lemma 3.1.8 this is an algebraic group). In Section 4.7 an algorithm will be indicated to find such a set Γ for an algebraic group in characteristic 0, for which we know a set of defining polynomials and its Lie algebra. The same section contains applications to centralizer and normalizer computations.

A fourth method that we mention here is to specify a semisimple (or somewhat more generally, reductive) algebraic group by a based root datum (see Section 5.4). In this way, a semisimple algebraic group is determined by two small integral matrices. So this is very concise. On the other hand, it can only be applied to algebraic groups of the mentioned type. Chapter 5 has algorithms for algebraic groups given by such a specification.

We remark that the questions that are of interest can differ with the way of specifying the group. If the group is given by a set of polynomial equations, it can be important to determine its Lie algebra, to see whether it is connected or not, whether it is solvable or not, and so on. On the other hand, if the group is semisimple and given by its root datum, the answers to most questions of that kind follow easily from the theory. In that case, of interest are other problems, like the determination of the dimensions of the irreducible representations (see Section 5.5).

3.14 Notes

Most of the material in this chapter is fairly standard. Our main references for the theory of algebraic groups are [Bor91], [Hum75], [Spr98], [TY05]. None of these books use the "naive" definition of algebraic group. However, because also with the more conceptual definition Theorem 3.5.4 is valid, both definitions yield the same class of groups.

Example 3.6.3 is due to Chevalley ([Che51], Section II.10). The treatment in Example 3.6.4 closely follows [Che51], Section II.7. The proof of Proposition 3.9.7 follows [BG06], Chapter 3.

Chapter 4

Algebraic Groups and Their Lie Algebras

The theme of this chapter is the correspondence between algebraic groups and Lie algebras, which we call the Lie correspondence. This correspondence works very well in characteristic 0, but is notably less powerful in positive characteristic. For this reason all base fields in this chapter will be assumed to be of characteristic 0 (unless otherwise stated). As in the previous chapters, K denotes an algebraically closed field, whereas $k, k', \dots$ denote not necessarily algebraically closed fields.

The first three sections contain a number of results on the Lie correspondence, as well as on algebraic Lie algebras (that is, Lie algebras that arise from algebraic groups). One of the cornerstones of this theory is the unique smallest algebraic group whose Lie algebra contains a given element (shown in Section 4.1). The remaining sections are all devoted to algorithmic problems (computing the smallest algebraic Lie algebra containing a given Lie algebra, computing defining polynomials of an algebraic group given its Lie algebra, computing the smallest algebraic group containing a given set of elements, computing a set of elements "generating" a given algebraic group and applications of that to centralizer and normalizer computations and computing the closure of an orbit of an algebraic group).

One more word about notation: we denote the elements of the Lie algebra of an algebraic group G as they lie in $K[G]^*$ by symbols like δ, γ (as in Chapter 3), but when the group is an algebraic subgroup of $\mathrm{GL}(n, K)$, and therefore the Lie algebra is viewed as a subalgebra of $\mathfrak{gl}(n, K)$, we prefer to denote its elements by symbols like $a, b, \dots$.

4.1 $G(\delta)$

Let G be an algebraic group, and $\delta \in \mathrm{Lie}(G)$. In this section we show that there exists a unique smallest algebraic subgroup $G(\delta) \subset G$ with the property that $\delta \in \mathrm{Lie}(G(\delta))$. We make use of the concepts and notation introduced in Section 3.4.

Consider $K[G]^*$ with the associative multiplication defined in Section 3.4.

In this algebra we can form the elements δ^n (with $\delta^0 = \varepsilon$). Set

$$J(\delta) = \{f \in K[G] \mid \delta^n(f) = 0 \text{ for all } n \geq 0\}.$$

Also from Section 3.4 we recall the map $\eta : \mathrm{Lie}(G) \to \mathrm{Der}_\ell(K[G])$, $\eta(\gamma) = D_\gamma$, where $D_\gamma(f) = f * \gamma$. We write $D_\delta^n = D_\delta \circ \cdots \circ D_\delta$ (n factors D_δ).

Lemma 4.1.1 *For $n \geq 0$, $f \in K[G]$ we have $\delta^n(f) = D_\delta^n(f)(e)$.*

Proof. We use induction on n. For $n = 0$ the statement is clear. The case $n = 1$ follows from the definition of the right convolution. Suppose $n > 1$ and write $\Delta(f) = \sum_i f_i \otimes f_i'$; then $D_\delta^n(f)(e) = D_\delta^{n-1}(f * \delta)(e) = \sum_i \delta(f_i')(D_\delta^{n-1}(f_i))(e) = \delta^n(f)$. In the second to last equality we used (3.1), and in the last equality we used induction. □

Lemma 4.1.2 $D_\delta(J(\delta)) \subset J(\delta)$.

Proof. Let $f \in J(\delta)$. Then by the previous lemma we get $\delta^n(D_\delta(f)) = D_\delta^n(D_\delta(f))(e) = D_\delta^{n+1}(f)(e) = \delta^{n+1}(f) = 0$. □

Lemma 4.1.3 *$J(\delta)$ is a prime ideal of $K[G]$.*

Proof. First we show that $\delta^n(fh) = 0$ for all $f \in J(\delta), h \in K[G]$, using induction on n. For $n = 0$ this is straightforward. For $n \geq 0$ we get, using Lemma 4.1.1,

$$\begin{aligned}\delta^{n+1}(fh) &= D_\delta^n(D_\delta(fh))(e) = D_\delta^n(D_\delta(f)h + fD_\delta(h))(e)\\ &= \delta^n(D_\delta(f)h) + \delta^n(fD_\delta(h)).\end{aligned}$$

By Lemma 4.1.2, $D_\delta(f) \in J(\delta)$. So both summands are zero by induction. We conclude that $J(\delta)$ is an ideal.

Now we show that $J(\delta)$ is prime. Let $f, h \in K[G]$ be such that neither lies in $J(\delta)$. Let s, t be the smallest integers with $\delta^s(f) \neq 0 \neq \delta^t(h)$. Then by the Leibniz rule for derivations we get

$$\begin{aligned}\delta^{s+t}(fh) = D_\delta^{s+t}(fh)(e) &= \left[\sum_{i=0}^{s+t} \binom{s+t}{i} D_\delta^i(f) D_\delta^{s+t-i}(h)\right](e)\\ &= \sum_{i=0}^{s+t} \binom{s+t}{i} \delta^i(f)\delta^{s+t-i}(h) = \binom{s+t}{s} \delta^s(f)\delta^t(h) \neq 0,\end{aligned}$$

and $fh \notin J(\delta)$. Therefore it is a prime ideal. □

Now set $G(\delta) = \{g \in G \mid f(g) = 0 \text{ for all } f \in J(\delta)\}$.

Lemma 4.1.4 *Let $f \in J(\delta)$ and $g \in G(\delta)$. Then $\lambda_g(f) \in J(\delta)$.*

Proof. Recall that D_δ is left invariant. So, using Lemma 4.1.1,

$$\delta^n(\lambda_g(f)) = D_\delta^n(\lambda_g(f))(e) = \lambda_g(D_\delta^n(f))(e) = D_\delta^n(f)(g).$$

But by Lemma 4.1.2 this is zero. □

Theorem 4.1.5 *$G(\delta)$ is a connected algebraic group, with $\delta \in \mathrm{Lie}(G(\delta))$. Moreover if $H \subset G$ is an algebraic subgroup with $\delta \in \mathrm{Lie}(H)$, then $G(\delta) \subset H$.*

Proof. Because $J(\delta)$ is prime, $G(\delta)$ is connected.

Since $f(e) = \delta^0(f) = 0$ for all $f \in J(\delta)$, we have $e \in G(\delta)$. Let $g, g' \in G(\delta)$, and $f \in J(\delta)$. Then $f(gg') = \lambda_g(f)(g')$, and by Lemma 4.1.4 this is zero. Therefore $gg' \in G(\delta)$. Also by Corollary 3.5.3 along with Lemma 4.1.4, λ_g is invertible on $J(\delta)$. Moreover, the inverse has to be $\lambda_{g^{-1}}$. So for $f \in J(\delta)$, $f(g^{-1}) = \lambda_{g^{-1}}(f)(e) = 0$. Hence $g^{-1} \in G(\delta)$. We conclude that $G(\delta)$ is a group.

Since $J(\delta) * \delta \subset J(\delta)$ (Lemma 4.1.2) we get $\delta \in \mathrm{Lie}(G(\delta))$ (Proposition 3.5.1). Suppose $\delta \in \mathrm{Lie}(H)$. Write $I = \mathcal{I}(H)$. Then by Proposition 3.5.1, $D_\delta^n(I) \subset I$. So for $f \in I$ we have $\delta^n(f) = D_\delta^n(f)(e) = 0$. It follows that $I \subset J(\delta)$ and hence $G(\delta) \subset H$. □

Corollary 4.1.6 *Let G be an algebraic group and $\mathfrak{h} \subset \mathrm{Lie}(G)$ be a subalgebra. Let $G(\mathfrak{h})$ denote the intersection of all algebraic subgroups H of G such that $\mathrm{Lie}(H)$ contains $\mathfrak{h}$. Then $G(\mathfrak{h})$ is a connected algebraic group, with $\mathfrak{h} \subset \mathrm{Lie}(G(\mathfrak{h}))$. If, moreover, there is a connected algebraic subgroup $H \subset G$ with $\mathrm{Lie}(H) = \mathfrak{h}$, then $H = G(\mathfrak{h})$.*

Proof. It is clear that $G(\mathfrak{h})$ is a connected algebraic subgroup of G. Let $\delta \in \mathfrak{h}$. Let $H' \subset G$ be an algebraic subgroup with $\mathfrak{h} \subset \mathrm{Lie}(H')$. Then also $\delta \in \mathrm{Lie}(H')$ and by the previous theorem, $G(\delta) \subset H'$. Therefore, $G(\delta) \subset G(\mathfrak{h})$. Hence $\mathrm{Lie}(G(\delta)) \subset \mathrm{Lie}(G(\mathfrak{h}))$, and we conclude $\mathfrak{h} \subset \mathrm{Lie}(G(\mathfrak{h}))$.

Let H be such that $\mathrm{Lie}(H) = \mathfrak{h}$. Then certainly $G(\mathfrak{h}) \subset H$. Then $\mathfrak{h} \subset \mathrm{Lie}(G(\mathfrak{h})) \subset \mathrm{Lie}(H) = \mathfrak{h}$. Hence $\mathrm{Lie}(G(\mathfrak{h})) = \mathrm{Lie}(H)$. Now we conclude by Theorem 1.3.4. □

4.2 The Lie correspondence

In this section we show some of the main results that make the Lie correspondence so effective in characteristic 0. Some of the highlights are Theorem

4.2.2 by which a connected algebraic group is uniquely determined by its Lie algebra and Corollary 4.2.8 (and its consequences) relating several properties of a rational representation to corresponding properties of its differential.

Theorem 4.2.1 *Let $\sigma : G \to H$ be a surjective morphism of algebraic groups. Then* $d\sigma : \mathrm{Lie}(G) \to \mathrm{Lie}(H)$ *is surjective as well.*

Proof. According to Theorem 1.4.3 there is a $g \in G$ such that $d_g\sigma : T_g(G) \to T_{\sigma(g)}(H)$ is surjective. The maps $\phi : G \to G$, $\phi(g') = gg'$ and $\psi : H \to H$, $\psi(h') = \sigma(g)h'$ are isomorphisms. So their differentials are isomorphisms as well (Corollary 1.2.4). Also, $\sigma\phi = \psi\sigma$. So by Lemma 1.2.3, $d_g\sigma \circ d_{e_G}\phi = d_{e_H}\psi \circ d_{e_G}\sigma$ (where e_G and e_H are the identity elements of G and H, respectively). This implies that $d_{e_G}\sigma$ is surjective. □

Theorem 4.2.2 *Let H_1, H_2 be algebraic subgroups of the algebraic group G.*

(i) *Suppose both H_1 and H_2 are connected. If* $\mathrm{Lie}(H_1) \subset \mathrm{Lie}(H_2)$, $H_1 \subset H_2$. *In particular, $H_1 = H_2$ if and only if* $\mathrm{Lie}(H_1) = \mathrm{Lie}(H_2)$.

(ii) $\mathrm{Lie}(H_1 \cap H_2) = \mathrm{Lie}(H_1) \cap \mathrm{Lie}(H_2)$.

Proof. Write $\mathfrak{h}_i = \mathrm{Lie}(H_i)$. Then $\mathfrak{h}_1 \subset \mathfrak{h}_2$ implies $G(\mathfrak{h}_1) \subset G(\mathfrak{h}_2)$ (notation as in Corollary 4.1.6). But by Corollary 4.1.6, $H_i = G(\mathfrak{h}_i)$. The second statement of (i) follows immediately from the first.

Since $\mathrm{Lie}(H_1 \cap H_2) \subset \mathrm{Lie}(H_i)$ we infer $\mathrm{Lie}(H_1 \cap H_2) \subset \mathrm{Lie}(H_1) \cap \mathrm{Lie}(H_2)$. But also $\mathfrak{h}_1 \cap \mathfrak{h}_2 \subset \mathrm{Lie}(H_i)$, and therefore $G(\mathfrak{h}_1 \cap \mathfrak{h}_2) \subset H_i$. So $G(\mathfrak{h}_1 \cap \mathfrak{h}_2) \subset H_1 \cap H_2$. Using Corollary 4.1.6: $\mathfrak{h}_1 \cap \mathfrak{h}_2 \subset \mathrm{Lie}(G(\mathfrak{h}_1 \cap \mathfrak{h}_2)) \subset \mathrm{Lie}(H_1 \cap H_2)$. □

Example 4.2.3 Let K be of characteristic $p > 0$ and G the algebraic group of Example 3.6.3. Set

$$H = \left\{ \begin{pmatrix} a & 0 & 0 \\ 0 & 1 & b \\ 0 & 0 & 1 \end{pmatrix} \mid a, b \in K, a \neq 0 \right\}.$$

Then H has the same Lie algebra as G, both G and H are connected, but they are not equal. This shows that Theorem 4.2.2 fails in non-zero characteristic.

Theorem 4.2.4 *Let $\sigma : G \to H$ be a morphism of algebraic groups. Then* $\mathrm{Lie}(\ker \sigma) = \ker(d\sigma)$.

Proof. Let $g \in G$. Then $g \in \ker \sigma$ if and only if $\sigma(g) = e_H$ (the identity element of H) if and only if $f(\sigma(g)) = f(e_H)$ for all $f \in K[H]$, which is the same as $\sigma^*(f)(g) - \sigma^*(f)(e_G) = 0$. So if we let A be the set of all $\sigma^*(f) - \sigma^*(f)(e_G)$ for $f \in K[H]$, then $\ker \sigma = \{g \in G \mid h(g) = 0 \text{ for all } h \in A\}$.

Let $\delta \in \mathrm{Lie}(\ker \sigma)$; then by Proposition 3.5.1, $\delta(h) = 0$ for all $h \in \mathcal{I}(\ker \sigma)$.

In particular for $f \in K[H]$ we have $0 = \delta(\sigma^*(f) - \sigma^*(f)(e_G)) = \delta(\sigma^*(f)) = \mathrm{d}\sigma(\delta)(f)$. So $\delta \in \ker \mathrm{d}\sigma$.

Let $\delta \in \mathrm{Lie}(G)$ and $f \in K[H]$. Then for $g \in G$ we calculate, using the left invariance of D_δ, and Lemma 4.1.1:

$$\begin{aligned} D_\delta(\sigma^*(f))(g) &= \lambda_g(D_\delta(\sigma^*(f))(e_G) = D_\delta(\lambda_g(\sigma^*(f)))(e_G) = \delta(\lambda_g(\sigma^*(f))) \\ &= \delta(\sigma^*(\lambda_{\sigma(g)} f)) = \mathrm{d}\sigma(\delta)(\lambda_{\sigma(g)} f). \end{aligned}$$

Let $\delta \in \ker(\mathrm{d}\sigma)$. Then the above implies that $D_\delta(\sigma^*(f)) = 0$. Therefore, $D_\delta(A) = 0$. Hence by Lemma 4.1.1, $A \subset J(\delta)$, implying $G(\delta) \subset \ker \sigma$. And that entails $\delta \in \mathrm{Lie}(\ker \sigma)$ by Theorem 4.1.5. □

Example 4.2.5 Let the notation be as in Example 3.9.2. In particular, $G = \mathrm{D}(n, K)$ and H is an algebraic subgroup of G. By Proposition 3.9.5, there are $e^1, \ldots, e^r \in \mathbb{Z}^n$ such that H is the intersection of the kernels of the χ_{e^i}, or, using the notation of Example 3.9.6, $H = \mathrm{D}(\Lambda)$, where $\Lambda \subset \mathbb{Z}^n$ is the lattice spanned by the e^i. From Theorems 4.2.2 and 4.2.4 it follows that $\mathrm{Lie}(H)$ is the intersection of the kernels of the differentials $\mathrm{d}\chi_{e^i}$. Observe that the Lie algebra of G consists of all diagonal matrices. We conclude that, writing $e^i = (e^i_1, \ldots, e^i_n)$, and setting

$$\mathfrak{d}(\Lambda) = \{\mathrm{diag}(\alpha_1, \ldots, \alpha_n) \mid \alpha_j \in K \text{ and } \sum_{j=1}^{n} e^i_j \alpha_j = 0 \text{ for } 1 \le i \le r\},$$

we have $\mathrm{Lie}(H) = \mathfrak{d}(\Lambda)$.

Lemma 4.2.6 *Let $\sigma : G \to H$ be a surjective morphism of algebraic groups. Let $H' \subset H$ be an algebraic subgroup, with Lie algebra $\mathfrak{h}' = \mathrm{Lie}(H')$. Set $G' = \sigma^{-1}(H')$. Then G' is an algebraic subgroup of G with $\mathrm{Lie}(G') = (\mathrm{d}\sigma)^{-1}(\mathfrak{h}')$.*

Proof. It is clear that G' is an algebraic subgroup of G. (It is defined, as subgroup of G, by the polynomials $\sigma^*(f)$, for $f \in \mathcal{I}(H')$.) So $\sigma : G' \to H'$ is a surjective morphism of algebraic groups. Hence by Theorem 4.2.1, $\mathrm{d}\sigma(\mathrm{Lie}(G')) = \mathfrak{h}'$. This implies that $(\mathrm{d}\sigma)^{-1}(\mathfrak{h}') = \mathrm{Lie}(G') + \ker \mathrm{d}\sigma$. But by Theorem 4.2.4, $\ker \mathrm{d}\sigma = \mathrm{Lie}(\ker \sigma) \subset \mathrm{Lie}(G')$ (as $\ker \sigma \subset G'$). □

Theorem 4.2.7 *Let $\sigma : G \to H$ be a morphism of algebraic groups and $H' \subset H$ an algebraic subgroup. Set $G' = \{g \in G \mid \sigma(g) \in H'\}$. Then $G' \subset G$ is an algebraic subgroup, and $\mathrm{Lie}(G') = \{\delta \in \mathrm{Lie}(G) \mid \mathrm{d}\sigma(\delta) \in \mathrm{Lie}(H')\}$.*

Proof. Set $\widetilde{H} = \sigma(G)$; then $\widetilde{H}$ is an algebraic subgroup of H (Proposition 3.1.9). Set $\widetilde{H}' = \widetilde{H} \cap H'$. Then $G' = \sigma^{-1}(\widetilde{H}')$. Lemma 4.2.6 says $\mathrm{Lie}(G') = \{\delta \in \mathrm{Lie}(G) \mid \mathrm{d}\sigma(\delta) \in \mathrm{Lie}(\widetilde{H}')\}$. But $\mathrm{Lie}(\widetilde{H}') = \mathrm{Lie}(\widetilde{H}) \cap \mathrm{Lie}(H')$

(Theorem 4.2.2) $= \mathrm{d}\sigma(\mathrm{Lie}(G)) \cap \mathrm{Lie}(H')$ (Theorem 4.2.1), proving the theorem. □

Corollary 4.2.8 *Let G be an algebraic group and $\rho : G \to \mathrm{GL}(V)$ a rational representation. Let $U \subset W$ be subspaces of V, and set $G' = \{g \in G \mid \rho(g)(v) \equiv v \bmod U$ for all $v \in W\}$. Then G' is an algebraic group and $\mathrm{Lie}(G') = \{\delta \in \mathrm{Lie}(G) \mid \mathrm{d}\rho(\delta)(W) \subset U\}$.*

Proof. Let $H = \mathrm{GL}(V)$ and $H' \subset H$ be the group of all $g \in \mathrm{GL}(V)$ with $g \cdot v \equiv v \bmod U$ for all $v \in W$. Now we use the preceding theorem, along with Example 3.6.10. □

Corollary 4.2.9 *Let G be an algebraic group, and $\rho : G \to \mathrm{GL}(V)$ a rational representation. Let U be a subspace of V and set $G' = \{g \in G \mid \rho(g)(U) = U\}$. Then G' is an algebraic group and $\mathrm{Lie}(G') = \{\delta \in \mathrm{Lie}(G) \mid \mathrm{d}\rho(\delta)(U) \subset U\}$.*

Proof. This follows from Corollary 4.2.8, by taking $W = U$. □

Corollary 4.2.10 *Let the notation be as in the preceding corollary. Suppose in addition that G is connected. Then U is stable under $\rho(G)$ if and only if U is stable under $\mathrm{d}\rho(\mathrm{Lie}(G))$.*

Proof. Set $G' = \{g \in G \mid \rho(g)U = U\}$. Then by Corollary 4.2.9, $\mathrm{Lie}(G') = \{\delta \in \mathrm{Lie}(G) \mid \mathrm{d}\rho(\delta)(U) \subset U\}$. Since G is connected we have $G = G'$ if and only if $\mathrm{Lie}(G) = \mathrm{Lie}(G')$ (Theorem 4.2.2). □

Corollary 4.2.11 *Let G be a connected algebraic group, and $\rho : G \to \mathrm{GL}(V)$ a rational representation. Let $v \in V$. Then $\rho(g)v = v$ for all $g \in G$ if and only if $\mathrm{d}\rho(\delta)v = 0$ for all $\delta \in \mathrm{Lie}(G)$.*

Proof. Let $G' \subset G$ be the subgroup consisting of $g \in G$ such that $\rho(g)v = v$. We apply Corollary 4.2.8 where W is the subspace spanned by v and $U = 0$. It follows that $\mathrm{Lie}(G')$ consists of the $\delta \in \mathrm{Lie}(G)$ with $\mathrm{d}\rho(\delta)v = 0$. We conclude in the same way as in the proof of Corollary 4.2.10. □

Proposition 4.2.12 *Let G be an algebraic group, and H an algebraic subgroup of G. Write $\mathfrak{g} = \mathrm{Lie}(G)$, $\mathfrak{h} = \mathrm{Lie}(H)$. If H is normal in G, then $\mathfrak{h}$ is an ideal of $\mathfrak{g}$. If moreover G and H are connected, and $\mathfrak{h}$ is an ideal in $\mathfrak{g}$, then H is normal in G.*

Proof. For $g \in G$, $\mathrm{Int}(g) : H \to H$ is an isomorphism. So its differential $\mathrm{Ad}(g)$ maps $\mathfrak{h}$ to $\mathfrak{h}$. We obtain a homomorphism $\mathrm{Ad} : G \to \mathrm{GL}(\mathfrak{h})$. Its differential is $d\mathrm{Ad} : \mathfrak{g} \to \mathfrak{gl}(\mathfrak{h})$. But by Theorem 3.8.1, $d\mathrm{Ad}(\delta)(\gamma) = [\delta, \gamma]$. It follows that $[\mathfrak{g}, \mathfrak{h}] \subset \mathfrak{h}$, and $\mathfrak{h}$ is an ideal of $\mathfrak{g}$.

Suppose $\mathfrak{h}$ is an ideal in $\mathfrak{g}$. Fix $g \in G$. Then by Corollary 4.2.10 it follows that $\mathrm{Ad}(g)$ maps $\mathfrak{h}$ to $\mathfrak{h}$. Furthermore, $\mathrm{Int}(g)$ maps H to gHg^{-1}. Also, $\mathrm{Lie}(gHg^{-1}) = \mathrm{Ad}(g)(\mathfrak{h})$ by Theorem 4.2.1. But the latter is equal to $\mathfrak{h}$. Therefore, Theorem 4.2.2 forces $H = gHg^{-1}$. □

4.3 Algebraic Lie algebras

A Lie algebra $\mathfrak{g} \subset \mathfrak{gl}(n, K)$ is called *algebraic* if there is an algebraic group $G \subset \mathrm{GL}(n, K)$ with $\mathfrak{g} = \mathrm{Lie}(G)$. In this section we collect a number of results on such Lie algebras and their corresponding algebraic groups. In the first subsection we look at the smallest algebraic Lie algebra containing a given Lie algebra. Section 4.3.2 is devoted to unipotent algebraic groups, and their Lie algebras. The final subsection discusses the structure of a general algebraic Lie algebra.

4.3.1 The algebraic hull

Let $\mathfrak{a} \subset \mathfrak{gl}(n, K)$ be a Lie algebra. By Theorem 4.2.2 there is a unique smallest algebraic Lie algebra $\mathfrak{g} \subset \mathfrak{gl}(n, K)$ that contains $\mathfrak{a}$. This Lie algebra is called the *algebraic hull* of $\mathfrak{a}$ and denoted by $\mathfrak{g}(\mathfrak{a})$.

Let $a \in \mathfrak{gl}(n, K)$; then, as in Section 4.1, by $G(a)$ we denote the smallest algebraic subgroup of $\mathrm{GL}(n, K)$ with a in its Lie algebra. By $\mathfrak{g}(a)$ we denote its Lie algebra. So $\mathfrak{g}(a)$ is the algebraic hull of the Lie algebra spanned by a. Lemma 4.3.1 immediately follows from Example 3.6.7. As an application we show a proposition which will be useful later.

Lemma 4.3.1 *Suppose a is nilpotent. Then $G(a) = \{\exp(ta) \mid t \in K\}$ and $\mathfrak{g}(a)$ is spanned by a.*

Proposition 4.3.2 *Let G be an algebraic group, and $\mathfrak{g} = \mathrm{Lie}(G)$. Let $\mathfrak{h}_1, \mathfrak{h}_1$ be two Cartan subalgebras of $\mathfrak{g}$. Then there is a $g \in G$ such that $\mathrm{Ad}(g)(\mathfrak{h}_1) = \mathfrak{h}_2$.*

Proof. By Theorem 2.5.8 we may assume that there is an $x \in \mathfrak{g}$ such that $\mathrm{ad}x$ is nilpotent, and $\exp(\mathrm{ad}x)(\mathfrak{h}_1) = \mathfrak{h}_2$. By Theorem 3.5.4 we may assume that $G \subset \mathrm{GL}(V)$ and consequently, $\mathfrak{g} \subset \mathfrak{gl}(V)$. Let $x = s + n$ be the Jordan decomposition of x. Then $\mathrm{ad}x = \mathrm{ad}s + \mathrm{ad}n$ is the Jordan decomposition of $\mathrm{ad}x$ (Proposition 2.2.5), whence $\mathrm{ad}s = 0$ so that we may assume that $s = 0$.

Then $g = \exp x \in G$ (Lemma 4.3.1, along with Theorem 4.2.2(i)). Lemma 2.3.1 now yields $\mathfrak{h}_2 = \exp(\mathrm{ad}x)(\mathfrak{h}_1) = \exp(x)\mathfrak{h}_1\exp(x)^{-1} = \mathrm{Ad}(g)(\mathfrak{h}_1)$. □

Lemma 4.3.3 *Let* $a = \mathrm{diag}(\alpha_1, \dots, \alpha_n)$ *be a diagonal matrix, where* $\alpha_i \in K$. *Set*

$$\Lambda = \{(e_1, \dots, e_n) \in \mathbb{Z}^n \mid \sum_{i=1}^n \alpha_i e_i = 0\}.$$

Then

$$G(a) = \{\mathrm{diag}(\beta_1, \dots, \beta_n) \mid \beta_i \in K \textit{ and } \prod_{i=1}^n \beta_i^{e_i} = 1 \textit{ for all } (e_1, \dots, e_n) \in \Lambda\},$$

$$\mathfrak{g}(a) = \{\mathrm{diag}(\gamma_1, \dots, \gamma_n) \mid \gamma_i \in K \textit{ and } \sum_{i=1}^n e_i\gamma_i = 0 \textit{ for all } (e_1, \dots, e_n) \in \Lambda\}.$$

Proof. Consider the group $G = \mathrm{D}(n, K)$ (see Example 3.9.2). Its Lie algebra consists of all diagonal matrices. Therefore $G(a) \subset G$. As seen in Example 3.9.6 there is a finite set $C \subset \mathbb{Z}^n$ with $G(a) = \mathrm{D}(C)$. Then $\mathrm{Lie}(G(a))$ consists of all $\mathrm{diag}(\gamma_1, \dots, \gamma_n)$ such that $\sum_i e_i\gamma_i = 0$, for all $e = (e_1, \dots, e_n) \in C$ (see Example 4.2.5). So $C \subset \Lambda$, whence $\mathrm{D}(\Lambda) \subset G(a)$.

In Example 4.2.5 we show that

$$\mathrm{Lie}(\mathrm{D}(\Lambda)) = \{\mathrm{diag}(\gamma_1, \dots, \gamma_n) \mid \sum_{i=1}^n e_i\gamma_i = 0 \text{ for all } (e_1, \dots, e_n) \in \Lambda\}.$$

So $a \in \mathrm{Lie}(\mathrm{D}(\Lambda))$, whence $G(a) \subset \mathrm{D}(\Lambda)$ (Theorem 4.1.5). It follows that $G(a) = \mathrm{D}(\Lambda)$, which proves the lemma. □

Theorem 4.3.4 *Let* A *be an algebra (not necessarily associative nor Lie). Let* $G = \mathrm{Aut}(A)$. *Then* G *is an algebraic subgroup of* $\mathrm{GL}(A)$ *with Lie algebra* $\mathrm{Der}(A)$ *(the algebra of all derivations of* A*).*

Proof. Set $L = \mathrm{Lie}(\mathrm{Aut}(A))$. First let $d \in \mathrm{Der}(A)$. We want to show that $d \in L$. Let $d = d_s + d_n$ be its Jordan decomposition. Then $d_s, d_n \in \mathrm{Der}(A)$ (Lemma 2.7.8). So it is enough to show that $d \in L$ whenever d is either nilpotent or semisimple. Suppose d is nilpotent. Then $G(d) = \{\exp(td) \mid t \in K\}$ is an algebraic subgroup of $\mathrm{Aut}(A)$ (see Section 2.3) and its Lie algebra is spanned by d (Lemma 4.3.1), whence $d \in L$. Secondly, suppose that d is semisimple. Let $a_1, \dots, a_m$ be a basis of A consisting of eigenvectors of A, and define λ_i by $d(a_i) = \lambda_i a_i$. Set

$$\Lambda = \{(e_1, \dots, e_m) \in \mathbb{Z}^m \mid \sum_{i=1}^m \lambda_i e_i = 0\}.$$

Then by Lemma 4.3.3,

$$G(d) = \{\mathrm{diag}(t_1, \dots, t_m) \mid \prod_{i=1}^{m} t_i^{e_i} = 1 \text{ for all } (e_1, \dots, e_m) \in \Lambda\}.$$

Define $c_{i,j}^k \in K$ by $a_i a_j = \sum_k c_{i,j}^k a_k$. A small calculation shows that $\lambda_k = \lambda_i + \lambda_j$ whenever $c_{i,j}^k \neq 0$. So for a $g \in G(d)$, $g = \mathrm{diag}(t_1, \dots, t_m)$, we have $t_k = t_i t_j$ whenever $c_{i,j}^k \neq 0$. But that implies that $g \in \mathrm{Aut}(A)$. So $G(d) \subset \mathrm{Aut}(A)$. Again we conclude that $d \in L$.

Now we use dual numbers (Section 1.2). Let $\phi = I_A + \varepsilon D$, where $D \in \mathfrak{gl}(A)$, and I_A is the identity on A. Then $\phi(ab) = \phi(a)\phi(b)$ is the same as $D(ab) = D(a)b + aD(b)$. Hence every element of L is a derivation of A. □

Lemma 4.3.5 *Let G be an algebraic group, and $G_1, \dots, G_s \subset G$ connected algebraic subgroups. Let $H \subset G$ be the connected algebraic subgroup of G generated by the G_i (see Proposition 3.2.3). Then* $\mathrm{Lie}(H)$ *is spanned by the subspaces* $\mathrm{Ad}(h)(\mathrm{Lie}(G_i))$, *where h runs through H.*

Proof. By Proposition 3.2.3 there are $i_1, \dots, i_n$ such that the product map $\sigma : G_{i_1} \times \cdots \times G_{i_n} \to H$ is surjective. Write $X = G_{i_1} \times \cdots \times G_{i_n}$. From Theorem 1.4.3 it follows that there is an $x = (g_{i_1}, \dots, g_{i_n}) \in X$ such that the differential $\mathrm{d}_x\sigma : T_x(X) \to T_{\sigma(x)}(H)$ is surjective. For $1 \leq k \leq n$ set $\bar{g}_k = g_{i_1} \cdots g_{i_k}$, and $Z_k = \bar{g}_k G_{i_k} \bar{g}_k^{-1}$. Furthermore, let $Z = Z_1 \times \cdots \times Z_n$. Now we define the following maps

$$\begin{aligned}
\psi &: X \to X, \quad (x_1, \dots, x_n) \mapsto (g_{i_1} x_1, \dots, g_{i_n} x_n),\\
\chi &: X \to Z, \quad (x_1, \dots, x_n) \mapsto (\bar{g}_1 x_1 \bar{g}_1^{-1}, \dots, \bar{g}_n x_n \bar{g}_n^{-1}),\\
\varphi &: Z \to H, \quad (z_1, \dots, z_n) \mapsto z_1 \cdots z_n,\\
\phi &: H \to H, \quad \phi(h) = h\sigma(x)^{-1}.
\end{aligned}$$

Then $\phi \circ \sigma \circ \psi = \varphi \circ \chi$. Write $\epsilon = (e, \dots, e) \in X$; then $\psi(\epsilon) = x$. Now ϕ and ψ are isomorphisms, so their differentials are bijective. Moreover, $\mathrm{d}_x\sigma$ is surjective. We conclude that $\mathrm{d}_\epsilon(\varphi \circ \chi) : T_\epsilon(X) \to \mathrm{Lie}(H)$ is surjective. Let $\delta = (\delta_1, \dots, \delta_n) \in T_\epsilon(X)$, i.e., $\delta_i \in \mathrm{Lie}(G_i)$. Then

$$\mathrm{d}_\epsilon(\varphi \circ \chi)(\delta) = \mathrm{d}_\epsilon\varphi(\mathrm{d}_\epsilon\chi(\delta)) = \sum_{k=1}^{n} \mathrm{Ad}(\bar{g}_k)(\delta_k).$$

(In the last equality we use Example 3.7.1 (the differential of the product map is the sum map), and the definition of the adjoint representation Ad.) It follows that $\mathrm{Lie}(H)$ is contained in the sum of all $\mathrm{Ad}(h)(\mathrm{Lie}(G_i))$. The other inclusion is obvious. □

Theorem 4.3.6 *Let the notation be as in the previous lemma.* $\mathrm{Lie}(H)$ *is generated by the* $\mathrm{Lie}(G_i)$ *(as Lie algebra).*

Proof. Set $\mathfrak{h} = \mathrm{Lie}(H)$. Let $\mathfrak{h}'$ be the Lie algebra generated by the $\mathrm{Lie}(G_i)$. Then $\mathfrak{h}' \subset \mathfrak{h}$. By Lemma 4.3.5 it is enough to show that $\mathrm{Ad}(h)(\mathfrak{h}') = \mathfrak{h}'$ for all $h \in H$. This is the same as $H \subset N_G(\mathfrak{h}') = \{g \in G \mid \mathrm{Ad}(g)(\mathfrak{h}') = \mathfrak{h}'\}$. We note that from Corollary 4.2.9 it follows that $N_G(\mathfrak{h}')$ is an algebraic subgroup of G. Furthermore, since $d\mathrm{Ad}(\delta) = \mathrm{ad}\delta$ (Theorem 3.8.1), we also have that $\mathrm{Lie}(N_G(\mathfrak{h}'))$ consists of all $\delta \in \mathrm{Lie}(G)$ with $[\delta, \mathfrak{h}'] \subset \mathfrak{h}'$. To show that $H \subset N_G(\mathfrak{h}')$ it suffices to show that $G_i \subset N_G(\mathfrak{h}')$. However, since the G_i are connected this follows from $\mathrm{Lie}(G_i) \subset \mathrm{Lie}(N_G(\mathfrak{h}'))$ (Theorem 4.2.2). In turn, that follows from $\mathrm{Lie}(G_i) \subset \mathfrak{h}'$. □

Corollary 4.3.7 *Let* $\mathfrak{g} \subset \mathfrak{gl}(n, K)$ *be a Lie algebra, and let* $a_1, \dots, a_r$ *be a set of generators of* $\mathfrak{g}$. *Then* $\mathfrak{g}$ *is algebraic if and only if* $\mathfrak{g}(a_i) \subset \mathfrak{g}$ *for* $1 \le i \le r$.

Proof. The "only if" is clear. Set $G_i = G(a_i)$, and let $H \subset \mathrm{GL}(n, K)$ be the algebraic group generated by the G_i. Then by Theorem 4.3.6, $\mathrm{Lie}(H)$ is generated by the $\mathfrak{g}(a_i)$. Hence $\mathrm{Lie}(H) = \mathfrak{g}$, and $\mathfrak{g}$ is algebraic. □

From Section 2.2 we recall that the associative algebra (with identity) generated by $a \in \mathfrak{gl}(n, K)$ is denoted $K[a]$.

Proposition 4.3.8 *Let* $a \in \mathfrak{gl}(n, K)$.

(i) $G(a)$ *and* $\mathfrak{g}(a)$ *are both contained in* $K[a]$.

(ii) *Suppose* a *has coefficients in* $k \subset K$. *Then* $G(a)$ *is defined over* k, *and* $\mathfrak{g}(a)$ *has a basis consisting of elements with coefficients in* k.

(iii) *Let* $g \in \mathrm{GL}(n, K)$ *and* $b = gag^{-1}$. *Then* $G(b) = gG(a)g^{-1}$ *and* $\mathfrak{g}(b) = g\mathfrak{g}(a)g^{-1}$.

Proof. Let H be the group of all invertible elements of $K[a]$. As seen in Example 3.6.9, H is algebraic and $\mathrm{Lie}(H) = K[a]$ (where the Lie multiplication is given by the commutator). Since $a \in K[a]$ we conclude that $G(a) \subset H$ and $\mathfrak{g}(a) \subset K[a]$.

Define δ_a as in Section 3.10, and D_{δ_a} as in Section 3.4. If a has coefficients in k then by (3.3), D_{δ_a} maps polynomials with coefficients in k to polynomials with coefficients in k. So by Lemma 4.1.1 the ideal $J(\delta_a)$ (see Section 4.1) is the solution space of a set of linear equations with coefficients in k. Hence it is generated by polynomials over k.

Consider the automorphism $\mathrm{Int}(g) : \mathrm{GL}(n, K) \to \mathrm{GL}(n, K)$ (Section 3.8). Set $H = \mathrm{Int}(g)(G(a))$. Then H is a connected algebraic subgroup of $\mathrm{GL}(n, K)$. Furthermore, by Theorem 4.2.1, $\mathrm{Lie}(H) = \mathrm{Ad}(g)(\mathrm{Lie}(G(a)) = g\mathfrak{g}(a)g^{-1}$. Hence $b \in \mathrm{Lie}(H)$, and therefore $G(b) \subset H$. It folows that $\mathfrak{g}(b) = \mathrm{Lie}(G(b)) \subset$

$\mathrm{Lie}(H) = g\mathfrak{g}(a)g^{-1}$. By analogous reasoning we get $\mathfrak{g}(a) \subset g^{-1}\mathfrak{g}(b)g$. Hence $\mathfrak{g}(b) = g\mathfrak{g}(a)g^{-1}$. By Theorem 4.2.2(i) we infer $G(b) = H$. □

4.3.2 Unipotent groups

A subgroup $G \subset \mathrm{GL}(n, K)$ is called *unipotent* if all of its elements are. An example of a unipotent group is $\mathrm{U}(n, K)$, which consists of all upper-triangular matrices in $\mathrm{GL}(n, K)$ with 1's on the diagonal. We note that $\mathrm{U}(n, K)$ is a connected algebraic group and its Lie algebra denoted $\mathfrak{u}(n, K)$ consists of all strictly upper-triangular matrices. A general theorem from group theory (see [Rob96], 8.1.10) says that for any subgroup S of $\mathrm{GL}(n, K)$ consisting of unipotent elements, there exists a $g \in \mathrm{GL}(n, K)$ such that gSg^{-1} is contained in $\mathrm{U}(n, K)$. (This holds more generally over any field.) In this section we collect some results on unipotent algebraic groups, their Lie algebras and the correspondence between them.

For a unipotent $a \in \mathrm{GL}(n, k)$ we define

$$\log(a) = \sum_{i=1}^{\infty}(-1)^{i-1}\frac{(a - I_n)^i}{i}, \tag{4.1}$$

where I_n denotes the $n \times n$ identity matrix (note that this is a finite sum).

Lemma 4.3.9 *Let $a \in \mathrm{GL}(n, k)$ be unipotent. Then $a = \exp(\log(a))$.*

Proof. For integers $r > 0$ we have

$$a^r = (I_n + a - I_n)^r = I_n + \sum_{i=1}^{r}\binom{r}{i}(a - I_n)^i = I_n + \sum_{i=1}^{\infty}\binom{r}{i}(a - I_n)^i.$$

Note that this last sum is finite because $\binom{r}{i} = 0$ if $i > r$. Now $\binom{z}{i}$ is a polynomial of degree i in z, with constant term equal to 0. It follows that $a^r = x_0 + rx_1 + r^2x_2 + \cdots + r^m x_m$, where m is the largest integer with $(a - I_n)^m \neq 0$, and the $x_i \in M_n(k)$ do not depend on r. Since the coefficient of the linear term of $\binom{z}{i}$ is $i^{-1}(-1)^{i-1}$, we get $x_1 = \log(a)$. Furthermore, $x_0 = I_n$.

Set $k' = k(T_1, T_2)$ (rational function field in two indeterminates). Set $p(T_1) = x_0 + T_1x_1 + \cdots + T_1^m x_m$. Then $p(T_1) \in M_n(k')$ and $p(r) = a^r$ for integers $r > 0$. Let $c_{s,t} : M_n(k') \to k'$ be the function that associates to a matrix its coefficient on position (s, t). Set $q(T_1, T_2) = c_{s,t}(p(T_1 + T_2) - p(T_1)p(T_2))$ (a polynomial in T_1, T_2). For integers $r_1, r_2 > 0$ we have $p(r_1 + r_2) = a^{r_1+r_2} = a^{r_1}a^{r_2} = p(r_1)p(r_2)$. Hence $q(r_1, r_2) = 0$ for all integers $r_1, r_2 > 0$. This implies that $q = 0$. (Indeed, we can view q as polynomial in T_2 with coefficients in $k[T_1]$. For all $r_1 > 0$, $q(r_1, T_2) = 0$, as it is a polynomial in one indeterminate with an infinite number of zeros. So all coefficients of q have an infinite number of zeros. It follows that they are all zero.) Now the coefficient of T_2 in q is $c_{s,t}(\sum_{i=1}^{m+1}(ix_i - x_{i-1}x_1)T_1^{i-1})$, where we set $x_{m+1} = 0$. But this is zero for all s, t. Hence $ix_i = x_{i-1}x_1$ for $i \geq 1$. It follows that $x_i = \frac{x_1^i}{i!}$ and $p(T_1) = \exp(T_1 x_1)$. In particular, $a = p(1) = \exp(\log(a))$. □

Proposition 4.3.10 *Let $a \in \mathrm{GL}(n, K)$ be unipotent and $G \subset \mathrm{GL}(n, K)$ be the smallest algebraic group containing a. Then it is of dimension 1 and connected. Moreover, $\mathrm{Lie}(G)$ is spanned by $\log(a)$.*

Proof. Set $x = \log(a)$. Then x is nilpotent, and by Lemma 4.3.1, $G(x) = \{\exp(tx) \mid t \in K\}$. So from Lemma 4.3.9 it follows that $a \in G(x)$. Hence $G \subset G(x)$. Let $p \in K[\mathrm{GL}(n, K)]$ be zero on G. Then $p(\exp(rx)) = p(a^r) = 0$ for integers $r > 0$. So $p(\exp(Tx))$ is a polynomial in T with an infinite number of zeros. Hence it is zero. Therefore, $p(\exp(tx)) = 0$ for all $t \in K$. So p vanishes on $G(x)$, whence $G(x) \subset G$. So $G(x) = G$, and therefore G is connected. Finally by Lemma 4.3.1, $\log(a)$ spans $\mathrm{Lie}(G)$. □

Corollary 4.3.11 *Let $G \subset \mathrm{GL}(n, K)$ be a unipotent algebraic group. Then G is connected.*

Proof. For $a \in G$ denote by G_a the smallest algebraic subgroup of $\mathrm{GL}(n, K)$ containing a. Then $G_a \subset G$ and by the previous proposition, G_a is connected. So G is generated by connected subgroups, and hence the corollary follows from Proposition 3.2.3. □

Lemma 4.3.12 *Let $x, x' \in M_n(K)$ be nilpotent such that $\exp(x) = \exp(x')$. Then $x = x'$.*

Proof. Set $a = \exp(x)$. Then for $r > 0$, $\exp(rx) = a^r = \exp(rx')$. Consider $\exp(Tx) - \exp(Tx')$; the coefficients of this matrix are polynomials in T with all integers $r > 0$ as zeros. Hence they are zero, and we get $\exp(tx) = \exp(tx')$ for all $t \in K$. So by Lemma 4.3.1, $G(x) = G(x')$. Hence $x' \in \mathrm{Lie}(G(x))$, which is spanned by x. We conclude that $x' = \alpha x$ where $\alpha \in K$. So if $x = 0$, then also $x' = 0$ and we are done. If $x \neq 0$, then there is a vector $v \in K^n$ with $xv \neq 0$, but $x^2 v = 0$. Then $av = v + xv$, but also $av = v + (\alpha x)v$. This implies that $\alpha = 1$. □

Corollary 4.3.13 *Let $x \in M_n(k)$ be nilpotent. Then $\log(\exp(x)) = x$.*

Proof. Set $x' = \log(\exp(x))$. Then by Lemma 4.3.9, $\exp(x') = \exp(x)$. So by the previous lemma, $x = x'$. □

Theorem 4.3.14 *Let $\mathfrak{g} \subset \mathfrak{gl}(n, K)$ be a Lie algebra. If $\mathfrak{g}$ consists of nilpotent elements, $\mathfrak{g}$ is algebraic, and the connected algebraic group with Lie algebra equal to $\mathfrak{g}$ is unipotent. Conversely, let $G \subset \mathrm{GL}(n, K)$ be a unipotent algebraic group with Lie algebra $\mathfrak{g}$. Then $\mathfrak{g}$ consists of nilpotent elements.*

Proof. The subalgebra spanned by each basis element of $\mathfrak{g}$ is algebraic (Lemma 4.3.1). Hence $\mathfrak{g}$ is algebraic (Corollary 4.3.7). Let G denote the connected algebraic group such that $\mathrm{Lie}(G) = \mathfrak{g}$. Let $V = K^n$ be the natural $\mathfrak{g}$-module. Let $0 = V_0 \subset V_1 \cdots \subset V_{s+1} = V$ be a composition series of V. By Lemma 2.4.2 we infer that $x \cdot V_{i+1} \subset V_i$ for $0 \leq i \leq s$, and all $x \in \mathfrak{g}$. Hence there is an $A \in \mathrm{GL}(n, K)$ such that AxA^{-1} is strictly upper triangular, for all $x \in \mathfrak{g}$. In other words, $A\mathfrak{g}A^{-1} \subset \mathfrak{u}(n, K)$, and therefore $AGA^{-1} \subset \mathrm{U}(n, K)$ (Theorem 4.2.2; note that the Lie algebra of AGA^{-1} is $A\mathfrak{g}A^{-1}$). In particular, G is unipotent.

As noted at the beginning of this section, G can be conjugated into $\mathrm{U}(n, K)$. The same conjugation maps $\mathfrak{g}$ into $\mathfrak{u}(n, K)$. So $\mathfrak{g}$ consists of nilpotent elements. □

Proposition 4.3.15 *Let $G \subset \mathrm{GL}(n, K)$ be a unipotent algebraic group with Lie algebra $\mathfrak{g}$. Let $x_1, \ldots, x_s$ be a basis of $\mathfrak{g}$. Then for $a \in G$ there are unique $p_1(a), \ldots, p_s(a) \in K$ with $a = \exp(\sum_{i=1}^{s} p_i(a)x_i)$. The functions $a \mapsto p_i(a)$ are contained in $K[G]$. They are algebraically independent over K, and $K[G] = K[p_1, \ldots, p_s]$.*

Proof. Let $a \in G$; then $\log(a)$ spans the Lie algebra of the smallest algebraic group that contains a (Proposition 4.3.10). So $\log(a) \in \mathfrak{g}$. Write $\log(a) = \sum_{i=1}^{s} p_i(a)x_i$. From (4.1) it follows that the p_i are contained in $K[G]$. Also, by Lemma 4.3.9, $a = \exp(\log(a))$; this implies that the coefficients of a are polynomials in $p_i(a)$. So the p_i generate $K[G]$. The transcendence degree of the field of fractions of $K[G]$ is $\dim \mathfrak{g}$ (see Section 1.3). Hence the p_i are algebraically independent over K. □

Let the notation be as in Proposition 4.3.15. This proposition, along with elimination techniques using Gröbner bases (see Section 1.6), leads to an immediate algorithm that, given a basis $x_1, \ldots, x_s$ of $\mathfrak{g}$, computes polynomial equations for G (as subgroup of $\mathrm{GL}(n, K)$). In Section 4.5 we will describe a different algorithm for the same purpose that avoids the use of Gröbner bases and is much more efficient.

4.3.3 The structure of algebraic Lie algebras

In this section we prove a theorem giving a lot of structural information about algebraic Lie algebras $\subset \mathfrak{gl}(n, K)$. The Lie correspondence makes it possible to obtain the corresponding information for an algebraic group.

Lemma 4.3.16 *Let $U \subset W \subset \mathfrak{gl}(n,K)$ be subspaces. Set $G = \{g \in \mathrm{GL}(n,K) \mid gwg^{-1} \equiv w \bmod U$ for all $w \in W\}$. Then G is an algebraic group and $\mathrm{Lie}(G) = \{a \in \mathfrak{gl}(n,K) \mid [a,W] \subset U\}$.*

Proof. Ths follows from Corollary 4.2.8, taking for ρ the adjoint representation of $\mathrm{GL}(n,K)$ and using Theorem 3.8.1. □

Lemma 4.3.17 *Let $\mathfrak{g} \subset \mathfrak{gl}(n,K)$ be a subalgebra. Let $\mathfrak{g}'$ denote its algebraic hull. Then*

(i) *Every ideal of $\mathfrak{g}$ is an ideal of $\mathfrak{g}'$.*

(ii) $[\mathfrak{g}',\mathfrak{g}'] = [\mathfrak{g},\mathfrak{g}]$.

(iii) *If $\mathfrak{g}$ is solvable, $\mathfrak{g}'$ is solvable.*

Proof. Let $\mathfrak{a} \subset \mathfrak{g}$ be an ideal. Set $\mathfrak{h}_1 = \{a \in \mathfrak{gl}(n,K) \mid [a,\mathfrak{a}] \subset \mathfrak{a}\}$. Then $\mathfrak{h}_1$ is algebraic by Lemma 4.3.16. Since $\mathfrak{g} \subset \mathfrak{h}_1$, we see that $\mathfrak{g}' \subset \mathfrak{h}_1$, and therefore $\mathfrak{a}$ is an ideal of $\mathfrak{g}'$.

Set $\mathfrak{h}_2 = \{a \in \mathfrak{gl}(n,K) \mid [a,\mathfrak{g}] \subset [\mathfrak{g},\mathfrak{g}]\}$. By Lemma 4.3.16 it is an algebraic Lie algebra. Since $\mathfrak{g} \subset \mathfrak{h}_2$ we get $\mathfrak{g}' \subset \mathfrak{h}_2$. This implies that $\mathfrak{g} \subset \mathfrak{h}_3 = \{a \in \mathfrak{gl}(n,K) \mid [a,\mathfrak{g}'] \subset [\mathfrak{g},\mathfrak{g}]\}$. But the latter is also an algebraic Lie algebra by Lemma 4.3.16. Hence $\mathfrak{g}' \subset \mathfrak{h}_3$, so that $[\mathfrak{g}',\mathfrak{g}'] \subset [\mathfrak{g},\mathfrak{g}]$. But evidently, $[\mathfrak{g},\mathfrak{g}] \subset [\mathfrak{g}',\mathfrak{g}']$.

Since $[\mathfrak{g},\mathfrak{g}] = [\mathfrak{g}',\mathfrak{g}']$, the derived series of $\mathfrak{g}'$ is equal to the one of $\mathfrak{g}$, apart from the first term. Hence if $\mathfrak{g}$ is solvable, so is $\mathfrak{g}'$. □

Proposition 4.3.18 *Let $\mathfrak{g} \subset \mathfrak{gl}(n,K)$ be a subalgebra and suppose $\mathfrak{g}$ is algebraic. Then its solvable radical is algebraic.*

Proof. Let $\mathfrak{r} \subset \mathfrak{g}$ denote the solvable radical of $\mathfrak{g}$. Let $\mathfrak{r}'$ be its algebraic hull. Then $\mathfrak{r}' \subset \mathfrak{g}$ as $\mathfrak{g}$ is algebraic. Moreover, by Lemma 4.3.17, $\mathfrak{r}'$ is solvable. Set $\mathfrak{h} = \{a \in \mathfrak{gl}(n,K) \mid [a,\mathfrak{g}] \subset \mathfrak{r}\}$. It is algebraic by Lemma 4.3.16, and contains $\mathfrak{r}$ as $\mathfrak{r}$ is an ideal of $\mathfrak{g}$. Hence $\mathfrak{r}' \subset \mathfrak{h}$. It follows that $[\mathfrak{r}',\mathfrak{g}] \subset \mathfrak{r} \subset \mathfrak{r}'$. So $\mathfrak{r}'$ is an ideal of $\mathfrak{g}$, and since it is solvable, $\mathfrak{r}' \subset \mathfrak{r}$. □

Proposition 4.3.19 *Let $\mathfrak{g} \subset \mathfrak{gl}(n,K)$ be an algebraic and solvable subalgebra. Let $\mathfrak{n} \subset \mathfrak{g}$ be the set of all nilpotent elements of $\mathfrak{g}$. Then $\mathfrak{n}$ is an ideal of $\mathfrak{g}$, and there exists an abelian subalgebra $\mathfrak{t} \subset \mathfrak{g}$, consisting of semisimple elements, such that $\mathfrak{g} = \mathfrak{t} \oplus \mathfrak{n}$ (direct sum of vector spaces). Moreover, any such $\mathfrak{t}$ is algebraic.*

Proof. Let $U = K^n$ be the natural $\mathfrak{g}$-module. Let $0 = U_0 \subset U_1 \cdots \subset U_s = U$ be a composition series of U. Set $\mathfrak{n}' = \{a \in \mathfrak{g} \mid aU_i \subset U_{i-1}$ for $1 \leq i \leq s\}$. Then $\mathfrak{n}'$ is an ideal of $\mathfrak{g}$ consisting of nilpotent elements. So $\mathfrak{n}' \subset \mathfrak{n}$. Since

U_i/U_{i-1} is an irreducible $\mathfrak{g}$-module, it follows that $\dim U_i/U_{i-1} = 1$ (Theorem 2.6.5). But then for an $a \in \mathfrak{n}$ it follows that $aU_i \subset U_{i-1}$. We conclude that $\mathfrak{n} = \mathfrak{n}'$.

Consider the set of all abelian subalgebras of $\mathfrak{g}$, consisting of semisimple elements. This set is not empty as $\{0\}$ belongs to it, so we can choose a maximal such subalgebra $\mathfrak{t}$. Also $\mathrm{ad}_{\mathfrak{g}}\mathfrak{t}$ is an abelian Lie algebra, and consists of semisimple elements. This implies that, as a $\mathfrak{t}$-module, $\mathfrak{g}$ is completely reducible (Theorem 2.12.3). Since $\mathfrak{n}$ is an ideal, $\mathfrak{n} \oplus \mathfrak{t}$ is a $\mathfrak{t}$-submodule. Hence there exists a $\mathfrak{t}$-submdule, $\mathfrak{t}' \subset \mathfrak{g}$, with $\mathfrak{g} = \mathfrak{n} \oplus \mathfrak{t} \oplus \mathfrak{t}'$.

We claim that $\mathfrak{t}' = 0$. Firstly, $[\mathfrak{t}, \mathfrak{t}'] \subset \mathfrak{t}'$. But $[\mathfrak{t}, \mathfrak{t}'] \subset [\mathfrak{g}, \mathfrak{g}] \subset \mathfrak{n}$ (the last inclusion follows from Theorem 2.6.5). So $[\mathfrak{t}, \mathfrak{t}'] = 0$. Let $a \in \mathfrak{t}'$, and $a = s + n$ be its Jordan decomposition. Since s and n can be written as polynomials in a (Proposition 2.2.3), both commute with $\mathfrak{t}$. Let $\mathfrak{t}_1$ be the subalgebra generated by s and $\mathfrak{t}$. As s commutes with $\mathfrak{t}$, $\mathfrak{t}_1$ is abelian and consists of semisimple elements (Corollary 2.2.2). So $\mathfrak{t}_1 = \mathfrak{t}$, whence $s \in \mathfrak{t}$. On the other hand, $n \in \mathfrak{n}$. It follows that $a \in \mathfrak{n} \oplus \mathfrak{t}$. Therefore $a = 0$ and the claim follows.

It remains to show that $\mathfrak{t}$ is algebraic. Let $a \in \mathfrak{t}$, and $a' \in \mathfrak{g}(a)$. Then $a' \in \mathfrak{g}$. Moreover, a' can be written as a polynomial in a (Proposition 4.3.8). Hence a' commutes with $\mathfrak{t}$. Also a' is semisimple (by Lemma 4.3.3 in combination with Proposition 4.3.8(iii)). As above, $a' \in \mathfrak{t}$. The conclusion is that $\mathfrak{g}(a) \subset \mathfrak{t}$, and therefore $\mathfrak{t}$ is algebraic (Corollary 4.3.7). □

Theorem 4.3.20 *Let $\mathfrak{g} \subset \mathfrak{gl}(n, K)$ be an algebraic subalgebra. Then as a vector space $\mathfrak{g}$ can be written as a direct sum, $\mathfrak{g} = \mathfrak{s} \oplus \mathfrak{t} \oplus \mathfrak{n}$, where*

(i) *the solvable radical $\mathfrak{r}$ of $\mathfrak{g}$ is equal to $\mathfrak{t} \oplus \mathfrak{n}$,*

(ii) *$\mathfrak{n}$ is the ideal consisting of all nilpotent elements of $\mathfrak{r}$; it is also the largest ideal of $\mathfrak{g}$ consisting entirely of nilpotent elements,*

(iii) *$\mathfrak{t}$ is an abelian subalgebra consisting of semisimple elements,*

(iv) *$\mathfrak{s}$ is a semisimple subalgebra and $[\mathfrak{s}, \mathfrak{t}] = 0$,*

(v) *$\mathfrak{s}$, $\mathfrak{t}$ and $\mathfrak{n}$ are all algebraic.*

Proof. Note that by Proposition 4.3.18, $\mathfrak{r}$ is algebraic. Let $\mathfrak{r} = \mathfrak{t} \oplus \mathfrak{n}$ be the decomposition of $\mathfrak{r}$ from Propostion 4.3.19. So (i) and (iii) are immediate. Let $U = K^n$ be the natural $\mathfrak{g}$-module. Let $0 = U_0 \subset U_1 \cdots \subset U_s = U$ be a composition series of U. Set $\mathfrak{n}' = \{a \in \mathfrak{g} \mid aU_i \subset U_{i-1} \text{ for } 1 \leq i \leq s\}$. Then $\mathfrak{n}'$ is the largest ideal of $\mathfrak{g}$ consisting of nilpotent elements (Lemma 2.4.2). As a Lie algebra it is nilpotent (since there is a basis of K^n with respect to which all elements of $\mathfrak{n}'$ are upper triangular), and therefore $\mathfrak{n}' \subset \mathfrak{r}$. Hence $\mathfrak{n}' \subset \mathfrak{n}$. Let $a \in \mathfrak{n}$. Let $\rho_i : \mathfrak{g} \to \mathfrak{gl}(U_i/U_{i-1})$ denote the induced representation. By Lemma 2.6.4 $\rho_i(a)$ lies in the centre of $\rho_i(\mathfrak{g})$, so that it spans an ideal of the latter algebra. By Lemma 2.4.2 it follows that $\rho_i(a) = 0$. We conclude that $\mathfrak{n} \subset \mathfrak{n}'$, proving (ii).

As in the proof of Proposition 4.3.19 we see that $\mathfrak{g}$ is a completely reducible $\mathfrak{t}$-module. So there is a subspace $\mathfrak{s}'$ of $\mathfrak{g}$ with $\mathfrak{g} = \mathfrak{n} \oplus \mathfrak{t} \oplus \mathfrak{s}'$ and $[\mathfrak{t}, \mathfrak{s}'] \subset \mathfrak{s}'$. But also $[\mathfrak{t}, \mathfrak{s}'] \subset \mathfrak{r}$ (as $\mathfrak{t} \subset \mathfrak{r}$). Hence $[\mathfrak{t}, \mathfrak{s}'] = 0$.

Now set $\mathfrak{h} = \{a \in \mathfrak{g} \mid [a, \mathfrak{t}] = 0\}$. Then $\mathfrak{h}$ is a subalgebra of $\mathfrak{g}$ and $\mathfrak{s}' \subset \mathfrak{h}$. Set $\mathfrak{r}' = \mathfrak{h} \cap \mathfrak{r}$. We claim that $\mathfrak{r}'$ is the solvable radical of $\mathfrak{h}$. It is certainly a solvable ideal of $\mathfrak{h}$. Moreover, $\mathfrak{h}/\mathfrak{r}'$ is isomorphic to $(\mathfrak{h} + \mathfrak{r})/\mathfrak{r}$. But $\mathfrak{h} + \mathfrak{r} \supset \mathfrak{s}' + \mathfrak{r} = \mathfrak{g}$. It follows that $\mathfrak{h}/\mathfrak{r}'$ is isomorphic to $\mathfrak{g}/\mathfrak{r}$, which is semisimple. Therefore $\mathfrak{r}'$ is the solvable radical of $\mathfrak{h}$.

Let $\mathfrak{h} = \mathfrak{s} \oplus \mathfrak{r}'$ be a Levi decomposition of $\mathfrak{h}$ (Theorem 2.7.10). Then $\mathfrak{g} = \mathfrak{r} \oplus \mathfrak{s}' \subset \mathfrak{r} + \mathfrak{h} = \mathfrak{r} \oplus \mathfrak{s}$, so that $\mathfrak{r} \oplus \mathfrak{s}$ is the Levi decomposition of $\mathfrak{g}$. Furthermore, $[\mathfrak{s}, \mathfrak{t}] = 0$ as $\mathfrak{s} \subset \mathfrak{h}$.

Finally, $\mathfrak{n}$ is algebraic by Theorem 4.3.14 and $\mathfrak{t}$ is algebraic by Proposition 4.3.19. Furthermore, $\mathfrak{s}$, being semisimple, is generated by nilpotent elements (for example by the root vectors corresponding to a root system). Hence by Lemma 4.3.1, Corollary 4.3.7, $\mathfrak{s}$ is algebraic. □

Now we consider the problem of computing the subalgebras and ideal promised by the theorem. The algorithm is based on the following lemma.

Lemma 4.3.21 *Let $\mathfrak{r}$, $\mathfrak{t}$, $\mathfrak{n}$ be as in Theorem 4.3.20.*

(i) *Set $\mathfrak{h} = \mathfrak{c}_{\mathfrak{r}}(\mathfrak{t})$, the centralizer of $\mathfrak{t}$ in $\mathfrak{r}$. Then $\mathfrak{h}$ is a Cartan subalgebra of $\mathfrak{r}$, and $\mathfrak{h} = \mathfrak{t} \oplus \mathfrak{c}_{\mathfrak{n}}(\mathfrak{t})$.*

(ii) *Let $\mathfrak{h}'$ be a Cartan subalgebra of $\mathfrak{r}$. Then there is an abelian subalgebra $\mathfrak{t}' \subset \mathfrak{r}$ consisting of semisimple elements such that $\mathfrak{r} = \mathfrak{t}' \oplus \mathfrak{n}$, and $\mathfrak{h}' = \mathfrak{c}_{\mathfrak{r}}(\mathfrak{t}') = \mathfrak{t}' \oplus \mathfrak{c}_{\mathfrak{n}}(\mathfrak{t}')$.*

(iii) *Let $a_1, \dots, a_r$ be a basis of $\mathfrak{h}'$ and let $a_i = s_i + n_i$ be the Jordan decomposition of a_i. Then $\mathfrak{t}'$ is spanned by $s_1, \dots, s_r$. Let $\mathfrak{r}_1(\mathfrak{h}')$ be the Fitting one component of $\mathfrak{r}$ with respect to the adjoint action of $\mathfrak{h}'$. Then $\mathfrak{n}$ is spanned by the n_i along with $\mathfrak{r}_1(\mathfrak{h}')$.*

Proof. Let $x \in \mathfrak{c}_{\mathfrak{r}}(\mathfrak{t})$, then we can write $x = y + z$, where $y \in \mathfrak{t}$ and $z \in \mathfrak{n}$. So for any $u \in \mathfrak{t}$ we get $0 = [u, x] = [u, y] + [u, z] = [u, z]$. It follows that $z \in \mathfrak{c}_{\mathfrak{r}}(\mathfrak{t})$. Hence $\mathfrak{h} = \mathfrak{t} \oplus \mathfrak{c}_{\mathfrak{n}}(\mathfrak{t})$. Since $\mathfrak{n}$ is nilpotent and $[\mathfrak{t}, \mathfrak{c}_{\mathfrak{n}}(\mathfrak{t})] = 0$ it also follows that $\mathfrak{h}$ is nilpotent.

Let $x \in \mathfrak{t}$. Then x is a semisimple linear transformation. By Proposition 2.2.5, $\mathrm{ad}x : \mathfrak{r} \to \mathfrak{r}$ is semisimple. Hence $\mathfrak{c}_{\mathfrak{r}}(\mathfrak{t}) = \mathfrak{r}_0(\mathfrak{t})$ (the Fitting zero component, see Section 2.5.2). But by Proposition 2.5.6, $\mathfrak{r}_0(\mathfrak{t})$ equals its normalizer in $\mathfrak{r}$. As $\mathfrak{h} = \mathfrak{r}_0(\mathfrak{t})$ is nilpotent, it follows that it is a Cartan subalgebra.

In order to prove (ii) we use a refinement of Theorem 2.5.8, whose proof we omit. Let D be the group of automorphisms of $\mathfrak{r}$ generated by $\exp(\mathrm{ad}x)$ for $x \in [\mathfrak{r}, \mathfrak{r}]$. There is a $\sigma \in D$ such that $\sigma(\mathfrak{h}) = \mathfrak{h}'$ ([Hum78], Theorem 16.2). Set $\mathfrak{t}' = \sigma(\mathfrak{t})$. Suppose that $\sigma = \exp(\mathrm{ad}x)$, where $x \in [\mathfrak{r}, \mathfrak{r}] \subset \mathfrak{n}$. But then $x \in \mathfrak{n}$ and $\mathfrak{t}' = (\exp x)\mathfrak{t}(\exp x)^{-1}$ (Lemma 2.3.1). So in this case $\mathfrak{t}'$ consists of semisimple elements. Since D is generated by automorphisms of this form, the same conclusion follows for general σ. The other properties of $\mathfrak{t}'$ follow from the analogous properties of $\mathfrak{h}$ and the fact that σ is an automorphism.

Let $h \in \mathfrak{h}'$; then by (ii) there are $s \in \mathfrak{t}'$ and $n \in \mathfrak{c}_{\mathfrak{n}}(\mathfrak{t}')$ with $h = s + n$. But then s is semisimple, n is nilpotent and $[s, n] = 0$. It follows that $h = s + n$ is the Jordan decomposition of h. In particular, all s_i lie in $\mathfrak{t}'$ and all n_i lie in $\mathfrak{c}_{\mathfrak{n}}(\mathfrak{t}')$. Also $h = \sum_i \alpha_i a_i = (\sum_i \alpha_i s_i) + (\sum_i \alpha_i n_i)$. This implies that $\sum_i \alpha_i s_i = s$ and $\sum_i \alpha_i n_i = n$. Therefore, the s_i span $\mathfrak{t}'$, and the n_i span $\mathfrak{c}_{\mathfrak{n}}(\mathfrak{t}')$. Note that $\mathfrak{r}_0(\mathfrak{h}') = \mathfrak{h}'$ as $\mathfrak{h}'$ is a Cartan subalgebra of $\mathfrak{r}$ (Proposition 2.5.7). Also $\mathfrak{r}_1(\mathfrak{h}') \subset [\mathfrak{r}, \mathfrak{r}] \subset \mathfrak{n}$. Now $\mathfrak{r} = \mathfrak{t}' \oplus \mathfrak{c}_{\mathfrak{n}}(\mathfrak{t}') \oplus \mathfrak{r}_1(\mathfrak{h}')$. The last two spaces are contained in $\mathfrak{n}$. Since we also have $\mathfrak{r} = \mathfrak{t}' + \mathfrak{n}$, $\mathfrak{n} = \mathfrak{c}_{\mathfrak{n}}(\mathfrak{t}') \oplus \mathfrak{r}_1(\mathfrak{h}')$, whence the last statement. □

The algorithm for obtaining the decomposition of Theorem 4.3.20 given a basis of $\mathfrak{g}$ is quite obvious. First we compute a basis of the solvable radical $\mathfrak{r}$ of $\mathfrak{g}$ (Section 2.6.3) and a Cartan subalgebra $\mathfrak{h}$ of $\mathfrak{r}$ (Section 2.5.3). Let $a_1, \dots, a_r$ be a basis of $\mathfrak{h}$, and compute the Jordan decomposition $a_i = s_i + n_i$ (Section 2.2). Then $\mathfrak{t}$ is the subspace spanned by $s_1, \dots, s_r$. Subsequently we compute the Fitting one component $\mathfrak{r}_1(\mathfrak{h})$ (Section 2.5.2). Then $\mathfrak{n}$ is the subspace spanned by $n_1, \dots, n_r$ and $\mathfrak{r}_1(\mathfrak{h})$. Finally, we set $\mathfrak{g}' = \mathfrak{c}_{\mathfrak{g}}(\mathfrak{t})$ and compute a semisimple subalgebra $\mathfrak{s}$ of $\mathfrak{g}'$, complementing the radical of $\mathfrak{g}'$ (Section 2.7.2).

Now we have a theorem that describes what this decomposition means for algebraic groups.

Theorem 4.3.22 *Let $G \subset \mathrm{GL}(n, K)$ be an algebraic group with Lie algebra $\mathfrak{g}$. Let $\mathfrak{s}$, $\mathfrak{t}$, $\mathfrak{n}$, $\mathfrak{r}$ be as in Theorem 4.3.20. Then $\mathfrak{r}$, $\mathfrak{n}$ are the Lie algebras of the radical $R(G)$ and the unipotent radical $R_u(G)$ respectively. If G is connected then $G = H \ltimes R_u(G)$ (semidirect product of algebraic groups), where H is the connected algebraic subgroup with Lie algebra $\mathfrak{s} \oplus \mathfrak{t}$.*

Proof. By Proposition 3.8.4, the Lie algebra of $R(G)$ is solvable. By Proposition 4.2.12 it is an ideal of $\mathfrak{g}$. So it is contained in $\mathfrak{r}$. Let R be the connected algebraic subgroup of G with Lie algebra $\mathfrak{r}$ (note that $\mathfrak{r}$ is algebraic by Proposition 4.3.18). Then R is normal in G by Proposition 4.2.12. Since $R_u(G)$ is the set of unipotent elements of $R(G)$, it follows that $\mathfrak{n} = \mathrm{Lie}(R_u(G))$.

Let G' be the smallest algebraic subgroup of G containing H and $R_u(G)$. Since $R_u(G)$ is normal in G we have that $R_u(G)H = HR_u(G)$. So by Proposition 3.2.3, $G' = HR_u(G)$. Furthermore, $\mathrm{Lie}(G')$ contains $\mathrm{Lie}(H) = \mathfrak{s} \oplus \mathfrak{t}$ and $\mathrm{Lie}(R_u(G)) = \mathfrak{n}$, and therefore $\mathrm{Lie}(G')$ contains $\mathrm{Lie}(G)$. The reverse inclusion is obvious, so by Theorem 4.2.2, $G = G'$. In particular, $G = HR_u(G)$.

The Lie algebra of $H \cap R_u(G)$ is trivial (Theorem 4.2.2), so this group is finite, and being unipotent as well, it has to be trivial.

Let V be a finite-dimensional subspace of $K[G]$, generating $K[G]$ and stable under the action of G (via $g \cdot f = \lambda_g(f)$; see Section 3.5). Let $V = V_1 \supset V_2 \supset \cdots \supset V_s = 0$ be a composition series of the G-module V. This means that the V_i are G-submodules, and the successive quotients V_i/V_{i+1} are irreducible

G-modules. Set $V' = V_1/V_2 \oplus \cdots \oplus V_{s-1}/V_s$, which is a G-module. Write $\mathfrak{h} = \mathrm{Lie}(H)$. Note that this is a reductive Lie algebra (Proposition 2.12.2). By Theorem 2.12.3, V is completely reducible as an $\mathfrak{h}$-module. So by Corollary 4.2.10, V is also completely reducible as an H-module. It follows that V and V' are isomorphic as H-modules. Let ρ, ρ' denote the representations of G associated with V, V'. Then there is a bijective linear map $\gamma : V' \to V$ such that $\gamma(\rho'(h)v) = \rho(h)\gamma(v)$ for $v \in V'$, $h \in H$.

Define $\alpha : G \to H$ by $\alpha(hu) = h$, for $h \in H$, $u \in R_u(G)$. This is a surjective group homomorphism. We show that it is a regular map.

As $R_u(G)$ is unipotent, it acts trivially on the quotients V_i/V_{i+1}. (Indeed, the space of all $\bar{v} \in V_i/V_{i+1}$ such that $g\bar{v} = \bar{v}$ for all $g \in R_u(G)$ is a non-zero G-submodule.) Therefore, $\rho' = \rho' \circ \alpha$. Hence for $g \in G$ we have

$$\gamma\rho'(g)\gamma^{-1} = \gamma\rho'\alpha(g)\gamma^{-1} = \rho(\alpha(g)).$$

Let $f \in V$, then by the definition of the G-action on V we obtain $f(\alpha(g)) = \rho(\alpha(g))(f)(e) = (\gamma\rho'(g)\gamma^{-1}(f))(e)$, where $e \in G$ denotes the identity element. Since ρ' is a rational representation of G it follows that $f \circ \alpha$ lies in $K[G]$. Because V generates $K[G]$ this holds for all $f \in K[G]$ and shows that α is a regular map.

Therefore the map $\beta : G \to R_u(G)$, $\beta(g) = \alpha(g)^{-1}g$ is regular as well. The map $H \times R_u(G) \to G$, $(h, u) \mapsto hu$ is regular, and its inverse is $g \mapsto (\alpha(g), \beta(g))$, which is regular as well. The conclusion is that $G = H \ltimes R_u(G)$. □

Remark 4.3.23 If G is not connected, let H be a maximal subgroup of G with Lie algebra $\mathfrak{s} \oplus \mathfrak{t}$. Mostow ([Mos56]) has shown that also in that case $G = HR_u(G)$. So also in this case we get a decomposition of G as a semidirect product of H and $R_u(G)$.

4.4 Computing the algebraic hull

In this section we consider the problem of constructing the algebraic hull of a given subalgebra $\mathfrak{a} \subset \mathfrak{gl}(n, K)$.

Let $a_1, \ldots, a_r$ be a set of Lie algebra generators of $\mathfrak{a}$ (for example a basis), and let $\mathfrak{g} \subset \mathfrak{gl}(n, K)$ be the Lie algebra generated by the $\mathfrak{g}(a_i)$ (notation as in Section 4.3.1). Then from Corollary 4.3.7 it follows that $\mathfrak{g}$ is algebraic. Hence the algebraic hull of $\mathfrak{a}$ is contained in $\mathfrak{g}$. On the other hand, every $\mathfrak{g}(a_i)$ has to be contained in the algebraic hull of $\mathfrak{a}$. We conclude that $\mathfrak{g}$ is the algebraic hull of $\mathfrak{a}$. So it suffices to construct $\mathfrak{g}(a_i)$ for all i. Let $a_i = s_i + n_i$ be the Jordan decomposition of a_i. Then by Theorem 3.10.2, $s_i, n_i \in \mathfrak{g}$. Hence $\mathfrak{g}$ is generated by the $\mathfrak{g}(s_i), \mathfrak{g}(n_i)$. Moreover, $\mathfrak{g}(n_i)$ is spanned by n_i (Lemma 4.3.1). So we are left with the problem of constructing $\mathfrak{g}(s)$, where $s \in \mathfrak{gl}(n, K)$ is a semisimple matrix. For this we suppose that s has coefficients in a field k which is a finite extension of $\mathbb{Q}$. This means that we can construct a field $k' \supset k$ containing the eigenvalues $\alpha_1, \dots, \alpha_n$ of s (basically by factorizing polynomials over number fields and extending the field if a factor is not linear).

By writing the α_i as linear combinations of the elements of a basis of k' over $\mathbb{Q}$ and solving a set of linear equations, we can compute a basis for

$$\Lambda_{\mathbb{Q}} = \{(e_1, \dots, e_n) \in \mathbb{Q}^n \mid \sum_{i=1}^{n} e_i \alpha_i = 0\}.$$

Now for $e = (e_1, \dots, e_n) \in \mathbb{Q}^n$ and $i \geq 0$ we set $\Delta_i(e) = \sum_{j=1}^{n} e_j \alpha_j^i$. As in the previous section we let $K[s]$ denote the associative algebra (with identity) over K generated by s. Let m be such that $s^0 (= I_n), s, \dots, s^m$ is a basis of $K[s]$ (m is one less than the degree of the minimal polynomial of s). Set

$$\Upsilon = \{(\gamma_0, \dots, \gamma_m) \in K^{m+1} \mid \sum_{i=0}^{m} \Delta_i(e)\gamma_i = 0 \text{ for all } e \in \Lambda_{\mathbb{Q}}\}.$$

Lemma 4.4.1

$$\mathfrak{g}(s) = \{\sum_{i=0}^{m} \gamma_i s^i \mid (\gamma_0, \dots, \gamma_m) \in \Upsilon\}.$$

Proof. Set $s' = \mathrm{diag}(\alpha_1, \dots, \alpha_n)$. Then there is a $u \in \mathrm{GL}(n, k')$ with $usu^{-1} = s'$. Since the minimal polynomials of s and s' coincide, $I_n, s', s'^2, \dots, s'^m$ is a basis of $K[s']$. Moreover, by Proposition 4.3.8(i), $\mathfrak{g}(s')$ is contained in $K[s']$.

Let $y = \sum_{i=0}^{m} \gamma_i s'^i \in K[s']$. Write $y(l,l)$ for the entry of y on position (l,l). Define Λ as in Lemma 4.3.3. By that lemma, $y \in \mathfrak{g}(s')$ if and only if for all $e = (e_1, \dots, e_n) \in \Lambda$ we have $\sum_l e_l y(l,l) = 0$. It is clear that in this statement we may replace Λ by $\Lambda_{\mathbb{Q}}$. Indeed, Λ is a subgroup of $\mathbb{Z}^n$ and hence it is finitely generated. Furthermore, a $\mathbb{Z}$-basis of Λ will also be a $\mathbb{Q}$-basis of $\Lambda_{\mathbb{Q}}$. Now $y(l,l) = \sum_{i=0}^{m} \gamma_i \alpha_l^i$, and hence $\sum_{l=1}^{n} e_l y(l,l) = \sum_{i=0}^{m} \Delta_i(e)\gamma_i$. It follows that $\mathfrak{g}(s') = \{\sum_{i=0}^{m} \gamma_i s'^i \mid (\gamma_0, \dots, \gamma_m) \in \Upsilon\}$. By Proposition 4.3.8(iii) $\mathfrak{g}(s) = u^{-1}\mathfrak{g}(s')u$ and hence $\mathfrak{g}(s) = \{\sum_{i=0}^{m} \gamma_i s^i \mid (\gamma_0, \dots, \gamma_m) \in \Upsilon\}$. □

Using this we can compute $\mathfrak{g}(s)$. Indeed, first we compute a basis of $\Lambda_{\mathbb{Q}}$. Then by solving a set of linear equations with coefficients in k' we obtain a basis of Υ. This immediately gives us a basis of $\mathfrak{g}(s)$.

Example 4.4.2 Let $s \in \mathfrak{gl}(4, \mathbb{Q})$ have minimal polynomial $T^4 + bT^2 + c$ with $D = b^2 - 4c$ not a square in $\mathbb{Q}$. Then the eigenvalues of s are $\alpha_1 = \alpha$, $\alpha_2 = -\alpha$, $\alpha_3 = \beta$, $\alpha_4 = -\beta$, where $\alpha^2 = \frac{1}{2}(-b + \sqrt{D})$ and $\beta^2 = \frac{1}{2}(-b - \sqrt{D})$. So α and β cannot be proportional over $\mathbb{Q}$ (otherwise α^2 and β^2 would be as well). Hence the α_i span a 2-dimensional subspace of K. So $\dim \Lambda = 2$, and is spanned by $e^1 = (1,1,0,0)$ and $e^2 = (0,0,1,1)$. Then $\Delta_0(e^1) = 2$, $\Delta_1(e^1) = \Delta_3(e^1) = 0$, $\Delta_2(e^1) = 2\alpha^2$. For e^2 we get the same except that $\Delta_2(e^2) = 2\beta^2$. So

$$\Upsilon = \left\{(\gamma_0, \dots, \gamma_3) \in K^3 \mid 2\gamma_0 + 2\alpha^2\gamma_2 = 2\gamma_0 + 2\beta^2\gamma_2 = 0\right\}.$$

As $\alpha^2 \neq \beta^2$, Υ consists of all vectors of the form $(0, \gamma_1, 0, \gamma_3)$. We conclude that $\mathfrak{g}(s)$ is spanned by s, s^3.

Example 4.4.3 Let $s \in \mathfrak{gl}(6, \mathbb{Q})$ have minimal polynomial $f = x^6 + 3x^5 - 5x^3 + 3x + 2$. Then f is irreducible over $\mathbb{Q}$ and the degree of the splitting field of f over $\mathbb{Q}$ is 48. This can still be handled quite easily by modern computer algebra systems. Using some computations in such a system, we find that $\Lambda_{\mathbb{Q}}$ is spanned by

$$(1,1,0,0,-1,-1),\ (0,0,1,1,-1,-1),$$

and Υ by

$$(1,0,0,0,0,0),\ (0,1,0,0,0,0),\ (0,0,1,0,-1,-\tfrac{2}{5}),\ (0,0,0,1,\tfrac{3}{2},\tfrac{3}{5}).$$

So $\mathfrak{g}(s)$ is spanned by $s^0 = I_6$, s, $s^2 - s^4 - \frac{2}{5}s^5$, $s^3 + \frac{3}{2}s^4 + \frac{3}{5}s^5$.

Remark 4.4.4 If s lies in $\mathfrak{gl}(n, \mathbb{Q})$ then an idea that presents itself is to use numerical approximations to the roots, $\hat{\alpha}_1, \dots, \hat{\alpha}_n$ to determine a basis of $\Lambda_{\mathbb{Q}}$. However, it can happen that there are integer vectors $(l_1, \dots, l_n)$, depending on the precision used, such that $\sum_i l_i\hat{\alpha}_i$ is zero (to the precision used) but that do not lie in $\Lambda_{\mathbb{Q}}$. (In other words, the span of the α_i in $\mathbb{C} \cong \mathbb{R}^2$ is not necessarily a lattice.) So it is not obvious how to make this work. In [FG07], an approach based on using also p-adic approximations to the roots is developed.

4.5 Computing defining polynomials for an algebraic group from its Lie algebra

Let $\mathfrak{g} \subset \mathfrak{gl}(n, K)$ be an algebraic Lie algebra. Let $G \subset \mathrm{GL}(n, K)$ be the connected algebraic subgroup with $\mathrm{Lie}(G) = \mathfrak{g}$, i.e., $G = G(\mathfrak{g})$. The question addressed here is how to find polynomials defining G as a subgroup of $\mathrm{GL}(n, K)$. We assume that $\mathfrak{g}$ is given by a basis consisting of elements of $\mathfrak{gl}(n, k)$, where k is a number field.

We first consider two subcases, namely $\mathfrak{g}$ consisting of nilpotent elements and $\mathfrak{g}$ consisting of commuting semisimple elements. Then we use these to formulate an algorithm for the general case.

4.5.1 Unipotent case

Here we let $\mathfrak{n} \subset \mathfrak{gl}(n, K)$ be a subalgebra consisting of nilpotent elements. Then $\mathfrak{n}$ is algebraic by Theorem 4.3.14, and nilpotent by Lemma 2.4.2. Furthermore, $G(\mathfrak{n})$ is unipotent (Theorem 4.3.14). Here we describe an algorithm for finding a set of polynomials defining $G(\mathfrak{n})$ as a subgroup of $\mathrm{GL}(n, K)$.

Example 4.5.1 Here we indicate the main idea of the algorithm using a 1-dimensional example. Let

$$x = \begin{pmatrix} 0 & 1 & 0 \\ 0 & 0 & 1 \\ 0 & 0 & 0 \end{pmatrix}, \text{ then } \exp(Tx) = \begin{pmatrix} 1 & T & \frac{1}{2}T^2 \\ 0 & 1 & T \\ 0 & 0 & 1 \end{pmatrix}.$$

Let $\mathfrak{n}$ be the Lie algebra spanned by x, then $G(\mathfrak{n})$ consists of $\exp(tx)$, $t \in K$ (Lemma 4.3.1). From the matrix $\exp(Tx)$ we see that we can "identify" T with x_{12}. Then we obtain the non-trivial polynomial equations by substitution: $x_{23} = T = x_{12}$ and $x_{13} = \frac{1}{2}T^2 = \frac{1}{2}x_{12}^2$. The main point is that we can always do something like this.

Proposition 4.5.2 *Let $\mathfrak{n} \subset \mathfrak{gl}(n, k)$ be a Lie algebra consisting of nilpotent matrices. Let $x_1, \dots, x_m$ be a basis of $\mathfrak{n}$ such that $x_i, \dots, x_m$ span a subalgebra $\mathfrak{n}_i$ of $\mathfrak{n}$, and such that $\mathfrak{n}_{i+1}$ is an ideal in $\mathfrak{n}_i$, for $1 \leq i \leq n$. Then every element of $G(\mathfrak{n})$ can be written as $\exp(t_1x_1) \cdots \exp(t_mx_m)$ for uniquely determined $t_i \in K$.*

Proof. The proof is by induction on m. The case $m = 1$ is covered by Proposition 4.3.15. So suppose $m > 1$. Then $\mathfrak{n}_2$ is an ideal of $\mathfrak{n}$, and by induction the result holds for $G(\mathfrak{n}_2)$.

We now use the *Baker-Campbell-Hausdorff formula.* We will not describe this formula in all details; for that see [Jac79], Section V.5. Applied to our situation it says the following. Let $x, y \in \mathfrak{n}$, then $\exp(x)\exp(y) = \exp(z)$, where

$$z = x + y + \tfrac{1}{2}[x, y] + \tfrac{1}{12}[[x, y], y] - \tfrac{1}{12}[[x, y], x] + \cdots,$$

which is an infinite series of longer and longer commutators of x and y. As $\mathfrak{n}$ is a nilpotent Lie algebra, all but a finite number of terms are zero.

Let $g \in G(\mathfrak{n})$, then according to Proposition 4.3.15 there are unique $\tau_i \in K$ with $g = \exp(\sum_{i=1}^n \tau_i x_i)$. By the Baker-Campbell-Hausdorff formula we have $\exp(-\tau_1 x_1)g = \exp(z)$, where $z \in \mathfrak{n}_2$. So by induction there are $t_2, \dots, t_m \in K$ with $\exp(z) = \exp(t_2x_2) \cdots \exp(t_mx_m)$. Therefore, with $t_1 = \tau_1$ we get $g = \exp(t_1x_1) \cdots \exp(t_mx_m)$.

In order to prove uniqueness, suppose $g = \exp(t_1x_1)h = \exp(t_1'x_1)h'$, where $t_1, t_1' \in K$, $h, h' \in G(\mathfrak{n}_2)$. Then $\exp((t_1 - t_1')x_1)$ lies in $G(\mathfrak{n}_2)$, and hence, by Proposition 4.3.15 it follows that $t_1 - t_1' = 0$. By induction we infer that the $t_2, \dots, t_m$ are uniquely determined as well. □

For our purposes we will need a basis that satisfies some additional properties. We suppose $\mathfrak{n}$ consists of strictly upper triangular matrices. (This is no essential restriction since we can always conjugate $\mathfrak{n}$ to an algebra with that property.) For a ring R we let $\mathcal{M}(R)$ be the associative algebra consisting of all strictly upper triangular matrices with coefficients in R. Moreover, for $r \geq 1$, $\mathcal{M}(R)^r$ will be the subalgebra spanned by the e_{ij} with $j-i \geq r$. We set $\mathfrak{n}(r) = \mathfrak{n} \cap \mathcal{M}(k)^r$. Then the $\mathfrak{n}(r)$ are ideals of $\mathfrak{n}$ with $[\mathfrak{n}(r), \mathfrak{n}(r')] \subset \mathfrak{n}(r+r')$. Let c be such that $\mathfrak{n}(c) \neq 0$ but $\mathfrak{n}(c+1) = 0$. Let $x_1, \dots, x_m$ be a basis of $\mathfrak{n}$ with the following properties. Firstly there are $1 = r_1 \leq r_2 \leq \cdots \leq r_c$ such that $x_{r_l}, x_{r_l+1}, \cdots, x_m$ is a basis of $\mathfrak{n}(l)$, for all l. Secondly, the elements $x_{r_l}, \dots, x_{r_{l+1}-1}$ are "echelonized": for each x_t with $r_l \leq t \leq r_{l+1}-1$ there is an (i,j) with $j-i = l$ such that $x_t(i,j) \neq 0$, but $x_s(i,j) = 0$ if $s \neq t$, and $r_l \leq s \leq r_{l+1}-1$. In the sequel we call such a basis an *adapted basis*. We note that it satisfies the requirements of Proposition 4.5.2.

Lemma 4.5.3 *Let $x_1, \dots, x_m$ be an adapted basis of $\mathfrak{n}$ and $T_1, \dots, T_m$ be indeterminates and set $A = \exp(T_1x_1) \cdots \exp(T_mx_m)$. Let $1 \leq s \leq m$ and r be such that $x_s \in \mathfrak{n}(r) \setminus \mathfrak{n}(r+1)$. Then there are p, q with $q - p = r$ and $A(p,q) = \alpha_s T_s + f_s(T_1, \dots, T_{s-1})$, where f_s is a polynomial in $s-1$ variables and $\alpha_s \in k$, $\alpha_s \neq 0$.*

Proof. For $1 \leq \nu \leq m$ we set $A_\nu = (\exp T_\nu x_\nu) \cdots (\exp T_m x_m)$. By descending induction we show that for $\nu = m, m-1, \dots, 1$ the following holds: if $q - p = t \geq 1$ then $A_\nu(p,q) \in k[T_1, \dots, T_{r_i-1}]$, where i is minimal such that $x_{r_i}, \dots, x_m$ belong to $\mathfrak{n}(t+1)$. For $\nu = m$ this clearly holds if we set $r_{c+1} = m+1$ (then we get $i = c+1$). Now supposing it holds for A_ν, we show it for $A_{\nu-1}$. Note that $A_{\nu-1}(p,q) = \sum_{j=1}^n (\exp T_{\nu-1}x_{\nu-1})(p,j)A_\nu(j,q)$, which equals

$$A_\nu(p,q) + \sum_{p<j<q} (\exp T_{\nu-1}x_{\nu-1})(p,j)A_\nu(j,q) + (\exp T_{\nu-1}x_{\nu-1})(p,q). \quad (4.2)$$

Let $p < j < q$. Then $q - j < q - p$, and therefore $A_\nu(j,q) \in k[T_1, \dots, T_{r_u-1}]$, where $u \leq i$. Let v be such that $x_{\nu-1} \in \mathfrak{n}(v) \setminus \mathfrak{n}(v+1)$. If $t \geq v$ then $x_{\nu-1} \notin \mathfrak{n}(t+1)$, hence $r_i > \nu - 1$ and $(\exp T_{\nu-1}x_{\nu-1})(p,q) \in k[T_1, \dots, T_{r_i-1}]$. On the other hand, if $t < v$ then $(\exp T_{\nu-1}x_{\nu-1})(p,q) = 0$. So the claim follows for $A_{\nu-1}$.

Now we show the statement of the lemma for A_ν instead of A and $\nu \leq s \leq m$. Again we use descending induction, the case $\nu = m$ being obvious. Again we use (4.2) for the induction step. If $\nu \leq s \leq m$, then by induction there are p, q with $p - q = r$ and $A_\nu(p,q) = \alpha_s T_s + p(T_1, \dots, T_{s-1})$. By the first part of the proof $A_\nu(j,q) \in k[T_1, \dots, T_{r_i-1}]$ with i minimal such that $x_{r_i}, \dots, x_m \in \mathfrak{n}(u+1)$, where $u = q - j < r$. But since $x_s \in \mathfrak{n}(u+1)$ it follows that $r_i - 1 < s$. Hence $A_{\nu-1}(k,q) = \alpha_s T_s + \tilde{f}(T_1, \dots, T_{s-1})$ where $\tilde{f}$ is a polynomial in $s-1$ variables. If $s = \nu - 1$, we let t be maximal with $x_{\nu-1}, \dots, x_t \in \mathfrak{n}(r)$ (so $x_{t+1}, \dots, x_m$ form a basis of $\mathfrak{n}(r+1)$). Then

$A_{\nu-1} = I_n + T_{\nu-1}x_{\nu-1} + \cdots + T_t x_t \bmod \mathcal{M}(k[T_1, \ldots, T_m])^{r+1}$. So by the echelon properties of the basis we can find p, q such that $x_{\nu-1}(p,q) = \alpha_{\nu-1} \neq 0$ and $x_i(p,q) = 0$ for $i = \nu+1, \ldots, t$. $\square$

Now we state the algorithm. As in Section 3.10 we write $R_n = K[x_{11}, x_{12}, \ldots, x_{nn}]$. For $f \in R_n$ and an $n \times n$ matrix A we write $f(A) = f(A(1,1), \ldots, A(n,n))$.

Algorithm 4.5.4 *Input: a basis of a Lie algebra* $\mathfrak{n} \subset \mathfrak{gl}(n,k)$ *consisting of nilpotent matrices.*
Output: defining polynomials for $G(\mathfrak{n})$.

1. *Compute an* $n \times n$ *matrix* U *such that* $\tilde{\mathfrak{n}} = U\mathfrak{n}U^{-1}$ *is in upper triangular form.*
2. *Compute an adapted basis* $x_1, \ldots, x_m$ *of* $\tilde{\mathfrak{n}}$.
3. *Compute* $A = \exp(T_1x_1) \cdots \exp(T_mx_m)$ *where* $T_1, \ldots, T_m$ *are indeterminates.*
4. *For* $1 \leq i \leq m$, *compute* s_i, t_i *with* $A(s_i, t_i) = \alpha_i T_i + f_i(T_1, \ldots, T_{i-1})$, *where* $\alpha_i \in k \setminus \{0\}$ *and* $f_i \in k[T_1, \ldots, T_{i-1}]$.
5. *Using the expressions obtained in the previous step compute polynomials* $P_i \in R_n$ *with* $T_i = P_i(A)$.
6. *Let* $\varphi : k[T_1, \ldots, T_m] \to R_n$ *be the homomorphism induced by the substitution* $T_i \mapsto P_i$.
7. *For* $1 \leq s, t \leq n$ *set* $\psi_{s,t} = x_{s,t} - \varphi(A(s,t))$.
8. *Let* $X = (x_{ij})$ *be the* $n \times n$ *matrix containing the indeterminates* x_{ij}. *Return the set consisting of* $\psi_{s,t}(UXU^{-1})$ *for* $1 \leq s, t \leq n$.

Comments: We note that all steps are computable. In order to compute U we consider the natural $\mathfrak{n}$-module, $V = k^n$. We let $V_1 = \{v \in V \mid \mathfrak{n} \cdot v = 0\}$; then $V_1 \neq 0$ by Proposition 2.4.1. For $l > 1$ we set $V_l = \{v \in V \mid \mathfrak{n} \cdot v \in V_{l-1}\}$; then $V_l \neq 0$ by Proposition 2.4.1 applied to the $\mathfrak{n}$-module V/V_{l-1}. It follows that there is an $r > 0$ such that $V = V_r$. We now construct a basis $v_1, \ldots, v_n$ of V such that $v_1, \ldots, v_{l_i}$ forms a basis of V_i where $1 \leq l_1 < l_2 < \cdots < l_r = m$. With respect to this basis, $\mathfrak{n}$ acts by upper triangular matrices. So we immediately get a matrix U with the required property.

Then computing an adapted basis of $\tilde{\mathfrak{n}}$ is straightforward. The s_i, t_i of Step 4 exist by Lemma 4.5.3, and can easily be found. We have $P_1 = \frac{1}{\alpha_1} x_{s_1,t_1}$. If the $P_1, \ldots, P_{i-1}$ are constructed, then $P_i = \frac{1}{\alpha_i}(x_{s_i,t_i} - f_i(P_1, \ldots, P_{i-1}))$. The remaining steps are straightforward to carry out.

Proposition 4.5.5 *Algorithm 4.5.4 is correct.*

Proof. We claim that the $\psi_{s,t}$ define $G(\tilde{\mathfrak{n}})$, i.e.,

$$G(\tilde{\mathfrak{n}}) = \{a \in \mathrm{GL}(n, K) \mid \psi_{s,t}(a) = 0 \text{ for all } s, t\}.$$

To see this, let $a \in G(\tilde{\mathfrak{n}})$. Then by Theorem 4.5.2 there are $\tau_1, \ldots, \tau_m \in K$ with $a = \exp(\tau_1 x_1) \cdots \exp(\tau_m x_m)$. This implies that $\tau_i = P_i(a)$. We define two evaluation maps: $e_1 : R_n \to K$ by $e_1(x_{st}) = a(s,t)$, and $e_2 : K[T_1, \ldots, T_m] \to K$ by $e_2(T_i) = \tau_i$. Then $e_1 \circ \varphi(T_i) = \tau_i = e_2(T_i)$. Hence $e_2 = e_1 \circ \varphi$. Now $\psi_{s,t}(a) = e_1(\psi_{s,t}) = a(s,t) - e_1(\varphi(A(s,t))) = a(s,t) - e_2(A(s,t)) = 0$. Hence a lies in the set on the right-hand side. Conversely, let $a \in \mathrm{GL}(n, K)$ be such that $\psi_{s,t}(a) = 0$ for all s, t. Again define $e_1 : R_n \to K$ by $e_1(x_{st}) = a(s,t)$. Then $\psi_{s,t}(a) = 0$ translates to $a(s,t) = e_1(\varphi(A(s,t))$. Hence $a = e_1(\varphi(A))$. Set $\tau_i = e_1(\varphi(T_i))$. Then $a = e_1(\varphi(A)) = \exp(\tau_1 x_1) \cdots \exp(\tau_m x_m)$. In other words, $a \in G(\tilde{\mathfrak{n}})$.

We note that $G(\tilde{\mathfrak{n}}) = UG(\mathfrak{n})U^{-1}$. Let $a \in \mathrm{GL}(n, K)$ then $a \in G(\mathfrak{n})$ if and only if $UaU^{-1} \in G(\tilde{\mathfrak{n}})$ if and only if $\psi_{s,t}(UaU^{-1}) = 0$ for all s, t. So the output of the algorithm is correct. □

Example 4.5.6 Let $\mathfrak{n} \subset \mathfrak{gl}(4, \mathbb{Q})$ have basis $x_1 = -e_{23}$, $x_2 = e_{12} - e_{34}$, $x_3 = e_{13} + e_{24}$, $x_4 = e_{14}$ (where e_{ij} denotes the 4×4 matrix with a 1 on position (i,j) and zeros elsewhere). We note that this is an adapted basis. Then

$$\exp(T_1 x_1) \cdots \exp(T_4 x_4) = \begin{pmatrix} 1 & T_2 & T_3 & T_4 + T_2 T_3 \\ 0 & 1 & -T_1 & T_3 + T_1 T_2 \\ 0 & 0 & 1 & -T_2 \\ 0 & 0 & 0 & 1 \end{pmatrix}.$$

So we get the substitution $T_1 \mapsto -x_{23}$, $T_2 \mapsto x_{12}$, $T_3 \mapsto x_{13}$, $T_4 \mapsto x_{14} - x_{12}x_{13}$. The polynomials that we find are the "obvious" ones, namely x_{ij} for $i > j$, $x_{ii} - 1$, along with $x_{34} + x_{12}$ and $x_{24} - x_{13} + x_{12}x_{23}$.

4.5.2 Diagonalizable case

Let $\mathfrak{a} \subset \mathfrak{gl}(n, K)$ be an algebraic Lie algebra consisting of commuting semisimple matrices. Here we consider the problem of computing a set of polynomials defining $G(\mathfrak{a})$. We assume that $\mathfrak{a}$ is given by a basis consisting of elements of $\mathfrak{gl}(n, k)$ where k is a finite extension of $\mathbb{Q}$.

Let $A \subset M_n(K)$ be the associative algebra with one generated by $\mathfrak{a}$. By A^* denote the set of all $a \in A$ with non-zero determinant. As seen in Example 3.6.9, A^* is an algebraic subgroup of $\mathrm{GL}(n, K)$ and $\mathfrak{a}$ is contained in its Lie algebra; therefore $G(\mathfrak{a}) \subset A^*$.

By an elementary result from linear algebra, the elements of $\mathfrak{a}$ are simultaneously diagonalizable. This means that there is an extension $k' \supset k$ and an $X \in \mathrm{GL}(n, k')$ such that $X\mathfrak{a}X^{-1}$ consists of diagonal matrices. Therefore XAX^{-1} also consists of diagonal matrices. It follows that A is a commutative semisimple associative algebra. As $G(\mathfrak{a}) \subset A^*$, we find that $G(\mathfrak{a})$ is diagonalizable.

We now indicate how to compute a k' and X as above. An $a_0 \in A$ that generates A (as algebra with one), is called a *splitting element.* By [Gra00], Lemma A.3.2, A has splitting elements, and moreover, a random element of A has a high probability of being a splitting element. (By following a proof of the primitive element theorem in number theory, it is also straightforward to devise a deterministic algorithm for constructing a splitting element of A.) So let $a_0 \in A$ be a splitting element, having coefficients in k. We construct the splitting field $k' \supset k$ of the minimal polynomial of a_0. By computing the eigenvalues and eigenvectors of a_0 we can compute an $X \in \mathrm{GL}(n, k')$ such that Xa_0X^{-1} is diagonal, implying that all XaX^{-1} are diagonal, for $a \in A$.

Let $a_1, \dots, a_r$ be a basis of $\mathfrak{a}$, and write $Xa_iX^{-1} = \mathrm{diag}(\alpha_{i1}, \dots, \alpha_{in})$. Define $\Lambda_\mathbb{Q}$ as the subspace of $\mathbb{Q}^n$ consisting of all $(e_1, \dots, e_n) \in \mathbb{Q}^n$ such that $\sum_j e_j\alpha_{ij} = 0$, for $1 \le i \le r$. By expressing elements of k' as linear combinations of the elements of a basis of k' over $\mathbb{Q}$, and solving a set of linear equations, we can compute a basis of the space $\Lambda_\mathbb{Q}$. Set $\Lambda = \Lambda_\mathbb{Q} \cap \mathbb{Z}^n$; a basis of Λ can be computed using the purification algorithm (see Section 6.2). By Example 4.2.5 and the construction of Λ, we have that $\mathrm{Lie}(\mathrm{D}(\Lambda)) = X\mathfrak{a}X^{-1}$. The latter is equal to $\mathrm{Lie}(G(X\mathfrak{a}X^{-1}))$. Since Λ is pure, $\mathrm{D}(\Lambda)$ is connected (Proposition 3.9.7). Hence by Theorem 4.2.2, we see that $G(X\mathfrak{a}X^{-1}) = \mathrm{D}(\Lambda)$. In the following we write $\mathfrak{b} = X\mathfrak{a}X^{-1}$, and $b_0 = Xa_0X^{-1} = \mathrm{diag}(\alpha_1, \dots, \alpha_n)$.

The algebra A is spanned by $I_n, a_0, a_0^2, \dots, a_0^t$, where $t + 1$ is the degree of the minimal polynomial of a_0. Now b_0 has the same characteristic polynomial as a_0. The minimal polynomial of a semisimple matrix is the square free part of its characteristic polynomial. Hence b_0 shares the minimal polynomial of a_0. So the algebra $K[b_0]$ generated by b_0 is spanned by $I_n, b_0, \dots, b_0^t$.

Let $y = \sum_{i=0}^t \delta_i b_0^i \in K[b_0]$. By $y(l, l)$ we denote the entry on position (l, l). Now $y \in G(\mathfrak{b})$ if and only if $\det(y) \neq 0$ and $\prod_l y(l, l)^{e_l} = 1$ for $e = (e_1, \dots, e_n)$ in a basis of Λ. Set $e_l' = e_l$ if $e_l \ge 0$, and $e_l' = 0$ otherwise. Also $e_l'' = e_l' - e_l$. Then $\prod_l y(l, l)^{e_l} = 1$ if and only if $\prod_l y(l, l)^{e_l'} = \prod_l y(l, l)^{e_l''}$. We substitute $y(l, l) = \sum_i \delta_i \alpha_l^i$. This yields a polynomial equation for the δ_i with coefficients in k'.

We do this for all e in a basis of Λ, and obtain polynomials $p_1, \dots, p_s \in k'[T_0, \dots, T_t]$, with the property that $\prod_l y(l, l)^{e_l} = 1$ for all $e \in \Lambda$ if and only if $p_i(\delta_0, \dots, \delta_t) = 0$ for $1 \le i \le s$. In other words, $y \in G(\mathfrak{b})$ if and only if $\det(y) \neq 0$ and $p_i(\delta_0, \dots, \delta_t) = 0$ for $1 \le i \le s$.

In the same way as in Proposition 4.3.8(iii) we infer that $G(\mathfrak{b}) = XG(\mathfrak{a})X^{-1}$. So $y \in G(\mathfrak{b})$ if and only if $X^{-1}yX \in G(\mathfrak{a})$. But $X^{-1}yX = \sum_i \delta_i a_0^i$. We conclude that an invertible $\sum_i \delta_i a_0^i \in G(\mathfrak{a})$ if and only if $p_j(\delta_0, \dots, \delta_t) = 0$ for $1 \le j \le s$.

Algorithm 4.5.7 *Input: a subalgebra $\mathfrak{a} \subset \mathfrak{gl}(n, K)$ consisting of commuting semisimple matrices, given by a basis whose elements have coefficients in k. Output: defining polynomials for $G(\mathfrak{a})$.*

1. *Compute a splitting element a_0 of the associative algebra with one A generated by $\mathfrak{a}$.*
2. *Construct the splitting field $k' \supset k$ of the minimal polynomial of a_0.*
3. *Compute a basis of Λ.*
4. *Construct polynomials $p_1, \dots, p_s \in k'[T_0, \dots, T_t]$ with the property that $\sum_{i=0}^{t} \delta_i a^i \in G(\mathfrak{a})$ if and only if $\det(y) \neq 0$ and $p_j(\delta_0, \dots, \delta_t) = 0$ for $1 \leq j \leq s$.*
5. *Set $M = \sum_{i=0}^{t} T_i a_0^i$ and let $\alpha_{ij}^l \in k$ be such that $T_l = \sum_{ij} \alpha_{ij}^l M(i,j)$.*
6. *Let x_{ij} for $1 \leq i, j \leq n$ be indeterminates and consider the substitution $T_l \mapsto \sum_{ij} \alpha_{ij}^l x_{ij}$. Let $\varphi : k'[T_0, \dots, T_t] \to R_n = k'[x_{11}, x_{12}, \dots, x_{nn}]$ be the corresponding ring homomorphism.*
7. *Let $g_1, \dots, g_r \in R_n$ be linear polynomials with the property that $x \in \mathfrak{gl}(n, K)$ lies in A if and only if $g_i(x) = 0$ for $1 \leq i \leq r$.*
8. *Return $\{g_1, \dots, g_r\} \cup \{\varphi(p_1), \dots, \varphi(p_s)\}$.*

Comments: We demonstrated above how to perform steps 1 to 4. Since the a_0^i for $0 \leq i \leq t$ are linearly independent, the coefficients α_{ij}^l in Step 5 exist (but they are not necessarily unique). The remaining steps are straightforward to carry out.

Proposition 4.5.8 *Algorithm 4.5.7 is correct.*

Proof. Let $c = (c_{ij}) \in \mathrm{GL}(n, K)$ be such that $g_i(c) = 0$ for all i. Then $c \in A$ so $c = \sum_i \delta_i a_0^i$. By substituting $T_l \mapsto \delta_l$ in the equation $T_l = \sum_{ij} \alpha_{ij}^k M(i,j)$ we see that $\delta_l = \sum_{i,j} \alpha_{ij}^l c_{ij}$. It follows that $c \in G(\mathfrak{a})$ if and only if

$$p_r\Big(\sum_{ij} \alpha_{ij}^0 c_{ij}, \dots, \sum_{ij} \alpha_{ij}^t c_{ij}\Big) = 0$$

for $1 \leq r \leq s$. But this is equivalent to $\varphi(p_r)(c) = 0$ for $1 \leq r \leq s$. We conclude that this algorithm returns defining polynomials for $G(\mathfrak{a})$. □

Example 4.5.9 Let $\mathfrak{a}$ be spanned by

$$a = \begin{pmatrix} 0 & 0 & 0 & -1 \\ 1 & 0 & 0 & 0 \\ 0 & 1 & 0 & 0 \\ 0 & 0 & 1 & 0 \end{pmatrix}.$$

In this case, the splitting element is just a. The minimal polynomial of a is T^4+1, so its eigenvalues are the primitive 8-th roots of unity. Let ζ denote a fixed primitive 8-th root of unity; then $b = XaX^{-1} = \mathrm{diag}(\zeta, \zeta^3, \zeta^5, \zeta^7)$ (here it does not matter what X looks like; it is enough to know that it exists). Furthermore, Λ is spanned by $(1,0,1,0)$, $(0,1,0,1)$. Let $y = \sum_{i=0}^{3} \delta_i b^i$. Then $y(1,1)y(3,3) = 1$ amounts to

$$\delta_0^2 + 2\delta_1\delta_3 - \delta_2^2 + (2\delta_0\delta_2 - \delta_1^2 + \delta_3^2)\zeta^2 = 1.$$

From $y(2,2)y(4,4) = 1$ we get

$$\delta_0^2 + 2\delta_1\delta_3 - \delta_2^2 - (2\delta_0\delta_2 - \delta_1^2 + \delta_3^2)\zeta^2 = 1.$$

The polynomial equations arising from this are equivalent to $T_0^2+2T_1T_3-T_2^2 = 1$ and $2T_0T_2 - T_1^2 + T_3^2 = 0$.

If $M = \sum_{i=0}^{3} T_i a^i$, then

$$M = \begin{pmatrix} T_0 & -T_3 & -T_2 & -T_1 \\ T_1 & T_0 & -T_3 & -T_2 \\ T_2 & T_1 & T_0 & -T_3 \\ T_3 & T_2 & T_1 & T_0 \end{pmatrix}.$$

We can, for instance, take the substitution $T_0 \mapsto x_{11}$, $T_1 \mapsto x_{21}$, $T_2 \mapsto x_{31}$, $T_3 \mapsto x_{41}$. The linear polynomials defining $K[a]$ can easily be read from M; they are: $x_{22} - x_{11}$, $x_{12} + x_{41}$, and so on. So we get these linear equations along with $x_{11}^2 + 2x_{21}x_{41} - x_{31}^2 - 1$ and $2x_{11}x_{31} - x_{21}^2 + x_{41}^2 = 0$.

From Example 4.4.2 we conclude that $\mathfrak{g}(a)$ is spanned by a, a^3. So $G(a)$ has dimension 2.

Remark 4.5.10 From Proposition 4.3.8 it follows that the ideal generated by the polynomials found by Algorithm 4.5.7 has a generating set consisting of polynomials with coefficients in k. This fact can also be established in a different way. Let $\mathcal{G} = \mathrm{Gal}(k'/k)$ and identify a $\sigma \in \mathcal{G}$ with an element (also denoted σ) of the symmetric group S_n by $\sigma(\alpha_i) = \alpha_{\sigma(i)}$. Then $e = (e_1, \dots, e_n) \in \Lambda$ if and only if $e^\sigma := (e_{\sigma^{-1}(1)}, \dots, e_{\sigma^{-1}(n)}) \in \Lambda$. Note that $\mathcal{G}$ acts by automorphisms on $k'[T_0, \dots, T_t]$ by stipulating $\sigma(T_i) = T_i$. Then σ maps the polynomials obtained from $e \in \Lambda$ to the polynomials obtained from e^σ. So the ideal in $k'[T_0, \dots, T_t]$ generated by the polynomials corresponding to all $e \in \Lambda$ is $\mathcal{G}$-stable. And that implies that this ideal has a basis defined over k (see [Win74], Theorem 3.2.5).

4.5.3 General case

Now let $\mathfrak{g} \subset \mathfrak{gl}(n,k)$ be an algebraic Lie algebra. As seen in Section 4.3.3 we can compute the decomposition

$$\mathfrak{g} = \mathfrak{s} \oplus \mathfrak{t} \oplus \mathfrak{n}$$

of Theorem 4.3.20. Here $\mathfrak{s}$ is semisimple and algebraic by Theorem 4.3.20.

Let $\mathfrak{h}$ be a Cartan subalgebra of $\mathfrak{s}$. Consider the root system of $\mathfrak{s}$ with respect to $\mathfrak{h}$. Let $\mathfrak{n}_-$, $\mathfrak{n}_+$ be the subalgebras of $\mathfrak{s}$ spanned by, respectively, the negative and positive root vectors. These subalgebras consist of nilpotent elements, and are therefore algebraic (Theorem 4.3.14). Also $\mathfrak{h}$ is algebraic (indeed, let $\mathfrak{h}'$ be the algebraic hull of $\mathfrak{h}$, then $\mathfrak{h}' \subset \mathfrak{s}$, and by Lemma 4.3.17(i), $\mathfrak{h}$ is normalized by $\mathfrak{h}'$, forcing $\mathfrak{h} = \mathfrak{h}'$ as $\mathfrak{h}$ equals its own normalizer). It follows that $\mathfrak{a} = \mathfrak{h} \oplus \mathfrak{t}$ is algebraic as well (Corollary 4.3.7). Moreover, it consists of commuting semisimple matrices (Proposition 2.11.5, Theorem 4.3.20). Using the algorithms of the previous subsections we can compute defining polynomials of $G(\mathfrak{n}_\pm)$, $G(\mathfrak{a})$ and $G(\mathfrak{n})$.

The map $\psi : \mathrm{GL}(n,K) \times \mathrm{GL}(n,K) \times \mathrm{GL}(n,K) \times \mathrm{GL}(n,K) \to \mathrm{GL}(n,K)$ given by $\psi(g_1,\dots,g_4) = g_1 \cdots g_4$ is regular. So by elimination methods (Section 1.6) we can compute polynomials defining the closure, denoted H, of $\psi(V)$, where $V = G(\mathfrak{n}_-) \times G(\mathfrak{a}) \times G(\mathfrak{n}_+) \times G(\mathfrak{n})$.

The differential of ψ is given by $\mathrm{d}\psi(x_1,\dots,x_4) = x_1 + \dots + x_4$ (Example 3.7.1). The tangent space (at $(I_n,\dots,I_n)$) of V is $T = \mathfrak{n}_- \oplus \mathfrak{a} \oplus \mathfrak{n}_+ \oplus \mathfrak{n}$ (Proposition 1.2.8). It follows that $\mathrm{d}\psi(T) = \mathfrak{g}$. So H is a closed subset of $G(\mathfrak{g})$, and the dimension of H is at least $\dim G(\mathfrak{g})$. It follows that $H = G(\mathfrak{g})$ (Theorem 1.3.4).

This algorithm uses some Gröber basis computations that can be quite difficult. The next examples give some idea of the possibilities and limitations of the algorithm. In both examples we let e_{ij} denote the 6×6 matrix with a 1 on position (i,j) and zeros elsewhere.

Example 4.5.11 Let $\mathfrak{g} = \mathfrak{n}^- \oplus \mathfrak{t} \oplus \mathfrak{n}^+ \subset \mathfrak{gl}(6,k)$, where $\mathfrak{t}$ is diagonalisable (in fact, diagonal) and spanned by

$$-e_{33} - e_{44} + e_{55} + e_{66},\ -e_{11} - e_{22} + e_{33} + e_{44},\ 2e_{11} + e_{33} - e_{44} + e_{55} - e_{66},\ 2e_{22} - e_{33} + e_{44} - e_{55} + e_{66},$$

$\mathfrak{n}^+$ is spanned by

$$x_{\alpha_1} = -e_{53} - e_{64},\ x_{\alpha_2} = -e_{31} - e_{42},\ x_{\alpha_3} = -e_{51} - e_{62},$$

and $\mathfrak{n}^-$ is spanned by

$$x_{-\alpha_1} = -e_{35} - e_{46},\ x_{-\alpha_2} = -e_{13} - e_{24},\ x_{-\alpha_3} = -e_{15} - e_{26}.$$

This Lie algebra is algebraic. Furthermore, it is reductive, with semisimple part of type A_2. The $x_{\pm\alpha_i}$ are the root vectors, with $\alpha_3 = \alpha_1 + \alpha_2$. It took about 11 seconds to compute defining polynomials for the corresponding group.

Example 4.5.12 Let $\mathfrak{g} = \mathfrak{n}_- \oplus \mathfrak{t} \oplus \mathfrak{n}_+ \subset \mathfrak{gl}(6,k)$, where $\mathfrak{t}$ is diagonalisable (in fact, diagonal) and spanned by

$$-e_{11} - e_{22} - 3e_{33} + e_{44} + e_{55} + 3e_{66},\ e_{11} - e_{22} - e_{33} - e_{44} + e_{55} - e_{66}.$$

$\mathfrak{n}_+$ is spanned by

$$x_\alpha = e_{15} + 2e_{24} - 3e_{32} - e_{46},$$

and $\mathfrak{n}_-$ is spanned by

$$x_{-\alpha} = -e_{23} + 2e_{42} + e_{51} - 3e_{64}.$$

This Lie algebra is algebraic. Furthermore, it is reductive, with semisimple part of type A_1. The computation of the defining polynomials of the corresponding group was terminated after 70 hours, when we found the program was using more than 30 GB of memory.

Remark 4.5.13 A different approach is the following. First construct a generating set of $\mathfrak{g}$ consisting entirely of elements that are nilpotent or semsisimple. Second, for each generating element a compute defining polynomials of $G(a)$. Third, put these groups together with the following algorithm.

Algorithm 4.5.14 *Input: connected algebraic subgroups* $G_1, G_2 \subset \mathrm{GL}(n, K)$. *Output: defining polynomials for the smallest algebraic subgroup of* $\mathrm{GL}(n, K)$ *containing* G_1, G_2.

1. *Let* G *be the trivial subgroup of* $\mathrm{GL}(n, K)$.
2. *Set* $G' := G$.
3. *Set* $G' := \overline{G'G_1}$.
4. *Set* $G' := \overline{G'G_2}$.
5. *If* $G' \neq G$ *then set* $G := G'$ *and return to 2. Otherwise return* G.

Here the closures $\overline{G'G_i}$ are computed with the usual Gröbner basis elimination algorithm. The correctness of the algorithm follows from Proposition 3.2.3.

The disadvantage of this approach, as opposed to the one above, is that many more Gröbner basis computations are needed, resulting in an algorithm that is more difficult to apply in practice. (Indeed, for the computation with the Lie algebra of Example 4.5.11 as input this algorithm needed 19669 seconds.) On the other hand, using this algorithm, we immediately see that the resulting group is defined over k if the G_i are.

4.6 The algebraic group generated by a given set

Here we look at the following problem: given a finite set $A \subset \mathrm{GL}(n, k)$, construct the smallest algebraic group $G \subset \mathrm{GL}(n, K)$ containing A. In contrast to the smallest algebraic group whose Lie algebra contains a given set of elements, this group is not necessarily connected: indeed, it could, for example, be finite. This leads to additional difficulties.

Throughout this section we use the following notation: for $a \in \mathrm{GL}(n,k)$ we denote the smallest algebraic subgroup of $\mathrm{GL}(n,K)$ containing it by G_a. Its connected component of the identity will be denoted G_a°.

Recall that algebraic groups are closed under the multiplicative Jordan decomposition (Theorem 3.10.4). So it is natural to first look at constructing G_a where a is unipotent or semisimple. However, instead of the full groups G_a we will describe how to construct their identity components, G_a°, along with bases of their Lie algebras because that is needed in the final algorithm.

4.6.1 Unipotent case

Let $u \in \mathrm{GL}(n,k)$ be unipotent; note that $a = \log(u)$ is nilpotent, and $G(a)$ is described by Lemma 4.3.1. From Lemma 4.3.9 we have $u = \exp(a)$. Therefore $u \in G(a)$, so $G_u \subset G(a)$. On the other hand, all powers $u^l = \exp(la)$, $l \in \mathbb{Z}$ lie in $G_u \cap G(a)$. But these form a dense subset of $G(a)$. So $G_u = G(a)$, and we can compute polynomial equations defining G_u by the algorithm given in Section 4.5.1. Note that $G(a)$ is connected and $\mathrm{Lie}(G_u)$ is spanned by a (see Example 3.6.7).

4.6.2 Semisimple case

Let $s \in \mathrm{GL}(n,k)$ be semisimple. We consider two cases.

In the first case s is diagonal. We use the notation of Examples 3.9.2, 3.9.6 and 4.2.5. In particular we recall that for $e \in \mathbb{Z}^n$ we define the character χ_e of $\mathrm{D}(n,K)$ by $\chi_e(\mathrm{diag}(\epsilon_1,\dots,\epsilon_n)) = \prod_{i=1}^n \epsilon_i^{e_i}$. Set

$$\Lambda_s = \{e \in \mathbb{Z}^n \mid \chi_e(s) = 1\}.$$

By the arguments used in Example 3.9.6 we see that $\mathrm{D}(\Lambda_s)$ is the smallest algebraic subgroup of $\mathrm{GL}(n,K)$ containing s. In other words, $G_s = \mathrm{D}(\Lambda_s)$. Let Λ_s' be the purification of Λ_s (Section 6.2). Then $\mathfrak{d}(\Lambda_s) = \mathfrak{d}(\Lambda_s')$. By Proposition 3.9.7 $\mathrm{D}(\Lambda_s')$ is connected. It follows that $G_s^\circ = \mathrm{D}(\Lambda_s')$.

In order to compute defining polynomials for $\mathrm{D}(\Lambda_s')$, we need to perform two tasks:

1. Compute a basis of Λ_s. In full generality this is rather difficult; but for the important case where s is defined over a number field, this can be done by the algorithm indicated in Remark 6.2.7.

2. Compute a basis of Λ_s' with the purification algorithm of Section 6.2.

Now we comment on the second case where s is not diagonal. We compute a field $k' \supset k$ containing the eigenvalues of s, and an $A \in \mathrm{GL}(n,k')$ such that $s' = AsA^{-1}$ is diagonal. By the algorithm outlined above we obtain a set of polynomials $P' \subset K[x_{ij}]$ defining $G_{s'}^{\circ} = AG_s^{\circ}A^{-1}$. Let X be the $n \times n$ matrix with $X(i,j) = x_{ij}$. For $p' \in P'$ set $p = p'(A^{-1}XA)$. Then the set of all p so obtained defines G_s°.

Alternatively, we can also proceed exactly as in Section 4.5.2, using the lattice $\Lambda'_{s'}$ and the associative algebras $K[s]$, $K[s']$. Using this approach we see that G_s° is defined over k (see Remark 4.5.10). In Example 4.6.4 this method is used.

Finally we remark that the Lie algebra of $G_{s'}$ consists of all matrices $\mathrm{diag}(\beta_1, \dots, \beta_n)$ such that $\sum_i e_i\beta_i = 0$ for $e = (e_1, \dots, e_n)$ in a basis of $\Lambda_{s'}$ (see Example 4.2.5). Furthermore, $\mathrm{Lie}(G_s) = A^{-1}\mathrm{Lie}(G_{s'})A$.

4.6.3 The algorithm

Here we simply state the algorithm, subsequently prove its correctness and illustrate it with an example.

Algorithm 4.6.1 *Input: a finite set $A \subset \mathrm{GL}(n,k)$, where all elements of A are either unipotent or semisimple.*
Output: defining polynomials for the smallest algebraic subgroup of $\mathrm{GL}(n,K)$ containing A.

1. *For each $a \in A$ compute $\mathfrak{h}_a = \mathrm{Lie}(G_a)$. Let $\mathfrak{h}$ be the Lie algebra generated by all $\mathfrak{h}_a$, $a \in A$.*

2. *While there is an $a \in A$ such that $a\mathfrak{h}a^{-1} \neq \mathfrak{h}$, replace $\mathfrak{h}$ by the Lie algebra generated by $\mathfrak{h}$ and $a\mathfrak{h}a^{-1}$.*

3. *Apply the algorithms of the previous section to obtain defining polynomials of $H = G(\mathfrak{h})$.*

4. *Let $G \subset \mathrm{GL}(n,K)$ denote the smallest algebraic group containing A. Let $\mathcal{A} \subset G/H$ be the group generated by the cosets aH, for $a \in A$. By repeatedly computing products of the aH, enumerate the group $\mathcal{A}$. For each newly formed element $bH \in \mathcal{A}$ do the following:*

   (a) Compute the semisimple and unipotent parts, s_b, u_b, of b. and $\mathfrak{g}_s = \mathrm{Lie}(G_{s_b})$, $\mathfrak{g}_u = \mathrm{Lie}(G_{u_b})$.

   (b) If at least one of $\mathfrak{g}_s$, $\mathfrak{g}_u$ is not contained in $\mathfrak{h}$, replace A by $A \cup \{s, u\}$ and return to 1.

5. *When we have enumerated all of $\mathcal{A}$, let $b_1, \dots, b_t \in \mathrm{GL}(n,K)$ be representatives of the cosets in $\mathcal{A}$. Compute polynomials defining the closed set $b_1H \cup \cdots \cup b_tH$, and return those.*

Proposition 4.6.2 *Algorithm 4.6.1 terminates and is correct.*

Proof. By Corollary 4.3.7 the Lie algebra $\mathfrak{h}$ as constructed in the first step is algebraic. Let $\widehat{H} \subset \mathrm{GL}(n,K)$ be the corresponding connected algebraic group. By Theorem 4.3.6, $\widehat{H}$ is the connected algebraic group generated by the G_a°, for $a \in A$.

The Lie algebra of the group $a\widehat{H}a^{-1}$ is $a\mathfrak{h}a^{-1}$. So by the same reasoning as above, after the second step terminates, $\mathfrak{h}$ is the Lie algebra of the smallest connected algebraic group that contains G_a° and is normalized by a for all $a \in A$.

After step 3., H is a connected algebraic subgroup of G. Moreover it is a normal subgroup of G. Indeed, let $p_1, \dots, p_r \in K[x_{ij}]$ generate the vanishing ideal of H. For $h \in H$ define $p_i^h \in K[x_{ij}, \frac{1}{\det(x_{ij})}]$ by $p_i^h(g) = p_i(ghg^{-1})$. Let $\mathcal{G}$ be the group generated by A. Then for a fixed $h \in H$, the p_i^h vanish on $\mathcal{G}$, and hence on $\overline{\mathcal{G}} = G$ (this equality follows from Lemma 3.1.8). This shows that H is normal in G. So G/H is a group, and $\mathcal{A}$ is a subgroup.

After a finite number of rounds $\mathfrak{h}$ is not changed further in step 4(b). Consider the situation at that point. Step 4(b) is not entered. Let $bH \in \mathcal{A}$; at some point in Step 4(a), b is considered. Since Step 4(b) is not entered, G_b° is contained in H. Hence there is an integer $l > 0$ such that $b^l \in H$. It follows that $\mathcal{A}$ consists of elements of finite order. By a theorem of Chevalley ([Che55b], Section V.3, Proposition 11), there is a rational representation $\rho : G \to \mathrm{GL}(m,K)$ with kernel H. Hence it induces an injective homomorphism $\bar{\rho} : G/H \to \mathrm{GL}(m,K)$. It follows that $\bar{\rho}(\mathcal{A})$ is a subgroup of $\mathrm{GL}(m,K)$ consisting of elements of finite order. However, it is known that a finitely generated subgroup of $\mathrm{GL}(m,K)$, whose elements have finite order, needs to be finite ([Kap95], Section II.2, Theorem G). We conclude that $\mathcal{A}$ is a finite group, and therefore Step 4 terminates and so does the algorithm.

When the algorithm terminates, $b_1H \cup \cdots \cup b_tH$ is a closed set, and moreover a group. Furthermore, this set is contained in any algebraic group containing the input set A. It follows that it coincides with G. □

Remark 4.6.3 Since this algorithm depends on the algorithms of Section 4.5.3 it will share their practical limitations (see Examples 4.5.11 and 4.5.12).

Example 4.6.4 Consider

$$a_1 = \begin{pmatrix} 0 & 0 & 0 & -1 \\ 1 & 0 & 0 & 0 \\ 0 & 1 & 0 & 0 \\ 0 & 0 & 1 & 0 \end{pmatrix}, \quad a_2 = \begin{pmatrix} 0 & 0 & 0 & 1 \\ 1 & 0 & 0 & 0 \\ 0 & 1 & 0 & 1 \\ 0 & 0 & 1 & 0 \end{pmatrix}.$$

Then a_1 has finite order (its order is 8), so $G_{a_1}^\circ = \{I_4\}$, and its Lie algebra is zero. The minimal polynomial of a_2 is $f = T^4 - T^2 - 1$. Its splitting field is $k' = \mathbb{Q}(\imath, \theta)$, where $\imath^2 = -1$ and θ is a root of f; it has degree 8 over $\mathbb{Q}$. The roots of f are $\pm\theta$, $\pm\imath(\theta - \theta^3)$. The lattice Λ of the multiplicative relations of the eigenvalues of a_2 is spanned by $(1,1,0,2)$, $(0,2,1,1)$, $(0,0,2,-2)$. (From the expressions for the eigenvalues given above it is easy to see that the corresponding multiplicative relations hold; it is a bit more work to show that they span the entire lattice.) Then Λ' (the purification of Λ) is spanned by $(1,0,0,1)$, $(0,1,0,1)$, $(0,0,1,-1)$. We follow the method of Section 4.5.2. Set $b = \mathrm{diag}(\theta, -\theta, \imath(\theta-\theta^3), -\imath(\theta-\theta^3))$. We want to compute the Lie algebra $\mathfrak{h}_2$ of G_{a_2}. For this we first consider the Lie algebra $\mathfrak{h}_b$ of G_b. Since G_b is contained in $K[b]$, the same holds for $\mathfrak{h}_b$. Let $y = \sum_{i=0}^{3} \delta_i b^i \in K[b]$, where $\delta_i \in K$. Then $y \in \mathfrak{h}_b$ if and only if $\sum_l e_l y(l,l) = 0$ for all $(e_1, \dots, e_4) \in \Lambda'$. The basis elements of Λ' yield the equations $y(1,1) + y(4,4) = 0$, $y(2,2) + y(4,4) = 0$ and $y(3,3) - y(4,4) = 0$. A short computation shows that this is equivalent to $\delta_1 = \delta_3 = 0$ and $\delta_2 = -2\delta_0$. It follows that $\mathfrak{h}_2$ consists of the elements $\sum_i \delta_i a_2^i$, where the δ_i satisfy these equations. Moreover,

$$\delta_0 + \delta_1 a_2 + \delta_2 a_2^2 + \delta_3 a_2^3 = \begin{pmatrix} \delta_0 & \delta_3 & \delta_2 & \delta_1+\delta_3 \\ \delta_1 & \delta_0 & \delta_3 & \delta_2 \\ \delta_2 & \delta_1+\delta_3 & \delta_0+\delta_2 & \delta_1+2\delta_3 \\ \delta_3 & \delta_2 & \delta_1+\delta_3 & \delta_0+\delta_2 \end{pmatrix}.$$

Therefore $\mathfrak{h}_2$ is spanned by

$$\begin{pmatrix} 1 & 0 & -2 & 0 \\ 0 & 1 & 0 & -2 \\ -2 & 0 & -1 & 0 \\ 0 & -2 & 0 & -1 \end{pmatrix}.$$

So we start with $\mathfrak{h} = \mathfrak{h}_2$. This algebra is not stable under conjugation with a_1; the Lie algebra generated by $\mathfrak{h}$ and $a_1 \mathfrak{h} a_1^{-1}$ is spanned by

$$\begin{pmatrix} 1 & 0 & -2 & 0 \\ 0 & 0 & 0 & 0 \\ -2 & 0 & -1 & 0 \\ 0 & 0 & 0 & 0 \end{pmatrix}, \quad \begin{pmatrix} 0 & 0 & 0 & 0 \\ 0 & 1 & 0 & -2 \\ 0 & 0 & 0 & 0 \\ 0 & -2 & 0 & -1 \end{pmatrix}.$$

These matrices commute and are semisimple. The computation of the defining polynomials for the group H takes 0.01 (s), using the algorithm of Section 4.5.2. We get

$$H = \left\{ \begin{pmatrix} x & 0 & y & 0 \\ 0 & s & 0 & t \\ y & 0 & x+y & 0 \\ 0 & t & 0 & s+t \end{pmatrix} \mid x^2 + xy - y^2 = 1,\ s^2 + st - t^2 = 1 \right\}.$$

The elements a_1^4, a_2^4, $(a_1a_2)^2$, $(a_1a_2^{-1})^2$ lie in H. The finitely presented group $\mathcal{G} = \langle g_1, g_2 \mid g_1^4, g_2^4, (g_1g_2)^2(g_1g_2^{-1})^2 \rangle$ has 16 elements. Furthermore, the cosets $a_1^i a_2^j H$, $0 \le i,j \le 3$, are all different. It follows that the group generated by the $a_i H$ is isomorphic to $\mathcal{G}$. We conclude that we get a disconnected algebraic group G, with $G^\circ = H$ and $G/G^\circ \cong \mathcal{G}$.

Remark 4.6.5 A related problem is to decide whether the smallest algebraic group H containing given $a_1, \dots, a_m \in \mathrm{GL}(n, K)$ is equal to a given algebraic group $G \subset \mathrm{GL}(n, K)$. With the algorithm for constructing H, this reduces to checking whether $G = H$, and that can be done by Gröbner basis computations. However, if the target group G happens to be connected, some shortcuts may be possible. If the groups G_{a_i} are connected, the $\mathfrak{h}$ computed in Step 1 of Algorithm 4.6.1 is already the Lie algebra of H, and we can decide by Theorem 4.2.2(i). (Indeed, $\mathfrak{h}$ is algebraic by Corollary 4.3.7; moreover, the connected algebraic group with Lie algebra $\mathfrak{h}$ contains all G_{a_i}, as they are connected.) If a_i is nilpotent, G_{a_i} is automatically connected (Corollary 4.3.11). If a_i is semisimple, as seen in Section 4.6.2, we can compute defining polynomials of the group $G^\circ_{a_i}$, allowing us to check whether $a_i \in G^\circ_{a_i}$ (which ensures that G_{a_i} is connected).

Remark 4.6.6 It is also possible to formulate an algorithm without using the Lie algebras of the various groups involved. First, defining polynomials of the G°_a are computed, and Algorithm 4.5.14 is used for computing the algebraic group H they generate. On the one hand, that algorithm would be less efficient. On the other hand, it can be made to work when the ground field is of characteristic $p > 0$ (see [DJK05]).

4.7 Generators, centralizers and normalizers

In this section we consider the problem converse to the one of Section 4.6: for an algebraic group $G \subset \mathrm{GL}(n, K)$ find a finite set of elements of G such that G is equal to the closure of the group generated by this set. We say that such a set *generates* G as an algebraic group, because the smallest algebraic group containing it is G. The group generated by the set will of course be smaller.

We describe an algorithm for finding a generating set and apply it to find algorithms for computing the centralizer and normalizer of an algebraic subgroup of an algebraic group.

4.7.1 Generating sets

First we supppose that G is connected and let $\mathfrak{g} \subset \mathfrak{gl}(n, K)$ be the Lie algebra of G. We assume that $\mathfrak{g}$ is given by a basis consisting of elements of $\mathfrak{gl}(n, k)$, where k is a subfield of K (typically this will be $\mathbb{Q}$, but it may also be a number field, for instance). Also we suppose that the given basis of $\mathfrak{g}$ consists of elements that are either nilpotent or semisimple (see Theorem 3.10.2).

Let $x \in \mathfrak{g}$ be either nilpotent or semisimple. As in Section 4.1 we denote the smallest algebraic group whose Lie algebra contains x by $G(x)$. We consider the problem of finding a finite set of elements generating $G(x)$ as an algebraic group. We denote the set that we find by $\Gamma(x)$. If x is nilpotent, this is easy: we can set $\Gamma(x) = \{\exp(x)\}$. (Indeed, $\exp(x)$ has infinite order, so the smallest algebraic group containing it has dimension at least 1; moreover, by Lemma 4.3.1, $\exp(x) \in G(x)$ and $G(x)$ has dimension 1, implying that $G(x)$ is generated, as algebraic group, by $\exp(x)$.)

If x is semisimple, then using the algorithms of Section 4.4 we compute the algebraic hull $\mathfrak{a}$ of x. We compute an $m \times m$ matrix X such that $\mathfrak{b} = X\mathfrak{a}X^{-1}$ consists of diagonal matrices and a basis of a pure lattice $\Lambda \subset \mathbb{Z}^m$ such that $G(\mathfrak{b}) = \mathrm{D}(\Lambda)$ (Section 4.5.2). As observed in Section 3.9, we can compute the lattice $\Lambda^\perp \subset \mathbb{Z}^m$. Let p denote the rank of $\Lambda^\perp$, then as seen in Section 3.9, we can construct a surjective morphism of algebraic groups $\varphi : \mathbb{G}_\mathrm{m}^p \to G(\mathfrak{b})$. This also yields a surjective morphism $\psi : \mathbb{G}_\mathrm{m}^p \to G(\mathfrak{a})$, by $\psi(\xi) = X^{-1}\varphi(\xi)X$. Let $\alpha \in K$ be an element of infinite order (for example, $\alpha = 2$). For $1 \leq i \leq p$ set $g_i = \psi(1, \dots, 1, \alpha, 1, \dots, 1)$, where the α is in the i-th argument. Let H_i be the group generated by $g_1, \dots, g_i$. Also define $\psi_i : (K^*)^i \to G$ by $\psi_i(z_1, \dots, z_i) = \psi(z_1, \dots, z_i, 1, \dots, 1)$. The image of ψ_i is an algebraic subgroup of G (Proposition 3.1.9), containing the closure of H_i. So no power of g_{i+1} is contained in this closure and hence the index of $\overline{H}_i$ in $\overline{H}_{i+1}$ is infinite. It follows that $\dim \overline{H}_{i+1} > \dim \overline{H}_i$. But $G(\mathfrak{a})$ is a connected algebraic group of dimension p containing all g_i. It follows that the algebraic group generated by the g_i is $G(\mathfrak{a})$. Also note that the latter is equal to $G(x)$. So in this case we set $\Gamma(x) = \{g_1, \dots, g_p\}$.

Now let Γ be the union of all sets $\Gamma(x)$, as x runs through the given basis of $\mathfrak{g}$. Let G' be the smallest algebraic group containing Γ. Then G' contains $G(x)$, for all x in the given basis of $\mathfrak{g}$. This implies that $\mathrm{Lie}(G')$ contains $\mathfrak{g}$. So by Theorem 4.2.2, $G \subset G'$. But also $\Gamma \subset G$, whence $G' \subset G$. It follows that $G = G'$.

If G is not connected, the problem is much more difficult. We first have to find the primary decomposition of $\mathcal{I}(G)$ (Remark 1.9.8), and for each primary ideal Q occurring in this decomposition, we have to find an element of $\mathcal{V}(Q)$. These tasks are difficult, but can in principle be carried out (possibly by extending the base field).

Remark 4.7.1 For algebraic groups of characteristic $p > 0$ it is not always possible to find a finite generating set. For example, if $G \subset \mathrm{GL}(n, \overline{\mathbb{F}}_p)$, any finite subset of G has the property that all of its elements have coefficients lying in a finite field. Hence such a subset generates a finite group, which is the smallest algebraic group containing the set.

4.7.2 The centralizer of an algebraic subgroup

Let $H \subset G \subset \mathrm{GL}(n, K)$ be algebraic groups, and let

$$Z_G(H) = \{g \in G \mid ghg^{-1} = h \text{ for all } h \in H\}$$

be its centralizer in G. Then $Z_G(H)$ is an algebraic subgroup of G. Here we describe how to compute a set of defining polynomials for it. We assume that we have the Lie algebras $\mathfrak{g} = \mathrm{Lie}(G)$, $\mathfrak{h} = \mathrm{Lie}(H)$ and defining polynomials of G and H. By the algorithm outlined in the previous subsection we can compute a set $\Gamma \subset H$, generating H as algebraic group.

Lemma 4.7.2 $Z_G(H) = \{g \in G \mid gh = hg \text{ for all } h \in \Gamma\}$.

Proof. Let M denote the right-hand side. Obviously $Z_G(H) \subset M$. Let $\langle\Gamma\rangle$ denote the group generated by Γ. Let $g \in M$, and for $1 \leq i, j \leq n$ define $p_{ij} \in K[G]$ by letting $p_{ij}(g')$ be the coefficient on position (i, j) of $gg'g^{-1} - g'$. Then the p_{ij} vanish on $\langle\Gamma\rangle$, and hence on its closure, which is H. We conclude that $g \in Z_G(H)$. □

This lemma yields an immediate algorithm for computing defining polynomials of $Z_G(H)$, as each condition $gh = hg$ is equivalent to a set of linear equations on the entries of g.

There also is a different approach, not using a generating set, and using only the Lie algebra. It is based on the following lemma.

Lemma 4.7.3 *Suppose H is connected. Then*

$$Z_G(H) = \{g \in G \mid gx = xg \text{ for all } x \in \mathfrak{h}\}.$$

Proof. For $g \in G$ define $\alpha_g : G \to G$ by $\alpha_g(h) = ghg^{-1}h^{-1}$. As seen in the proof of Lemma 3.8.3, $\mathrm{d}\alpha_g(x) = gxg^{-1} - x$. Moreover, by Theorem 4.2.4 we see that $\mathrm{Lie}(\ker \alpha_g) = \ker \mathrm{d}\alpha_g$. Now $g \in Z_G(H)$ if and only if $H \subset \ker \alpha_g$. As H is connected, that is the same as $\mathfrak{h} \subset \ker \mathrm{d}\alpha_g$ (Theorem 4.2.2), whence the result. □

Observe that it is enough to have the condition $gx = xg$ for all x in a basis of $\mathfrak{h}$. So, if G is connected, the lemma gives a finite number of linear equations that a $g \in G$ has to satisfy to lie in $Z_G(H)$. These linear equations are exactly the defining polynomials of $Z_G(H)$.

Example 4.7.4 Let $H = \left\{ \begin{pmatrix} x & 0 \\ 0 & y \end{pmatrix} \mid xy = 1 \right\} \subset G = \mathrm{GL}(2, K)$. Here we can take Γ just consisting of $h = \begin{pmatrix} 2 & 0 \\ 0 & \frac{1}{2} \end{pmatrix}$. Let $g = \begin{pmatrix} x_{11} & x_{12} \\ x_{21} & x_{22} \end{pmatrix}$. Then $gh = hg$ is equivalent to $\frac{1}{2}x_{12} = 2x_{12}$, $\frac{1}{2}x_{21} = 2x_{21}$. We conclude that the centralizer $Z_G(H)$ is the group of all diagonal matrices.

The Lie algebra of H is spanned by $\begin{pmatrix} 1 & 0 \\ 0 & -1 \end{pmatrix}$. Using that we arrive at the same conclusion.

4.7.3 The normalizer of an algebraic subgroup

Let the notation be as in the previous subsection. The group

$$N_G(H) = \{g \in G \mid ghg^{-1} \in H \text{ for all } h \in H\}$$

is called the *normalizer* of H in G. This is an algebraic subgroup of G (a fact that is perhaps not immediately clear; it will follow from the lemma below).

For $h \in H$ define $\varphi_h : G \to G$ by $\varphi_h(g) = ghg^{-1}$. This is a regular map, so the preimage $\varphi_h^{-1}(H)$ is a closed subset of G.

Lemma 4.7.5 *Let H be generated, as an algebraic group, by the set Γ. Then $N_G(H)$ is the intersection of the $\varphi_h^{-1}(H)$, where h runs through Γ.*

Proof. Let N denote the intersection in question. Obviously $N_G(H) \subset N$.

Observe that if g lies in the intersection of $\varphi_{h_1}^{-1}(H)$ and $\varphi_{h_2}^{-1}(H)$, then $ghg^{-1} \in H$ for all h in the group generated by h_1, h_2. Let $\langle\Gamma\rangle$ denote the group generated by Γ. Then for $g \in N$ it follows that $g\langle\Gamma\rangle g^{-1} \subset H$.

Let $g \in G$. We claim that $\overline{g\langle\Gamma\rangle g^{-1}} = g\overline{\langle\Gamma\rangle}g^{-1}$. First note that $g\overline{\langle\Gamma\rangle}g^{-1}$ is closed and contains $g\langle\Gamma\rangle g^{-1}$, and hence it contains $\overline{g\langle\Gamma\rangle g^{-1}}$ as well. Let $p \in K[G]$ vanish on $g\langle\Gamma\rangle g^{-1}$. Define $p_g \in K[G]$ by $p_g(g') = p(gg'g^{-1})$. Then p_g vanishes on $\langle\Gamma\rangle$, and hence on $\overline{\langle\Gamma\rangle}$. It follows that p vanishes on $g\overline{\langle\Gamma\rangle}g^{-1}$. This implies the reverse inclusion.

Now for $g \in N$ we conclude: $gHg^{-1} = g\overline{\langle\Gamma\rangle}g^{-1} = \overline{g\langle\Gamma\rangle g^{-1}} \subset \overline{H} = H$, whence $g \in N_G(H)$. □

In order to apply this lemma, we need to compute defining polynomials for the closed sets $\varphi_h^{-1}(H)$, $h \in \Gamma$. On a theoretical level this is straightforward, as $\varphi_h^{-1}(H)$ is defined by $\varphi_h^*(p)$, where p runs through a set of defining polynomials of H. However, $\varphi_h^*(p)(g) = p(ghg^{-1})$. In this equation we substitute the matrix (x_{ij}) for g (where we write $K[\mathrm{GL}(n,K)] = K[x_{ij}, \frac{1}{\det(x_{ij})}]$). The main problem is that we need to write the symbolic inverse of (x_{ij}). For small n this is readily done, but for increasing n it becomes very difficult (it is necessary to write the determinant of (x_{ij}), for example).

Example 4.7.6 Let H, G be as in Example 4.7.4. Let h, g also be as in that example. Then

$$ghg^{-1} = \frac{1}{x_{11}x_{22} - x_{12}x_{21}} \begin{pmatrix} 2x_{11}x_{22} - \frac{1}{2}x_{12}x_{21} & -\frac{3}{2}x_{11}x_{12} \\ \frac{3}{2}x_{21}x_{22} & \frac{1}{2}x_{11}x_{22} - 2x_{12}x_{21} \end{pmatrix}.$$

A set of defining polynomials of H consists of $p_1 = x_{12}$, $p_2 = x_{21}$ and $p_3 = x_{11}x_{22} - 1$. So

$$\varphi_h^*(p_1) = \frac{1}{\det(x_{ij})}(-\tfrac{3}{2}x_{11}x_{12}),\ \varphi_h^*(p_2) = \frac{1}{\det(x_{ij})}(\tfrac{3}{2}x_{21}x_{22}),$$

$$\varphi_h^*(p_3) = \frac{1}{\det(x_{ij})^2}(-\tfrac{9}{4}x_{11}x_{12}x_{21}x_{22}).$$

It follows that $N_G(H)$ is defined by $x_{11}x_{12}$, $x_{21}x_{22}$. So this group has two connected components: one consisting of the diagonal matrices (this is $Z_G(H)$, see Example 4.7.4) and one consisting of the antidiagonal matrices.

4.8 Orbit closures

Let G be an algebraic group and $\rho : G \to \mathrm{GL}(V)$ a rational representation. For a $v \in V$ we consider its orbit $\rho(G)v$ and the closure $\overline{\rho(G)v}$; see Section 3.12.

Concerning the orbit closures two questions come to mind:

A Can we find polynomials defining $\overline{\rho(G)v}$?

B Given $w \in V$, can we decide whether $w \in \overline{\rho(G)v}$?

We remark that a positive answer to A would immediately yield a positive answer to B. Furthermore, if we have a method for B, we can also decide whether $w \in \rho(G)v$, as that is equivalent to $w \in \overline{\rho(G)v}$ and $v \in \overline{\rho(G)w}$.

Note that $G \to \rho(G)v$, $g \mapsto \rho(g)v$ is a regular map. So if G is given by a set of defining polynomials and ρ is given by $\rho_{ij} \in K[G]$ such that $(\rho_{ij}(g))$ is the matrix of $\rho(g)$ with respect to a fixed basis of V, we can compute defining polynomials for $\overline{\rho(G)v}$ by elimination methods (Section 1.6). This gives an answer to question A, and hence also to B. This works over fields of any characteristic. However, due to the need to compute Gröbner bases, this algorithm is of limited use in practice.

If the base field is of characteristic 0, $G \subset \mathrm{GL}(m, K)$ is connected and the Lie algebra $\mathfrak{g}$ of G is available (or easily computable), a slightly different approach is possible. First we compute a parametrized dense subset of G, as follows. Use the notation of Section 4.5.3. Let $x_1, \dots, x_r$ be a basis of $\mathfrak{n}$; then by Proposition 4.3.15, the map $\mathbb{A}^r \to G(\mathfrak{n})$, $(t_1, \dots, t_r) = \exp(t_1x_1 + \cdots + t_rx_r)$ is surjective. In the same way we obtain surjective maps $\mathbb{A}^s \to G(\mathfrak{n}_-)$ and $\mathbb{A}^s \to G(\mathfrak{n}_+)$. Furthermore, as seen in the previous section, we can compute a surjective morphism $\mathbb{G}_\mathrm{m}^p \to G(\mathfrak{a})$. Composing all of this with the multiplication map as in Section 4.5.3, we obtain a dominant regular map $\mathbb{A}^{r+2s} \times \mathbb{G}_\mathrm{m}^p \to G$. Composing that with ρ we have a dominant regular map $\mathbb{A}^{r+2s} \times \mathbb{G}_\mathrm{m}^p \to \rho(G)v$. Again by elimination techniques we can compute defining polynomials for the closure of the image of this map, which is the same as $\overline{\rho(G)v}$.

Remark 4.8.1 In full generality this works only in characteristic 0. However, using Theorem 5.3.9 it is not difficult to extend this approach to semisimple algebraic groups over fields of positive characteristic.

The main advantage of this approach, compared to the previous one, is that it can be used when just the Lie algebra of G is known, and not a set of defining polynomials of G. Its main disadvantage is that it can only be applied to connected algebraic groups.

Example 4.8.2 Let G be the 10-dimensional group with Lie algebra $\mathfrak{g}$ from Example 4.5.11. Let $V = K^6$ be the natural G-module and $v_1, \dots, v_6$ denote the elements of its standard basis. Let $v = v_5 + v_6$. Write $K[V] = K[z_1, \dots, z_6]$. Computing defining polynomials for the closure $\overline{Gv}$, using a set of defining polynomials of G and the general method outlined above, took 0.07 (s), and the resulting polynomials were

$$z_1z_4 - z_2z_3$$
$$z_1z_6 - z_2z_5$$
$$z_3z_6 - z_4z_5.$$

In particular we see that the orbit is not closed as 0 lies in its closure, but not in the orbit itself.

In order to apply the second method, we have

$$\exp(x_1x_{\alpha_1} + x_2x_{\alpha_2} + x_3x_{\alpha_3}) = \begin{pmatrix} 1 & 0 & 0 & 0 & 0 & 0 \\ 0 & 1 & 0 & 0 & 0 & 0 \\ -x_2 & 0 & 1 & 0 & 0 & 0 \\ 0 & -x_2 & 0 & 1 & 0 & 0 \\ \frac{1}{2}x_1x_2 - x_3 & 0 & -x_1 & 0 & 1 & 0 \\ 0 & \frac{1}{2}x_1x_2 - x_3 & 0 & -x_1 & 0 & 1 \end{pmatrix},$$

$$\exp(y_1x_{-\alpha_1} + y_2x_{-\alpha_2} + y_3x_{-\alpha_3}) = \begin{pmatrix} 1 & 0 & -y_2 & 0 & \frac{1}{2}y_1y_2 - y_3 & 0 \\ 0 & 1 & 0 & -y_2 & 0 & \frac{1}{2}y_1y_2 - y_3 \\ 0 & 0 & 1 & 0 & -y_1 & 0 \\ 0 & 0 & 0 & 1 & 0 & -y_1 \\ 0 & 0 & 0 & 0 & 1 & 0 \\ 0 & 0 & 0 & 0 & 0 & 1 \end{pmatrix}.$$

The subalgebra $\mathfrak{t}$ is diagonal, so we do not need to diagonalize. We use the same notation as in the description of the algorithm above. The lattice Λ is spanned by $(1, -1, 0, 0, -1, 1)$, $(0, 0, 1, -1, -1, 1)$, and the lattice $\Lambda^{\perp}$ is spanned by

$$(1, 0, 0, -1, 0, -1), (0, 1, 0, 1, 0, 1), (0, 0, 1, 1, 0, 0), (0, 0, 0, 0, 1, 1).$$

So the morphism $\varphi : \mathbb{G}_{\mathrm{m}}^4 \to G(\mathfrak{t})$ maps $(s_1, \dots, s_4)$ to

$$T(s_1, s_2, s_3, s_4) = \mathrm{diag}(s_1, s_2, s_3, s_1^{-1}s_2s_3, s_4, s_1^{-1}s_2s_4).$$

Now $\exp(y_1x_{-\alpha_1} + y_2x_{-\alpha_2} + y_3x_{-\alpha_3}) \cdot T(s_1, \dots, s_4) \cdot \exp(x_1x_{\alpha_1} + x_2x_{\alpha_2} + x_3x_{\alpha_3}) \cdot v$ is equal to

$$(\tfrac{1}{2}y_1y_2s_4 - y_3s_4, \tfrac{1}{2}y_1y_2s_1^{-1}s_2s_4 - y_3s_1^{-1}s_2s_4, -y_1s_4, -y_1s_1^{-1}s_2s_4, s_4, s_1^{-1}s_2s_4).$$

It took 0.01 (s) to compute the algebraic relations between the coordinates of this vector, yielding the same polynomials as with the first method.

4.9 Notes

The main results of Sections 4.2, 4.3 are due to Chevalley ([Che51], [Che55b]). One of the main technical tools he used is a formal exponential. In [Bor91] some of these results are also proved, but using quotients of algebraic groups by normal subgroups, and not a formal exponential. Our treatment uses the results of Section 4.1, which is based on [Hoc81], [TY05]. Most of the proof of Theorem 4.3.22 is taken from [Hoc81].

The algorithm of Section 4.4 can be found in [FG07]. The algorithm of Sections 4.5 is based on [Gra09]. However, in that paper the approach based on Algorithm 4.5.14 is used.

The problem of computing the smallest algebraic group containing a given set of matrices was first considered in [DJK05]. In that paper an algorithm similar to the one outlined in Remark 4.6.6 is given.

The approach to computing the closure of an orbit by using a parametrised dense subset of the group was first proposed by Popov ([Pop09]). In this paper one more algorithm for question B of Section 4.8 is given. This algorithm reduces the question to a set of linear equations. However, the numbers of equations and unknowns in these equations are so large that it is very difficult to use this algorithm in practice. A second difficulty is the need to compute the degree of the algebraic group, which is not an easy problem either. For these reasons we have not included this algorithm here.

Chapter 5

Semisimple Algebraic Groups

Without doubt, one of the main achievements in the theory of algebraic groups is the classification of the semisimple ones, due to the work of Chevalley ([Che05]). This classification has many intriguing aspects, one being its independence on the characteristic of the ground field: the semisimple algebraic groups over a given algebraically closed field k are, up to isomorphism, parametrized by the semisimple root data. The existence of an algebraic group with a given root datum was not shown by Chevalley in [Che05]. One way of proving that was initiated in his famous "Tôhuko" paper ([Che55a]), describing a uniform construction of the semisimple algebraic groups of adjoint type. In his lecture notes ([Ste67]), Steinberg extended Chevalley's approach, giving a construction of an arbitrary semisimple algebraic group.

In the first part of this chapter we follow Steinberg's approach to the construction of the semisimple algebraic groups. In Section 5.1 we study a certain class of "abstract" groups $\mathcal{G}$ having a set of generators satisfying a number of relations. The Chevalley groups defined in Section 5.2 belong to that class, so all results of Section 5.1 carry over to them. In Section 5.3 we show that when the base field is algebraically closed, these groups are semisimple algebraic. We show a number of their properties: a basis of the Lie algebra is given and a characterization of the irreducible representations is obtained.

Another amazing fact concerning the semisimple algebraic groups is that we can parametrize their elements via the Bruhat decomposition. In Section 5.1 it is shown for the abstract groups $\mathcal{G}$, then in Section 5.2 it is shown in strengthened form for the Chevalley groups. In the second part of the chapter, concerned with algorithms, it also plays an important role. The Bruhat decomposition shows that every element of a semisimple algebraic group can be expressed as a "normal word", and Section 5.6 has an algorithm for computing the normal word representing the product of two elements, each given as a normal word. Section 5.7 describes an algorithm that, given a highest weight representation of a semisimple algebraic group and a matrix lying in the image of that representation, gives a preimage of that matrix expressed as a normal word.

As mentioned before, Section 5.3, has a characterization of the irreducible representations of a semisimple algebraic group. However, if the base field is of characteristic $p > 0$, it is by no means clear what the dimension of such a representation is. This is a question to which computers have been applied since the late 1960's. In Section 5.5 we describe the main algorithm used in this area.

In Section 5.8 we show, among other things, that the Bruhat decomposition generalizes to reductive algebraic groups in characteristic 0. Finally in Section 5.9 we look at Dynkin's alorithm for listing the regular semisimple subalgebras of a semisimple Lie algebra up to conjugacy by the adjoint group.

Throughout this chapter, for a field k we denote by k^* the set of non-zero elements of k.

5.1 Groups defined by certain generators and relations

In this section we study a class of groups, which we call C-groups that have a set of generators parametrized by the Cartesian product of a root system and a field. These generators are required to satisfy a specific set of relations.

Let Φ be a root system, with fixed set of positive roots Φ^+, and corresponding basis of simple roots $\Delta = \{\alpha_1, \dots, \alpha_\ell\}$. We use the notation relative to root systems and Weyl groups introduced in Chapter 2. In particular, W denotes the Weyl group of Φ, generated by the reflections s_α for $\alpha \in \Phi$.

Let k be a field, and $\mathcal{G}$ be a group generated by elements $\xi_\alpha(t)$, for $\alpha \in \Phi$ and $t \in k$. For $\alpha \in \Phi$, $t \in k$, $t \neq 0$, we define the elements

$$\omega_\alpha(t) = \xi_\alpha(t)\xi_{-\alpha}(-t^{-1})\xi_\alpha(t),\ \eta_\alpha(t) = \omega_\alpha(t)\omega_\alpha(1)^{-1},\ \varpi_\alpha = \omega_\alpha(1).$$

For $\alpha, \beta \in \Phi$, $\alpha \neq -\beta$ we let $I_{\alpha,\beta} = \{(i,j) \mid i,j \in \mathbb{Z}_{>0},\ i\alpha + j\beta \in \Phi\}$. For $g, h \in \mathcal{G}$ we let $[g,h] = ghg^{-1}h^{-1}$ denote their commutator. We suppose that the following relations hold:

$$\xi_\alpha(0) = 1 \text{ (for } \alpha \in \Phi), \tag{5.1}$$

$$\xi_\alpha(t)\xi_\alpha(u) = \xi_\alpha(t+u) \text{ (for } \alpha \in \Phi,\ t,u \in k), \tag{5.2}$$

$$[\xi_\alpha(t), \xi_\beta(u)] = \prod_{(i,j)\in I_{\alpha,\beta}} \xi_{i\alpha+j\beta}(c_{i,j}^{\alpha,\beta} t^i u^j) \text{ (where } c_{i,j}^{\alpha,\beta} \in \mathbb{Z}, \text{ and the product is taken in a fixed order and } \alpha \neq -\beta), \tag{5.3}$$

$$\varpi_\alpha \eta_\beta(t) \varpi_\alpha^{-1} = \eta_{s_\alpha(\beta)}(t) \text{ (for } \alpha, \beta \in \Phi,\ t \in k^*), \tag{5.4}$$

$$\varpi_\alpha \xi_\beta(t) \varpi_\alpha^{-1} = \xi_{s_\alpha(\beta)}(c_{\alpha,\beta} t) \text{ (for } \alpha, \beta \in \Phi,\ t \in k \text{ and } c_{\alpha,\beta} = \pm 1, \text{ is such that } c_{\alpha,\beta} = c_{\alpha,-\beta}), \tag{5.5}$$

$$\eta_\alpha(t)\xi_\beta(u)\eta_\alpha(t)^{-1} = \xi_\beta(t^{\langle \beta, \alpha^\vee \rangle} u) \text{ (for } t \in k^*, u \in k, \alpha, \beta \in \Phi). \tag{5.6}$$

Definition 5.1.1 *We say that a group $\mathcal{G}$ generated by $\xi_\alpha(t)$ for $\alpha \in \Phi$, $t \in k$, and satisfying the relations* (5.1) *to* (5.6)*, is a C-group. The set of generators $\xi_\alpha(t)$ is said to be a C-generating set of $\mathcal{G}$.*

For $\alpha \in \Phi$ we set $\mathcal{X}_\alpha = \{\xi_\alpha(t) \mid t \in k\}$. By (5.1), (5.2), $\mathcal{X}_\alpha$ is an abelian subgroup of $\mathcal{G}$ (and a homomorphic image of the additive group of k).

A subset $S \subset \Phi$ is said to be *closed* if $\alpha, \beta \in S$ and $\alpha + \beta \in \Phi$ imply that $\alpha + \beta \in S$. A subset I of a closed set S is called an *ideal* if $\alpha \in I$, $\beta \in S$ and $\alpha + \beta \in S$ imply that $\alpha + \beta \in I$. For example, $S = \Phi^+$ is closed and when $\alpha \in \Delta$, then $I = S \setminus \{\alpha\}$ is an ideal in it.

Furthermore, for $S \subset \Phi$ we denote the subgroup of $\mathcal{G}$ generated by all $\mathcal{X}_\alpha$ for $\alpha \in S$, by $\mathcal{X}_S$.

Lemma 5.1.2 *Let $S \subset \Phi^+$ be closed, and $I \subset S$ an ideal. Then $\mathcal{X}_I$ is a normal subgroup of $\mathcal{X}_S$.*

Proof. Let $\alpha \in S$, $\beta \in I$; by (5.3) we have

$$\xi_\alpha(t)\xi_\beta(u)\xi_\alpha(t)^{-1} = \left(\prod_{(i,j)\in I_{\alpha,\beta}} \xi_{i\alpha+j\beta}(c_{i,j}^{\alpha,\beta} t^i u^j) \right) \xi_\beta(u).$$

If $(i,j) \in I_{\alpha,\beta}$, then $i\alpha + j\beta \in I$. The desired conclusion follows. □

Lemma 5.1.3 *Let $S \subset \Phi^+$ be closed and choose any ordering of S. Then every element of $\mathcal{X}_S$ can be written as a product $\prod_{\alpha\in S} \xi_\alpha(t_\alpha)$ where $t_\alpha \in k$ and the product is taken in the chosen ordering.*

Proof. First assume that the chosen ordering is height-compatible (i.e., roots of smaller height are smaller). Let $\alpha \in S$ be its smallest element. Then $I = S \setminus \{\alpha\}$ is an ideal in S. By the previous lemma, $\mathcal{X}_S = \mathcal{X}_\alpha \mathcal{X}_I$, and the proof for this case is finished by induction.

The general case follows from the special case, along with a group theoretic fact: let $G_1, \ldots, G_s$ be subgroups of a group G such that $G = G_1 \cdots G_s$, and $G_i \cdots G_s$ is a normal subgroup of G for $1 \leq i \leq s$; then for any permutation π of $1, \ldots, s$ we have $G = G_{\pi(1)} \cdots G_{\pi(s)}$. (This is shown by induction on s; the details are left to the reader.) □

Now we define a number of subgroups of $\mathcal{G}$. Let $\mathcal{U}$ be the subgroup of $\mathcal{G}$ generated by all $\mathcal{X}_\alpha$ for $\alpha \in \Phi^+$. Let $\mathcal{H}$ be the subgroup generated by all $\eta_\alpha(t)$ for $\alpha \in \Phi$, $t \in k^*$. Let $\mathcal{B}$ be the subgroup generated by $\mathcal{U}$ and $\mathcal{H}$. Let $\mathcal{N}$ be the subgroup of $\mathcal{G}$ generated by the $\omega_\alpha(t)$ for $\alpha \in \Phi$, $t \in k^*$.

Lemma 5.1.4 (i) *$\mathcal{U}$ is a normal subgroup of $\mathcal{B}$ and $\mathcal{B} = \mathcal{U}\mathcal{H}$.*

(ii) *$\mathcal{H}$ is a normal subgroup of $\mathcal{N}$.*

(iii) *There is a surjective homomorphism $\vartheta : W \to \mathcal{N}/\mathcal{H}$ such that $\vartheta(s_\alpha) = \mathcal{H}\omega_\alpha(t)$ for all $\alpha \in \Phi$ (where the latter coset is independent of t).*

Proof. By (5.6), $\mathcal{U}$ is normalized by the $\eta_\alpha(t)$. So it is normalized by all generators of $\mathcal{B}$, whence $\mathcal{U}$ is normal in $\mathcal{B}$. The second statement of (i) is a direct consequence.

By (5.4), $\mathcal{H}$ is normalized by the ϖ_α, and hence also by $\omega_\alpha(t) = \eta_\alpha(t)\varpi_\alpha$. So (ii) follows.

In order to prove (iii) we first consider the group $\widetilde{W}$ generated by the symbols σ_α, for $\alpha \in \Phi$, subject to the relations $\sigma_\alpha^2 = 1$, $\sigma_\alpha\sigma_\beta\sigma_\alpha = \sigma_{s_\alpha(\beta)}$, for all $\alpha, \beta \in \Phi$. The generators s_α of W satisfy these relations (see Section 2.8.3). So there is a surjective group homomorphism $\phi : \widetilde{W} \to W$, with $\phi(\sigma_\alpha) = s_\alpha$. The claim is that ϕ is an isomorphism. By arguments similar to those used for Lemma 2.8.19 and Theorem 2.8.20 we see that $\widetilde{W}$ is generated by $A = \{\sigma_\alpha \mid \alpha \in \Delta\}$. Write $\sigma_i = \sigma_{\alpha_i}$. Let $\sigma_{i_1} \cdots \sigma_{i_m}$ lie in $\ker\phi$, and choose such an element with a minimal m. Then $s_{i_1} \cdots s_{i_m}$ is equal to 1, and hence has length 0. So there is an $r \geq 2$ such that $\mathcal{L}(s_{i_1} \cdots s_{i_{r-1}}) = r - 1$ and $\mathcal{L}(s_{i_1} \cdots s_{i_r}) = r - 2$. This implies that $s_{i_1} \cdots s_{i_{r-1}}(\alpha_{i_r}) < 0$ (Corollary 2.8.25). As in the proof of Lemma 2.8.21 we get a t such that $(s_{i_{t+1}} \cdots s_{i_{r-1}})s_{i_r}(s_{i_{r-1}} \cdots s_{i_{t+1}}) = s_{s_{i_{t+1}} \cdots s_{i_{r-1}}(\alpha_{i_r})} = s_{\gamma_t} = s_{i_t}$, but $(\sigma_{i_{t+1}} \cdots \sigma_{i_{r-1}})\sigma_{i_r}(\sigma_{i_{r-1}} \cdots \sigma_{i_{t+1}}) = \sigma_{i_t}$, and by substituting we find an expression of shorter length lying in $\ker\phi$, which is a contradiction.

We note that $\mathcal{H}\omega_\alpha(t) = \mathcal{H}\omega_\alpha(t)\varpi_\alpha^{-1}\varpi_\alpha = \mathcal{H}\varpi_\alpha$. So the coset is independent of t. Write $\hat{w}_\alpha = \mathcal{H}\omega_\alpha(t)$. Then $\omega_\alpha(1), \omega_\alpha(-1) \in \hat{w}_\alpha$, so that $1 \in \hat{w}_\alpha^2$, whence $\hat{w}_\alpha^2 = \mathcal{H} \cdot 1$. Furthermore,

$$\begin{aligned}
\varpi_\alpha\varpi_\beta\varpi_\alpha^{-1} &= \varpi_\alpha\xi_\beta(1)\xi_{-\beta}(-1)\xi_\beta(1)\varpi_\alpha^{-1} \\
&= \varpi_\alpha\xi_\beta(1)\varpi_\alpha^{-1}\varpi_\alpha\xi_{-\beta}(-1)\varpi_\alpha^{-1}\varpi_\alpha\xi_\beta(1)\varpi_\alpha^{-1} \\
&= \xi_{s_\alpha(\beta)}(c_{\alpha,\beta})\xi_{-s_\alpha(\beta)}(-c_{\alpha,\beta})\xi_{s_\alpha(\beta)}(c_{\alpha,\beta}) \quad \text{(by (5.5))} \\
&= \omega_{s_\alpha(\beta)}(c_{\alpha,\beta}) = \omega_{s_\alpha(\beta)}(1)^{c_{\alpha,\beta}} = \varpi_{s_\alpha(\beta)}^{c_{\alpha,\beta}}.
\end{aligned}$$

And as $c_{\alpha,\beta} = \pm 1$ and $\varpi_\alpha\varpi_\beta\varpi_\alpha^{-1} \in \hat{w}_\alpha\hat{w}_\beta\hat{w}_\alpha^{-1}$ we see that $\hat{w}_\alpha\hat{w}_\beta\hat{w}_\alpha^{-1} = \hat{w}_{s_\alpha(\beta)}$. So the $\hat{w}_\alpha$ satisfy the defining relations of $\widetilde{W}$. Therefore there is a surjective group homomorphism $\widetilde{W} \to \mathcal{N}/\mathcal{H}$, mapping $\sigma_\alpha \mapsto \hat{w}_\alpha$. Composing this with the isomorphism $W \to \widetilde{W}$ yields ϑ. □

For $w \in W$ we let $\dot{w} \in \mathcal{N}$ be such that $\vartheta(w) = \mathcal{H}\dot{w}$. Then the sets $\dot{w}\mathcal{B}$ and $\mathcal{B}\dot{w}$ do not depend on the choice of $\dot{w}$. Indeed, let $\dot{w}' \in \mathcal{N}$ be such that $\dot{w}' = h\dot{w}$ for some $h \in \mathcal{H}$, then $\dot{w}'\mathcal{B} = h\dot{w}\mathcal{B} = \dot{w}\dot{w}^{-1}h\dot{w}\mathcal{B} = \dot{w}\mathcal{B}$ by Lemma 5.1.4(ii).

Lemma 5.1.5 *Let $\alpha \in \Phi$, $w \in W$. Then $\dot{w}\mathcal{X}_\alpha\dot{w}^{-1} = \mathcal{X}_{w(\alpha)}$.*

Proof. Let $w = s_{i_1} \cdots s_{i_r}$ be a reduced expression. Write $\varpi_i = \varpi_{\alpha_i}$. Since $\vartheta(s_i) = \mathcal{H}\varpi_i$, we have that $\dot{w} = h\varpi_{i_1} \cdots \varpi_{i_r}$, for some $h \in \mathcal{H}$. By (5.6), $h\mathcal{X}_\alpha h^{-1} = \mathcal{X}_\alpha$. By (5.5), $\varpi_i\mathcal{X}_\alpha\varpi_i^{-1} = \mathcal{X}_{s_i(\alpha)}$. The lemma follows. □

Lemma 5.1.6 *Let $\alpha \in \Delta$. Then $\mathcal{B} \cup \mathcal{B}\dot{s}_\alpha\mathcal{B}$ is a subgroup of $\mathcal{G}$.*

Proof. Set $R = \mathcal{B} \cup \mathcal{B}\dot{s}_\alpha\mathcal{B}$. As $\dot{s}_\alpha\mathcal{B} = \dot{s}_\alpha^{-1}\mathcal{B}$ we see that R is closed under inversion.

Note that $\xi_{-\alpha}(t) = \xi_\alpha(t^{-1})\xi_\alpha(-t^{-1})\xi_{-\alpha}(t)\xi_\alpha(-t^{-1})\xi_\alpha(t^{-1})$, which equals $\xi_\alpha(t^{-1})\omega_\alpha(-t^{-1})\xi_\alpha(t^{-1})$ so that $\xi_{-\alpha}(t) \in \mathcal{B}\dot{s}_\alpha\mathcal{B} \subset R$. Now $\mathcal{B} = \mathcal{U}\mathcal{H}$ by Lemma 5.1.4, and by Lemma 5.1.3 we have $\mathcal{U} = \mathcal{X}_\alpha\mathcal{X}_J$, where $J = \Phi^+ \setminus \{\alpha\}$. So $\dot{s}_\alpha\mathcal{B}\dot{s}_\alpha = \dot{s}_\alpha\mathcal{B}\dot{s}_\alpha^{-1} = \dot{s}_\alpha\mathcal{X}_\alpha\mathcal{X}_J\mathcal{H}\dot{s}_\alpha^{-1} = \dot{s}_\alpha\mathcal{X}_\alpha\dot{s}_\alpha^{-1}\dot{s}_\alpha\mathcal{X}_J\dot{s}_\alpha^{-1}\dot{s}_\alpha\mathcal{H}\dot{s}_\alpha^{-1}$. By Lemma 5.1.5, $\dot{s}_\alpha\mathcal{X}_\alpha\dot{s}_\alpha^{-1} = \mathcal{X}_{-\alpha}$, and $\dot{s}_\alpha\mathcal{X}_J\dot{s}_\alpha^{-1} = \mathcal{X}_J$ (Lemma 2.8.18). Furthermore, $\dot{s}_\alpha\mathcal{H}\dot{s}_\alpha^{-1} = \mathcal{H}$ by (5.4). It follows that $\dot{s}_\alpha\mathcal{B}\dot{s}_\alpha \subset \mathcal{X}_{-\alpha}\mathcal{U}\mathcal{H} = \mathcal{X}_{-\alpha}\mathcal{B} \subset R\mathcal{B} \subset R$. This implies that R is closed under multiplication. □

Lemma 5.1.7 *Let $w \in W$, $\alpha \in \Delta$ and $v = ws_\alpha$. We have*

(i) $\mathcal{B}\dot{w}\mathcal{B} \cdot \mathcal{B}\dot{s}_\alpha\mathcal{B} \subset \mathcal{B}\dot{w}\mathcal{B} \cup \mathcal{B}\dot{v}\mathcal{B}$,

(ii) *and if $w(\alpha) > 0$ then even $\mathcal{B}\dot{w}\mathcal{B} \cdot \mathcal{B}\dot{s}_\alpha\mathcal{B} \subset \mathcal{B}\dot{v}\mathcal{B}$.*

Proof. We first prove (ii). Let $J = \Phi^+ \setminus \{\alpha\}$. As in the proof of the previous lemma we have $\mathcal{B} = \mathcal{X}_\alpha\mathcal{X}_J\mathcal{H}$. Hence $\mathcal{B}\dot{w}\mathcal{B} \cdot \mathcal{B}\dot{s}_\alpha\mathcal{B} = \mathcal{B}\dot{w}\mathcal{X}_\alpha\mathcal{X}_J\mathcal{H}\dot{s}_\alpha\mathcal{B} = \mathcal{B}\dot{w}\mathcal{X}_\alpha\dot{w}^{-1} \cdot \dot{w}\dot{s}_\alpha \cdot \dot{s}_\alpha^{-1}\mathcal{X}_J\dot{s}_\alpha\dot{s}_\alpha^{-1}\mathcal{H}\dot{s}_\alpha\mathcal{B}$. By Lemma 5.1.5, $\dot{w}\mathcal{X}_\alpha\dot{w}^{-1} = \mathcal{X}_{w(\alpha)} \subset \mathcal{B}$ (as $w(\alpha) > 0$). Using also that $\dot{s}_\alpha\mathcal{X}_J\dot{s}_\alpha^{-1} = \mathcal{X}_J$, $\dot{s}_\alpha\mathcal{H}\dot{s}_\alpha^{-1} = \mathcal{H}$, noted also in the previous proof, we conclude that $\mathcal{B}\dot{w}\mathcal{B} \cdot \mathcal{B}\dot{s}_\alpha\mathcal{B} \subset \mathcal{B}\dot{w}\dot{s}_\alpha\mathcal{B} = \mathcal{B}\dot{v}\mathcal{B}$.

Now we only need to show (i) when $w(\alpha) < 0$. Then $v(\alpha) > 0$ and $w = vs_\alpha$. Using (ii) and Lemma 5.1.6 we infer $\mathcal{B}\dot{w}\mathcal{B} \cdot \mathcal{B}\dot{s}_\alpha\mathcal{B} = \mathcal{B}\dot{v}\dot{s}_\alpha\mathcal{B} \cdot \mathcal{B}\dot{s}_\alpha\mathcal{B} \subset \mathcal{B}\dot{v}\mathcal{B} \cdot \mathcal{B}\dot{s}_\alpha\mathcal{B} \cdot \mathcal{B}\dot{s}_\alpha\mathcal{B} \subset \mathcal{B}\dot{v}\mathcal{B} \cdot (\mathcal{B} \cup \mathcal{B}\dot{s}_\alpha\mathcal{B}) = \mathcal{B}\dot{v}\mathcal{B} \cup \mathcal{B}\dot{v}\mathcal{B} \cdot \mathcal{B}\dot{s}_\alpha\mathcal{B} \subset \mathcal{B}\dot{v}\mathcal{B} \cup \mathcal{B}\dot{w}\mathcal{B}$. □

Proposition 5.1.8 *$\mathcal{G}$ is generated by all $\mathcal{X}_\alpha$ and ϖ_α for $\alpha \in \Delta$.*

Proof. By (5.5), $\varpi_\alpha\mathcal{X}_\beta\varpi_\alpha^{-1} = \mathcal{X}_{s_\alpha(\beta)}$. So the proposition follows by Lemma 2.8.19 and Theorem 2.8.20. □

Proposition 5.1.9 *$\mathcal{G}$ is the union of the sets $\mathcal{B}\dot{w}\mathcal{B}$ as w runs through W.*

Proof. Let M be the union of all $\mathcal{B}\dot{w}\mathcal{B}$ for $w \in W$. If $w = 1$ then $\mathcal{B}\dot{w}\mathcal{B} = \mathcal{B}$, so that M contains $\mathcal{X}_\alpha$ for all $\alpha > 0$, along with $\mathcal{H}$. By Lemma 5.1.4(iii), there is an $h \in \mathcal{H}$ such that $h\dot{s}_\alpha = \varpi_\alpha$. It follows that $\varpi_\alpha \in M$ for all $\alpha \in \Phi$. So by Proposition 5.1.8, M contains a set of generators of $\mathcal{G}$. Furthermore, by Lemma 5.1.7, M is closed under multiplication by these generators. As M is also closed under inversion, M is a subgroup of $\mathcal{G}$, so is equal to $\mathcal{G}$. □

Proposition 5.1.10 *Let $w \in W$. As in Section 2.8.3 set $\Phi_w = \Phi^+ \cap w^{-1}(-\Phi^+)$ (i.e., the set of all positive roots sent to negative roots by w). Set $\mathcal{U}_w = \mathcal{X}_{\Phi_w}$. Then $\mathcal{B}\dot{w}\mathcal{B} = \mathcal{B}\dot{w}\mathcal{U}_w$.*

Proof. Set $\Psi = \Phi^+ \cap w^{-1}(\Phi^+)$ and $\widetilde{\mathcal{U}}_w = \mathcal{X}_\Psi$. Then Φ^+ is the disjoint union of Φ_w and Ψ, and hence $\mathcal{U} = \widetilde{\mathcal{U}}_w \mathcal{U}_w$ (Lemma 5.1.3). Using Lemma 5.1.4(i) we see that $\mathcal{B}\dot{w}\mathcal{B} = \mathcal{B}\dot{w}\widetilde{\mathcal{U}}_w\mathcal{U}_w\mathcal{H} = \mathcal{B}\dot{w}\widetilde{\mathcal{U}}_w\dot{w}^{-1}\dot{w}\mathcal{U}_w\mathcal{H}$. Now $\dot{w}\widetilde{\mathcal{U}}_w\dot{w}^{-1} \subset \mathcal{B}$ by Lemma 5.1.5. Hence $\mathcal{B}\dot{w}\mathcal{B} = \mathcal{B}\dot{w}\mathcal{U}_w\mathcal{H}$.

As Φ_w is closed, $\mathcal{U}_w$ is a subgroup of $\mathcal{B}$, normalized by $\mathcal{H}$ (by (5.6)). Therefore, $\mathcal{U}_w\mathcal{H} = \mathcal{H}\mathcal{U}_w$. Now $\mathcal{B}\dot{w}\mathcal{H} = \mathcal{B}\dot{w}\mathcal{H}\dot{w}^{-1}\dot{w} = \mathcal{B}\mathcal{H}\dot{w}$ (5.4) $= \mathcal{B}\dot{w}$. It follows that $\mathcal{B}\dot{w}\mathcal{B} = \mathcal{B}\dot{w}\mathcal{U}_w$. □

5.2 Chevalley groups

This section is devoted to the Chevalley groups constructed from a representation of a semisimple Lie algebra and a field. The first two subsections have some preparatory material. In Section 5.2.3 we define the Chevalley groups, and show that they are C-groups, so that all results of the previous section can be applied to them. Section 5.2.4 is concerned with the Bruhat decomposition, which is a refinement of Propositions 5.1.9 and 5.1.10. In the final subsection we show that a Chevalley group can be defined by a *presentation*: a (possibly infinite) set of generators required to satisfy a specific set of relations. First, however, we fix some notation to be used throughout this section.

Let $\mathfrak{g}$ be a semisimple Lie algebra over $\mathbb{C}$. Let $\mathfrak{h} \subset \mathfrak{g}$ be a fixed Cartan subalgebra, and Φ be the corresponding root system. Fix a set of positive roots Φ^+ with corresponding basis of simple roots $\Delta = \{\alpha_1, \dots, \alpha_\ell\}$. From Section 2.10.3 we recall that $\mathcal{U}_\mathbb{Z}$ denotes an integral form of the universal enveloping algebra of $\mathfrak{g}$. We let x_α for $\alpha \in \Phi$, $h_1, \dots, h_\ell$ denote a fixed Chevalley basis of $\mathfrak{g}$. We use the basis of $\mathcal{U}_\mathbb{Z}$ consisting of the integral monomials constructed in Section 2.10.3, starting from this Chevalley basis. Also we let $\mathcal{U}_\mathbb{Z}^+, \mathcal{U}_\mathbb{Z}^0, \mathcal{U}_\mathbb{Z}^-$ be the subalgebras of $\mathcal{U}_\mathbb{Z}$ generated respectively, by the $x_\alpha^{(m)}$, $\alpha > 0$, by the $\binom{h_i}{r}$, $1 \le i \le \ell$, $r \ge 0$, and by the $x_\alpha^{(m)}$, $\alpha < 0$. Then the form of the basis elements of $\mathcal{U}_\mathbb{Z}$ immediately implies that $\mathcal{U}_\mathbb{Z} = \mathcal{U}_\mathbb{Z}^- \mathcal{U}_\mathbb{Z}^0 \mathcal{U}_\mathbb{Z}^+$.

For $\alpha, \beta \in \Phi$, $\alpha \neq \pm\beta$, define $N_{\alpha,\beta} \in \mathbb{Z}$ by $[x_\alpha, x_\beta] = N_{\alpha,\beta} x_{\alpha+\beta}$ (where $N_{\alpha,\beta} = 0$ if $\alpha+\beta \notin \Phi$). By Theorem 2.9.13, if $\alpha+\beta \in \Phi$, then $N_{\alpha,\beta} = \pm(r+1)$, where r is the maximal integer with $\alpha - r\beta \in \Phi$.

We denote the $\mathbb{Z}$-span of the fixed Chevalley basis by $\mathfrak{g}_\mathbb{Z}$. The $\mathbb{Z}$-span of $h_1, \dots, h_\ell$ will be denoted $\mathfrak{h}_\mathbb{Z}$.

We use the concepts and notation relative to the representation theory of $\mathfrak{g}$ contained in Section 2.11.

5.2.1 Admissible lattices

Let V be a finite-dimensional $\mathfrak{g}$-module. With respect to addition, V is an abelian group, and a finitely generated subgroup of it is called a *lattice* in V. A lattice Λ in V is said to be *admissible* if it is the $\mathbb{Z}$-span of a basis of V and it is invariant under $\mathcal{U}_{\mathbb{Z}}$ (i.e., $\mathcal{U}_{\mathbb{Z}} \cdot \Lambda \subseteq \Lambda$). In order to start our study of admissible lattices in V we need a technical result on polynomials.

Let $R = \mathbb{Q}[T_1, \ldots, T_\ell]$ denote the polynomial ring over $\mathbb{Q}$ in ℓ indeterminates. For $f \in R$ and $l \geq 1$ we define

$$\binom{f}{l} = \frac{f(f-1)\cdots(f-l+1)}{l!},$$

which also lies in R. The zeros of $\binom{f}{l}$ are the $(a_1, \ldots, a_\ell) \in \mathbb{Q}^\ell$ such that $f(a_1, \ldots, a_\ell) = i$, with $0 \leq i \leq l-1$. Also, for $a = (a_1, \ldots, a_\ell) \in \mathbb{Z}^\ell$ and $l \geq 1$ define

$$\psi_{a,l} = \prod_{i=1}^{\ell} \binom{T_i - a_i + l}{l}\binom{-T_i + a_i + l}{l}.$$

Lemma 5.2.1 *Let $a = (a_1, \ldots, a_\ell) \in \mathbb{Z}^\ell$, and $S \subset \mathbb{Z}^\ell$ a finite set not containing a. For $l \geq 1$, let B_l be the set of all $(b_1, \ldots, b_l) \in \mathbb{Z}^l$ such that $|b_i - a_i| \leq l$. Let l be such that $S \subset B_l$, and set $f = \psi_{a,l}$. Then $f(a) = 1$, $f(b) = 0$ for all $b \in S$.*

Proof. These are straightforward verifications. In fact, $f(b) = 0$ for all $b \in B \setminus \{a\}$. □

Lemma 5.2.2 *Let $\Lambda \subset V$ be an admissible lattice. For a weight μ of V, set $\Lambda_\mu = \Lambda \cap V_\mu$. Then Λ is the direct sum of all Λ_μ, where μ runs over the weights of V.*

Proof. For an integral weight $\mu \in \mathfrak{h}^*$ write $a(\mu) = (\mu(h_1), \ldots, \mu(h_\ell))$. Fix a weight μ_0 of V, and write $a_0 = a(\mu_0)$. By the previous lemma we find a polynomial $f = \psi_{a_0,l}$ such that $f(a_0) = 1$, $f(a(\mu)) = 0$ for all other weights μ of V. Set $u = f(h_1, \ldots, h_\ell)$. By the formulas given in Section 2.10.3, we see that $u \in \mathcal{U}_{\mathbb{Z}}^0$ (for that, note also $\binom{-h}{l} = (-1)^l\binom{h+l-1}{l}$). Moreover, u acts on V and it maps all weight spaces to 0, except V_{μ_0} on which it acts as the identity. It follows that $\Lambda_{\mu_0} \subset \Lambda$, and the lemma is proved. □

Lemma 5.2.3 *Let $\lambda \in \mathfrak{h}^*$ be a dominant integral weight. Let $V(\lambda)$ denote the irreducible highest weight module over $\mathfrak{g}$ of highest weight λ. Fix a non-zero $v_\lambda \in V(\lambda)_\lambda$. Then $\Lambda = \mathcal{U}_{\mathbb{Z}}^- \cdot v_\lambda$ is an admissible lattice in $V(\lambda)$.*

Proof. Obviously Λ is a subgroup of $V(\lambda)$. Furthermore, only finitely many monomials $\prod_{\alpha>0} x_{-\alpha}^{(k_\alpha)}$ fail to kill v_λ, so Λ is finitely generated.

We show that Λ is $\mathcal{U}_\mathbb{Z}$-invariant. Let $u \in \mathcal{U}_\mathbb{Z}$, and $\xi = v^- \cdot v_\lambda \in \Lambda$, where $v^- \in \mathcal{U}_\mathbb{Z}^-$. Then $u \cdot \xi = (uv^-) \cdot v_\lambda$, and uv^- is a sum of integral monomials $u^- u^0 u^+$, where $u^- \in \mathcal{U}_\mathbb{Z}^-$, $u^0 \in \mathcal{U}_\mathbb{Z}^0$, $u^+ \in \mathcal{U}_\mathbb{Z}^+$. Now $u^+ \cdot v_\lambda = 0$ unless $u^+ = 1$. Furthermore $u^0 \cdot v_\lambda$ is equal to an integer multiple of v_λ. It follows that $u \cdot \xi \in \Lambda$.

There is a basis $\{v_1, \dots, v_m\}$ of $V(\lambda)$ such that $v_1 = v_\lambda$ and $u \cdot v_i = \sum_j q_{ij} v_j$ with $q_{ij} \in \mathbb{Q}$, for all $u \in \mathcal{U}_\mathbb{Z}$. Indeed, we can take the elements $u^- \cdot v_\lambda$, where u^- runs through the integral monomials in $\mathcal{U}_\mathbb{Z}^-$ not divisible by a leading monomial of the Gröbner basis constructed in Section 2.11.4. A set of vectors with coefficients in $\mathbb{Q}$ is linearly independent over $\mathbb{C}$ if and only if it is linearly independent over $\mathbb{Q}$ if and only if it is linearly independent over $\mathbb{Z}$. This implies that Λ has rank equal to $\dim V(\lambda)$. □

This lemma, the classification of the irreducible $\mathfrak{g}$-modules (Theorem 2.11.7), and Weyl's theorem (Theorem 2.7.6), immediately imply the following.

Corollary 5.2.4 *Let V be a finite-dimensional $\mathfrak{g}$-module. Then V has an admissible lattice.*

Now let $V = V(\lambda)$ be an irreducible highest weight module over $\mathfrak{g}$, given by a basis, along with a method to compute the action of an $x \in \mathfrak{g}$ on a $v \in V$ (for example, as found with the algorithms of Section 2.11.4). We consider the problem to find a basis of an admissible lattice in V. The first method that comes to mind for this is directly based on the proof of Lemma 5.2.3. Let $\{v_1, \dots, v_n\}$ be the given basis of V, and assume that v_1 is a highest weight vector. For convenience we assume that for each x in the given Chevalley basis of $\mathfrak{g}$, $x \cdot v_i$ is a linear combination of the v_j with coefficients in $\mathbb{Q}$ (any basis consisting of elements $u^- \cdot v_\lambda$, where u^- are integral monomials in $\mathcal{U}_\mathbb{Z}^-$, has this property). There are only finitely many monomials $u^- = \prod_{\alpha>0} x_{-\alpha}^{(m_\alpha)}$ in $\mathcal{U}_\mathbb{Z}^-$ that fail to kill v_1. For each such u^- compute $u^- \cdot v_1 = \sum_i \tau_i v_i$. All coefficient vectors $(\tau_1, \dots, \tau_n)$ that we obtain in this way form the rows of a matrix A. Set $B = dA$, where d is an integer such that all entries of B are integral. Compute the Hermite normal form H of B (see [Sim94], Chapter 8). Then the non-zero rows of H form a basis of the lattice spanned by the rows of B. Therefore, the non-zero rows of $\frac{1}{d}H$ form a basis of the lattice spanned by the rows of A. We transform these rows to elements of V, which then form a basis of an admissible lattice.

This method can be rather inefficient, because the number of monomials u^- considered can be much larger than the dimension of V. Here we also briefly describe an algorithm that uses the path model (Section 2.11.3), which uses a number of elements of $\mathcal{U}_\mathbb{Z}^-$ equal to $\dim V$. Secondly, it immediately gives a basis of the admissible lattice $\mathcal{U}_\mathbb{Z}^- \cdot v_\lambda$, without a need for linear algebra calculations over the integers.

Let $\Pi(\lambda)$ and the crystal graph Γ_λ be as in Section 2.11.3. Fix a reduced expression $\hat{w}_0 = s_{i_1} \cdots s_{i_t}$ of the longest element w_0 of the Weyl group. Let $\pi \in \Pi(\lambda)$ and define a sequence of non-negative integers n_l and paths $\pi_l \in \Pi(\lambda)$ as follows. We set $\pi_0 = \pi$. For $l \geq 1$ we let n_l be maximal such that $e_{\alpha_{i_l}}^{n_l} \pi_{l-1} \neq 0$, and we set $\pi_l = e_{\alpha_{i_l}}^{n_l}$. Following [Lit98], we call $\eta_\pi = (n_1, \ldots, n_t)$ an *adapted string* of the path π. We note that for a given π and $\hat{w}_0$, the corresponding adapted string can easily be read from Γ_λ (see Remark 2.11.15).

For $\pi \in \Pi(\lambda)$ let $\eta_\pi = (n_1, \ldots, n_t)$ and set $v_\pi = x_{-\alpha_{i_1}}^{(n_1)} \cdots x_{-\alpha_{i_t}}^{(n_t)} \cdot v_\lambda$. Then the set of vectors $\{v_\pi \mid \pi \in \Pi(\lambda)\}$ forms a basis of $\mathcal{U}_\mathbb{Z}^- \cdot v_\lambda$ (we do not prove this here, but refer to [Lit98], Theorem 10.1(ii)).

Example 5.2.5 Let $\mathfrak{g}$ be the simple Lie algebra of type A_2, with multiplication table as in Example 2.9.14. Let $\lambda = 2\lambda_1$. Using the method of Section 2.11.4 it is straightforward to construct $V(\lambda)$. Writing $y_i = x_{-\alpha_i}$, $1 \leq i \leq 3$, the Gröbner basis of $I(\lambda)$ is

$$\{y_1^{(3)}, y_2, y_1^{(2)} y_3, y_1 y_3^{(2)}, y_3^{(3)}\},$$

and a basis of $V(\lambda)$ is

$$B = \{v_\lambda, y_1 \cdot v_\lambda, y_3 \cdot v_\lambda, y_1^{(2)} \cdot v_\lambda, y_1 y_3 \cdot v_\lambda, y_3^{(2)} \cdot v_\lambda\}.$$

Example 2.11.16 has the crystal graph for this case. Using $\hat{w}_0 = s_1 s_2 s_1$, the adapted strings are read from Γ_λ. They are $(0,0,0)$, $(1,0,0)$, $(2,0,0)$, $(0,1,1)$, $(1,1,1)$, $(0,2,2)$. This means that a basis of $\mathcal{U}_\mathbb{Z}^- \cdot v_\lambda$ is formed by v_λ, $y_1 \cdot v_\lambda$, $y_1^{(2)} \cdot v_\lambda$, $y_2 y_1 \cdot v_\lambda = -y_3 \cdot v_\lambda$, $y_1 y_2 y_1 \cdot v_\lambda = -y_1 y_3 \cdot v_\lambda$, and $y_2^{(2)} y_1^{(2)} \cdot v_\lambda = y_3^{(2)} \cdot v_\lambda$.

5.2.2 Chevalley's commutator formula

In this section we work with the ring $\mathcal{U}_\mathbb{Z}[\![\tau, \upsilon]\!]$ of formal power series in the indeterminates τ, υ with coefficients in $\mathcal{U}_\mathbb{Z}$. This ring consists of infinite sequences $(a_{ij}\tau^i \upsilon^j)_{i,j \geq 0}$, with $a_{ij} \in \mathcal{U}_\mathbb{Z}$. It is convenient to represent such a sequence as a formal sum

$$\sum_{i,j \geq 0} a_{ij} \tau^i \upsilon^j.$$

The addition in $\mathcal{U}_\mathbb{Z}[\![\tau, \upsilon]\!]$ is defined componentwise, whereas for the multiplication we have

$$\left(\sum_{i,j \geq 0} a_{ij} \tau^i \upsilon^j\right) \left(\sum_{k,l \geq 0} b_{kl} \tau^k \upsilon^l\right) = \sum_{m,n \geq 0} \left(\sum_{\substack{i+k=m \\ j+l=n}} a_{ij} b_{kl}\right) \tau^m \upsilon^n.$$

(Note that the summation defining the coefficient of $\tau^m \upsilon^n$ is finite, and hence defines an element of $\mathcal{U}_\mathbb{Z}$.)

Let $a \in \mathbb{Z}[\tau, \upsilon]$ be a polynomial with constant term equal to 0. Then for $\gamma \in \Phi$ we set

$$\exp(ax_\gamma) = \sum_{l=0}^{\infty} x_\gamma^{(l)} a^l,$$

which is a well-defined element of $\mathcal{U}_\mathbb{Z}[\![\tau, \upsilon]\!]$. A short calculation shows that $\exp(ax_\gamma)\exp(-ax_\gamma) = 1$. Indeed, the product is equal to

$$\sum_{m \geq 0} \left(\sum_{l=0}^{m} (-1)^l \binom{m}{l} \right) x_\gamma^{(m)} a^m,$$

and $\sum_{l=0}^{m} (-1)^l \binom{m}{l} = 0$ for $m \geq 1$.

In the sequel $\mathfrak{g}_\mathbb{Z}[\tau, \upsilon]$ denotes the $\mathbb{Z}[\tau, \upsilon]$-span of $\mathfrak{g}_\mathbb{Z}$ inside $\mathcal{U}_\mathbb{Z}[\![\tau, \upsilon]\!]$. It is a Lie ring over $\mathbb{Z}[\tau, \upsilon]$.

Let $b \in \mathbb{Z}[\tau, \upsilon]$ and $\gamma \in \Phi$. Define the $\mathbb{Z}[\tau, \upsilon]$-linear map $D_{bx_\gamma} : \mathfrak{g}_\mathbb{Z}[\tau, \upsilon] \to \mathfrak{g}_\mathbb{Z}[\tau, \upsilon]$ by $D_{bx_\gamma}(y) = b[x_\gamma, y]$. Then $D_{bx_\gamma}(x_\delta) = N_{\gamma,\delta} b x_{\gamma+\delta}$, $D^2_{bx_\gamma}(x_\delta) = N_{\gamma,\delta} N_{\gamma,\gamma+\delta} b^2 x_{2\gamma+\delta}$, $D^3_{bx_\gamma}(x_\delta) = N_{\gamma,\delta} N_{\gamma,\gamma+\delta} N_{\gamma,2\gamma+\delta} b^3 x_{3\gamma+\delta}$ and hence $D^4_{bx_\gamma}(x_\delta) = 0$ (by Lemma 2.8.9). Furthermore, by Theorem 2.9.13 we see that if $3\gamma + \delta$ is a root, then, as $\gamma - \delta \notin \Phi$, $N_{\gamma,\delta} = \pm 1$, $N_{\gamma,\gamma+\delta} = \pm 2$, $N_{\gamma,2\gamma+\delta} = \pm 3$. Also, if $3\gamma + \delta \notin \Phi$, but $2\gamma + \delta \in \Phi$ then if $\gamma - \delta \in \Phi$ we have $N_{\gamma,\delta} = \pm 2$, $N_{\gamma,\gamma+\delta} = \pm 3$, and otherwise $N_{\gamma,\delta} = \pm 1$, $N_{\gamma,\gamma+\delta} = \pm 2$. In all cases we see that $N_{\gamma,\delta} N_{\gamma,\gamma+\delta}$ is divisible by 2 and $N_{\gamma,\delta} N_{\gamma,\gamma+\delta} N_{\gamma,2\gamma+\delta}$ is divisible by 6. Since also $D_{bx_\gamma}(h_i) = -\langle \gamma, \alpha_i^\vee \rangle b x_\gamma$, so that $D^2_{bx_\gamma}(h_i) = 0$, we conclude that $\exp(D_{bx_\gamma})$ maps $\mathfrak{g}_\mathbb{Z}[\tau, \upsilon]$ into itself.

Lemma 5.2.6 *Let $\alpha, \beta \in \Phi$, $\alpha \neq \pm\beta$. Let $a \in \mathbb{Z}[\tau, \upsilon]$ have zero constant term. Then*

$$\exp(ax_\beta) x_\alpha \exp(-ax_\beta) = \exp(D_{ax_\beta})(x_\alpha).$$

Proof. Consider the sets $R_{\alpha,\beta}$ from Section 2.10.3. If $R_{\alpha,\beta} = \{\alpha, \beta\}$, then x_α and x_β commute, and the lemma is obvious. If $R_{\alpha,\beta} = \{\alpha, \beta, \alpha + \beta\}$ then, using Proposition 2.10.7, $x_\beta^{(m)} x_\alpha = x_\alpha x_\beta^{(m)} - N_{\alpha,\beta} x_\beta^{(m-1)} x_{\alpha+\beta}$. It follows that $\exp(ax_\beta) x_\alpha = x_\alpha \exp(ax_\beta) - aN_{\alpha,\beta} x_{\alpha+\beta} \exp(ax_\beta)$, whence $\exp(ax_\beta) x_\alpha \exp(-ax_\beta) = x_\alpha - aN_{\alpha,\beta} x_{\alpha+\beta} = \exp(D_{ax_\beta})(x_\alpha)$. If $R_{\alpha,\beta} = \{\alpha, \beta, \alpha + \beta, \alpha + 2\beta\}$ then, again using Proposition 2.10.7, we see that

$$\begin{aligned} x_\beta^{(m)} x_\alpha &= x_\alpha x_\beta^{(m)} - N_{\alpha,\beta} x_\beta^{(m-1)} x_{\alpha+\beta} + \tfrac{1}{2} N_{\alpha,\beta} N_{\beta,\alpha+\beta} x_\beta^{(m-2)} x_{\alpha+2\beta} \\ &= x_\alpha x_\beta^{(m)} - N_{\alpha,\beta} x_{\alpha+\beta} x_\beta^{(m-1)} - \tfrac{1}{2} N_{\alpha,\beta} N_{\beta,\alpha+\beta} x_{\alpha+2\beta} x_\beta^{(m-2)}. \end{aligned}$$

Hence $\exp(ax_\beta) x_\alpha \exp(-ax_\beta) = x_\alpha - aN_{\alpha,\beta} x_{\alpha+\beta} - \frac{1}{2} a^2 N_{\alpha,\beta} N_{\beta,\alpha+\beta} x_{\alpha+2\beta} = \exp(D_{ax_\beta})(x_\alpha)$. In the remaining cases the lemma is proved analogously. □

The formula in the next theorem is called the *Chevalley commutator formula*.

Theorem 5.2.7 *Let $\alpha, \beta \in \Phi$, $\alpha \neq \pm\beta$. Define $I_{\alpha,\beta}$ as in Section 5.1, and order the elements of $I_{\alpha,\beta}$ according to height (where the height of (i,j) is $i+j$). Then in $\mathcal{U}_{\mathbb{Z}}[\![\tau, \upsilon]\!]$ we have*

$$[\exp(\tau x_\alpha), \exp(\upsilon x_\beta)] = \prod_{(i,j)\in I_{\alpha,\beta}} \exp(c_{i,j}^{\alpha,\beta} \tau^i \upsilon^j x_{i\alpha+j\beta}),$$

where $[A, B] = ABA^{-1}B^{-1}$ denotes the commutator of A and B, the product is taken using the chosen ordering, and $c_{i,j}^{\alpha,\beta} \in \mathbb{Z}$. More precisely, $c_{1,1}^{\alpha,\beta} = N_{\alpha,\beta}$, $c_{1,2}^{\alpha,\beta} = \frac{1}{2}N_{\alpha,\beta}N_{\beta,\alpha+\beta}$, $c_{2,1}^{\alpha,\beta} = \frac{1}{2}N_{\alpha,\beta}N_{\alpha,\alpha+\beta}$, $c_{1,3}^{\alpha,\beta} = \frac{1}{6}N_{\alpha,\beta}N_{\beta,\alpha+\beta}N_{\beta,\alpha+2\beta}$, $c_{2,3}^{\alpha,\beta} = -\frac{1}{6}N_{\alpha,\beta}N_{\beta,\alpha+\beta}N_{\beta,\alpha+2\beta}N_{\alpha,\alpha+3\beta}$, $c_{3,1}^{\alpha,\beta} = \frac{1}{6}N_{\alpha,\beta}N_{\alpha,\alpha+\beta}N_{\alpha,2\alpha+\beta}$, $c_{3,2}^{\alpha,\beta} = \frac{1}{3}N_{\alpha,\beta}N_{\alpha,\alpha+\beta}N_{\alpha,2\alpha+\beta}N_{\beta,3\alpha+\beta}$.

Proof. Write $c_{i,j}$ instead of $c_{i,j}^{\alpha,\beta}$. Let $B = \prod \exp(-c_{i,j}\tau^i\upsilon^j x_{i\alpha+j\beta})$, where the product is taken in the order *opposite* the chosen one. Set

$$\begin{aligned}\psi(\tau, \upsilon) &= [\exp(\tau x_\alpha), \exp(\upsilon x_\beta)]B \\ &= \exp(\tau x_\alpha)\exp(\upsilon x_\beta)\exp(-\tau x_\alpha)\exp(-\upsilon x_\beta)B.\end{aligned}$$

We use the derivation $\tau\frac{\mathrm{d}}{\mathrm{d}\tau}$ of $\mathcal{U}_{\mathbb{Z}}[\![\tau, \upsilon]\!]$. We show that $\tau\frac{\mathrm{d}}{\mathrm{d}\tau}\psi(\tau, \upsilon) = A\psi(\tau, \upsilon)$, and that the given values of the $c_{i,j}$ imply that $A = 0$. This yields $\psi(\tau, \upsilon) = \psi(0, \upsilon) = 1$, whence the theorem.

First of all, $\tau\frac{\mathrm{d}}{\mathrm{d}\tau}\psi(\tau, \upsilon)$ is a sum of three terms:

1. $\tau x_\alpha \psi(\tau, \upsilon)$,
2. $\exp(\tau x_\alpha)\exp(\upsilon x_\beta)(-\tau x_\alpha)\exp(-\tau x_\alpha)\exp(-\upsilon x_\beta)B$,
3. $\exp(\tau x_\alpha)\exp(\upsilon x_\beta)\exp(-\tau x_\alpha)\exp(-\upsilon x_\beta)\tau\frac{\mathrm{d}}{\mathrm{d}\tau}B$.

In order to rewrite the second and third term into the required form (i.e., as multiples of $\psi(\tau, \upsilon)$), we use commutation relations that follow from Lemma 5.2.6. In the second term, the first step is to substitute $\exp(\upsilon x_\beta)(-\tau x_\alpha) = -\tau\exp(D_{\upsilon x_\beta})(x_\alpha)\exp(\upsilon x_\beta)$. The remaining steps depend on the specific form of the set $I_{\alpha,\beta}$.

The possibilities for $I_{\alpha,\beta}$ follow from the sets $R_{\alpha,\beta}$ listed in Section 2.10.3. If $I_{\alpha,\beta} = \emptyset$, then x_α, x_β commute and $[\exp(\tau x_\alpha), \exp(\upsilon x_\beta)] = 1$. If $I_{\alpha,\beta} = \{(1,1)\}$, then $\exp(D_{\upsilon x_\beta})(x_\alpha) = x_\alpha - \upsilon N_{\alpha,\beta}x_{\alpha+\beta}$. So the second term is $-\tau(x_\alpha - \upsilon N_{\alpha,\beta}x_{\alpha+\beta})\psi(\tau, \upsilon)$. Furthermore, since $B = \exp(-c_{1,1}\tau\upsilon x_{\alpha+\beta})$, the third term is

$$[\exp(\tau x_\alpha), \exp(\upsilon x_\beta)](-c_{1,1}\tau\upsilon x_{\alpha+\beta})B = -c_{1,1}\tau\upsilon x_{\alpha+\beta}\psi(\tau, \upsilon).$$

After summing and using $c_{1,1} = N_{\alpha,\beta}$ we obtain $A = 0$.

In the third case $I_{\alpha,\beta} = \{(1,1), (1,2)\}$. Here $\exp(D_{\upsilon x_\beta})(x_\alpha) = x_\alpha - \upsilon N_{\alpha,\beta}x_{\alpha+\beta} - \frac{1}{2}\upsilon^2 N_{\alpha,\beta}N_{\beta,\alpha+\beta}x_{\alpha+2\beta}$. All terms in this sum commute with x_α so the second term is

$$(-\tau x_\alpha + \tau\upsilon N_{\alpha,\beta}x_{\alpha+\beta} + \tfrac{1}{2}\tau\upsilon^2 N_{\alpha,\beta}N_{\beta,\alpha+\beta}x_{\alpha+2\beta})\psi(\tau, \upsilon).$$

Since $B = \exp(-c_{1,2}\tau\upsilon^2 x_{\alpha+2\beta})\exp(-c_{1,1}\tau\upsilon x_{\alpha+\beta})$, $\tau\frac{\mathrm{d}}{\mathrm{d}\tau}B$ is the sum of two terms:

$$C_1 = (-c_{1,2}\tau\upsilon^2 x_{\alpha+2\beta})B,$$
$$C_2 = \exp(-c_{1,2}\tau\upsilon^2 x_{\alpha+2\beta})(-c_{1,1}\tau\upsilon x_{\alpha+\beta})\exp(-c_{1,1}\tau\upsilon x_{\alpha+\beta}).$$

Since $x_{\alpha+2\beta}$ commutes with x_α, x_β we have $[\exp(\tau x_\alpha), \exp(\upsilon x_\beta)]C_1 = (-c_{1,2}\tau\upsilon^2 x_{\alpha+2\beta})\psi(\tau,\upsilon)$. By Lemma 5.2.6,

$$\begin{aligned}\exp(-\upsilon x_\beta)x_{\alpha+\beta} &= \exp(D_{-\upsilon x_\beta})(x_{\alpha+\beta})\exp(-\upsilon x_\beta)\\ &= (x_{\alpha+\beta} - \upsilon N_{\beta,\alpha+\beta}x_{\alpha+2\beta})\exp(-\upsilon x_\beta),\end{aligned}$$

and for $\exp(\upsilon x_\beta)x_{\alpha+\beta}$ we have a very similar expression with different signs. This leads to the cancellation of two terms of the form $\pm\upsilon N_{\beta,\alpha+\beta}x_{\alpha+2\beta}\psi(\tau,\upsilon)$. It follows that $[\exp(\tau x_\alpha), \exp(\upsilon x_\beta)]C_2 = -c_{1,1}\tau\upsilon x_{\alpha+\beta}\psi(\tau,\upsilon)$. Summing shows that

$$A = \tau\upsilon(N_{\alpha,\beta} - c_{1,1})x_{\alpha+\beta} + \tau\upsilon^2(\tfrac{1}{2}N_{\alpha,\beta}N_{\beta,\alpha+\beta} - c_{1,2})x_{\alpha+2\beta} = 0.$$

The arguments for the remaining cases are very similar. Therefore we leave them to the reader. □

Remark 5.2.8 It is straightforward to see that the $c_{i,j}^{\alpha,\beta}$ in the previous theorem are all integers. In fact, they are are equal to ± 1, except $c_{3,2}^{\alpha,\beta}$, which is equal to ± 2.

5.2.3 Chevalley groups

Let V be a finite-dimensional faithful $\mathfrak{g}$-module, and let Λ be an admissible lattice in V. (The hypothesis of faithfulness simply means that $\mathfrak{g}$ has no direct summand acting trivially.) Let k be a field, and set $V^k = k \otimes_{\mathbb{Z}} \Lambda$. Let μ be a weight of V, then by Λ_μ we denote $\Lambda \cap V_\mu$, and set $V_\mu^k = k \otimes_{\mathbb{Z}} \Lambda_\mu$. Then V^k is the direct sum of the various V_μ^k (Lemma 5.2.2).

Let $v_1, \dots, v_m$ be a basis of Λ; then $1 \otimes v_1, \dots, 1 \otimes v_m$ is a basis of the k-vector space V^k. Let $\mathcal{U}_k$ denote $k \otimes_{\mathbb{Z}} \mathcal{U}_{\mathbb{Z}}$. Let $t \in k$ and $\alpha \in \Phi$, then $t^n \otimes x_\alpha^{(n)} \in \mathcal{U}_k$ corresponds to an endomorphism of V^k by $t^n \otimes x_\alpha^{(n)} \cdot \zeta \otimes v = (t^n\zeta) \otimes (x_\alpha^{(n)} v)$. So if M is the matrix of $x_\alpha^{(n)}$ with respect to the basis $v_1, \dots, v_m$, then $t^n M$ is the matrix of $t^n \otimes x_\alpha^n$ with respect to the basis $1 \otimes v_1, \dots, 1 \otimes v_m$. (As M has integer entries, it can be viewed as a matrix with coefficients in k, therefore it makes sense to talk about $t^n M$.) Note that $x_\alpha^{(n)}$ acts as zero for n large enough (see the proof of Lemma 2.11.3). Therefore $\sum_{n=0}^{\infty} t^n \otimes x_\alpha^{(n)}$ corresponds to a well-defined element of $\mathrm{End}(V^k)$, which is denoted by $x_\alpha(t)$. By the next lemma it even lies in $\mathrm{GL}(V^k)$. We let G be the group generated by $x_\alpha(t)$, for $t \in k$, $\alpha \in \Phi$. This is called the *Chevalley group* defined by V

and k (later we will see that this group does not depend on the choice of Λ). It is the objective of this section to show that G is a C-group.

Example 5.2.9 Let $\mathfrak{g}$ and $V = V(\lambda)$ be as in Example 5.2.5 where a basis B of V is given and an admissible lattice is computed. It is seen that the admissible lattice is spanned by B. Using the Gröbner basis given in the example and the commutation relations in $\mathcal{U}_\mathbb{Z}$, it is straightforward to obtain the matrices of elements of $\mathfrak{g}$ with respect to B:

$$x_{\alpha_1}(t) = \begin{pmatrix} 1 & 2t & 0 & t^2 & 0 & 0 \\ 0 & 1 & 0 & t & 0 & 0 \\ 0 & 0 & 1 & 0 & t & 0 \\ 0 & 0 & 0 & 1 & 0 & 0 \\ 0 & 0 & 0 & 0 & 1 & 0 \\ 0 & 0 & 0 & 0 & 0 & 1 \end{pmatrix}, \quad x_{-\alpha_1}(t) = \begin{pmatrix} 1 & 0 & 0 & 0 & 0 & 0 \\ t & 1 & 0 & 0 & 0 & 0 \\ 0 & 0 & 1 & 0 & 0 & 0 \\ t^2 & 2t & 0 & 1 & 0 & 0 \\ 0 & 0 & t & 0 & 1 & 0 \\ 0 & 0 & 0 & 0 & 0 & 1 \end{pmatrix}.$$

Lemma 5.2.10 *Let $\alpha, \beta \in \Phi$, $\alpha \neq \pm\beta$ and $t, u \in k$. Then*

(i) *$x_\alpha(0)$ is the identity on V^k,*

(ii) *$x_\alpha(t)x_\alpha(u) = x_\alpha(t+u)$,*

(iii) *$[x_\alpha(t), x_\beta(u)] = \prod_{(i,j)\in I_{\alpha,\beta}} x_{i\alpha+j\beta}(c_{i,j}^{\alpha,\beta} t^i u^j)$, where the $c_{i,j}^{\alpha,\beta}$ are as in Theorem 5.2.7 and the product is taken in a height-compatible order.*

Proof. The first statement is obvious. Consider, as in the previous section, the ring $\mathcal{U}_\mathbb{Z}[\![\tau, \upsilon]\!]$. Let R be the subring consisting of those $\xi = \sum_{i,j} a_{ij}\tau^i\upsilon^j$, such that only finitely many a_{ij} act non-trivially in V. Define $\sigma : R \to \mathrm{End}(V^k)$ by $\sigma(\xi)(1 \otimes v) = \sum_{i,j} t^i u^j \otimes a_{ij}v$. Then R is a ring homomorphism. Applying σ to the relation $\exp(\tau x_\alpha)\exp(\upsilon x_\alpha) = \exp((\tau+\upsilon)x_\alpha)$ yields (ii). Similarly, (iii) follows from Theorem 5.2.7. □

Example 5.2.11 Let $\mathfrak{g}$ be the Lie algebra of type A_2, so that the root system has positive roots $\alpha_1, \alpha_2, \alpha_3 = \alpha_1 + \alpha_2$. Using the multiplication table of Example 2.9.14 we see that $[x_{\alpha_2}(t), x_{\alpha_1}(u)] = x_{\alpha_3}(tu)$, $[x_{\alpha_3}(t), x_{\alpha_1}(u)] = [x_{\alpha_3}(t), x_{\alpha_2}(u)] = 1$. We can rewrite these relations as follows

$$\begin{aligned} x_{\alpha_2}(t)x_{\alpha_1}(u) &= x_{\alpha_1}(u)x_{\alpha_2}(t)x_{\alpha_3}(tu) \\ x_{\alpha_3}(t)x_{\alpha_1}(u) &= x_{\alpha_1}(u)x_{\alpha_3}(t) \\ x_{\alpha_3}(t)x_{\alpha_2}(u) &= x_{\alpha_2}(u)x_{\alpha_3}(t). \end{aligned}$$

For $\alpha \in \Phi$ and $t \in k^*$ we set

$$w_\alpha(t) = x_\alpha(t)x_{-\alpha}(-t^{-1})x_\alpha(t),\ h_\alpha(t) = w_\alpha(t)w_\alpha(1)^{-1} \text{ and } \bar{w}_\alpha = w_\alpha(1).$$

Also we set $\mathfrak{g}_k = k \otimes_\mathbb{Z} \mathfrak{g}_\mathbb{Z}$, $\mathfrak{h}_k = k \otimes_\mathbb{Z} \mathfrak{h}_\mathbb{Z}$ and let $\rho_k : \mathfrak{g}_k \to \mathfrak{gl}(V^k)$ be the representation of $\mathfrak{g}_k$ afforded by V^k. Remark 2.9.9 provides an action of the Weyl group W on $\mathfrak{h}$. This action leaves $\mathfrak{h}_\mathbb{Z}$ invariant, and therefore W acts on $\mathfrak{h}_k$ as well.

Example 5.2.12 Let the notation be as in Example 5.2.9. Using the matrices of $x_{\alpha_1}(t)$, $x_{-\alpha_1}(t)$ given there, we obtain

$$w_{\alpha_1}(t) = \begin{pmatrix} 0 & 0 & 0 & t^2 & 0 & 0 \\ 0 & -1 & 0 & 0 & 0 & 0 \\ 0 & 0 & 0 & 0 & t & 0 \\ t^{-2} & 0 & 0 & 0 & 0 & 0 \\ 0 & 0 & -t^{-1} & 0 & 0 & 0 \\ 0 & 0 & 0 & 0 & 0 & 1 \end{pmatrix},$$

and $h_{\alpha_1}(t) = \mathrm{diag}(t^2, 1, t, t^{-2}, t^{-1}, 1)$.

Lemma 5.2.13 *Let $\alpha \in \Phi$, $t \in k^*$ and $h \in \mathfrak{h}_k$. Then $w_\alpha(t)\rho_k(h)w_\alpha(t)^{-1} = \rho_k(s_\alpha(h))$.*

Proof. First assume that k is of characteristic 0. Set $h_\alpha = [x_\alpha, x_{-\alpha}]$. Then $\alpha(h_\alpha) = 2$ (Remark 2.9.15). If $h \in h_k$ is such that $\alpha(h) = 0$, then h and x_α commute, so that both sides of the equation equal $\rho_k(h)$. Hence it suffices to show the lemma when $h = h_\alpha$. Since k is of characteristic 0, $x_\alpha(t) = \exp(t\rho_k(x_\alpha))$. By applying Lemma 2.3.1 three times, $w_\alpha(t)\rho_k(h_\alpha)w_\alpha(t)^{-1} = -\rho_k(h_\alpha)$, which is equal to $\rho_k(s_\alpha(h_\alpha))$.

For the general case we fix a basis of Λ, yielding a corresponding basis of V^k, and we represent all linear maps with respect to that basis. Consider the ring of Laurent polynomials $\mathbb{Z}[\tau, \tau^{-1}]$. Let $X_{n,\alpha}(\tau)$ be τ^n times the matrix of $x_\alpha^{(n)}$ with respect to the fixed basis of Λ, so $X_{n,\alpha}(\tau)$ has entries in $\mathbb{Z}[\tau]$. Furthermore, set $X_\alpha(\tau) = \sum_{n\geq 0} X_{n,\alpha}(\tau)$. Next consider $M_\alpha = X_\alpha(\tau)X_{-\alpha}(-\tau^{-1})X_\alpha(\tau)$, $\overline{M}_\alpha = X_\alpha(-\tau)X_{-\alpha}(\tau^{-1})X_\alpha(-\tau)$. Then we obtain the matrix of $w_\alpha(t)$ from M_α by mapping $\tau \mapsto t$. Similarly, we obtain the matrix of $w_\alpha(t)^{-1}$ from $\overline{M}_\alpha$. For $\tilde{h} \in \mathfrak{h}_\mathbb{Z}$ we let $\rho_\mathbb{Z}(\tilde{h})$ denote the matrix of $\tilde{h}$ acting on Λ. Then $M_\alpha \rho_\mathbb{Z}(\tilde{h})\overline{M}_\alpha = \rho_\mathbb{Z}(s_\alpha(\tilde{h}))$, is an equation that holds when we specialize τ to any non-zero value of a field of characteristic 0. So the equation holds over $\mathbb{Z}[\tau, \tau^{-1}]$. Therefore, the lemma is proved over fields of any characteristic. □

Lemma 5.2.14 *Let $\alpha \in \Phi$ and μ be a weight of V and $v \in V_\mu^k$. Then there is a $v' \in V_{s_\alpha(\mu)}^k$ such that $w_\alpha(t)v = t^{-\langle \mu, \alpha^\vee \rangle}v'$ for all $t \in k^*$.*

Proof. As in the proof of the previous lemma we may assume that the characteristic of k is 0. Set $v'' = w_\alpha(t)v$. As $x_\alpha(t) = \exp(t\rho_k(x_\alpha))$, $v'' = \sum_{i\in\mathbb{Z}} t^i v_i$, where $v_i \in V_{\mu+i\alpha}$ (as there are only finitely many weights, this sum is finite). Let $h \in \mathfrak{h}_k$; then $hv'' = hw_\alpha(t)v = w_\alpha(t)w_\alpha(t)^{-1}hw_\alpha(t)v = w_\alpha(t)s_\alpha(h)v$ (by Lemma 5.2.13) $= \mu(s_\alpha(h))w_\alpha(t)v = s_\alpha(\mu)(h)v''$. It follows that v'' is a weight vector of weight $s_\alpha(\mu)$. So the only i for which v_i is non-zero is $i = -\langle \mu, \alpha^\vee \rangle$. □

Lemma 5.2.15 *We have $w_\alpha(t)\rho_k(x_\beta)w_\alpha(t)^{-1} = c_{\alpha,\beta}t^{-\langle\beta,\alpha^\vee\rangle}\rho_k(x_{s_\alpha(\beta)})$, for all $\alpha, \beta \in \Phi$, where $c_{\alpha,\beta} = \pm 1$ is independent of t and $c_{\alpha,\beta} = c_{\alpha,-\beta}$. More precisely, the $c_{\alpha,\beta}$ are as follows. Firstly, $c_{\alpha,\alpha} = c_{\alpha,-\alpha} = -1$. For $\alpha, \beta \in \Phi$, $\alpha \neq \pm\beta$ let r be maximal such that $\beta - r\alpha \in \Phi$. Define $\varepsilon(\alpha,\beta)$ by $N_{\alpha,\beta} = \varepsilon(\alpha,\beta)(r+1)$. If $\langle\beta,\alpha^\vee\rangle > 0$, set $\gamma = -\alpha$ and $\gamma = \alpha$ otherwise. Set $m = |\langle\beta,\alpha^\vee\rangle|$. Then $c_{\alpha,\beta} = (-1)^r \prod_{i=0}^{m-1} \varepsilon(\gamma, i\gamma + \beta)$.*

Proof. As for the previous lemmas it suffices to prove this for k of characteristic 0. Consider the adjoint representation of $\mathfrak{g}$ and define $W_\alpha(t)$ in the same way as $w_\alpha(t)$. Then using Lemma 2.3.1 we see that $w_\alpha(t)\rho_k(x_\beta)w_\alpha(t)^{-1} = \rho_k(W_\alpha(t)(x_\beta))$. By Lemma 5.2.14 this is equal to $\rho_k(c_{\alpha,\beta}t^{-\langle\beta,\alpha^\vee\rangle}x_{s_\alpha(\beta)})$. Now $W_\alpha(1)$ is an automorphism of $\mathfrak{g}_\mathbb{Z}$, mapping $x_\alpha \mapsto c_{\alpha,\beta}x_{s_\alpha(\beta)}$. In particular, it preserves the space spanned by $x_\beta, x_{s_\alpha(\beta)}$. So the determinant of the restriction of $W_\alpha(1)$ to that space is ± 1, forcing $c_{\alpha,\beta} = \pm 1$.

Furthermore, by Lemma 5.2.13, setting $h_\beta = [x_\beta, x_{-\beta}]$,

$$\begin{aligned}\rho_k(h_{s_\alpha(\beta)}) &= w_\alpha(1)\rho_k(h_\beta)w_\alpha(1)^{-1}\\ &= [w_\alpha(1)\rho_k(x_\beta)w_\alpha(1)^{-1}, w_\alpha(1)\rho_k(x_{-\beta})w_\alpha(1)^{-1}]\\ &= c_{\alpha,\beta}c_{\alpha,-\beta}[\rho_k(x_{s_\alpha(\beta)}), \rho_k(x_{-s_\alpha(\beta)})] = c_{\alpha,\beta}c_{\alpha,-\beta}\rho_k(h_{s_\alpha(\beta)}),\end{aligned}$$

yielding $c_{\alpha,\beta} = c_{\alpha,-\beta}$.

In order to compute $c_{\alpha,\beta}$ it is sufficient to work out the coefficient of $x_{s_\alpha(\beta)}$ in $\exp(\mathrm{ad}x_\alpha)\exp(-\mathrm{ad}x_{-\alpha})\exp(\mathrm{ad}x_\alpha)(x_\beta)$. This is a straightforward case by case verification, where each case depends on the values of r and $\langle\beta,\alpha^\vee\rangle$. Here we deal with one case, leaving the others to the reader. Suppose $r = 0$ and $\langle\beta,\alpha^\vee\rangle = -2$. Then $s_\alpha(\beta) = \beta + 2\alpha$. Furthermore,

$$\begin{aligned}\exp(-\mathrm{ad}x_{-\alpha})\exp(\mathrm{ad}x_\alpha)(x_\beta) &= x_\beta + N_{\alpha,\beta}(x_{\alpha+\beta} - N_{-\alpha,\alpha+\beta}x_\beta) +\\ &\tfrac{1}{2}N_{\alpha,\beta}N_{\alpha,\alpha+\beta}(x_{2\alpha+\beta} - N_{-\alpha,2\alpha+\beta}x_{\alpha+\beta} + \tfrac{1}{2}N_{-\alpha,2\alpha+\beta}N_{-\alpha,\alpha+\beta}x_\beta).\end{aligned}$$

The coefficient of $x_{2\alpha+\beta}$ in $\exp(\mathrm{ad}x_\alpha)\exp(-\mathrm{ad}x_{-\alpha})\exp(\mathrm{ad}x_\alpha)(x_\beta)$ is then

$$\begin{aligned}&\tfrac{1}{2}N_{\alpha,\beta}N_{\alpha,\alpha+\beta}(1 - N_{-\alpha,\alpha+\beta}N_{\alpha,\beta} + \tfrac{1}{4}N_{\alpha,\beta}N_{\alpha,\alpha+\beta}N_{-\alpha,2\alpha+\beta}N_{-\alpha,\alpha+\beta})\\ &+ N_{\alpha,\alpha+\beta}(N_{\alpha,\beta} - \tfrac{1}{2}N_{\alpha,\beta}N_{\alpha,\alpha+\beta}N_{-\alpha,2\alpha+\beta}) + \tfrac{1}{2}N_{\alpha,\beta}N_{\alpha,\alpha+\beta}.\end{aligned}$$

By evaluating the Jacobi identity involving $x_{-\alpha}$, x_α and x_β we find that $N_{-\alpha,\alpha+\beta}N_{\alpha,\beta} = 2$. By evaluating the Jacobi identity involving $x_{-\alpha}$, x_α and $x_{\alpha+\beta}$ we find that $N_{\alpha,\alpha+\beta}N_{-\alpha,2\alpha+\beta} = N_{-\alpha,\alpha+\beta}N_{\alpha,\beta} = 2$. Substituting these values we find that the coefficient of $x_{2\alpha+\beta}$ is equal to $\frac{1}{2}N_{\alpha,\beta}N_{\alpha,\alpha+\beta}$ as required. □

Lemma 5.2.16 *Let the notation be as in the previous lemma. We have $c_{\alpha,\beta} = c_{-\alpha,-s_\alpha(\beta)}$ for all $\alpha, \beta \in \Phi$.*

Proof. We may suppose that $\alpha \neq \pm\beta$.

Let $\delta_1, \delta_2, \delta_3$ be three roots such that $\delta_1 + \delta_2 + \delta_3 = 0$. Using the Jacobi identity it can be shown that

$$\frac{N_{\delta_1,\delta_2}}{(\delta_3,\delta_3)} = \frac{N_{\delta_2,\delta_3}}{(\delta_1,\delta_1)} = \frac{N_{\delta_3,\delta_1}}{(\delta_2,\delta_2)}$$

(see [Car72], Theorem 4.1.2(ii)). It follows that

$$\varepsilon(\delta_1, \delta_2) = -\varepsilon(\delta_1, \delta_3). \tag{5.7}$$

Write $\beta' = -s_\alpha(\beta)$, and let r' be maximal with $\beta' - r'(-\alpha) \in \Phi$. Since $\beta' = -\beta + \langle \beta, \alpha^\vee \rangle \alpha$, it follows that $r' = r - \langle \beta, \alpha^\vee \rangle$.

If $\langle \beta, \alpha^\vee \rangle = 0$, the lemma is trivial. Suppose $\langle \beta, \alpha^\vee \rangle < 0$. Then $c_{\alpha,\beta} = e \prod_{i=0}^{m-1} \varepsilon(\alpha, i\alpha + \beta)$ (with $e = (-1)^r$), whereas after some rewriting we see that $c_{-\alpha,-s_\alpha(\beta)} = e' \prod_{j=1}^{m} \varepsilon(\alpha, -j\alpha - \beta)$ (with $e' = (-1)^{r'}$). Now, $e' = (-1)^m e$, and using (5.7) we see that $\varepsilon(\alpha, i\alpha + \beta) = -\varepsilon(\alpha, -(i+1)\alpha - \beta)$. The lemma is proved in this case. When $\langle \beta, \alpha^\vee \rangle > 0$, it is proved by an analogous argument. □

Lemma 5.2.17 *Let $\alpha \in \Phi$. Then $h_\alpha(t)$ acts on V_μ^k as multiplication by $t^{\langle \mu, \alpha^\vee \rangle}$.*

Proof. Observe that $w_\alpha(t)^{-1} = w_\alpha(-t)$, whence $h_\alpha(t) = w_\alpha(-t)^{-1} w_\alpha(-1)$. Let $v \in V_\mu^k$, and let $v' \in V_{s_\alpha(\mu)}^k$ be such that $w_\alpha(t)v = t^{-\langle \mu, \alpha^\vee \rangle} v'$ (Lemma 5.2.14). Then $w_\alpha(t)^{-1} v' = t^{\langle \mu, \alpha^\vee \rangle} v$ and $w_\alpha(-1)v = (-1)^{-\langle \mu, \alpha^\vee \rangle} v'$. It follows that $w_\alpha(-t)^{-1} w_\alpha(-1) v = t^{\langle \mu, \alpha^\vee \rangle} v$. □

Proposition 5.2.18 *Let $\alpha, \beta \in \Phi$. Then*

(i) $\bar{w}_\alpha h_\beta(t) \bar{w}_\alpha^{-1} = h_{s_\alpha(\beta)}(t)$ *for* $t \in k^*$,

(ii) $\bar{w}_\alpha x_\beta(t) \bar{w}_\alpha^{-1} = x_{s_\alpha(\beta)}(c_{\alpha,\beta} t)$, *for* $t \in k$, *with* $c_{\alpha,\beta}$ *as in Lemma 5.2.15,*

(iii) $h_\alpha(t) x_\beta(s) h_\alpha(t)^{-1} = x_\beta(t^{\langle \beta, \alpha^\vee \rangle} s)$, *for* $t \in k^*$, $s \in k$.

Proof. As in the proof of Lemma 5.2.13 it suffices to prove these assertions assuming that k is of characteristic 0.

Let μ be a weight of V and $v \in V_\mu^k$. By Lemma 5.2.14, $\bar{w}_\alpha^{-1} v \in V_{s_\alpha(\mu)}^k$ (note that $\bar{w}_\alpha^{-1} = w_\alpha(-1)$). Hence $h_\beta(t) \bar{w}_\alpha^{-1} v = t^{\langle s_\alpha(\mu), \beta^\vee \rangle} \bar{w}_\alpha^{-1} v$ (Lemma 5.2.17). So $\bar{w}_\alpha h_\beta(t) \bar{w}_\alpha^{-1} v = t^{\langle s_\alpha(\mu), \beta^\vee \rangle} v$. Now $\langle s_\alpha(\mu), \beta^\vee \rangle = \langle \mu, s_\alpha(\beta)^\vee \rangle$. Therefore by Lemma 5.2.17 we have (i).

By Lemma 5.2.15, $\bar{w}_\alpha \rho_k(x_\beta) \bar{w}_\alpha^{-1} = c_{\alpha,\beta} \rho_k(x_{s_\alpha(\beta)})$. So (ii) follows by exponentiating.

For (iii) we let $H_\alpha(t)$ be defined in the same way as $h_\alpha(t)$, but with respect to the adjoint representation. Using Lemma 2.3.1 three times we see that

$h_\alpha(t)\rho_k(x_\beta)h_\alpha(t)^{-1} = \rho_k(H_\alpha(t)(x_\beta))$. By Lemma 5.2.17, the latter is equal to $t^{\langle\beta,\alpha^\vee\rangle}\rho_k(x_\beta)$. The statement of the proposition now follows by exponentiating both sides. □

Corollary 5.2.19 *G is a C-group and the set consisting of $x_\alpha(t)$ for $\alpha \in \Phi$, $t \in k$ is a C-generating set of G.*

5.2.4 Bruhat decomposition

Let the notation be as in the previous section. For $\alpha \in \Phi$ let $X_\alpha = \{x_\alpha(t) \mid t \in k\}$, which is a subgroup of G. Let $U = U^+$, U^- be the subgroups of G generated by all X_α for $\alpha \in \Phi^+$ and $\alpha \in \Phi^-$ respectively. Let H be the subgroup of G generated by all $h_\alpha(t)$, $\alpha \in \Phi$, $t \in k^*$ and B be the subgroup of G generated by U and H. Finally, let N be the subgroup generated by $w_\alpha(t)$, $\alpha \in \Phi$, $t \in k^*$. By Corollary 5.2.19, all results of Section 5.1 concerning the groups $\mathcal{G}$, $\mathcal{X}_\alpha$, $\mathcal{U}$, $\mathcal{B}$, $\mathcal{H}$ and $\mathcal{N}$ hold for the groups G, X_α, U, B, H and N. It is the objective of this section to strengthen some of the statements in Section 5.1 (most importantly the decomposition of Propositions 5.1.9 and 5.1.10) by using the construction of G from a representation of $\mathfrak{g}$.

Lemma 5.2.20 *Let $S \subset \Phi^+$ be closed, and let X_S be the subgroup of G generated by X_α for $\alpha \in S$. Then every element of X_S can be written uniquely as $\prod_{\alpha\in S} x_\alpha(t_\alpha)$, where $t_\alpha \in k$ and the product is taken in any fixed order.*

Proof. In view of Lemma 5.1.3 we only have to prove uniqueness. Let $g = \prod_{\alpha\in S} x_\alpha(t_\alpha)$ and $\alpha_1 \in S$ be of minimal height. Let $v \in V_\mu^k$ be such that $x_{\alpha_1}v \neq 0$, where μ is some weight of V. Then $gv = v + t_{\alpha_1}x_{\alpha_1}v + v_1$, where $x_{\alpha_1}v \in V_{\mu+\alpha_1}^k$ and v_1 is a sum of weight vectors of weights not equal to μ or $\mu + \alpha_1$. It follows that t_{α_1} is uniquely determined by g. Furthermore, $I = S \setminus \{\alpha_1\}$ is an ideal in S. Therefore, X_I is normal in X_S (Lemma 5.1.2), so that $g' = x_{\alpha_1}(-t_{\alpha_1})g$ lies in X_I. The proof is finished by induction. □

Consider the partial order $\preceq$ on the weights defined in Section 2.11. List the weights of V in an order compatible with $\preceq$ from large to small. We take a basis of V^k consisting of weight vectors, arranged according to the order of the weights, with respect to which we define the matrix of an element of G. Then the elements of U^+ are upper triangular and unipotent, the elements of U^- are lower triangular and unipotent, and the elements of H are diagonal (Lemma 5.2.17).

Lemma 5.2.21 *Let $\alpha, \beta \in \Phi$. Then $X_\alpha = X_\beta$ if and only if $\alpha = \beta$.*

Proof. If α and β have the same sign, then this follows from the previous lemma (note that with a different choice of ordering of the roots we can have Φ^- as the set of positive roots, so a similar statement as in the lemma also holds when $S \subset \Phi^-$). If they have different sign, one of X_α, X_β is upper triangular unipotent, whereas the other is lower triangular unipotent. The lemma follows also in this case. □

Lemma 5.2.22 *The surjective homomorphim $\vartheta : W \to N/H$, with $\vartheta(s_\alpha) = Hw_\alpha(t)$ (Lemma 5.1.4(iii)), is an isomorphism.*

Proof. It remains to show that ϑ is injective. Let $w = s_{i_1} \cdots s_{i_r}$ lie in $\ker \vartheta$. Then $h = w_{\alpha_1}(1) \cdots w_{\alpha_r}(1)$ lies in H. By Proposition 5.2.18(ii), $hx_\beta(t)h^{-1} = x_{w\beta}(\pm t)$, for $\beta \in \Phi$, $t \in k$. So $hX_\beta h^{-1} = X_{w\beta}$. On the other hand, $hX_\beta h^{-1} = X_\beta$ by Proposition 5.2.18(iii). By Lemma 5.2.21 it now follows that $w\beta = \beta$ for all $\beta \in \Phi$, whence $w = 1$. □

Theorem 5.2.23 *For each $w \in W$ fix a $\dot{w} \in N$ with $\vartheta(w) = H\dot{w}$. For $w \in W$, let Φ_w be as in Proposition 5.1.10, and U_w be the subgroup of G generated by $x_\alpha(t)$ for $\alpha \in \Phi_w$, $t \in k$. Then*

(i) *G is the disjoint union of the sets $B\dot{w}B$ for $w \in W$,*

(ii) *every element in $B\dot{w}B$ can uniquely be written as $b\dot{w}u$, where $b \in B$, $u \in U_w$.*

Proof. By Proposition 5.1.9 we only have to prove disjointness. Let $w_1, w_2 \in W$ be such that $B\dot{w}_1B = B\dot{w}_2B$. Using induction on $\mathcal{L}(w_1)$ we show that this implies that $w_1 = w_2$. If $\mathcal{L}(w_1) = 0$, then $w_1 = 1$ and $B\dot{w}_1B = B$, so that $\dot{w}_2 \in B$. As a consequence, $\dot{w}_2B\dot{w}_2^{-1} = B$. But $\dot{w}_2\mathcal{X}_\alpha\dot{w}_2^{-1} = \mathcal{X}_{w_2(\alpha)}$ by Lemma 5.1.5. Using Lemma 5.2.21, it follows that $w_2(\alpha) > 0$ for all $\alpha \in \Phi^+$. That implies that $\mathcal{L}(w_2) = 1$ (Lemma 2.8.24).

Now suppose $\mathcal{L}(w_1) > 0$. Then there is an $\alpha \in \Delta$ such that $\mathcal{L}(w_1s_\alpha) < \mathcal{L}(w_1)$. Set $v_i = w_is_\alpha$ for $i = 1, 2$. Then $\dot{v}_1 \in B\dot{w}_1B \cdot B\dot{s}_\alpha B = B\dot{w}_2B \cdot B\dot{s}_\alpha B \subset B\dot{w}_2B \cup B\dot{v}_2B$ (Lemma 5.1.7(i)) $= B\dot{w}_1B \cup B\dot{v}_2B$. We have $\dot{v}_1 \notin B\dot{w}_1B$, as otherwise $B\dot{v}_1B = B\dot{w}_1B$ and $v_1 = w_1$ (induction), forcing $s_\alpha = 1$. It follows that $\dot{v}_1 \in B\dot{v}_2B$. Hence $B\dot{v}_1B = B\dot{v}_2B$, and therefore $v_1 = v_2$ by induction, whence $w_1 = w_2$.

For (ii) we only need to prove uniqueness in view of Proposition 5.1.10. Suppose $b\dot{w}u = b'\dot{w}u'$ where $b, b' \in B$, $u, u' \in U_w$. Then $b^{-1}b' = \dot{w}u(u')^{-1}\dot{w}^{-1}$. Now $u(u')^{-1} \in U_w$. But $\dot{w}X_\alpha\dot{w}^{-1} = X_{w\alpha}$ (Lemma 5.1.5). Since $w(\alpha) < 0$ for all $\alpha \in \Phi_w$, we have $\dot{w}u(u')^{-1}\dot{w}^{-1} \in U^-$. So $b^{-1}b'$ is both upper triangular and lower triangular unipotent; hence it is 1. It follows that $b' = b$ and $u = u'$. □

The decomposition of G as the union of the disjoint sets $B\dot{w}B$ is called the *Bruhat decomposition*.

Corollary 5.2.24 *Let the notation be as in the previous theorem. Every element of G has a unique expression as $u_1 h\dot{w}u_2$ where $u_1 \in U$, $u_2 \in U_w$ and $h \in H$.*

Proof. By Lemma 5.1.4(i), $B = UH$. An element of $U \cap H$ is both unipotent upper triangular and diagonal, and must be 1, so the corollary follows from Theorem 5.2.23(ii). □

Example 5.2.25 It follows that G is the disjoint union of the sets $B\dot{w}U_w$. Let $\mathfrak{g}$ be of type A_2, as in Example 5.2.11. Then, for instance, for $w = s_1 s_2$, we have $\Phi_w = \{\alpha_2, \alpha_3\}$ and every element of $B\dot{w}U_w$ can be written uniquely as $x_{\alpha_1}(t_1)x_{\alpha_2}(t_2)x_{\alpha_3}(t_3)h\dot{w}x_{\alpha_2}(t_4)x_{\alpha_3}(t_5)$, where $h \in H$, $t_i \in k$.

Proposition 5.2.26 *The centre of G is contained in H.*

Proof. Let $g \in G$ lie in the centre of G. By Corollary 5.2.24 there are unique $u_1 \in U$, $h \in H$, $w \in W$, $u_2 \in U_w$ with $g = u_1 h\dot{w}u_2$. Suppose $w \neq 1$. Then there is an $\alpha > 0$ such that $w\alpha < 0$. Hence $\alpha \in \Phi_w$ so that $X_\alpha \subset U_w$. As $gx_\alpha(t) = x_\alpha(t)g$ we have $u_1 h\dot{w}(u_2 x_\alpha(t)) = (x_\alpha(t)u_1)h\dot{w}u_2$ contrary to the uniqueness part of Corollary 5.2.24. It follows that $w = 1$, whence $g \in B$.

Let w_0 be the longest element of W. This element maps all positive roots to negative ones. So there is a $\preceq$-compatible order of the weights of V such that w_0 reverses the list of those weights. By Lemma 5.2.14, $\dot{w}_0$ maps V_μ^k to $V_{w_0(\mu)}^k$. It follows that there is a basis of V^k where the weights of the basis elements occur in the given order, from large to small, such that $\dot{w}_0$ reverses this basis. Hence $\dot{w}_0 g\dot{w}_0^{-1}$ is lower triangular. But also $\dot{w}_0 g\dot{w}_0^{-1} = g$ is upper triangular. Therefore, g is diagonal, and as $g \in B$, it follows that $g \in H$. □

Proposition 5.2.27 *Suppose k is infinite. Then $H = Z_G(H)$ and $N = N_G(H)$.*

Proof. Let $g \in N_G(H)$ and write $g = u_1 h\dot{w}u_2$ as in Corollary 5.2.24. Let $h_1 \in H$ and set $h_2 = gh_1g^{-1}$, then also $h_2 \in H$. Hence $u_1 h\dot{w}u_2 h_1 = h_2 u_1 h\dot{w}u_2$, or

$$(u_1)(h\dot{w}h_1\dot{w}^{-1})\dot{w}(h_1^{-1}u_2h_1) = (h_2u_1h_2^{-1})(h_2h)\dot{w}(u_2). \tag{5.8}$$

By Corollary 5.2.24 it follows that $h\dot{w}h_1\dot{w}^{-1} = h_2h$ so that $\dot{w}h_1\dot{w}^{-1} = h_2$.

Now let $g \in Z_G(H)$. Then in the above argument we have $h_2 = h_1$ and it follows that $\dot{w} \in Z_G(H)$. But by Proposition 5.2.18(i) we see that $\dot{w}h_\beta(t)\dot{w}^{-1} = h_{w(\beta)}(t)$ for all $t \in k$, $\beta \in \Phi$. It follows that $h_\beta(t) = h_{w(\beta)}(t)$ for all t and β. Using Lemma 5.2.17, and the fact that k is infinite, we infer that this is the same as $\langle \mu, \beta^\vee \rangle = \langle \mu, w(\beta)^\vee \rangle$ for all weights μ of V. Lemma 5.2.29 now implies that $w(\beta) = \beta$ for all β, whence $w = 1$. From Proposition 5.2.18(iii), Lemma 5.2.20, it follows that if a $u \in U$ commutes with all elements of H, then $u = 1$. But (5.8) entails also that $u_1 = h_1 u_1 h_1^{-1}$, and $u_2 = h_1^{-1} u_2 h_1$ for all $h_1 \in H$. The conclusion is that $Z_G(H) = H$.

If $g \in N_G(H)$, we see that $\dot{w}^{-1} g h_1 (\dot{w}^{-1} g)^{-1} = h_1$ for all $h_1 \in H$. In other words, $\dot{w}^{-1} g \in Z_G(H)$. So $g = \dot{w} h_3$ for some $h_3 \in H$ and therefore $g \in N$. □

5.2.5 Presentation of G

Let G be as in Section 5.2.3. Let $\mathcal{G}$ be a C-group with C-generating set consisting of $\xi_\alpha(t)$ for $\alpha \in \Phi$, $t \in k$. Relative to $\mathcal{G}$ we use the notation introduced in Section 5.1. So we have the elements $\omega_\alpha(t)$, $\eta_\alpha(t)$, ϖ_α, and subgroups $\mathcal{U}$, $\mathcal{H}$ and so on.

Lemma 5.2.28 *Suppose $\phi(\xi_\alpha(t)) = x_\alpha(t)$ extends to a surjective group homomorphism $\mathcal{G} \to G$. For $w \in W$ fix a $\dot{w} \in \mathcal{N}$ such that $\vartheta(w) = \mathcal{H}\dot{w}$ (with ϑ as in Lemma 5.1.4).*

- (i) *Every element of $\mathcal{U}$ can be written uniquely as $\prod_{\alpha>0} \xi_\alpha(t_\alpha)$ where $t_\alpha \in k$, and the product is taken in any fixed order.*
- (ii) *Every element of $\mathcal{G}$ can uniquely be written as $u_1 h \dot{w} u_2$ where $u_1 \in \mathcal{U}$, $h \in \mathcal{H}$, $u_2 \in \mathcal{U}_w$.*
- (iii) *Let $Z(\mathcal{G})$ denote the centre of $\mathcal{G}$. Then $\ker\phi \subset Z(\mathcal{G}) \subset \mathcal{H}$.*

Proof. In view of Lemma 5.1.3 we only need to show uniqueness in (i). If $\prod_{\alpha>0} \xi_\alpha(t_\alpha) = \prod_{\alpha>0} \xi_\alpha(t'_\alpha)$, by applying ϕ and Lemma 5.2.20 we see that $t_\alpha = t'_\alpha$ for all α.

In view of Lemma 5.1.4(i) and Proposition 5.1.10, in (ii) we also only need to show uniqueness. Suppose $u_1 h \dot{w} u_2 = \tilde{u}_1 \tilde{h} \dot{w} \tilde{u}_2$. By applying ϕ and Corollary 5.2.24, we infer that $\phi(u_2) = \phi(\tilde{u}_2)$ and $\phi(u_1 h) = \phi(\tilde{u}_1 \tilde{h})$. Again by Corollary 5.2.24 the latter implies that $\phi(u_1) = \phi(\tilde{u}_1)$. By (i), $\phi : \mathcal{U} \to U$ is an isomorphism, whence $u_1 = \tilde{u}_1$, $u_2 = \tilde{u}_2$. Hence $h\dot{w} = \tilde{h}\dot{w}$ and we have $h = \tilde{h}$ as well.

Now let $g = u_1 h \dot{w} u_2 \in \ker\phi$. By applying ϕ, Corollary 5.2.24 and (i) we obtain $u_1 = u_2 = 1$, $w = 1$, so that $g \in \mathcal{H}$. So there are $\beta_1, \dots, \beta_m \in \Phi$ with $g = \eta_{\beta_1}(t_1) \cdots \eta_{\beta_m}(t_m)$, where $t_i \in k^*$. Let $\gamma \in \Phi$; then by (5.6), $g \xi_\gamma(s) g^{-1} = \xi_\gamma(s \prod_{i=1}^m t_i^{\langle \gamma, \beta_i^\vee \rangle})$. By applying ϕ and using Lemma 5.2.20 we see that $\prod_i t_i^{\langle \gamma, \beta_i^\vee \rangle} = 1$. In turn that implies that g lies in the centre of $\mathcal{G}$.

Now let $g \in Z(\mathcal{G})$. Proposition 5.2.26 implies $\phi(g) \in H$. Hence $g = hz$ where $h \in \mathcal{H}$, $z \in \ker\phi$. Above it is shown that $\ker\phi \subset \mathcal{H}$ and we conclude that $g \in \mathcal{H}$. □

We let P denote the weight lattice spanned by the fundamental weights $\lambda_1, \dots, \lambda_\ell$ (see Lemma 2.8.27). Let $Q \subset P$ denote the root lattice spanned by the simple roots $\alpha_1, \dots, \alpha_\ell$. Furthermore, L_V denotes the lattice generated by all weights of V. This is the additive subgroup of P generated by all $\mu \in P$ such that $V_\mu \neq 0$.

Lemma 5.2.29 *We have $Q \subset L_V \subset P$.*

Proof. Let $\alpha \in \Phi$. As V is a faithful module, there is a weight vector v of weight μ such that $x_\alpha \cdot v \neq 0$. But then $\mu + \alpha$ is also a weight of V and $\alpha = (\mu + \alpha) - \mu \in L_V$. □

Proposition 5.2.30 *Let $\alpha, \beta \in \Phi$. Let the integers n_i^α be as in Theorem 2.9.13.*

(i) $h_\alpha(tu) = h_\alpha(t)h_\alpha(u)$ *for all* $t, u \in k^*$.

(ii) $h_\alpha(t)h_\beta(u) = h_\beta(u)h_\alpha(t)$ *for all* $t, u \in k^*$.

(iii) $h_\alpha(t) = \prod_{i=1}^{\ell} h_{\alpha_i}(t^{n_i^\alpha})$ *for all* $t \in k^*$.

(iv) $\prod_{i=1}^{\ell} h_{\alpha_i}(t_i) = 1$ *if and only if* $\prod_{i=1}^{\ell} t_i^{\langle \mu, \alpha_i^\vee \rangle} = 1$ *for all* $\mu \in L_V$.

(v) *The centre of G consists of $\prod_{i=1}^{\ell} h_{\alpha_i}(t_i)$ such that $\prod_i t_i^{\langle \beta, \alpha_i^\vee \rangle} = 1$ for all $\beta \in Q$. In particular the centre is finite.*

Proof. By Lemma 5.2.17, $h_\alpha(t)$ acts on V_μ^k as multiplication by $t^{\langle \mu, \alpha^\vee \rangle}$. This immediately implies (i), (ii) and (iv). For (iii) we note that $\langle \mu, \alpha^\vee \rangle = \sum_{i=1}^{\ell} n_i^\alpha \langle \mu, \alpha_i^\vee \rangle$.

For the last statement note that $\prod_{i=1}^{\ell} h_{\alpha_i}(t_i)$ commutes with $x_\beta(u)$ if and only if $\prod_i t_i^{\langle \beta, \alpha_i^\vee \rangle} = 1$ by Proposition 5.2.18. □

Now let $\mathcal{G}$ be the group generated by the symbols $\xi_\alpha(t)$, for $\alpha \in \Phi$, $t \in k$, subject to a number of relations which we now describe. First we impose the relations (5.1) to (5.6), where the $c_{i,j}^{\alpha,\beta}$ in (5.3) are as in G, the product is taken in the same order as in G (Lemma 5.2.10) and the $c_{\alpha,\beta}$ in (5.5) are as in G (Lemma 5.2.15). Second, we impose the relations from Proposition 5.2.30(i) through (iv) with $\eta_\alpha(t)$ in place of $h_\alpha(t)$. Since all these relations hold in G (with $x_\alpha(t)$ in place of $\xi_\alpha(t)$), there is a surjective homomorphism $\phi : \mathcal{G} \to G$ with $\phi(\xi_\alpha(t)) = x_\alpha(t)$.

Theorem 5.2.31 *ϕ is an isomorphism.*

Proof. By Theorem 5.2.28, $\ker \phi \subset \mathcal{H}$. But $\phi : \mathcal{H} \to H$ is an isomorphism. □

Corollary 5.2.32 *Let G' be a second Chevalley group constructed using the same data as for G, but with a possibly different $\mathfrak{g}$-module V'. Let $x'_\alpha(t)$ denote* *the generators of G'. Suppose $L_V \supset L_{V'}$. Then there is a surjective homomorphism $\varphi : G \to G'$ with $\varphi(x_\alpha(t)) = x'_\alpha(t)$. Furthermore, $\ker\varphi$ is contained in the centre of G, and if $L_{V'} = L_V$ then φ is an isomorphism.*

Proof. By Theorem 5.2.31, the relations satisfied by the $x_\alpha(t)$ form a subset of those satisfied by the $x'_\alpha(t)$. Hence φ exists. By Lemma 5.2.28, $\ker\varphi \subset Z(G) \subset H$. But if $L_{V'} = L_V$ then $H' \cong H$ (Proposition 5.2.30; here $H' \subset G'$ is defined as H in G) and φ induces an isomorphism $H' \to H$. So in that case $\ker\varphi$ is trivial. □

First we note that this corollary implies that G depends on L_V and k, but *not* on the choice of admissible lattice.

If $L_V = Q$, G is called *adjoint*. If $L_V = P$, G is said to be *universal* or *simply connected*. It is possible to realize any lattice L with $Q \subset L \subset P$ as L_V. For example, if $V = \mathfrak{g}$ is the adjoint module, $L_V = Q$. If V is the direct sum of the irreducible modules with the fundamental weights as highest weights, then $L_V = P$. Let G_{ad}, G_{sc} denote the adjoint and universal groups respectively. If V is any faithful $\mathfrak{g}$-module, and G denotes the corresponding Chevalley group, by the previous corollary there are surjective homomorphisms $\sigma : G_{\mathrm{sc}} \to G$ and $\pi : G \to G_{\mathrm{ad}}$.

Example 5.2.33 Using Proposition 5.2.30 we see that, if G is simply connected, every element of H can be written uniquely as $h_{\alpha_1}(r_1)\cdots h_{\alpha_\ell}(r_\ell)$, with $r_i \in k^*$.

Let $\mathfrak{g}$ be of type A_2 as in Example 5.2.11. Let G be simply connected, and $w = s_1 s_2$ as in Example 5.2.25. Then every element of $B\dot{w}U_w$ can be written uniquely as $x_{\alpha_1}(t_1)x_{\alpha_2}(t_2)x_{\alpha_3}(t_3)h_{\alpha_1}(r_1)h_{\alpha_2}(r_2)\dot{w}x_{\alpha_1}(t_4)x_{\alpha_3}(t_5)$ where $r_i \in k^*$, $t_i \in k$.

Example 5.2.34 Let again $\mathfrak{g}$ be of type A_2 and $V = V(\lambda)$, with $\lambda = 2\lambda_1$, as in Example 5.2.5. Let G be the Chevalley group constructed using this module. The weights of V are $2\lambda_1$, λ_2, $\lambda_1 - \lambda_2$, $-2\lambda_1 + 2\lambda_2$, $-\lambda_1$, $-2\lambda_2$. It follows that $L_V = P$, so that G is simply connected.

5.3 Semisimple algebraic groups

Let k be an algebraically closed field. Let G be a Chevalley group constructed from a faithful $\mathfrak{g}$-module V and using the field k and the admissible lattice Λ. We first show that G is a semisimple algebraic group. Then we look at some properties of G and its representations. We freely use the notation introduced in the previous sections.

Theorem 5.3.1 *G is a semisimple (hence connected) algebraic group.*

Proof. The entries of the matrix of $x_\alpha(t)$ are polynomials in t. So the map $t \mapsto x_\alpha(t)$ mapping $\mathbb{G}_a \to G$, is regular. As $\mathbb{G}_a$ is a connected algebraic group, so is its image X_α (Lemma 5.2.20, Lemma 1.1.6). By Proposition 3.2.3 we conclude that G is a connected algebraic group.

Write $R = R(G)$, the radical of G. This is the unique maximal connected solvable normal algebraic subgroup of G. Let $\pi : G \to G_{\text{ad}}$ be a surjective homomorphism of G to the adjoint group (see previous section). Then $\pi(R)$ is a normal solvable subgroup of G_{ad}. Decomposing $\mathfrak{g}$ as a direct sum of simple ideals yields a decomposition of G_{ad} as a product $G_{\text{ad}} = G_1 \cdots G_m$, where each G_i is the adjoint group corresponding to a simple ideal of $\mathfrak{g}$. It can be shown that each G_i is a simple group ([Car72], Theorem 11.1.2, or [Ste67], Theorem 5). So $\pi(R)$ is a product of some of the G_i's. But those are not solvable. It follows that $\pi(R)$ is trivial. Hence R is contained in the centre of G (Corollary 5.2.32), and therefore it is finite (Proposition 5.2.30(v)). As R is connected, it has to be trivial. □

Lemma 5.3.2 *Set $\tilde{\mathfrak{h}} = \{h \in \mathfrak{h} \mid \mu(h) \in \mathbb{Z}\}$ for all weights μ of V. Let $\tilde{h}_1, \dots, \tilde{h}_\ell$ be a basis of the $\mathbb{Z}$-module $\tilde{\mathfrak{h}}$. Then the elements $\rho_k(x_\alpha)$ for $\alpha \in \Phi$ and $\rho_k(\tilde{h}_i)$, $1 \le i \le \ell$ are linearly independent over k.*

Proof. It can be shown that the $\mathbb{Z}$-span of the x_α and $\tilde{h}_i$ is exactly the stabilizer of Λ, that is, it is equal to $\{x \in \mathfrak{g} \mid x \cdot \Lambda \subset \Lambda\}$ (see [Hum78], Section 27.2).

Choose a basis of V^k coming from a basis of Λ, and represent elements $\rho(x)$ for $x \in \mathfrak{g}_k$ by matrices with respect to that basis. Then the given elements all have matrices with coefficients in the prime field. So if there is a linear dependency over k, there is a linear dependency over its prime field. That in turn yields integers n_α, m_i such that $x = \sum_\alpha n_\alpha x_\alpha + \sum_i m_i \tilde{h}_i$ has a matrix relative to the chosen basis of Λ, having entries all divisible by p where p is the characteristic of k and $\frac{1}{p}x$ stabilizes Λ, contrary to what is said above. □

Example 5.3.3 Let $\mathfrak{g}$ be of type B_2, and let G be adjoint. Let $C = \begin{pmatrix} 2 & -2 \\ -1 & 2 \end{pmatrix}$ be the Cartan matrix, and α_1, α_2 the simple roots. Then $h \in \tilde{\mathfrak{h}}$ if and only if $\alpha_1(h), \alpha_2(h) \in \mathbb{Z}$. Writing $h = ah_1 + bh_2$, this amounts to $2a_1 - 2a_2, -a_1 + 2a_2 \in \mathbb{Z}$. And that means that h_1, $\frac{1}{2}h_2$ is a basis of $\tilde{\mathfrak{h}}$.

In the following we write $\Phi^+ = \{\alpha_1, \dots, \alpha_s\}$, and assume that $\text{ht}(\alpha_i) \le \text{ht}(\alpha_j)$ if $i < j$.

Proposition 5.3.4 *U is an algebraic subgroup of G. The map $\sigma : \mathbb{G}_a^s \to U$ given by $\sigma(t_1, \dots, t_s) = x_{\alpha_1}(t_1) \cdots x_{\alpha_s}(t_s)$ is an isomorphism of algebraic* *groups. Its differential maps* $(t_1,\dots,t_s)$ *to* $\sum_i t_i\rho_k(x_{\alpha_i})$. Lie$(U)$ *is spanned by* $\rho_k(x_{\alpha_i})$, $1\le i\le s$.

Proof. Obviously σ is regular and bijective (the latter follows from Lemma 5.2.20). We show that σ has a regular inverse.

Choose a basis $B=\{v_1,\dots,v_n\}$ of V^k coming from a basis of Λ and consisting of weight vectors. We represent elements of End(V^k) by matrices with respect to this basis, thus identifying GL(V^k) and GL(n,k). Let a_{ij}, $1\le i,j\le n$ be the matrix coordinate functions. They are regular on G.

For $\alpha\in\Phi^+$ write $x_\alpha\cdot v_i=\sum_{j=1}^n \zeta_{ij}^\alpha v_j$, $\zeta_{ij}\in\mathbb{Z}$. Furthermore, choose i_α, j_α such that $\zeta_{i_\alpha j_\alpha}^\alpha\neq 0$.

Let $g=\prod_{r=1}^s x_{\alpha_r}(t_r)$. Fix an $\alpha=\alpha_m\in\Phi^+$, and write $i=i_\alpha$, $j=j_\alpha$ and $\zeta=\zeta_{i_\alpha j_\alpha}^\alpha$. We are interested in the coefficient of v_j in $g\cdot v_i$. Let μ be the weight of v_i; then the weight of v_j is $\mu+\alpha$. If $r>m$ then $x_{\alpha_r}(t_r)\cdot v_i=v_i+\bar v$, where $\bar v$ is a sum of basis vectors of weights $\mu+q\alpha_r$, $q\ge 1$ and $\bar v$ will not contribute to the coefficient of v_j. Furthermore, $x_{\alpha_m}(t_m)\cdot v_i=v_i+t_m\zeta v_j+\bar v$, where again $\bar v$ does not contribute to the coefficient of v_j. It follows that the coefficient of v_j in $g\cdot v_i$ is $\zeta t_m+p_{ij}(t_1,\dots,t_{m-1})$, where $p_{ij}\in\mathbb{Z}[T_1,\dots,T_{m-1}]$. But this is also equal to $a_{ji}(g)$. By induction there are polynomials $f_1,\dots,f_{m-1}$ in the a_{lq} such that $t_r=f_r(g)$ for $1\le r\le m-1$. So we also find such a polynomial f_m for t_m. We conclude that the inverse of σ is a regular map.

Define $\sigma_r:\mathbb{G}_\mathrm{a}\to U$ by $\sigma_r(t)=x_\alpha(t)$. Then $\mathrm{d}\sigma_r(t)=t\rho_k(x_{\alpha_r})$. This implies the statement on the differential of σ (in view of Example 3.7.1 and Lemma 1.2.3). Finally, since σ is an isomorphism of algebraic groups, its differential is an isomorphism of Lie algebras, whence the last statement. □

Remark 5.3.5 Instead of Φ^+ we can also consider closed subsets S of it, or of Φ^-, and look at the group U_S. The same proof shows that there is an isomorphism $\mathbb{G}_\mathrm{a}^{|S|}\to U_S$, and that Lie$(U_S)$ is spanned by $\rho_k(x_\alpha)$ for $\alpha\in S$.

Let $\tilde h_i$, $1\le i\le \ell$ be as in Lemma 5.3.2. For $t\in k^*$ define $\tilde h_i(t)\in$ GL(V^k) by $\tilde h_i(t)\cdot v=t^{\mu(\tilde h_i)}v$ for $v\in V_\mu^k$.

Proposition 5.3.6 *H is a connected algebraic subgroup of G containing $\tilde h_i(t)$ for $1\le i\le\ell$, $t\in k^*$. The map $\tau:\mathbb{G}_\mathrm{m}^\ell\to H$ given by $\tau(t_1,\dots,t_\ell)=\tilde h_1(t_1)\cdots\tilde h_\ell(t_\ell)$ is an isomorphism of algebraic groups. Its differential maps $(t_1,\dots,t_\ell)$ to $\sum_i t_i\rho_k(\tilde h_i)$. The Lie algebra* Lie$(H)$ *is spanned by $\rho_k(\tilde h_i)$ for $1\le i\le\ell$.*

Proof. The map $\mathbb{G}_\mathrm{m}\to H$, $t\mapsto h_\alpha(t)$ is a homomorphism of algebraic groups (Proposition 5.2.30(i)). In the same way as in the proof of Theorem 5.3.1 we see that H is a connected algebraic group.

Write $h_i(t)=h_{\alpha_i}(t)\in H$. We have $\tilde h_i=\sum_{j=1}^\ell q_{ij}h_j$, where $q_{ij}\in\mathbb{Q}$. Hence $\mu(\tilde h_i)=\sum_j q_{ij}\langle\mu,\alpha_j^\vee\rangle$, so in view of Lemma 5.2.17, we see that $\tilde h_i(t)=\prod_j h_j(t^{q_{ij}})\in H$. (Note that $t^{q_{ij}}\in k$ as k is algebraically closed.)

Also $h_i = \sum_{j=1}^{\ell} n_{ij}\tilde{h}_j$, with $n_{ij} \in \mathbb{Z}$. Let $\tilde{t}_1, \dots, \tilde{t}_\ell \in k^*$ be given and consider the equations $\prod_{i=1}^{\ell} t_i^{n_{ij}} = \tilde{t}_j$, $1 \le j \le \ell$. As the matrix (n_{ij}) is non-singular, this has a solution with $t_i \in k^*$. Indeed, by elementary operations (multiplying two equations, taking the inverse) we can transform the equation system into an equivalent one, where the matrix (n_{ij}) is triangular with non-zero diagonal entries. Let μ be a weight of V. Then

$$\prod_{i=1}^{\ell} t_i^{\langle \mu, \alpha_i^\vee \rangle} = \prod_{i=1}^{\ell} t_i^{\mu(h_i)} = \prod_{j=1}^{\ell} \tilde{t}_j^{\mu(\tilde{h}_j)}.$$

In view of Lemma 5.2.17 we conclude that for $v \in V_\mu^k$ we have $\prod_i h_i(t_i) \cdot v = \prod_j \tilde{h}_j(\tilde{t}_j) \cdot v$, whence τ is surjective.

It is obvious that τ is a regular map. We show that it also has a regular inverse. Let $\mu_1, \dots, \mu_\ell$ be the basis of L_V with $\mu_i(\tilde{h}_j) = \delta_{ij}$ and $P(V)$ denote the set of weights of V. Write $\mu_i = \sum_{\mu \in P(V)} n_{i,\mu}\mu$. Then

$$t_i = \prod_{\mu \in P(V)} \left(\prod_{j=1}^{\ell} t_j^{\mu(\tilde{h}_j)} \right)^{n_{i,\mu}}$$

which implies that τ has a regular inverse.

The differential of $\tau_i : \mathbb{G}_\mathrm{m} \to G$, $\tau_i(t) = \tilde{h}_i(t)$ maps $t \in k$ to $t\rho_k(\tilde{h}_i)$. So the final statements follow in the same way as in the proof of the previous proposition. □

The next is a result of matrix theory. Here we will not go into the proof, but refer to [Hou75], Section 1.4.

Lemma 5.3.7 *Let A, L, D, R be $n \times n$ matrices with coefficients in a field such that L and R are lower triangular, respectively upper triangular, unipotent, D is diagonal and non-singular, and $A = LDR$. Write $A = (a_{ij})$, $L = (l_{ij})$, $D = (d_{ij})$, $R = (r_{ij})$. For $i \ge p$, let $\delta_{p,i}$, $\delta_{i,p}$ denote the determinants of the $p \times p$ matrices formed by, respectively, the entries of rows $1, \dots, p$ and columns $1, \dots, p-1, i$ of A, and the entries of rows $1, \dots, p-1, i$ and columns $1, \dots, p$ of A. Then for $i \ge p$ we have $l_{ip} = \delta_{ip}/\delta_{pp}$, $d_{pp} = \delta_{p,p}/\delta_{p-1,p-1}$ and $r_{pi} = \delta_{pi}/\delta_{pp}$.*

Proposition 5.3.8 *For $w \in W$, fix a $\dot{w} \in N$ such that $\vartheta(w) = H\dot{w}$ (see Lemma 5.2.22). Let Y_w denote the closure of $B\dot{w}B$. Then*

(i) *$B\dot{w}B$ is open in Y_w,*

(ii) $\dim Y_w = |\Phi^+| + \ell + \mathcal{L}(w)$.

Proof. If we let the group $B \times B$ act on G by $(b, b') \cdot g = bg(b')^{-1}$, then $B\dot{w}B$ is the orbit of $\dot{w}$, and hence is open in its closure (Lemma 3.12.1).

Set $\Phi_w = \Phi^+ \cap w^{-1}(-\Phi^+)$, and $U_w = X_{\Phi_w}$, as in Theorem 5.2.23. Then $B\dot{w}U_w = B\dot{w}U_w\dot{w}^{-1}\dot{w}$. Set $U_w^- = U_{w(\Phi_w)}$, which is a subgroup of U^-. By Lemma 5.1.5, $\dot{w}U_w\dot{w}^{-1} = U_w^-$. Let k' be the function field over k in the indeterminates T_α, for $\alpha \in \Phi$, T_i, $1 \le i \le \ell$. Set

$$g_0 = \prod_{\alpha>0} x_\alpha(T_\alpha) \prod_{i=1}^{\ell} \tilde{h}_i(T_i) \prod_{\alpha\in w(\Phi_w)} x_\alpha(T_\alpha).$$

Since, with respect to a well-chosen basis, U is upper triangular unipotent, H is diagonal and U_w^- is lower triangular unipotent, using Lemma 5.3.7 and the construction of the inverses in the proofs of Propositions 5.3.4 and 5.3.6, we find rational functions in the entries of g_0 that have values T_α, for $\alpha > 0$, T_i for $1 \le i \le \ell$, and T_α for $\alpha \in w(\Phi_w)$. Therefore $k(g_0)$ (notation as in Section 1.3.1) is the rational function field in these indeterminates. Obviously, g_0 is a generic point of the closure $\overline{BU_w^-}$ of BU_w^-, and hence $\dim \overline{BU_w^-} = s + \ell + |\Phi_w|$ (Proposition 1.3.5). By Theorem 5.2.23, $B\dot{w}B = BU_w^-\dot{w}$. Therefore, $Y_w = \overline{BU_w^-}\,\dot{w}$ and both closed sets have the same dimension. Finally, by Lemma 2.8.24, $|\Phi_w| = \mathcal{L}(w)$. □

Theorem 5.3.9 (i) *U^-HU is open in G,*

(ii) *Lie(G) is spanned by $\rho_k(x_\alpha)$ for $\alpha \in \Phi$, along with $\rho_k(\tilde{h}_i)$, $1 \le i \le \ell$, where $\tilde{h}_i$ is as in Lemma 5.3.2.*

Proof. For $w \in W$ let $\dot{w} \in N$ be as in the previous proposition. Let w_0 be the longest element of W (see Section 2.8.3). Since $|\Phi_w| = \mathcal{L}(w)$ (Lemma 2.8.24) by Proposition 5.3.8, the closure of $B\dot{w}_0B$ has the highest dimension among all $B\dot{w}B$ (and this dimension is $s + \ell + s = \dim \mathfrak{g}$). Because G is connected (Theorem 5.3.1) and also equals the union of the various $B\dot{w}B$ (Theorem 5.2.23), we conclude that G is equal to the closure of $B\dot{w}_0B$.

Finally, $B\dot{w}_0B = U\dot{w}_0B = \dot{w}_0\dot{w}_0^{-1}U\dot{w}_0B = \dot{w}_0U^-B$ (Lemma 5.1.5). As $B\dot{w}_0B$ is open in G (Proposition 5.3.8(i)), the same holds for $U^-HU = U^-B$.

We have just seen that G has dimension equal to $\dim \mathfrak{g}$. By Propositions 5.3.4 and 5.3.6 the given elements are contained in Lie(G). By Lemma 5.3.2 they span a Lie algebra of dimension $\dim \mathfrak{g}$. It follows that this Lie algebra is Lie(G). □

Corollary 5.3.10 *Consider the set-up of Corollary 5.2.32, but with k algebraically closed, so G and G' are semisimple algebraic groups. The homomorphism φ is a morphism of algebraic groups.*

Proof. We need to show that φ is a regular map. First we consider the restriction of φ to the open set U^-HU of G. Since, relative to the appropriate basis, U^- and U are lower triangular, respectively, upper triangular, unipotent and H is diagonal, by Lemma 5.3.7, there is a rational map $\tau : \mathrm{GL}(V^k) \to \mathrm{GL}(V^k) \times \mathrm{GL}(V^k) \times \mathrm{GL}(V^k)$ such that $\tau(u^-hu) = (u^-, h, u)$, for $u^- \in U^-$, $u \in U$, $h \in H$. As seen in the proof of Proposition 5.3.4, there are regular maps $\eta^-, \eta : \mathrm{GL}(V^k) \to \mathbb{G}_a^s$ such that for all $u \in U$ we have if $\eta(u) = (t_1, \dots, t_s)$, then $u = \prod_i x_{\alpha_i}(t_i)$, and similarly for η^- and U^-. An $h \in H$ can be written $h = \prod_{i=1}^{\ell} h_i(t_i)$ (Proposition 5.2.30). The diagonal entries of this element are $\prod_i t_i^{\langle \mu, \alpha_i^\vee \rangle}$, where μ ranges over the weights of V. Similarly, an $h' \in H'$ has diagonal entries $\prod_i t_i^{\langle \mu', \alpha_i^\vee \rangle}$, where μ' ranges over the weights of V'. As $L_{V'} \subset L_V$ every μ' can be expressed as a sum of μ's. So every diagonal entry of h' is a monomial in the entries of h. It follows that there exists a rational map $\sigma : \mathrm{GL}(V^k) \to \mathrm{GL}(V^k)$, defined on U^-HU such that $\sigma(g) = \varphi(g)$ for all $g \in U^-HU$.

For $w \in W$ let $\dot{w} \in N$ be as in Proposition 5.3.8. By the above, there is a rational map $\sigma_w : \mathrm{GL}(V^k) \to \mathrm{GL}(V^k)$, defined on $\dot{w}U^-HU$, such that $\sigma_w(g) = \varphi(g)$ for all $g \in \dot{w}U^-HU$. Furthermore, letting $\widetilde{U}_w = X_\Psi$, where Ψ is as in the proof of Proposition 5.1.10, and using Lemma 5.1.5, we see that $B\dot{w}B = UH\dot{w}B = \dot{w}\dot{w}^{-1}U\dot{w}B = \dot{w}\dot{w}^{-1}U_w\dot{w}\dot{w}^{-1}\widetilde{U}_w\dot{w}B \subset \dot{w}U^-UB = \dot{w}U^-B$, implying that the open sets $\dot{w}U^-HU$ cover G. The conclusion is that G is the union of a finite number of open sets, and restricted to each of these φ is equal to a rational function. It is well-known that this implies that φ is regular (see, e.g., [Sha94], Section I.3, Theorem 4). □

By Proposition 5.3.6, H is a connected algebraic subgroup of G. Furthermore, it is diagonalizable, and hence a torus. We consider the group of its characters $X^*(H)$ (see Section 3.9). Let $\mu \in L_V$ and we define $\chi_\mu \in X^*(H)$ by

$$\chi_\mu\left(\prod_{i=1}^{\ell} h_i(t_i)\right) = \prod_{i=1}^{\ell} t_i^{\langle \mu, \alpha_i^\vee \rangle}.$$

By Proposition 5.2.30(iv), this is well-defined.

Lemma 5.3.11 *Let $\mu \in L_V$, and $\gamma \in \Phi$. Then $\chi_\mu(h_\gamma(t)) = t^{\langle \mu, \gamma^\vee \rangle}$.*

Proof. By Proposition 5.2.30(iii), $h_\gamma(t) = \prod_i h_i(t^{n_i^\gamma})$. From the definition of n_i^γ (Theorem 2.9.13), it follows that $\sum_i n_i^\gamma \langle \mu, \alpha_i^\vee \rangle = \langle \mu, \gamma^\vee \rangle$. By putting these observations together with the definition of χ_μ, we obtain the required statement. □

Also we define an action of W on $X^*(H)$ by $w \cdot \chi(h) = \chi(\dot{w}h\dot{w}^{-1})$, where $h \in H$ and, as usual, $\dot{w} \in N$ is such that $\vartheta(w) = H\dot{w}$ (Lemma 5.2.22). Since H is normal in N and is abelian, this definition does not depend on the choice of $\dot{w}$.

Proposition 5.3.12 *The map $\mu \mapsto \chi_\mu$ is a W-equivariant isomorphism of abelian groups $L_V \to X^*(H)$.*

Proof. The map is obviously a group homomorphism, and if χ_μ is trivial, $\prod_i t_i^{\langle \mu, \alpha_i^\vee \rangle} = 1$ for all $t_i \in k^*$, implying that $\mu = 0$ so that the map is injective. Let $\mu_1, \dots, \mu_m$ be the weights of V. Then the diagonal entries of $\prod_i h_i(t_i)$ are $\chi_{\mu_j}(\prod_i h_i(t_i))$. Since for $\chi \in X^*(H)$ we have that $\chi(\prod_i h_i(t_i))$ is a monomial in these diagonal entries, it follows that χ is a monomial in the χ_{μ_i}. Hence the map is surjective as well.

Let $\alpha, \beta \in \Phi$ and choose $\dot{s}_\alpha = \bar{w}_\alpha$ (by definition this is $w_\alpha(1)$). Using Proposition 5.2.18(i) we then see that $s_\alpha \cdot \chi_\mu(h_\beta(t)) = \chi_\mu(h_{s_\alpha(\beta)}(t)) = t^{\langle \mu, s_\alpha(\beta)^\vee \rangle}$ by Lemma 5.3.11. By the same lemma, $\chi_{s_\alpha(\mu)}(h_\beta(t)) = t^{\langle s_\alpha(\mu), \beta^\vee \rangle}$. As $\langle s_\alpha(\mu), \beta^\vee \rangle = \langle \mu, s_\alpha(\beta)^\vee \rangle$ we infer that the map is W-equivariant. □

In the following we identify elements of L_V and $X^*(H)$. For example, a root in Φ may be viewed as a character in $X^*(H)$.

Let $\theta : G \to \mathrm{GL}(\widehat{V})$ be a finite-dimensional rational representation of G. Most of the time we will write $g \cdot v$ (or gv) instead of $\theta(g)v$. For $\mu \in X^*(H)$ we set

$$\widehat{V}_\mu = \{v \in \widehat{V} \mid h \cdot v = \mu(h)v \text{ for all } h \in H\}$$

which is called the *weight space* of $\widehat{V}$ corresponding to μ. If $\widehat{V}_\mu \neq 0$ then μ is said to be a weight of $\widehat{V}$, and the elements of V_μ are called *weight vectors* of weight μ.

Lemma 5.3.13 *Let μ be a weight of $\widehat{V}$ and $v \in \widehat{V}_\mu$. Let $\alpha \in \Phi$. Then there are $v_i \in \widehat{V}_{\mu+i\alpha}$, for $i \geq 1$ such that $x_\alpha(t) \cdot v = v + \sum_{i \geq 1} t^i v_i$ for all $t \in k$.*

Proof. The map $t \mapsto \theta(x_\alpha(t))$ is a regular map $\mathbb{G}_a \to \mathrm{GL}(\widehat{V})$. So there are $v_i \in \widehat{V}$, for $i \geq 0$ such that $x_\alpha(t)v = \sum_{i \geq 0} t^i v_i$, for all $t \in k$. Let $h \in H$. Then $hx_\alpha(t)h^{-1} = x_\alpha(\alpha(h)t)$, by Proposition 5.2.18(iii) (see also Lemma 5.3.11). Therefore $hx_\alpha(t)v = \sum_{i \geq 0} (\alpha(h)t)^i \mu(h) v_i$. On the other hand, this is equal to $\sum_{i \geq 0} t^i h \cdot v_i$. Since k is an infinite field, if $v_i \neq 0$, then $hv_i = \alpha(h)^i \mu(h) v_i$ so that $v_i \in \widehat{V}_{\mu+i\alpha}$. Finally, setting $t = 0$ yields $v_0 = v$. □

Proposition 5.3.14 *Suppose $\widehat{V}$ is an irreducible G-module.*

(i) *There is a unique weight λ such that $u \cdot v = v$ for all $u \in U$ and $v \in \widehat{V}_\lambda$.*

(ii) *As a* U^-*-module,* $\widehat{V}$ *is generated by any non-zero element of* $\widehat{V}_\lambda$.

(iii) *Every weight of* $\widehat{V}$ *has the form* $\lambda - \sum_{\alpha\in\Phi^+} m_\alpha\alpha$, *where* $m_\alpha \in \mathbb{Z}_{\geq 0}$.

(iv) $\widehat{V}$ *is the direct sum of its weight spaces.*

(v) $\dim \widehat{V}_\lambda = 1$.

(vi) $\langle\lambda, \alpha_i^\vee\rangle \geq 0$ *for* $1 \leq i \leq \ell$; *in other words,* λ *is dominant.*

Proof. As H is an abelian group, there is at least one non-zero weight space $\widehat{V}_\mu$. Let $v \in \widehat{V}_\mu$ be non-zero. If there is an $\alpha \in \Phi^+$ such that $x_\alpha(t)\cdot v \neq v$ for a $t \in k$, by Lemma 5.3.13, there is a non-zero weight space $\widehat{V}_{\mu+i\alpha}$ for some $i \geq 1$. Continuing, we obtain a sequence of weight spaces $\widehat{V}_{\mu_i}$ such that $\mu_i \prec \mu_{i+1}$. As $\widehat{V}$ is finite-dimensional, this sequence must terminate. Hence λ as in (i) exists.

Let $v_\lambda \in \widehat{V}_\lambda$ be non-zero. Then $U^-HU \cdot kv_\lambda = U^- \cdot kv_\lambda$. But U^-HU is open in G (Theorem 5.3.9(i)), so any linear polynomial vanishing on $U^- \cdot kv_\lambda$ vanishes on $G \cdot kv_\lambda$. The latter is equal to $\widehat{V}$, and so is the former.

We have just seen that $\widehat{V} = U^- \cdot kv_\lambda$. Together with Lemma 5.3.13, this implies (iii) and (iv).

Let $v_1, v_2 \in \widehat{V}_\lambda$ be linearly independent. Then $\widehat{V} = U^- \cdot kv_1 = U^- \cdot kv_2$. But $v_2 \notin U^- \cdot kv_1$, which is a contradiction. It follows that $\dim \widehat{V}_\lambda = 1$.

Consider $v = \bar{w}_{\alpha_i} \cdot v_\lambda$; using Proposition 5.2.18(i), and Lemma 5.3.11, we see that $h_\beta(t)\cdot v = \bar{w}_{\alpha_i} h_{s_i(\beta)}(t)\cdot v_\lambda = t^{\langle\lambda, s_i(\beta)^\vee\rangle}\bar{w}_{\alpha_i}v_\lambda = t^{\langle s_i(\lambda),\beta^\vee\rangle}v$. So v is a weight vector of weight $s_i(\lambda)$. Together with (iii) this shows (vi). □

The weight λ of this proposition is called the *highest weight* of $\widehat{V}$. A non-zero $v_\lambda \in \widehat{V}_\lambda$ is called a *highest weight vector.*

Theorem 5.3.15 *Let* $\lambda \in L_V$ *be dominant (in other words,* $\langle\lambda, \alpha_i\rangle \geq 0$ *for* $1 \leq i \leq \ell$*). Then there exists a unique irreducible rational representation* $\theta : G \to \mathrm{GL}(L(\lambda))$ *of highest weight* λ.

Proof. Let M be an irreducible highest weight module over $\mathfrak{g}$ of highest weight λ. Then by Proposition 2.11.1, $L_M \subset L_V$. Choose an admissible lattice in M, and let G' be the corresponding Chevalley group over k. Corollary 5.3.10 yields a surjective morphism of algebraic groups $\varphi : G \to G' \subset \mathrm{GL}(M^k)$. Since the lattice of M is the direct sum of its weight spaces (Lemma 5.2.2), M^k contains a non-zero vector v_λ such that $u \cdot v_\lambda = v_\lambda$ for all $u \in U$. Let V' be the G-submodule of M^k generated by v_λ. Let V'' be a maximal proper G-submodule of V', not containing v_λ (it is the sum of all proper G-submodules of V'). Set $L(\lambda) = V'/V''$. Then $L(\lambda)$ is an irreducible rational G-module of highest weight λ.

Suppose we have two irreducible rational G-modules V_1 and V_2 of highest weight λ, with respective highest weight vectors v_1 and v_2. Let $V_3 \subset V_1 \oplus V_2$

be the G-module generated by $v_1 + v_2$. In the same way as in the proof of Proposition 5.3.14, the weight space of weight λ in V_3 is spanned by $v_1 + v_2$. In particular, $v_2 \notin V_3$, so that by the irreducibility of V_2 we have $V_2 \cap V_3 = 0$. Now let $\pi_1 : V_1 \oplus V_2 \to V_1$ be the projection onto the first factor. Then $\pi_1(v_1 + v_2) = v_1$, whence $\pi_1(V_3) = V_1$. The kernel of π_1 is V_2. We conclude that π_1 maps V_3 bijectively onto V_1, so these are isomorphic G-modules. Similarly, V_3 and V_2 are isomorphic. □

5.4 Root data

Here we briefly explain how to specify a semisimple algebraic group in a way suitable for computation. Unlike for semisimple Lie algebras, it is not enough to give a root system: we must give a lattice L_V as well. The most obvious idea would be to specify a semisimple algebraic group by giving a root system Φ and a lattice X with $Q \subset X \subset P$ (where Q, P are the root and weight lattices). However, in the literature a semisimple (and, more generally, a reductive) algebraic group is usually specified by its *root datum.* Here we describe the equivalent concept of *based root datum,* which is also useful for computation with the group (for an example see Section 5.6), not just for specifying the group.

Definition 5.4.1 *A* based root datum *is a quadruple* $(X, A, X^\vee, B)$*, where* $X, X^\vee$ *are free* $\mathbb{Z}$*-modules of rank* m*, and* A *and* B *are* $\ell \times m$ *integral matrices (where* $\ell \leq m$*) such that* $C = AB^T$ *is a Cartan matrix. The based root datum is said to be* semisimple *if* $m = \ell$.

As the name suggests, semisimple algebraic groups are given by semisimple based root data (whereas a general based root datum corresponds to a reductive algebraic group). Here all our based root data will be semisimple.

Let $(X, A, X^\vee, B)$ be a semisimple based root datum. We now construct a number of objects that will play an important role in the following. Let $\mu_1, \dots, \mu_\ell$, $\mu_1^\vee, \dots, \mu_\ell^\vee$ be fixed bases of $X, X^\vee$. Define a $\mathbb{Z}$-bilinear map (a *pairing*) $\langle\ ,\ \rangle : X \times X^\vee \to \mathbb{Z}$ by $\langle \mu_i, \mu_j^\vee \rangle = \delta_{ij}$. We identify elements of X and $X^\vee$ and their coefficient vectors relative to these bases. Let $\mathcal{V} = \mathbb{R} \otimes_{\mathbb{Z}} X$, and $\alpha_1, \dots, \alpha_\ell$ be the elements of X corresponding to the rows of A. Use the Cartan matrix $C = AB^T$ to define an inner product on $\mathcal{V}$ using the basis $\alpha_1, \dots, \alpha_\ell$ (see Section 2.8.1), and let Φ be the root system in $\mathcal{V}$ constructed from C and the α_i (Algorithm 2.8.14). Then $\Delta = \{\alpha_1, \dots, \alpha_\ell\}$ is a basis of simple roots. For $\alpha \in \Phi$ set $\bar{\alpha} = \frac{2\alpha}{(\alpha,\alpha)}$, as in Theorem 2.9.13. As in that theorem we define integers n_i^α by $\bar{\alpha} = \sum_i n_i^\alpha \bar{\alpha}_i$. Let $\alpha_1^\vee, \dots, \alpha_\ell^\vee$ be the elements of $X^\vee$ corresponding to the rows of B. For $\alpha \in \Phi$ we set $\alpha^\vee = \sum_i n_i^\alpha \alpha_i^\vee$.

Note that for $\alpha, \beta \in \Phi$ we have two definitions of the symbol $\langle \alpha, \beta^\vee \rangle$. The first one has it equal to $\frac{2(\alpha,\beta)}{(\beta,\beta)}$, the second one uses the pairing above and the element $\beta^\vee \in X^\vee$. However, since the inner product on $\mathcal{V}$ is defined using C, these two definitions agree. That immediately implies that, for $\alpha \in \Phi$ we have that Φ is stable under the map $s_\alpha : X \to X$, defined by $s_\alpha(x) = x - \langle x, \alpha^\vee \rangle \alpha$. A short calculation shows that $s_\alpha(\beta) = \beta - \langle \alpha, \beta^\vee \rangle \bar{\alpha}$. It follows that $\Phi^\vee$ is stable under the map $s_\alpha^\vee : X^\vee \to X^\vee$, $s_\alpha^\vee(y) = y - \langle \alpha, y \rangle \alpha^\vee$. These considerations imply that the quadruple $(X, \Phi, X^\vee, \Phi^\vee)$ is a root datum (we do not define this concept here, as it plays no role in what follows, but refer to [Spr98]).

Define $\lambda_1, \dots, \lambda_\ell \in \mathcal{V}$ by $\langle \lambda_i, \alpha_j^\vee \rangle = \delta_{ij}$. They are called the *fundamental weights* of the based root datum (compare Section 2.8.3). Let Q, P be the free $\mathbb{Z}$-modules spanned by $\alpha_1, \dots, \alpha_\ell$ and $\lambda_1, \dots, \lambda_\ell$. These are called, respectively, the root and weight lattices of the based root datum. Let $\mathfrak{g}$ denote the complex semisimple Lie algebra with root system Φ. Since $Q \subset X \subset P$, there is a faithful $\mathfrak{g}$-module V with $L_V = X$. Then the semisimple algebraic group G, over the algebraically closed field k that corresponds to the given based root datum is constructed from V and an admissible lattice in it.

Example 5.4.2 A based root datum $(X, A, X^\vee, B)$ is said to be *simply connected* if $X = P$. This means that $A = C$ (the Cartan matrix) and B is the $\ell \times \ell$ identity matrix.

The based root datum is called *adjoint* if $X = Q$. In that case, A is the $\ell \times \ell$ identity matrix and $B = C^T$.

Example 5.4.3 In this example we give a based root datum that is neither adjoint nor simply connected. Let $\ell = 3$ and

$$A = \begin{pmatrix} 1 & 0 & 0 \\ 0 & 1 & 0 \\ 1 & 2 & 2 \end{pmatrix}, \quad B = \begin{pmatrix} 2 & -1 & 0 \\ -1 & 2 & -2 \\ 0 & -1 & 2 \end{pmatrix}, \quad C = AB^T = \begin{pmatrix} 2 & -1 & 0 \\ -1 & 2 & -1 \\ 0 & -1 & 2 \end{pmatrix}.$$

Denoting, as before, the basis elements of X, $X^\vee$ by μ_i, $\mu_i^\vee$ $(1 \le i \le 3)$ respectively, we have $\alpha_1^\vee = 2\mu_1^\vee - \mu_2^\vee$, $\alpha_2^\vee = -\mu_1^\vee + 2\mu_2^\vee - 2\mu_3^\vee$, $\alpha_3^\vee = -\mu_2^\vee + 2\mu_3^\vee$, whence

$$\begin{aligned} \lambda_1 &= \mu_1 + \mu_2 + \tfrac{1}{2}\mu_3 \\ \lambda_2 &= \mu_1 + 2\mu_2 + \mu_3 \\ \lambda_3 &= \mu_1 + 2\mu_2 + \tfrac{3}{2}\mu_3. \end{aligned}$$

By inverting this we see that $\mu_1 = 2\lambda_1 - \lambda_2$, $\mu_2 = -\lambda_1 + 2\lambda_2 - \lambda_3$, $\mu_3 = -2\lambda_2 + 2\lambda_3$. By a Hermite normal form computation we see that X is also spanned by $\lambda_1 + \lambda_3$, λ_2, $2\lambda_3$. These weights are all dominant, so if we take V to be the direct sum of the irreducible modules with these highest weights, $L_V = X$. Alternatively, we can take V to be the irreducible module with highest weight equal to one of them.

5.5 Irreducible representations in positive characteristic

Let G be a Chevalley group over an algebraically closed field k. Let $L(\lambda)$ be an irreducible G-module of highest weight λ (Theorem 5.3.15). Here we are concerned with determining the character of $L(\lambda)$ (i.e., the weights of $L(\lambda)$ and the dimensions of the corresponding weight spaces), given only the dominant weight λ. As seen in the proof of Theorem 5.3.15, $L(\lambda)$ is constructed from the irreducible $\mathfrak{g}$-module M with highest weight λ. This module gives rise to the group $G' \subset \mathrm{GL}(M^k)$, and using a surjective homomorphism $G \to G'$, we obtain a representation of G from which $L(\lambda)$ is obtained. It is clear that $L(\lambda)$ has the same character when viewed as a G-module, or as a G'-module. Therefore we lose nothing if we assume that $G = G'$. So we let $V = V(\lambda)$ be an irreducible $\mathfrak{g}$-module of highest weight λ, and G the Chevalley group constructed using an admissible lattice Λ in V. We let $L(\lambda)$ be the irreducible G-module constructed from $V^k = k \otimes_{\mathbb{Z}} \Lambda$. The question is what the character of $L(\lambda)$ is. From Proposition 5.5.1 it will follow that if k is of characteristic 0, then $L(\lambda) = V^k$. In that case the character is the same as the character of V (which can be determined using the methods indicated in Section 2.11.3). Therefore, we will be concerned with the case where the characteristic of k is $p > 0$.

There is a large body of work on this problem (see, for example, the book by Jantzen, [Jan03]). Here we will look at the computational side and describe an algorithm that, given λ, computes the character of $L(\lambda)$.

We recall that $\mathcal{U}_k = k \otimes \mathcal{U}_{\mathbb{Z}}$, $\mathfrak{g}_k = k \otimes_{\mathbb{Z}} \mathfrak{g}_{\mathbb{Z}}$, and that ρ_k denotes the representation of $\mathfrak{g}_k$ afforded by V^k. Note that ρ_k extends in a natural way to a representation of $\mathcal{U}_k$, and of G.

Proposition 5.5.1 *A subspace $W \subset V^k$ is a G-submodule if and only if it is a $\mathcal{U}_k$-submodule.*

Proof. There is an integer $N > 0$ such that for all $\alpha \in \Phi$ we have $\rho_k(x_\alpha^{(N+1)}) = 0$. So by definition, for $t \in k$,

$$\rho_k(x_\alpha(t)) = \sum_{i=0}^{N} t^i \rho_k(x_\alpha^{(i)}). \tag{5.9}$$

From this it follows that if W is a $\mathcal{U}_k$-submodule, it is also a G-submodule. Now suppose W is a G-submodule. Choose $t_0, \ldots, t_N \in k$ distinct. Because the Vandermonde matrix (t_i^j) is invertible, (5.9) allows us to express $\rho_k(x_\alpha^{(i)})$ for $0 \le i \le N$ in terms of the $\rho_k(x_\alpha(t_j))$. Therefore each $\rho_k(x_\alpha^{(i)})$ stabilizes W. Because $\mathcal{U}_{\mathbb{Z}}$ is generated by the $x_\alpha^{(i)}$, it follows that W is a $\mathcal{U}_k$-submodule. □

Let $\Lambda = \mathcal{U}_{\mathbb{Z}}^{-} \cdot v_\lambda$, where $v_\lambda \in V$ is a fixed weight vector of weight λ. By Lemma 5.2.3, Λ is an admissible lattice in V. In the following we also denote by v_λ the element $1 \otimes v_\lambda \in V^k$. Let V' be the G-submodule of V^k generated by v_λ. By Proposition 5.5.1, $V' = V^k$. As in the proof of Theorem 5.3.15, let V'' be a maximal proper G-submodule of V^k, not containing v_λ. As seen in the mentioned proof, $L(\lambda) = V^k/V''$ is the irreducible G-module of highest weight λ. Now we give a description of V'', leading to an algorithm to compute the character of $L(\lambda)$.

An anti-involution of an associative algebra A is a bijective linear map $\varphi : A \to A$, of order 2 such that $\varphi(ab) = \varphi(b)\varphi(a)$ for all $a, b \in A$.

Lemma 5.5.2 *Let $\mathcal{U}(\mathfrak{g})$ be the universal enveloping algebra of $\mathfrak{g}$ (recall that $\mathfrak{g}$ is a Lie algebra over $\mathbb{C}$). The fixed Chevalley basis that we are using has elements x_α ($\alpha \in \Phi$), $h_1, \dots, h_\ell$. There is a unique anti-involution $\sigma : \mathcal{U}(\mathfrak{g}) \to \mathcal{U}(\mathfrak{g})$ with $\sigma(x_\alpha^m) = x_{-\alpha}^m$ and $\sigma(h_i) = h_i$, $1 \le i \le \ell$.*

Proof. From Section 2.10.1 we recall that $\mathcal{U}(\mathfrak{g}) = k\langle X\rangle/I$, where X is the set of symbols x_α ($\alpha \in \Phi$), $h_1, \dots, h_\ell$, and I is generated by the relations corresponding to the multiplication table (2.2). On $k\langle X\rangle$ we uniquely define an anti-involution σ by requiring $\sigma(x_\alpha) = x_{-\alpha}$, $\sigma(h_i) = h_i$ and $\sigma(ab) = \sigma(b)\sigma(a)$ for all a, b. This σ maps the generators of I into I. This is obvious, except possibly for the one corresponding to the relation $[x_\alpha, x_\beta] = N_{\alpha,\beta}x_{\alpha+\beta}$. Here we have to use the fact that $N_{\alpha,\beta} = -N_{-\alpha,-\beta}$, which can be shown with a lengthy argument using the Jacobi identity (see [Car72], Theorem 4.1.2(iii)). It follows that $\sigma(I) = I$, so that σ induces an anti-involution of $\mathcal{U}(\mathfrak{g})$ with the required properties. □

Let σ be the anti-involution of the previous lemma. It is obvious that σ maps $\mathcal{U}_{\mathbb{Z}}$ to itself, so it induces an anti-involution of $\mathcal{U}_{\mathbb{Z}}$ and also of $\mathcal{U}_k$ that we also denote by σ. A bilinear form (,) on V^k is said to be *contravariant* if

$$(\rho_k(u)v_1, v_2) = (v_1, \rho_k(\sigma(u))v_2) \text{ for all } v_1, v_2 \in V^k, u \in \mathcal{U}_k. \tag{5.10}$$

Lemma 5.5.3 *There is unique contravariant bilinear form (,) on V^k with $(v_\lambda, v_\lambda) = 1$. Weight spaces corresponding to different weights are orthogonal under this form.*

Proof. Let $\beta_1, \dots, \beta_n$ be the elements of Φ^+ listed in a fixed order. For $\underline{i} = (i_1, \dots, i_n)$ define

$$e_{\underline{i}} = x_{\beta_1}^{(i_1)} \cdots x_{\beta_n}^{(i_n)} \text{ and } f_{\underline{i}} = x_{-\beta_1}^{(i_1)} \cdots x_{-\beta_n}^{(i_n)}.$$

Then the $e_{\underline{i}}$ and $f_{\underline{i}}$, as $\underline{i}$ runs over $\mathbb{Z}_{\ge 0}^n$, form bases of $\mathcal{U}_{\mathbb{Z}}^+$ and $\mathcal{U}_{\mathbb{Z}}^-$ respectively. Also, for $\underline{i}$ as above we write $\underline{i}^t = (i_n, \dots, i_1)$. Then $\sigma(f_{\underline{i}}) = e_{\underline{i}^t}$. There is a basis of V^k consisting of elements of the form $\rho_k(f_{\underline{i}})v_\lambda$. Now

$(\rho_k(f_{\underline{i}})v_\lambda, \rho_k(f_{\underline{j}})v_\lambda) = (v_\lambda, \rho_k(e_{\underline{i}^t} f_{\underline{j}})v_\lambda)$. We write $e_{\underline{i}^t} f_{\underline{j}}$ as a linear combination of monomials of the form $u^- u^0 u^+$, where $u^\pm \in \mathcal{U}_\mathbb{Z}^\pm$, $u^0 \in \mathcal{U}_\mathbb{Z}^0$. If $u^+ \neq 1$ then $\rho_k(u^- u^0 u^+)v_\lambda = 0$. If $u^+ = 1$ and $u^- \neq 1$, then $(v_\lambda, \rho_k(u^- u^0)v_\lambda)$ is a scalar times $(\rho_k(\sigma(u^-))v_\lambda, v_\lambda)$ which is zero. So $(\rho_k(f_{\underline{i}})v_\lambda, \rho_k(f_{\underline{j}})v_\lambda)$ can only be non-zero if in the linear combination there are monomials in $\mathcal{U}_\mathbb{Z}^0$ (i.e., with $u^- = u^+ = 1$). This can only happen if $\sum_s i_s\beta_s = \sum_t j_t\beta_t$, proving the second assertion of the lemma. Furthermore, for a $u^0 \in \mathcal{U}_\mathbb{Z}^0$, we have that $\rho_k(u^0)v_\lambda$ is a scalar, depending only on u^0 and λ, times v_λ. It follows that $(\rho_k(f_{\underline{i}})v_\lambda, \rho_k(f_{\underline{j}})v_\lambda)$ is uniquely determined. □

Proposition 5.5.4 *Let $(\ ,\ )$ be the contravariant form on V^k with $(v_\lambda, v_\lambda) = 1$. The radical of this form is the unique maximal proper G-submodule of V^k.*

Proof. We use the fact that G-submodules are the same as $\mathcal{U}_k$-submodules (Proposition 5.5.1). Let $W \subset V^k$ be a proper $\mathcal{U}_k$-submodule. Then W does not contain v_λ. Arguing as in the proof of Lemma 5.2.2, it is seen that W is the direct sum of the various $W \cap V_\mu^k$. Let $u \in \mathcal{U}_k$, and $w \in W$. Then $\rho_k(\sigma(u))w \in W$, and it lies in a sum of weight spaces whose weights are not equal to λ. Therefore, by the second statement of the previous lemma, $(\rho_k(u)v_\lambda, w) = (v_\lambda, \rho_k(\sigma(u))w) = 0$, and w lies in the radical of the form. Since the form is contravariant, its radical is a $\mathcal{U}_k$-submodule, and it does not contain v_λ as $(v_\lambda, v_\lambda) = 1$. Therefore it is a proper $\mathcal{U}_k$-submodule. □

Let the notation be as in the previous proposition. Let R denote the radical of the form $(\ ,\ )$. As noted in the proof of the previous proposition, R is the direct sum of the spaces $R_\mu = R \cap V_\mu^k$. We fix a weight μ of V^k and outline a procedure to compute $\dim V_\mu^k/R_\mu$. We need to do this only for dominant μ, in view of Lemma 5.2.14. (Note that $w_\alpha(t)$ maps R_μ to $R_{s_\alpha(\mu)}$.)

We use the notation introduced in the proof of Lemma 5.5.3. Let $\underline{i}^1, \dots, \underline{i}^m$ be such that the $f_{\underline{i}^a} v_\lambda \in V$, for $1 \leq a \leq m$, span $V_\mu \cap \Lambda$. (Note that these elements are not required to be linearly independent.) As observed in the proof of Lemma 5.5.3, $e_{(\underline{i}^a)^t} f_{\underline{i}^b} v_\lambda = n_{ab} v_\lambda$ where n_{ab} is an integer. This integer can be computed by rewriting $e_{(\underline{i}^a)^t} f_{\underline{i}^b}$ as a linear combination of monomials $u^- u^0 u^+$ using the commutation relations in Section 2.10.3, and noting that $\binom{h_i}{j} \cdot v_\lambda = \binom{\langle \lambda, \alpha_i^\vee \rangle}{j} v_\lambda$. Write $\bar{n}_{ab} = n_{ab} \cdot 1$, where $1 \in k$ denotes the unity element in k. Let $v = \sum_a \zeta_a \rho_k(f_{\underline{i}^a})v_\lambda$ be an element of V_μ^k, where $\zeta_a \in k$. Using the contravariance of the form we see that $v \in R_\mu$ if and only if $\sum_a \bar{n}_{ab}\zeta_a = 0$. It follows that $\dim V_\mu/R_\mu$ is equal to the rank of the matrix $(\bar{n}_{ab})$.

This can be carried out for a fixed characteristic p. In order to obtain an overview of the characteristics for which $\dim V_\mu^k < \dim V_\mu$, we compute the matrix (n_{ab}) over $\mathbb{Z}$, and then its elementary divisors $d_1, \dots, d_r$ (these are the non-zero diagonal elements of the Smith form of the matrix; see Section 6.2). Then the rank of the matrix $(\bar{n}_{ab})$ is equal to the number of d_i such that p does not divide d_i.

Example 5.5.5 Let $\mathfrak{g}$ be of type A_1 with basis h, e, f (as in Example 2.1.4). Let $\lambda = r\lambda_1$; then the admissible lattice Λ is spanned by $f^{(s)}v_\lambda$, $0 \le s \le r$. Using Lemma 2.10.10, $(f^{(s)}v_\lambda, f^{(s)}v_\lambda) = \binom{r}{s}$. So if the characteristic p divides this binomial coefficient, $f^{(s)}v_\lambda$ lies in the radical of the form, otherwise it does not. Take, for example, $r = 6$. The binomial coefficients that occur are $1, 6, 15, 20, 15, 6, 1$. Hence $\dim L(\lambda) = 4$ when $p = 2, 5$, $\dim L(\lambda) = 3$ if $p = 3$ and $\dim L(\lambda) = 7$ otherwise.

Remark 5.5.6 It can be shown that the form (,) is symmetric ([Jan03], II 8.17), so we only have to compute a bit more than half of the matrix (n_{ab}).

For a given μ, there are several ways to obtain a set of sequences $\underline{i}^a$ such that the $f_{\underline{i}^a}v_\lambda \in V$ for $1 \le a \le m$ span $\Lambda_\mu = V_\mu \cap \Lambda$. The simplest is to take all sequences $(i_1, \ldots, i_n)$ such that $\lambda - \sum_s i_s\beta_s = \mu$ (where the β_s are as in the proof of Lemma 5.5.3). This is the approach taken in [Bur71], [BW71]. However, as $\dim V$ increases, an enormous number of monomials may have to be considered because in general the number of these sequences is much higher than $\dim V_\mu$. In [GS88] an improvement is given, using a simple root α_i such that $\langle \lambda, \alpha_i^\vee \rangle = 0$, which reduces the number of monomials. In [Lüb01], another idea is proposed. The form (,) is non-degenerate on V (as V is irreducible), and therefore the rank of the matrix (n_{ab}) is equal to $m_\mu = \dim V_\mu$. Furthermore, there is a combinatorial algorithm to compute the determinant of the matrix of the restriction of the form to Λ_μ. Here we do not go into details, but refer to [Jan03], II 8.17(3). Now the algorithm enumerates sequences $\underline{i}^a$ such that $\lambda - \sum_s i_s^a\beta_s = \mu$ and gradually builds the matrix (n_{ab}). The process stops when we have a matrix of the correct rank and the product of whose elementary divisors is equal to the computed determinant. (Note that we need to check this last property because some set of $\underline{i}^a$ could yield a matrix of the correct rank, while the $f_{\underline{i}^a}$ are not a basis of Λ_μ.) Using an implementation based on this idea, Lübeck ([Lüb01]) has computed extensive tables containing the dimension of $L(\lambda)$ for simple $\mathfrak{g}$ of ranks ≤ 11 such that $\dim L(\lambda)$ is limited by some bound (for example, for A_{10}, B_{10}, C_{10} this bound is 10,000, whereas for E_8 it is 100,000). The cited paper also has a theorem giving the dimension of $L(\lambda)$ for classical groups of larger rank ℓ, and again the dimension of $L(\lambda)$ limited by a bound (which is $\ell^3/8$ for A_ℓ, and ℓ^3 otherwise).

Another idea, which to the best of my knowledge has not been tested in practice, comes from the theory of canonical bases of quantum groups. A quantum group is an associative algebra $\mathcal{U}_q$ over the ring $\mathbb{Z}[q, q^{-1}]$, where q is an indeterminate. It is a deformation of $\mathcal{U}_\mathbb{Z}$, which, roughly speaking, means that for $q \to 1$, the quantum group becomes isomorphic to $\mathcal{U}_\mathbb{Z}$. Lusztig ([Lus90]) introduced a special basis of a subalgebra $\mathcal{U}_q^-$ (which is a deformation of $\mathcal{U}_\mathbb{Z}^-$), called the canonical basis that has a lot of remarkable properties. Later Kashiwara ([Kas91]) also gave a construction of the canonical basis using independent methods. Setting $q \to 1$ transforms the canonical basis of $\mathcal{U}_q^-$ to the canonical basis of $\mathcal{U}_\mathbb{Z}^-$. Here we briefly indicate how this applies to our setting but without talking about quantum groups and therefore omitting all proofs. For an introduction to quantum groups and canonical bases we refer to [Jan96] and [Lus93]. For simplicity we assume that $\mathfrak{g}$ is simply laced (i.e., of type A_ℓ, D_ℓ, or E_ℓ).

Fix a reduced expression $\hat{w}_0 = s_{j_1} \cdots s_{j_n}$ of the longest element w_0 in the Weyl group. For $1 \leq i \leq n$ set $\beta_i = s_{j_1} \cdots s_{j_{i-1}}(\alpha_{j_i})$. Then the β_i exhaust all positive roots. Again we use the notation $f_{\underline{i}}$, as in the proof of Lemma 5.5.3, but using this ordering of the positive roots. We say that $f_{\underline{i}}$ is a $\hat{w}_0$-monomial to underline the dependency on the chosen reduced expression.

Let $\hat{w}_0'$ be a different reduced expression for w_0. We define an operator $R_{\hat{w}_0}^{\hat{w}_0'}$ that transforms a $\hat{w}_0$-monomial to a $\hat{w}_0'$-monomial. By Proposition 2.8.26, $\hat{w}_0$ can be transformed to $\hat{w}_0'$ by a sequence of elementary relations. We recall that an elementary relation consists of replacing $s_{j_a}s_{j_{a+1}}$ by $s_{j_{a+1}}s_{j_a}$ if $\langle \alpha_{j_a}, \alpha_{j_{a+1}}^\vee \rangle = 0$ or replacing $s_{j_a}s_{j_{a+1}}s_{j_a}$ (assuming that $j_{a+2} = j_a$) by $s_{j_{a+1}}s_{j_a}s_{j_{a+1}}$ if $\langle \alpha_{j_a}, \alpha_{j_{a+1}}^\vee \rangle = -1$ (note that these are the only possibilities, since $\mathfrak{g}$ is of simply laced type). Suppose $\hat{w}_0'$ is obtained from $\hat{w}_0$ by one such relation. Let $f_{\underline{i}}$ be a $\hat{w}_0$-monomial, $\underline{i} = (i_1, \ldots, i_n)$. If the relation is of the first type above, $\underline{i}'$ is obtained from $\underline{i}$ by replacing i_a, i_{a+1} by i_{a+1}, i_a. If it is of the second type, we set $\eta = \min(i_a, i_{a+2})$ and $\underline{i}'$ is obtained by replacing i_a, i_{a+1}, i_{a+2} by i'_a, i'_{a+1}, i'_{a+2}, where $i'_a = i_{a+1} + i_{a+2} - \eta$, $i'_{a+1} = \eta$, $i'_{a+2} = i_a + i_{a+1} - \eta$. Then $f_{\underline{i}'} = R_{\hat{w}_0}^{\hat{w}_0'}(f_{\underline{i}})$. (See [Lus93], Chapter 42 for an explanation of this procedure; see [Lus92], [Gra02] for what to do if $\mathfrak{g}$ is not simply laced). If $\hat{w}_0'$ is obtained from $\hat{w}_0$ using more elementary relations, we iterate this procedure.

Let $\alpha \in \Delta$ be a simple root. Then we define the *Kashiwara operator* $\widetilde{F}_\alpha$ on the set of monomials $f_{\underline{i}}$. Let $\hat{w}_0'$ be a reduced expression for w_0 starting with s_α. (Note that this means that the first root in the listing of positive roots constructed from $\hat{w}_0'$ is α.) Let $f_{\underline{i}}$ be a $\hat{w}_0$-monomial and set $f_{\underline{i}'} = R_{\hat{w}_0}^{\hat{w}_0'}(f_{\underline{i}})$. Let $\underline{i}''$ be equal to $\underline{i}'$, except that $i''_1 = i'_1 + 1$. Let $f_{\underline{i}''}$ be the corresponding $\hat{w}_0'$-monomial and set $f_{\underline{i}'''} = R_{\hat{w}_0'}^{\hat{w}_0}(f_{\underline{i}''})$. Then $\widetilde{F}_\alpha(f_{\underline{i}}) = f_{\underline{i}'''}$.

Let S be the set of adapted strings coming from the crystal graph Γ_λ (see Section 5.2.1). For $\eta = (l_1, \ldots, l_n) \in S$ set $f_\eta = \widetilde{F}_{\alpha_{j_1}}^{l_1} \cdots \widetilde{F}_{\alpha_{j_n}}^{l_n}(1)$. Let

$$\mathbf{B} = \{b_{\underline{i}} \mid \underline{i} = (i_1, \ldots, i_n) \in (\mathbb{Z}_{\geq 0})^n\}$$

denote the canonical basis of $\mathcal{U}_{\mathbb{Z}}^-$. So the basis elements are indexed in the same way as the basis of $\mathcal{U}_{\mathbb{Z}}^-$ consisting of the monomials in the $x_{-\beta_i}$. The canonical basis has the following properties:

1. There is a decomposition $\mathbf{B} = \mathbf{B}_\lambda \cup \mathbf{B}^\lambda$ where $\{b \cdot v_\lambda \mid b \in \mathbf{B}_\lambda\}$ is a basis of $V(\lambda)$ and $b \cdot v_\lambda = 0$ for all $b \in \mathbf{B}^\lambda$ ([Jan96], Theorem 11.16). (To appreciate the power of this, note that we have only one $\mathbf{B}$, but for each λ there is such a decomposition.)

2. We have $b_{\underline{i}} = f_{\underline{i}} + \sum_{\underline{i}' >_{\text{lex}} \underline{i}} \zeta_{\underline{i}',\underline{i}} f_{\underline{i}'}$ where $\zeta_{\underline{i}',\underline{i}} \in \mathbb{Z}$. (For this we refer to [Gra02], Proposition 5.1; here $\underline{i}' >_{\text{lex}} \underline{i}$ if $i'_l > i_l$, where l is minimal such that $i'_l \neq i_l$.)

3. $\mathbf{B}_\lambda = \{b_{\underline{i}} \mid \underline{i}$ is such that $f_{\underline{i}} = f_\eta$ for an $\eta \in S\}$ ([Gra02], Lemma 6.1).

By combining these statements, we see that the $f_\eta \cdot v_\lambda$, as η runs over S, form a basis of Λ, and we can use these monomials in the algorithm for computing the character of $L(\lambda)$. We illustrate this procedure in the next example.

Example 5.5.7 Let $\mathfrak{g}$ be of type A_2 with multiplication table as in Example 2.9.14. Let $\lambda = 2\lambda_1 + 2\lambda_2$. The dominant weights occurring in $V = V(\lambda)$ are λ, $3\lambda_1$, $3\lambda_2$, $\lambda_1 + \lambda_2$, 0. The picture below shows the top part of the crystal graph containing all paths ending in a dominant weight. Here we list these paths, with brackets indicating the corresponding end weight: π_1 (λ), π_2 ($3\lambda_2$), π_3 ($3\lambda_1$), π_5, π_6 ($\lambda_1 + \lambda_2$), $\pi_{12}, \pi_{13}, \pi_{16}$ (0).

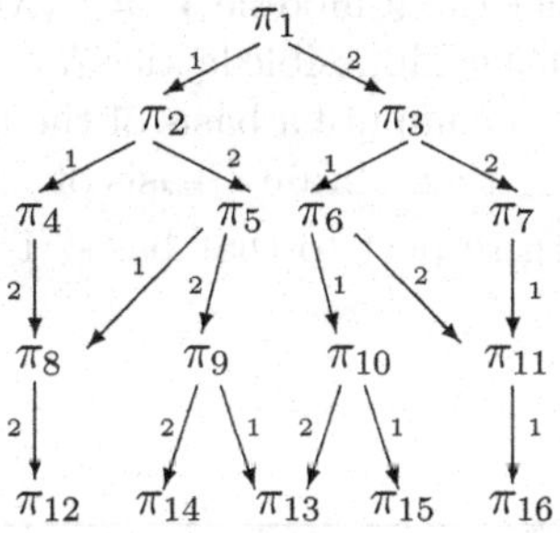

We choose $\hat{w}_0 = s_1 s_2 s_1$ and let η_i denote the adapted string corresponding to π_i. Then $\eta_1 = (0,0,0)$, $\eta_2 = (1,0,0)$, $\eta_3 = (0,1,0)$, $\eta_5 = (0,1,1)$, $\eta_6 = (1,1,0)$, $\eta_{12} = (0,2,2)$, $\eta_{13} = (1,2,1)$, $\eta_{16} = (2,2,0)$.

The ordering of positive roots corresponding to $\hat{w}_0$ is $\alpha_1, \alpha_3, \alpha_2$. We write $x_i = x_{\alpha_i}$ and $y_i = x_{-\alpha_i}$. There is only one other reduced expression for w_0, which is $\hat{w}'_0 = s_2 s_1 s_2$. Applying the Kashiwara operators, we get $f_{\eta_1} = 1$, $f_{\eta_2} = y_1$, $f_{\eta_3} = y_2$, $f_{\eta_5} = y_3$, $f_{\eta_6} = y_1 y_2$, $f_{\eta_{12}} = y_3^{(2)}$, $f_{\eta_{13}} = y_1 y_3 y_2$, $f_{\eta_{16}} = y_1^{(2)} y_2^{(2)}$.

We have $(y_1 \cdot v_\lambda, y_1 \cdot v_\lambda) = (v_\lambda, x_1 y_1 \cdot v_\lambda) = (v_\lambda, (y_1 x_1 + h_1) \cdot v_\lambda) = 2$. Therefore, $y_1 \cdot v_\lambda$ lies in the radical R if and only if the characteristic p is 2. We find the same for $y_2 \cdot v_\lambda$. Next we turn to $\mu = \lambda_1 + \lambda_2$. We have $x_3 y_3 = y_3 x_3 + h_1 + h_2$, so that $(f_{\eta_5} v_\lambda, f_{\eta_5} v_\lambda) = 4$. Furthermore, $x_2 x_1 y_3 = -y_1 x_1 + y_2 x_2 + y_3 x_1 x_2 + y_3 x_3 + h_2$, yielding $(f_{\eta_6} v_\lambda, f_{\eta_5} v_\lambda) = 2$. Finally,

$$x_2 x_1 y_1 y_2 = y_1 y_2 x_1 x_2 + y_1 y_2 x_3 + y_1 h_2 x_1 + 2 y_2 x_2 + y_2 h_1 x_2 + h_1 h_2 + h_2,$$

from which it follows that $(f_{\eta_6} v_\lambda, f_{\eta_6} v_\lambda) = 6$. So, in view of Remark 5.5.6 the matrix of the form restricted to Λ_μ is $\left(\begin{smallmatrix} 4 & 2 \\ 2 & 6 \end{smallmatrix}\right)$, which has Smith normal form $\left(\begin{smallmatrix} 2 & 0 \\ 0 & 10 \end{smallmatrix}\right)$. It follows that $\dim R_\mu = 2$ if $p = 2$, $\dim R_\mu = 1$ if $p = 5$, and $\dim R_\mu = 0$ for all other characteristics.

Let $\mu = 0$. Using similar computations we find that the matrix of the form restricted to Λ_μ is $\left(\begin{smallmatrix} 6 & 6 & 1 \\ 6 & 16 & 6 \\ 1 & 6 & 6 \end{smallmatrix}\right)$, which has elementary divisors $1, 5, 40$. Hence for $p = 2$ we have $\dim R_\mu = 1$, for $p = 5$ we have $\dim R_\mu = 2$ and $\dim R_\mu = 0$ otherwise.

Taking into account that the orbits of the Weyl group on the weights $3\lambda_1$, $3\lambda_2$ have three elements, whereas the orbit of $\lambda_1 + \lambda_2$ has six elements and the orbit of 0 has one element, and $\dim V = 27$, we conclude that $\dim L(\lambda) = 8$ if $p = 2$, $\dim L(\lambda) = 19$ if $p = 5$ and $\dim L(\lambda) = 27$ for all other characteristics. We also know the weights and the dimensions of the corresponding weight spaces in each case.

Remark 5.5.8 The algorithms described here can be extended to construct a basis of $L(\lambda)$ along with a function that computes the matrix of a given element (say $x_\alpha(t)$) of G with respect to that basis. Indeed, with the algorithm from Section 2.11.4 we construct the $\mathfrak{g}$-module $V = V(\lambda)$. As seen in Section 5.2.1 we can then find a basis of the admissible lattice $\Lambda = \mathcal{U}_\mathbb{Z}^- \cdot v_\lambda$. With the methods given in this section we can compute a basis of the radical $R \subset V^k = k \otimes \Lambda$ of the contravariant form. We then obtain a basis of V^k/R, and we can compute the matrix M_l of $x_\alpha^{(l)}$ with respect to that basis. Then the matrix of $x_\alpha(t)$ is equal to $\sum_{l \geq 0} t^l M_l$.

5.6 Multiplication in a semisimple algebraic group

Corollary 5.2.24 yields a normal form of elements in a semisimple algebraic group. We consider the following problem: given two elements in normal form, compute the normal form of their product and the normal form of the inverse of one of them. In this section we show an algorithm for doing that. In the context of this problem, we view the generators of the group as symbols (rather than as endomorphisms acting on a vector space). The elements of the group are formal words in these symbols, and we use relations to rewrite a word that is not in normal form to one that is. For this reason, the formal set-up is as follows. Let $R = (X, A, X^\vee, B)$ be a semisimple based root datum. As in Section 5.4, let $\mu_1, \ldots, \mu_\ell$, $\mu_1^\vee, \ldots, \mu_\ell^\vee$ be the given bases of X, $X^\vee$. Let k be an algebraically closed field and G be a semisimple algebraic group over k corresponding to R. Let $\Phi \subset X$ be the root system corresponding to R, with basis of simple roots $\Delta = \{\alpha_1, \ldots, \alpha_\ell\}$. Fix a choice for the signs $\varepsilon(\alpha, \beta)$ of the constants $N_{\alpha,\beta}$, $\alpha, \beta \in \Phi$ such that (2.2) defines a semisimple complex Lie algebra $\mathfrak{g}$. Let $\mathcal{G}$ be the group generated by the symbols $\xi_\alpha(t)$ (for $t \in k$, $\alpha \in \Phi$), subject to the relations (5.1) to (5.6), where the $c_{i,j}^{\alpha,\beta}$ in (5.3) are computed as in Lemma 5.2.10 (and the product is taken in the same order) and the $c_{\alpha,\beta}$ in (5.5) are as in Lemma 5.2.15. We then have the elements and subgroups of $\mathcal{G}$ as defined in Section 5.1. We also impose the relations of Proposition 5.2.30(i) through (iv), with $\eta_\alpha(t)$ in place of $h_\alpha(t)$. By Theorem 5.2.31, mapping $\xi_\alpha(t) \mapsto x_\alpha(t)$ extends to an isomorphism $\phi : \mathcal{G} \to G$.

We use the notation $\bar{\alpha} = \frac{2\alpha}{(\alpha,\alpha)}$; see Theorem 2.9.13.

We first look at the subgroup $\mathcal{H}$, generated by the $\eta_\alpha(t)$. By the relations of Proposition 5.2.30 it follows that $\mathcal{H}$ is also generated by $\eta_i(t) = \eta_{\alpha_i}(t)$, $1 \le i \le \ell$. Moreover, each element of $\mathcal{H}$ can be written as $\eta_1(t_1)\cdots\eta_\ell(t_\ell)$. However, the relations of Proposition 5.2.30(iv) make it difficult to obtain a normal form of the elements of $\mathcal{H}$ this way. Therefore, we define a different set of generators of $\mathcal{H}$, along the lines of Proposition 5.3.6. The coefficients q_{ij} used in the proof of that proposition can be chosen so that $(q_{ij}) = B^{-1}$. Therefore, in the following we let $(q_{ij})_{1\le i,j\le \ell} = B^{-1}$ and we set $\tilde{\eta}_i(t) = \prod_{j=1}^{\ell} \eta_j(t^{q_{ij}})$. Here the $t^{q_{ij}}$ are defined as follows. Note that $(q_{ij}) = \frac{1}{d}(m_{ij})$ where d, m_{ij} are integers. Fix $s \in k$ such that $s^d = t$, and let $t^{q_{ij}} = s^{m_{ij}}$ for all i, j. A short calculation based on the relation in Proposition 5.2.30(iv) shows that $\tilde{\eta}_i(t)$ does not depend on the choice of s.

Lemma 5.6.1 *Let $\tau : (k^*)^\ell \to \mathcal{H}$ be defined by $\tau(t_1,\dots,t_\ell) = \prod_{j=1}^{\ell} \tilde{\eta}_i(t_i)$. Then τ is an isomorphism of groups.*

Proof. By the relations of Proposition 5.2.30(i) and (ii), τ is a group homomorphism. We show that it is surjective. Fix an r, $1 \le r \le \ell$. Let $(m_1,\dots,m_\ell)$ be the r-th row of B. Then $(\sum_{i=1}^{\ell} m_i q_{ij})_{1\le j\le \ell}$ is equal to the r-th row of $B(q_{ij})$. By the choice of q_{ij}, that is the identity matrix. It follows that $\prod_{i=1}^{\ell} \tilde{\eta}_i(t^{m_i}) = \eta_r(t)$. For injectivity, suppose $\prod_{i=1}^{\ell} \tilde{\eta}_i(t_i) = 1$. Inserting the definition of $\tilde{\eta}_i$ and using the relation in Proposition 5.2.30(i), we see that $1 = \prod_{j=1}^{\ell} \eta_j(\prod_{i=1}^{\ell} t_i^{q_{ij}})$. By the relation in Proposition 5.2.30(iv) and taking $\mu = \mu_m$, this yields

$$1 = \prod_{j=1}^{\ell}\prod_{i=1}^{\ell} t_i^{\langle \mu_m, \alpha_j^\vee\rangle q_{ij}} = \prod_{i=1}^{\ell} t_i^{\sum_j \langle \mu_m, \alpha_j^\vee\rangle q_{ij}}$$

but by the choice of the q_{ij} the latter product is simply equal to t_m. □

This yields the normal form of the elements of $\mathcal{H}$ which are uniquely written as $\prod_i \tilde{\eta}_i(t_i)$, and the product of two such elements is simply formed by taking the products of the arguments of each $\tilde{\eta}_i$. The downside is that the $\tilde{\eta}_i$ in general do not satisfy the relations of Section 5.1. We first look at (5.4). For a $y = \sum_{i=1}^{\ell} m_i \mu_i^\vee \in X^\vee$ we write $\tilde{\eta}_y(t) = \prod_{i=1}^{\ell} \tilde{\eta}_i(t^{m_i})$. From Section 5.4 we recall that for $\alpha \in \Phi$ we have the map $s_\alpha^\vee : X^\vee \to X^\vee$ with $s_\alpha^\vee(y) = y - \langle \alpha, y\rangle \alpha^\vee$.

Lemma 5.6.2 *For $y \in X^\vee$ we have*

$$\varpi_{\alpha_r} \tilde{\eta}_y(t) \varpi_{\alpha_r}^{-1} = \tilde{\eta}_{s_{\alpha_r}^\vee(y)}(t).$$

Proof. First we consider $y = \mu_i^\vee$, so that $\tilde{\eta}_y(t) = \tilde{\eta}_i(t)$. Using the definition of $\tilde{\eta}_i$, and (5.4), we see that $\varpi_{\alpha_r}\tilde{\eta}_i(t)\varpi_{\alpha_r}^{-1} = \prod_{j=1}^{\ell} \eta_{s_{\alpha_r}(\alpha_j)}(t^{q_{ij}})$. Also, with $\overline{s_{\alpha_r}(\alpha_j)} = \bar{\alpha}_j - \langle \alpha_r, \alpha_j^\vee\rangle\bar{\alpha}_r$ (noted also in Section 5.4) and the relation in Proposition 5.2.30(iii), $\eta_{s_{\alpha_r}(\alpha_j)}(u) = \eta_j(u)\eta_r(u^{-\langle\alpha_r,\alpha_j^\vee\rangle})$. Therefore

$$\varpi_{\alpha_r}\tilde{\eta}_i(t)\varpi_{\alpha_r}^{-1} = \tilde{\eta}_i(t)\eta_r(t^{-\sum_{j=1}^{\ell} q_{ij}\langle\alpha_r,\alpha_j^\vee\rangle}).$$

As $\sum_{j=1}^{\ell} q_{ij}\langle\alpha_r,\alpha_j^\vee\rangle = C(q_{ij})^T(r,i)$ and $C = AB^T$, using the choice of q_{ij}, we have $\sum_{j=1}^{\ell} q_{ij}\langle\alpha_r,\alpha_j^\vee\rangle = A(r,i)$. Furthermore, as seen in the proof of Lemma 5.6.1, $\eta_r(u) = \prod_{i=1}^{\ell} \tilde{\eta}_i(u^{B(r,i)})$. Noting that $s_{\alpha_r}^\vee(\mu_i^\vee) = \mu_i^\vee - \langle\alpha_r,\mu_i^\vee\rangle\alpha_r^\vee$, $\alpha_r^\vee = \sum_i B(r,i)\mu_i^\vee$, $A(r,i) = \langle\alpha_r,\mu_i^\vee\rangle$, we arrive at the desired result in this case. The general case now follows by writing $y = \sum_i m_i\mu_i^\vee$, so that $\tilde{\eta}_y(t) = \prod_i \tilde{\eta}_i(t^{m_i})$. □

Lemma 5.6.3 *For $\beta \in \Phi$ we have*

$$\tilde{\eta}_i(t)\xi_\beta(u)\tilde{\eta}_i(t)^{-1} = \xi_\beta(t^{\langle\beta,\mu_i^\vee\rangle}u).$$

Proof. Using the definition of $\tilde{\eta}_i$ and (5.6), we infer that the left-hand side is equal to $\xi_\beta(t^a u)$, where $a = \sum_{j=1}^{\ell}\langle\beta,\alpha_j^\vee\rangle q_{ij}$. By writing β as a linear combination of simple roots and noting $C(q_{ij})^T = A$ (as in the proof of the previous lemma), the lemma is proved. □

For $1 \le i \le \ell$ set $n_i = \varpi_{\alpha_i}^{-1}$ (here we could also take ϖ_{α_i}, but later we will see that it is easier to work with $\varpi_{\alpha_i}^{-1}$). Let W denote the Weyl group of Φ. For each $w \in W$ fix a reduced expression $w = s_{i_1}\cdots s_{i_r}$, and set $\dot{w} = n_{i_1}\cdots n_{i_r}$. Then $\vartheta(w) = \mathcal{H}\dot{w}$ (notation as in Lemma 5.1.4(iii)).

Lemma 5.6.4 *For $\gamma \in \Phi$ we write $n_\gamma = \varpi_\gamma^{-1}$. Let $\alpha, \beta \in \Delta$, and let m be the order of $s_\alpha s_\beta$ in W. Then $n_\alpha n_\beta \cdots = n_\beta n_\alpha \cdots$, where both sides have m factors.*

Proof. Note that α and β are the simple roots of a root subsystem Ψ of Φ. If Ψ is of type $A_1 + A_1$, then $m = 2$. Furthermore, n_α, n_β commute by (5.3), as $n_\alpha = \xi_\alpha(-1)\xi_{-\alpha}(1)\xi_\alpha(-1)$, and similarly for n_β.

In the proof of Lemma 5.1.4(iii) we see that $\varpi_\alpha\varpi_\beta\varpi_\alpha^{-1} = \omega_{s_\alpha(\beta)}(c_{\alpha,\beta})$. By definition, the latter is $\eta_{s_\alpha(\beta)}(c_{\alpha,\beta})n_{s_\alpha(\beta)}^{-1}$. Therefore, as $c_{\alpha,\beta} = \pm 1$, we have $n_\alpha^{-1}n_\beta n_\alpha = n_{s_\alpha(\beta)}\eta_{s_\alpha(\beta)}(c_{\alpha,\beta})$. Furthermore, we note that $\eta_\alpha(-1) = \omega_\alpha(-1)\omega_\alpha(1)^{-1} = n_\alpha^2$.

We leave the case where Ψ is of type A_2 to the reader, and continue with the case where Ψ is of type B_2 and $\langle\alpha,\beta^\vee\rangle = -2$, $\langle\beta,\alpha^\vee\rangle = -1$. Then $n_\beta n_\alpha n_\beta n_\alpha = n_\beta^2 n_{\alpha+2\beta}\eta_{\alpha+2\beta}(c_{\beta,\alpha})n_\alpha$. We use (5.4), the fact that $n_{\alpha+2\beta}$ and

n_α commute (see the argument for the case where Ψ is of type $A_1 + A_1$), and $n_\beta^2 = \eta_\beta(-1)$ to "move" the n_α in this expression to the front, and obtain

$$n_\beta n_\alpha n_\beta n_\alpha = n_\alpha \eta_{\alpha+\beta}(-1) n_{\alpha+2\beta} \eta_{\alpha+2\beta}(c_{\beta,\alpha}).$$

Here $\overline{\alpha+\beta} = 2\bar{\alpha} + \bar{\beta}$, so that, by the relation in Proposition 5.2.30(iii), $\eta_{\alpha+\beta}(t) = \eta_\alpha(t^2)\eta_\beta(t)$, whence $\eta_{\alpha+\beta}(-1) = \eta_\beta(-1) = n_\beta^2$. The lemma is proved in this case.

This leaves the case where Ψ is of type G_2. We assume that $\langle \alpha, \beta^\vee \rangle = -1$, $\langle \beta, \alpha^\vee \rangle = -3$. Then $n_\alpha n_\beta n_\alpha = n_\alpha^2 n_{3\alpha+\beta} \eta_{3\alpha+\beta}(c_{\alpha,\beta})$ and $n_\beta n_\alpha n_\beta = n_\beta^2 n_{\alpha+\beta} \eta_{\alpha+\beta}(c_{\beta,\alpha})$. We must show that these two elements commute. We have $n_\alpha^2 n_{3\alpha+\beta} = n_{3\alpha+\beta}\eta_{2\alpha+\beta}(-1)$ and $n_\beta^2 n_{\alpha+\beta} = n_{\alpha+\beta}\eta_{3\alpha+2\beta}(-1)$ and must show that $A = n_{3\alpha+\beta}\eta_{2\alpha+\beta}(-1)\eta_{3\alpha+\beta}(c_{\alpha,\beta})$ and $B = n_{\alpha+\beta}\eta_{3\alpha+2\beta}(-1)\eta_{\alpha+\beta}(c_{\beta,\alpha})$ commute. We have

$$\begin{aligned} AB &= n_{3\alpha+\beta} n_{\alpha+\beta} \eta_\alpha(-1) \eta_{3\alpha+\beta}(c_{\alpha,\beta}) \eta_{3\alpha+2\beta}(-1) \eta_{\alpha+\beta}(c_{\beta,\alpha}), \\ BA &= n_{\alpha+\beta} n_{3\alpha+\beta} \eta_\beta(-1) \eta_{\alpha+\beta}(c_{\beta,\alpha}) \eta_{2\alpha+\beta}(-1) \eta_{3\alpha+\beta}(c_{\alpha,\beta}). \end{aligned}$$

Because $n_{3\alpha+\beta}$ and $n_{\alpha+\beta}$ commute, $AB = BA$ amounts to $\eta_\alpha(-1)\eta_{3\alpha+2\beta}(-1) = \eta_\beta(-1)\eta_{2\alpha+\beta}(-1)$. By some calculations in $\mathcal{H}$ based on (5.4), we see that both sides are trivial. □

This lemma, together with Proposition 2.8.26, implies that $\dot{w}$ does not depend on the chosen reduced expression for w. Furthermore, it follows that, setting $v = ws_{\alpha_i}$,

$$\begin{cases} \dot{w} n_i = \dot{v} & \text{if } \mathcal{L}(v) > \mathcal{L}(w) \\ \dot{w} n_i = \eta_{v(\alpha_i)}(-1)\dot{v} & \text{if } \mathcal{L}(v) < \mathcal{L}(w). \end{cases} \tag{5.11}$$

Fix an order of the positive roots $\beta_1, \ldots, \beta_n$. Also, for each $w \in W$ fix an order of the elements of $\Phi_w = \{\alpha \in \Phi^+ \mid w(\alpha) \in \Phi^-\}$. By Lemma 5.2.28, every element of $\mathcal{G}$ can be written uniquely as $u_1 h \dot{w} u_2$, where $u_1 \in \mathcal{U}$, $h \in \mathcal{H}$, $u_2 \in \mathcal{U}_w$. Furthermore, u_1 can be written uniquely as $\xi_{\beta_1}(t_1) \cdots \xi_{\beta_n}(t_n)$, with $t_i \in k$. Analogously, u_2 can uniquely be written as a product of $x_\alpha(t_\alpha)$, where the $\alpha \in \Phi_w$ are taken in the chosen order. By Lemma 5.6.1, h can be written uniquely as a product $\tilde{\eta}_1(t_1) \cdots \tilde{\eta}_\ell(t_\ell)$, with $t_i \in k^*$. If u_1, u_2, h are thus written, we say that the element $u_1 h \dot{w} u_2$ is a *normal word*. The problem is to rewrite the product of two normal words as a normal word. We deal with a few special cases, namely where the second normal word is a $u \in \mathcal{U}$, an $\tilde{\eta}_i(s)$ ($1 \le i \le \ell$, $s \in k^*$), or a n_i ($1 \le i \le \ell$). If we can handle each of these special cases, we can deal with the general case as well.

First we consider rewriting an element $\xi = \xi_{\beta_{i_1}}(t_{i_1}) \cdots \xi_{\beta_{i_m}}(t_{i_m})$ to a normal word. The various procedures that can be used are called *collection algorithms* for $\mathcal{U}$ (analogous to collection in nilpotent groups; see for example, [HEO05], Section 8.1.3).

We call a pair of indices (i_a, i_b) with $a < b$ and $i_a > i_b$ an *inversion* of ξ. If ξ has no inversions, it is a normal word. The *height* of the inversion is the maximum height of β_{i_a}, β_{i_b}. Let N be the maximal height of a positive root. To ξ we associate the N-tuple $\mathbf{h}(\xi) = (\iota_1, \ldots, \iota_N)$, where ι_j is the number of inversions of ξ of height j. Suppose ξ is not a normal word, and choose an inversion (i_a, i_b) with $b = a + 1$. Writing $\alpha = \beta_{i_a}$, $\beta = \beta_{i_b}$, we have, from (5.3), $\xi_\alpha(t)\xi_\beta(s)\xi_\alpha(-t)\xi_\beta(-s) = \prod_{(i,j)\in I_{\alpha,\beta}} \xi_{i\alpha+j\beta}(c_{i,j}^{\alpha,\beta} t^i s^j)$. Rewriting, and substituting $t \mapsto -t$, $u \mapsto -s$ we have

$$\xi_\alpha(t)\xi_\beta(s) = \xi_\beta(s)\xi_\alpha(t) \prod_{(i,j)\in I_{\alpha,\beta}} \xi_{i\alpha+j\beta}((-1)^{i+j} c_{i,j}^{\alpha,\beta} t^i s^j).$$

In ξ we substitute the right-hand side of this expression (with $t = t_{i_a}$, $s = t_{i_b}$) for the left-hand side, obtaining a new word ξ'. It is clear that $\mathbf{h}(\xi') <_{\text{lex}} \mathbf{h}(\xi)$. (Writing $\mathbf{h}(\xi') = (\iota_1', \ldots, \iota_N')$, this means that $\iota_j' < \iota_j$ where j is minimal such that $\iota_j \neq \iota_j'$.) When performing this repeatedly, after a finite number of steps we find a normal word.

In general, there is more than one inversion (i_a, i_{a+1}). For choosing which one to treat several strategies are possible. We could take the leftmost one, the rightmost one, the one of smallest height, the one of greatest height, and so on. Here we do not go into the question of which choice is best, but refer to [CHM08], where a few approaches are compared.

To rewrite a product of two normal words in $\mathcal{U}$ to a normal word, a symbolic approach is possible as well. By treating the t_i, s_i as symbols and performing the collection algorithm symbolically, it is possible to obtain polynomials p_i of $2n$ variables such that

$$\xi_{\beta_1}(t_1)\cdots\xi_{\beta_n}(t_n)\xi_{\beta_1}(s_1)\cdots\xi_{\beta_n}(s_n) = \xi_{\beta_1}(p_1(t_i, s_i))\cdots\xi_{\beta_n}(p_n(t_i, s_i)),$$

where $p_j(t_i, s_i) = p_j(t_1, \ldots, t_n, s_1, \ldots, s_n)$. Multiplication in $\mathcal{U}$ thus becomes a question of evaluating these polynomials. We illustrate this in an example.

Example 5.6.5 Let the root system be of type A_2, as in Example 5.2.11. Using the relations in that example we see that

$$\xi_{\alpha_1}(t_1)\xi_{\alpha_2}(t_2)\xi_{\alpha_3}(t_3)\xi_{\alpha_1}(s_1)\xi_{\alpha_2}(s_2)\xi_{\alpha_3}(s_3) = \xi_{\alpha_1}(t_1 + s_1)\xi_{\alpha_2}(t_2 + s_2)\xi_{\alpha_3}(t_3 + s_3 + t_2 s_1).$$

Now we treat the first case: multiplying $u_1 h\dot{w} u_2$ by a $u \in \mathcal{U}$. We list the positive roots in an order that has the roots not in Φ_w first, and then the roots in Φ_w, in the order fixed at the start. We perform collection on the element $u_2 u$, using this ordering of the positive roots. We obtain $u_2 u = u' u''$, where u' does not involve any $\xi_\beta(s)$, for $\beta \in \Phi_w$, and $u'' \in \mathcal{U}_w$. To commute u' and $\dot{w}$ we repeatedly use $n_i \xi_\beta(t) = \xi_{s_{\alpha_i}(\beta)}(c_{\alpha_i,\beta} t) n_i$, which follows from (5.5), to obtain $u''' \in \mathcal{U}$ with $\dot{w} u' = u''' \dot{w}$. (Note that for every α such that ξ_α is involved in u', we have $w(\alpha) \in \Phi^+$.) We then repeatedly use the relation from Lemma

5.6.3 to compute a u'''' with $hu''' = u''''h$. Finally, we perform the collection algorithm on u_1u'''' yielding u_1'. The result is $u_1'h\dot{w}u''$.

In the second case we consider multiplying $u_1h\dot{w}u_2$ by $\tilde{\eta}_i(t)$. By repeatedly using $\xi_\beta(s)\tilde{\eta}_i(t) = \tilde{\eta}_i(t)\xi_\beta(t^{-\langle\beta,\mu_i^\vee\rangle}s)$, which follows from Lemma 5.6.3, we get a $u' \in \mathcal{U}_w$ such that $u_2\tilde{\eta}_i(t) = \tilde{\eta}_i(t)u'$. From Lemma 5.6.2 we have $n_r\tilde{\eta}_y(t) = \tilde{\eta}_{s^\vee_{\alpha_r}(y)}(t)n_r$. Using this repeatedly we find $y \in X^\vee$ such that $\dot{w}\tilde{\eta}_i(t) = \tilde{\eta}_y(t)\dot{w}$. Finally we rewrite the product $h\tilde{\eta}_y(t)$ to a normal word in $\mathcal{H}$.

The third case, multiplying by n_i, is trickier. For this we first use the collection algorithm to find $u' \in \mathcal{U}_{\Phi_w\setminus\{\alpha_i\}}$ with $u_2 = \xi_{\alpha_i}(t)u'$. Repeatedly using (5.5) we find $u'' \in \mathcal{U}$ such that $u'n_i = n_iu''$. Now we distinguish two subcases. The first has $t \neq 0$. Then $\alpha_i \in \Phi_w$, so that $w(\alpha_i) \in \Phi^-$. By using the definitions we see that $\eta_{\alpha_i}(t) = \xi_{\alpha_i}(t)\xi_{-\alpha_i}(-t^{-1})\xi_{\alpha_i}(t)n_i$. So by (5.6), $\xi_{\alpha_i}(t)n_i = \xi_{-\alpha_i}(t^{-1})\eta_{\alpha_i}(t)\xi_{\alpha_i}(-t^{-1})$. Hence, writing $u''' = \xi_{\alpha_i}(-t^{-1})u''$ we have $gn_i = u_1h\dot{w}\xi_{-\alpha_i}(t^{-1})\eta_{\alpha_i}(t)u'''$. By repeatedly using $n_j\xi_\beta(s) = \xi_{s_{\alpha_j}(\beta)}(c_{\alpha_j,\beta}s)n_j$, which follows from (5.5), we find an $\epsilon = \pm 1$ such that $\dot{w}\xi_{-\alpha_i}(t^{-1}) = \xi_{-w(\alpha_i)}(\epsilon t^{-1})\dot{w}$. Using the relation in Lemma 5.6.3 we obtain an $s \in k$ such that $h\xi_{-w(\alpha_i)}(\epsilon t^{-1}) = \xi_{-w(\alpha_i)}(s)h$. Let $u_1' = u_1\xi_{-w(\alpha_i)}(s)$ (note that $-w(\alpha_i) \in \Phi^+$, so that $u_1' \in \mathcal{U}$). By (5.4) we have $\dot{w}\eta_{\alpha_i}(t) = \eta_{w(\alpha_i)}(t)\dot{w}$. By writing $\eta_{w(\alpha_i)}(t)$ as a product of $\tilde{\eta}_j(s_j)$'s, we find $h' \in \mathcal{H}$ such that $u_1h\dot{w}\xi_{-\alpha_i}(t^{-1})\eta_{\alpha_i}(t) = u_1'h'\dot{w}$. By using the procedure in the first case above, multiplying $u_1'h'\dot{w}$ by u''', yields the result.

The second subcase occurs when $t = 0$. Then by (5.11), $\dot{w}n_i = h'\dot{v}$ where $h' \in \mathcal{H}$. So we multiply $u_1hh'\dot{v}$ by u'', again using the procedure of the first case.

Example 5.6.6 Let the root system be of type A_2 and the root datum of adjoint type. For $\mathfrak{g}$ we use the multiplication table of Example 2.9.14. We multiply the elements $g_1 = \xi_{\alpha_2}(2)\tilde{\eta}_1(3)n_1n_2\xi_{\alpha_2}(-1)\xi_{\alpha_3}(\frac{1}{2})$ and $g_2 = \xi_{\alpha_1}(1)\xi_{\alpha_2}(-2)\tilde{\eta}_2(4)n_1\xi_{\alpha_1}(-2)$. We have $\Phi_{s_{\alpha_1}s_{\alpha_2}} = \{\alpha_2, \alpha_1+\alpha_2\}$, so that the ordering of positive roots defined in the first case above is simply α_1, α_2, $\alpha_3 = \alpha_1 + \alpha_2$. By Example 5.6.5,

$$\xi_{\alpha_2}(-1)\xi_{\alpha_3}(\tfrac{1}{2}) \cdot \xi_{\alpha_1}(1)\xi_{\alpha_2}(-2) = \xi_{\alpha_1}(1)\xi_{\alpha_2}(-3)\xi_{\alpha_3}(-\tfrac{1}{2}).$$

Furthermore, $c_{\alpha_2,\alpha_1} = 1$, $c_{\alpha_1,\alpha_2} = c_{\alpha_1,\alpha_3} = -1$, so that $n_1n_2\xi_{\alpha_1}(1) = \xi_{\alpha_2}(-1)n_1n_2$. Also, $\tilde{\eta}_1(3)\xi_{\alpha_2}(-1) = \xi_{\alpha_2}(3^{\langle\alpha_2,\mu_1^\vee\rangle}\cdot -1)\tilde{\eta}_1(3) = \xi_{\alpha_2}(-1)\tilde{\eta}_1(3)$. Hence $g_1 \cdot \xi_{\alpha_1}(1)\xi_{\alpha_2}(-2) = \xi_{\alpha_2}(1)\tilde{\eta}_1(3)n_1n_2\xi_{\alpha_2}(-3)\xi_{\alpha_3}(-\frac{1}{2})$.

Now we multiply $\xi_{\alpha_2}(1)\tilde{\eta}_1(3)n_1n_2\xi_{\alpha_2}(-3)\xi_{\alpha_3}(-\frac{1}{2})$ and $\tilde{\eta}_2(4)$. First, $\xi_{\alpha_2}(-3)\xi_{\alpha_3}(-\frac{1}{2})\tilde{\eta}_2(4) = \tilde{\eta}_2(4)\xi_{\alpha_2}(-\frac{3}{4})\xi_{\alpha_3}(-\frac{1}{8})$. Since $s^\vee_{\alpha_1}s^\vee_{\alpha_2}(\mu_2^\vee) = -\mu_1^\vee$, we have $n_1n_2\tilde{\eta}_2(4) = \tilde{\eta}_1(\frac{1}{4})n_1n_2$, so that the result of this multiplication is $\xi_{\alpha_2}(1)\tilde{\eta}_1(\frac{3}{4})n_1n_2\xi_{\alpha_2}(-\frac{3}{4})\xi_{\alpha_3}(-\frac{1}{8})$.

Next, we multiply the previous result by n_1. We have $\xi_{\alpha_2}(-\frac{3}{4})\xi_{\alpha_3}(-\frac{1}{8})n_1 = n_1\xi_{\alpha_3}(\frac{3}{4})\xi_{\alpha_2}(\frac{1}{8})$ and find the element $\xi_{\alpha_2}(1)\tilde{\eta}_1(\frac{3}{4})n_1n_2n_1\xi_{\alpha_2}(\frac{1}{8})\xi_{\alpha_3}(\frac{3}{4})$.

Finally, we multiply this by $\xi_{\alpha_1}(-2)$. Again using Example 5.6.5 we find that the result is equal to $\xi_{\alpha_2}(1)\tilde{\eta}_1(\frac{3}{4})n_1n_2n_1\xi_{\alpha_1}(-2)\xi_{\alpha_2}(\frac{1}{8})\xi_{\alpha_3}(\frac{1}{2})$.

A similar problem involves computing the normal word equal to the inverse of a given $g = u_1 h\dot{w} u_2$. Note that $g^{-1} = u_2^{-1}\dot{w}^{-1}h^{-1}u_1^{-1}$. We have $n_i^4 = \eta_{\alpha_i}(1) = 1$, so $n_i^{-1} = n_i^3 = \eta_{\alpha_i}(-1)n_i$. We can use that to compute a $h' \in \mathcal{H}$ such that $\dot{w}^{-1} = h'\dot{v}$, where $v = w^{-1}$. It is straightforward to compute the inverses of the other terms of g^{-1}. Subsequently we rewrite $u_2^{-1}\dot{w}^{-1}h^{-1}u_1^{-1}$ to a normal word, using the same techniques used for computing the product.

5.7 From matrix to word

In the previous section we saw how to multiply elements of $\mathcal{G}$ that are given as normal words. However, there are many problems concerning those words, that cannot easily be solved by working inside $\mathcal{G}$ only. As example, consider the multiplicative Jordan decomposition of an element (Section 3.10.2). For applications like that it is important to be able to transform normal words to matrices (with respect to a given representation of $\mathcal{G}$) and transform matrices back to normal words. The first of these problems is not difficult (see below). In this section we focus on the second problem. We restrict discussion to representations that are quotients of highest weight representations.

Let the notation be as in the previous section and $\lambda \in X$ be a dominant weight. Let V be the highest weight module over $\mathfrak{g}$ of highest weight λ. As usual, we assume that V is a faithful $\mathfrak{g}$-module. We use V to construct the Chevalley group G' over k. By Corollary 5.2.32, there exists a surjective homomorphism $G \to G'$, so that V^k is also a G-module. In much the same way as in Remark 5.5.8 we construct a G-module $\widehat{V}$ which is a quotient of V^k. So $\widehat{V}$ could be V^k, $L(\lambda)$, or a module between them. Composing the corresponding representation of G with the isomorphism $\phi : \mathcal{G} \to G$, we obtain a representation $\theta : \mathcal{G} \to \mathrm{GL}(\widehat{V})$. Here we consider the problem to construct the inverse of θ, that is, given $A \in \theta(\mathcal{G})$, find an element $g \in \mathcal{G}$ given as a normal word in the sense of the previous section such that $\theta(g) = A$. We note that such a g is not necessarily unique, as θ can have a kernel contained in $\mathcal{H}$. By construction, we know all the weights of $\widehat{V}$, and we have bases of the corresponding weight spaces.

The first half of this section is technical preparation for the algorithm described in the second half.

Set $[\ell] = \{1, \dots, \ell\}$. For $J \subset [\ell]$ we let W_J be the subgroup of W generated by s_{α_i}, for $i \in J$. This is a reflection subgroup of W and the Weyl group of the root subsystem of Φ with basis of simple roots $\{\alpha_i \mid i \in J\}$. Now we define a sequence of weights $\bar{\mu}_i$ and subsets $J_i \subset [\ell]$. We let $\bar{\mu}_0 = \lambda$, $J_0 = [\ell]$, $J_1 = \{i \in [\ell] \mid \langle \lambda, \alpha_i^\vee \rangle = 0\}$. Suppose $\bar{\mu}_0, \dots, \bar{\mu}_{m-1}$, $J_0, \dots, J_m$ have been determined, with $J_m \neq \emptyset$. Let $r \in [\ell] \setminus J_m$ be such that there are $i \in J_m$ with $\langle \alpha_r, \alpha_i^\vee \rangle \neq 0$ (note that such r exist as $J_m \subset J_1$ and V is a faithful $\mathfrak{g}$-module). Let $t \leq m-1$ be such that $r \in J_t \setminus J_{t+1}$. Then we set $\bar{\mu}_m = s_r(\bar{\mu}_t)$,

$J_{m+1} = \{i \in J_m \mid \langle \bar{\mu}_m, \alpha_i^\vee \rangle = 0\}$. The sequence stops when we find s with $J_{s+1} = \emptyset$. In the sequel we fix the sets J_m, the weights $\bar{\mu}_m$, and the number s. We define Φ_m to be the root subsystem of Φ spanned by the α_i for $i \in J_m$ and set $\Phi_m^+ = \Phi_m \cap \Phi^+$.

Lemma 5.7.1 (i) $J_{m+1} \subsetneq J_m$,

(ii) $\langle \bar{\mu}_m, \alpha_i^\vee \rangle \geq 0$ *for* $i \in J_m$,

(iii) $W_{J_m} = \{w \in W \mid w(\bar{\mu}_i) = \bar{\mu}_i$ *for* $0 \leq i \leq m-1\}$,

(iv) $\bar{\mu}_m + \alpha$ *is not a weight of* $\widehat{V}$ *for* $\alpha \in \Phi_m$,

(v) $\bar{\mu}_m - \alpha$ *is not a weight of* $\widehat{V}$ *for* $\alpha \in \Phi_{m+1}$.

Proof. For the first three statements we use induction on m. They trivially hold for $m = 0$. Suppose $m \geq 1$. Then $J_m = \{i \in J_{m-1} \mid \langle \bar{\mu}_{m-1}, \alpha_i^\vee \rangle = 0\}$. We have $s_r(\bar{\mu}_t) = \bar{\mu}_t - a\alpha_r$, with $a = \langle \bar{\mu}_t, \alpha_r^\vee \rangle$. As $r \in J_t$ and $t < m$, by induction it follows that $a \geq 0$. But since $r \notin J_{t+1}$, a is non-zero and has to be positive. Let $i \in J_m$. Then also $i \in J_{t+1}$, whence $\langle \bar{\mu}_t, \alpha_i^\vee \rangle = 0$. Therefore, $\langle s_r(\bar{\mu}_t), \alpha_i^\vee \rangle = -a\langle \alpha_r, \alpha_i^\vee \rangle \geq 0$. So (i) and (ii) hold for m. By induction, $\{w \in W \mid w(\bar{\mu}_i) = \bar{\mu}_i \text{ for } 0 \leq i \leq m-1\} = \{w \in W_{J_{m-1}} \mid w(\bar{\mu}_{m-1})) = \bar{\mu}_{m-1}\}$. Because $\langle \bar{\mu}_{m-1}, \alpha_i^\vee \rangle \geq 0$ for $i \in J_{m-1}$ and $W_{J_{m-1}}$ is the Weyl group of Φ_{m-1}, the latter group is generated by the s_{α_i} for $i \in J_m$ (Proposition 2.8.31).

For the fourth statement we note that $\bar{\mu}_m = \lambda - \nu$, where ν is a linear combination, with non-negative integral coefficients of α_i with $i \notin J_m$. Therefore, $\bar{\mu}_m + \alpha$, with $\alpha \in \Phi_m$, cannot be a weight of $\widehat{V}$ (see Proposition 2.11.1).

For (v) we note that $\bar{\mu}_m + \alpha$ is not a weight of $\widehat{V}$ by (iv). Furthermore, $\langle \bar{\mu}_m, \alpha^\vee \rangle = 0$ by definition of J_{m+1}. As s_α inverts the string of weights $\bar{\mu}_m + i\alpha$, $i \in \mathbb{Z}$, we conclude that $\bar{\mu}_m - \alpha$ is not a weight of $\widehat{V}$ either. □

Example 5.7.2 Let the root system be of type A_2 and $\lambda = 2\lambda_1$. Then $J_0 = \{1, 2\}$, $J_1 = \{2\}$, $J_3 = \emptyset$, and $\bar{\mu}_0 = 2\lambda_1$, $\bar{\mu}_1 = s_{\alpha_1}(\bar{\mu}_1) = -2\lambda_1 + 2\lambda_2$.

Let $w \in W$. By Proposition 2.8.32 we can uniquely write $w = w^1 w_1$, where w_1 is the unique shortest element in $W_{J_1} w$ and $w^1 \in W_{J_1}$. By Remark 2.8.33, $\mathcal{L}(w) = \mathcal{L}(w^1) + \mathcal{L}(w_1)$. As W_{J_1} is the Weyl group of Φ_1, we can apply the proposition again and find $w^1 = w^2 w_2$ where w_2 is the unique shortest element in $W_{J_2} w^1$, $w^2 \in W_{J_2}$ and $\mathcal{L}(w^1) = \mathcal{L}(w^2) + \mathcal{L}(w_2)$. Continuing we find a decomposition $w = w_s \cdots w_1$ with $w_s \cdots w_i \in W_{J_{i-1}}$, and w_i is the unique shortest element of $W_{J_i} w_s \cdots w_i$, for $1 \leq i \leq s$. Furthermore, $\mathcal{L}(w) = \mathcal{L}(w_1) + \cdots + \mathcal{L}(w_s)$. In this section we call this the *J-decomposition* of w.

Lemma 5.7.3 *Let* $w = w_s \cdots w_1$ *be the J-decomposition of a* $w \in W$. *Let* $1 \leq m \leq s$ *and set* $\nu = w_m^{-1} \bar{\mu}_{m-1}$. *Then* w_m^{-1} *is the unique shortest element of* $W_{J_{m-1}}$ *mapping* $\bar{\mu}_{m-1}$ *to* ν.

Proof. Let $v \in W_{J_{m-1}}$ be such that $v\bar{\mu}_{m-1} = \nu$. Then $v^{-1}w_m^{-1}$ stabilizes $\bar{\mu}_{m-1}$, and hence lies in W_{J_m} (Lemma 5.7.1(iii)). Hence $W_{J_m}v^{-1} = W_{J_m}w_m$, proving the lemma. □

Lemma 5.7.4 *Let $w \in W$ have J-decomposition $w = w_s \cdots w_1$. Let $\mathcal{U}_w$ be as in Proposition 5.1.10. Then $\dot{w}\mathcal{U}_w = \dot{w}_s\mathcal{U}_{w_s} \cdots \dot{w}_1\mathcal{U}_{w_1}$.*

Proof. Let $v \in W$, and let $v = s_{i_1} \cdots s_{i_r}$ be reduced. Write $v' = s_{i_1} \cdots s_{i_{r-1}}$. Using $\Phi_v = s_{i_r}(\Phi_{v'}) \dot{\cup} \{\alpha_{i_r}\}$ (see the proof of Lemma 2.8.24), we see that $\Phi_v = \{\alpha_{i_r}, s_{i_r}(\alpha_{i_{r-1}}), \ldots, s_{i_r} \cdots s_{i_2}(\alpha_{i_1})\}$. This implies that if $v = v_1v_2$ with $\mathcal{L}(v) = \mathcal{L}(v_1) + \mathcal{L}(v_2)$, then $\Phi_v = v_2^{-1}(\Phi_{v_1}) \dot{\cup} \Phi_{v_2}$.

Let $w^1 = w_s \cdots w_2$, as above. Then $\mathcal{U}_w = \mathcal{U}_{w_1^{-1}(\Phi_{w^1})}\mathcal{U}_{w_1}$. By substituting $w^{-1}(\alpha)$ for α in Lemma 5.1.5 we obtain $\mathcal{U}_{w_1^{-1}(\Phi_{w^1})} = \dot{w}_1^{-1}\mathcal{U}_{w^1}\dot{w}_1$. Therefore, $\dot{w}\mathcal{U}_w = \dot{w}^1\mathcal{U}_{w^1}\dot{w}_1\mathcal{U}_{w_1}$. The proof is finished by induction. □

Lemma 5.7.5 *Let w_0 be the longest element of W and $w_0 = w_{0,s} \cdots w_{0,1}$ be its J-decomposition. We have $\mathcal{U} = \mathcal{U}_{w_{0,1}^{-1}} \cdots \mathcal{U}_{w_{0,s}^{-1}}$.*

Proof. Let $v \in W$. We have $\Phi^+ \cap v^{-1}(\Phi^+) = \Phi^+ \cap v^{-1}w_0(\Phi^-) = \Phi_{w_0v}$. Also, $\Phi = v^{-1}(\Phi^-) \dot{\cup} v^{-1}(\Phi^+)$. Hence Φ^+ is the disjoint union of $\Phi^+ \cap v^{-1}(\Phi^-) = \Phi_v$ and $\Phi^+ \cap v^{-1}(\Phi^+) = \Phi_{w_0v}$. Write $w_0 = w_0^1 w_{0,1}$, with $w_0^1 \in W_{J_1}$. With $v = w_{0,1}^{-1}$ we obtain $\Phi^+ = \Phi_{w_{0,1}^{-1}} \dot{\cup} \Phi_{w_0^1}$. By Remark 2.8.33, w_0^1 is the longest element of W_{J_1}. By induction we find that Φ^+ is the disjoint union of $\Phi_{w_{0,i}^{-1}}$, $1 \le i \le s$. In view of Lemma 5.1.3, this finishes the proof. □

The previous two lemmas, along with Lemma 5.2.28, imply that $\mathcal{G}$ has the following modified Bruhat decomposition:

$$\mathcal{G} = \bigcup_{w \in W} \mathcal{U}_{w_{0,1}^{-1}} \cdots \mathcal{U}_{w_{0,s}^{-1}} \mathcal{H}\dot{w}_s\mathcal{U}_{w_s} \cdots \dot{w}_1\mathcal{U}_{w_1}, \tag{5.12}$$

where for $w \in W$, $w = w_s \cdots w_1$ is its J-decomposition.

Lemma 5.7.6 *Let $0 \le m \le s$ and $w \in W$ with J-decomposition $w = w_s \cdots w_1$.*

(i) *Let $m < j \le s$ and $\alpha \in \Phi_{w_j}$ or $\alpha \in \Phi_{w_{0,j}^{-1}}$. Then $\xi_\alpha(t)$ acts as the identity on $\widehat{V}_{\bar{\mu}_m}$.*

(ii) *Let $m + 1 < j \le s$ and $\alpha \in \Phi_{w_j}$ or $\alpha \in \Phi_{w_{0,j}^{-1}}$. Then $\xi_{-\alpha}(t)$ acts as the identity on $\widehat{V}_{\bar{\mu}_m}$.*

(iii) *Let $m < j \le s$ and $0 \le i \le m - 1$. Then $\dot{w}_j$ acts as multiplication by a non-zero scalar on $\widehat{V}_{\bar{\mu}_i}$.*

Proof. If $j > m$, then $w_j \in W_{J_m}$. From the characterization of Φ_v in the proof of Lemma 5.7.4, it follows that $\Phi_{w_j} \subset \Phi_m$. Therefore, by Lemma 5.7.1(iv), $\xi_\alpha(t) \cdot v = v$ for $\alpha \in \Phi_{w_j}$ and $v \in \widehat{V}_{\bar{\mu}_m}$. The same argument applies when $\alpha \in \Phi_{w_{0,j}^{-1}}$. An analogous reasoning, now using Lemma 5.7.1(v), shows (ii).

For $i \leq m-1$ we have $w_j(\bar{\mu}_i) = \bar{\mu}_i$ by Lemma 5.7.1(iii). Therefore, $\dot{w}_j$ acts as multiplication by a non-zero scalar on $\widehat{V}_{\bar{\mu}_i}$ (Lemma 5.2.14). □

Proposition 5.7.7 *Let $\alpha \in \Phi^+$ and $\mathcal{G}_\alpha$ denote the subgroup of $\mathcal{G}$ generated by $\xi_\alpha(t)$, $\xi_{-\alpha}(t)$ for $t \in k$. Write $n_\alpha = \varpi_\alpha^{-1}$. Let ν be a weight of $\widehat{V}$ such that $\nu - \alpha$ is not a weight of $\widehat{V}$. Let $v_\nu \in \widehat{V}_\nu$ be non-zero, and write $\xi_\alpha(t) \cdot v_\nu = \sum_{i \geq 0} t^i v_{\nu+i\alpha}$ (Lemma 5.3.13). Set $r = -\langle \nu, \alpha^\vee \rangle$. Then $v_{\nu+i\alpha} = 0$ for $i > r$ and $v_{\nu+r\alpha} = (-1)^r n_\alpha \cdot v_\nu$.*

Proof. Let M denote the $\mathfrak{g}$-module used to construct G. Consider the semisimple Lie algebra $\mathfrak{sl}(2,\mathbb{C})$, with basis h, e, f as in Example 2.1.4. When using its natural 2-dimensional module to construct the corresponding Chevalley group, the result is $\mathrm{SL}(2,k)$. Mapping $e \mapsto x_\alpha$, $f \mapsto x_{-\alpha}$, $h \mapsto h_\alpha = [x_\alpha, x_{-\alpha}]$ yields an injective homomorphsim $\mathfrak{sl}(2,\mathbb{C}) \to \mathfrak{g}$ (Remark 2.9.15). This way, M becomes a $\mathfrak{sl}(2,\mathbb{C})$-module and again we can construct the corresponding Chevalley group, which is naturally a subgroup of G. As $\mathrm{SL}(2,k)$ is simply connected, Corollary 5.2.32 yields a homomorphism $\mathrm{SL}(2,k) \to G$. By composing it with ϕ^{-1} we get a homomorphism $\delta : \mathrm{SL}(2,k) \to \mathcal{G}_\alpha$, mapping $\left(\begin{smallmatrix} 1 & t \\ 0 & 1 \end{smallmatrix}\right) \mapsto \xi_\alpha(t)$, $\left(\begin{smallmatrix} 1 & 0 \\ u & 1 \end{smallmatrix}\right) \mapsto \xi_{-\alpha}(u)$. Using this homomorphism yields an action of $\mathrm{SL}(2,k)$ on $\widehat{V}$.

Let $V(v_\nu)$ denote the $\mathrm{SL}(2,k)$-module generated by v_ν. Then $V(v_\nu)$ is a sum of 1-dimensional weight spaces of weights $\nu + i\alpha$, $i \geq 0$. Let r_0 be maximal such that $V(v_\nu)$ has a weight space of weight $\nu + r_0\alpha$. Using Lemma 5.2.14 we see that s_α preserves the set of weights of the form $\nu + i\alpha$. This implies that $s_\alpha(\nu) = \nu + r_0\alpha$, so that $r_0 = -\langle \nu, \alpha^\vee \rangle = r$. Furthermore, by Lemma 5.2.14, n_α maps v_ν to a non-zero weight vector of weight $\nu + r\alpha$. Hence $V(v_\nu)$ is also generated, as an $\mathrm{SL}(2,k)$-module, by $n_\alpha \cdot v_\nu$, which is a highest weight vector of weight r.

Now let N_r denote the $(r+1)$-dimensional k-vector space consisting of the homogeneous polynomials of degree r in the indeterminates X and Y. This is an $\mathrm{SL}(2,k)$-module by $\left(\begin{smallmatrix} a & b \\ c & d \end{smallmatrix}\right) \cdot f(X,Y) = f(aX + cY, bX + dY)$. Furthermore, N_r is a highest weight module over $\mathrm{SL}(2,k)$, of highest weight r and X^r is a highest weight vector. So mapping $X^r \mapsto n_\alpha \cdot v_\nu$ extends to a surjective homomorphism of $\mathrm{SL}(2,k)$-modules. We multiply this homomorphism by a scalar to obtain a surjective homomorphism $\eta : N_r \to V(v_\nu)$, with $\eta(Y^r) = v_\nu$.

Set $\hat{n} = \left(\begin{smallmatrix} 0 & -1 \\ 1 & 0 \end{smallmatrix}\right)$. Then $\delta(\hat{n}) = n_\alpha$. We have $\hat{n} \cdot Y^r = (-1)^r X^r$, and the coefficient of X^r in $\left(\begin{smallmatrix} 1 & t \\ 0 & 1 \end{smallmatrix}\right) \cdot Y^r$ is t^r. Applying η proves the proposition. □

Let $\Omega = W \cdot \lambda$, the W-orbit of λ. Then all $\bar{\mu}_i \in \Omega$. Furthermore, $\dim \widehat{V}_\mu = 1$ for $\mu \in \Omega$ (Lemma 5.2.14). For $\mu \in \Omega$ we fix a non-zero $v_\mu \in \widehat{V}_\mu$.

Consider a basis of $\widehat{V}$ consisting of weight vectors and containing v_μ for $\mu \in \Omega$. Let $v \in \widehat{V}$ and write v as a linear combination of the elements of this basis. If c is the coefficient of a v_μ ($\mu \in \Omega$) in this expression, we say c is *the coefficient of* v_μ *in* v.

We let $(\ ,\ )$ denote the unique contravariant form on $\widehat{V}$, with $(v_\lambda, v_\lambda) = 1$ (Lemma 5.5.3; it is induced by the contravariant form on V^k, as $\widehat{V}$ is the quotient of V^k by a subspace of the radical of that form). In the sequel it will be necessary to compute (v_μ, v_μ), for $\mu \in \Omega$ by the methods of Section 5.5.

We claim that there is an automorphism $\varphi : \mathcal{G} \to \mathcal{G}$, with $\varphi(\xi_\alpha(t)) = \xi_{-\alpha}(-t)$. For this we must show that φ preserves the relations satisfied by the $\xi_\alpha(t)$. For (5.1) and (5.2) this is obvious. For (5.3) we must show that $(-1)^i(-1)^j c_{i,j}^{-\alpha,-\beta} = -c_{i,j}^{\alpha,\beta}$. But that follows from the explicit formulas in Theorem 5.2.7, along with the fact that $N_{-\alpha,-\beta} = -N_{\alpha,\beta}$ (also noted in the proof of Lemma 5.5.2). Note that $\varphi(\omega_\alpha(t)) = \omega_{-\alpha}(-t)$, $\varphi(\varpi_\alpha) = \varpi_{-\alpha}^{-1}$, $\varphi(\eta_\alpha(t)) = \omega_{-\alpha}(-t)\varpi_{-\alpha} = \eta_{-\alpha}(-t)\varpi_{-\alpha}^2 = \eta_{-\alpha}(t)$ (as $\varpi_{-\alpha}^2 = \eta_{-\alpha}(-1)$). This implies that φ preserves (5.4) and (5.6) and the relations following from Proposition 5.2.30. Finally, from Lemma 5.2.16 it follows that φ preserves (5.5).

We let $\tau : \mathcal{G} \to \mathcal{G}$ be the anti-automorphism defined by $\tau(g) = \varphi(g)^{-1}$. This means that $\tau(\xi_\alpha(t)) = \xi_{-\alpha}(t)$.

Now we consider the elements $x_\alpha(t) \in G$. Let M be as in the proof of Proposition 5.7.7. As in Section 5.5, this yields a representation ρ_k of the algebra $\mathcal{U}_k$. Then $x_\alpha(t) = \sum_{j\geq 0} t^j \rho_k(x_\alpha^{(j)})$, which is a finite sum. Let $\sigma : \mathcal{U}_k \to \mathcal{U}_k$ be the anti-automorphism of Section 5.5. Then $\sum_{j\geq 0} t^j \rho_k(\sigma(x_\alpha^{(j)})) = \sum_{j\geq 0} t^j \rho_k(x_{-\alpha}^{(j)}) = x_{-\alpha}(t) = \phi(\tau(\xi_\alpha(t)))$. Therefore, by (5.10) we have

$$(\theta(g)v_1, v_2) = (v_1, \theta(\tau(g))v_2) \text{ for all } v_1, v_2 \in \widehat{V}, g \in \mathcal{G}. \tag{5.13}$$

Let $\mu, \nu \in \Omega$. The coefficient of v_μ in $\theta(\tau(g))v_\nu$ is equal to $(v_\mu, \theta(\tau(g))v_\nu)/(v_\mu, v_\mu)$, which by (5.13) is equal to $(\theta(g)v_\mu, v_\nu)/(v_\mu, v_\mu)$. In turn, that is equal to the coefficient of v_ν in $\theta(g)v_\mu$, times $\frac{(v_\nu,v_\nu)}{(v_\mu,v_\mu)}$. We conclude that without computing $\tau(g)$ explicitly, we can compute the coefficient of v_μ in $\theta(\tau(g))v_\nu$.

Now we return to the situation at the beginning of this section: we have $A \in \theta(\mathcal{G})$, and want $g \in \mathcal{G}$ such that $\theta(g) = A$. According to (5.12) we can write

$$g = u_1 \cdots u_s h \dot{w}_s u_s' \cdots \dot{w}_1 u_1',$$

where $u_i \in \mathcal{U}_{w_{0,i}^{-1}}$, $u_i' \in \mathcal{U}_{w_i}$. Let $m \geq 1$ be such that we know the elements u_i, u_i', $\dot{w}_i$, for $i < m$. (So, initially, we have $m = 1$.) Multiplying by the inverses of $\theta(u_i)$, $\theta(u_i')$, $\theta(\dot{w}_i)$, in the appropriate manner, we find A' such that $A' = \theta(g')$, with $g' = u_m \cdots u_s h \dot{w}_s u_s' \cdots \dot{w}_m u_m'$. We now want to find u_m, u_m', $\dot{w}_m$. We have $(A')^{-1} = \theta((g')^{-1})$, and $(g')^{-1} =$

$(u'_m)^{-1}\dot{w}_m^{-1}\cdots(u'_s)^{-1}\dot{w}_s^{-1}h^{-1}u_s^{-1}\cdots u_m^{-1}$. By Lemma 5.7.6 we now see that

$$(A')^{-1}v_{\bar{\mu}_{m-1}} = a\theta((u'_m)^{-1})v_{w_m^{-1}\bar{\mu}_{m-1}}, \text{ with } a \in k^*.$$

Write $\nu = w_m^{-1}\bar{\mu}_{m-1}$. By Lemma 5.3.13 we infer that $\theta((u'_m)^{-1})v_\nu$ is equal to v_ν plus a linear combination of weight vectors of weights $\succ \nu$. Therefore, as a first step we compute $v = (A')^{-1}v_{\bar{\mu}_{m-1}}$ and write it as a linear combination of weight vectors. Let ν be the lowest weight (with respect to $\prec$) occurring in this expression. Then $\nu = w_m^{-1}\bar{\mu}_{m-1}$. But by Lemma 5.7.3, w_m^{-1} is exactly the shortest element of $W_{J_{m-1}}$ with this property. Since we have computed ν, we also find w_m^{-1}, and therefore w_m.

Next, we compute Φ_{w_m}, and order its elements $\gamma_1, \dots, \gamma_q$ according to height (from small to large). Write $(u'_m)^{-1} = \xi_{\gamma_1}(t_1)\cdots\xi_{\gamma_q}(t_q)$. Let $p \geq 1$ be such that we know t_i for $i < p$ (so initially, $p = 1$). We show how to obtain t_p. Set $v' = \theta(\xi_{\gamma_{p-1}}(-t_{p-1})\cdots\xi_{\gamma_1}(-t_1))v$, so that $v' = a\theta(\xi_{\gamma_p}(t_p)\cdots\xi_{\gamma_q}(t_q))v_\nu$. Now $\theta(\xi_{\gamma_{p+1}}(t_{p+1})\cdots\xi_{\gamma_q}(t_q))v_\nu = v_\nu + v''$, where v'' is a linear combination of weight vectors of weights that are equal to ν plus a non-trivial linear combination of the γ_i for $p+1 \leq i \leq q$. It follows that $v' = a\theta(\xi_{\gamma_p}(t_p))v_\nu + v'''$, and the weights of the weight vectors occurring in the first summand are of the form $\nu + i\gamma_p$, $i \geq 0$, while no such weight vectors occur in v'''. Hence a is the coefficient of v_ν in v'. Moreover, by Proposition 5.7.7, the coefficient of $n_{\gamma_p} \cdot v_\nu$ in v' is $a(-1)^r t_p^r$, where $r = -\langle \nu, \gamma_p^\vee \rangle$. So we can compute $\hat{t} = t_p^r$. Then t_p is the r-th root of $\hat{t}$ with the property that the coefficient of $n_{\gamma_p} \cdot v_\nu$ in $\theta(\xi_{\gamma_p}(-t_p))v'$ is 0.

Now set $g'' = g'(u'_m)^{-1}$. As τ is an anti-automorphism, $\tau((g'')^{-1})$ equals

$$\tau(u_m^{-1})\cdots\tau(u_s^{-1})\tau(h^{-1})\tau(\dot{w}_s^{-1})\tau((u'_s)^{-1})\cdots\tau(\dot{w}_{m+1}^{-1})\tau((u'_{m+1})^{-1})\tau(\dot{w}_m^{-1}).$$

We have $\varphi(n_i) = \varpi_{-\alpha_i}$. Hence, if $w = s_{i_1}\cdots s_{s_r}$ is reduced, $\tau(\dot{w}^{-1}) = \varphi(\dot{w}) = \varpi_{-\alpha_{i_1}}\cdots\varpi_{-\alpha_{i_r}} \in \mathcal{H}\dot{w}$. Therefore $\tau(\dot{w}_m^{-1})v_\nu$ is a non-zero scalar multiple of $v_{\bar{\mu}_{m-1}}$. By Lemma 5.7.6 (ii) and (iii), it follows that $\tau((g'')^{-1})v_\nu$ is a non-zero scalar multiple of $\tau(u_m^{-1})v_{\bar{\mu}_{m-1}}$.

Let $\gamma_1, \dots, \gamma_q$ be the elements of $\Phi_{w_{0,m}^{-1}}$, sorted according to height. Write $u_m^{-1} = \xi_{\gamma_q}(t_q)\cdots\xi_{\gamma_1}(t_1)$. Then $\tau(u_m^{-1}) = \xi_{-\gamma_1}(t_1)\cdots\xi_{-\gamma_q}(t_q)$. Let $p \geq 1$ be such that we know t_i for $i < p$ (so initially, $p = 1$). We show how to obtain t_p. Set $g''' = \xi_{\gamma_{p-1}}(t_{p-1})\cdots\xi_{\gamma_1}(t_1)g''$ and $\hat{u}_m = \xi_{\gamma_p}(-t_p)\cdots\xi_{\gamma_q}(-t_q)$ so that, as before, $\theta(\tau((g''')^{-1}))v_\nu$ is a non-zero scalar multiple of $\theta(\tau(\hat{u}_m^{-1}))v_{\bar{\mu}_{m-1}}$. Moreover, $\theta(\tau(\hat{u}_m^{-1}))v_{\bar{\mu}_{m-1}}$ is equal to $\theta(\xi_{-\gamma_p}(t_p))v_{\bar{\mu}_{m-1}} + z$, where the first summand is a linear combination of weight vectors of weights $\bar{\mu}_{m-1} - i\gamma_p$, $i \geq 0$, whereas no such weight vectors occur in z. By a result similar to Proposition 5.7.7 (whose statement and proof we leave to the reader; note that $\gamma_p \in \Phi_{m-1}$ so that $\bar{\mu}_{m-1} + \gamma_p$ is not a weight of $\widehat{V}$ by Lemma 5.7.1(iv)), it follows that the coefficient of $n_{\gamma_p} \cdot v_{\bar{\mu}_{m-1}}$ in $\theta(\xi_{-\gamma_p}(t_p))v_{\bar{\mu}_{m-1}}$ is t_p^r, where $r = \langle \bar{\mu}_{m-1}, \gamma_p^\vee \rangle$. Summarizing, let $a \in k^*$ be such that $\theta(\tau((g''')^{-1}))v_\nu = a\theta(\tau(\hat{u}_m^{-1}))v_{\bar{\mu}_{m-1}}$. Then a and at_p^r are respectively the coefficients of $v_{\bar{\mu}_{m-1}}$

and $n_{\gamma_p} \cdot v_{\bar{\mu}_{m-1}}$ in $\theta(\tau((g''')^{-1}))v_\nu$. We have seen that these coefficients can be computed using the contravariant form. In particular, a is equal to the coefficient of v_ν in $\theta((g''')^{-1})v_{\bar{\mu}_{m-1}}$ times $\frac{(v_\nu, v_\nu)}{(v_{\bar{\mu}_{m-1}}, v_{\bar{\mu}_{m-1}})}$. Similarly, at_p^r is equal to the coefficient of v_ν in $\theta((g''')^{-1})(n_{\gamma_p} \cdot v_{\bar{\mu}_{m-1}})$ times $\frac{(v_\nu, v_\nu)}{(n_{\gamma_p} \cdot v_{\bar{\mu}_{m-1}}, n_{\gamma_p} \cdot v_{\bar{\mu}_{m-1}})}$. Furthermore, setting $A'' = A'\theta(u'_m)^{-1}$, $A''' = \theta(\xi_{\gamma_{p-1}}(t_{p-1}) \cdots \xi_{\gamma_1}(t_1))A''$, we have $\theta((g''')^{-1}) = (A''')^{-1}$. It follows that we can compute $\hat{t} = t_p^r$, and let t_p be the r-th root of $\hat{t}$ such that with $A'''' = \theta(\xi_{\gamma_p}(t_p))A'''$ the coefficient of v_ν in $(A'''')^{-1}(n_{\gamma_p} \cdot v_{\bar{\mu}_{m-1}})$ is 0.

After performing the above steps for $m = 1, \dots, s$, we know the u_i, u'_i and w_i. After multiplying A in the correct way by the inverses of these elements we obtain a matrix D with the property that $D = \theta(h)$ for some $h \in \mathcal{H}$. As we use a basis of $\widehat{V}$ consisting of weight vectors, D is a diagonal matrix acting on each weight space as multiplication by a scalar. We now recall from Section 5.4 that $\mu_1, \dots, \mu_\ell$, $\mu_1^\vee, \dots, \mu_\ell^\vee$ denote the standard bases of X and $X^\vee$, respectively. Using the definition of $\tilde{\eta}_i$ of the previous section, we see that $\tilde{\eta}_i(t)$ acts as multiplication by $t^{\langle \mu, \mu_i^\vee \rangle}$ on $\widehat{V}_\mu$. Let $\nu_1, \dots, \nu_p$ be the weights of $\widehat{V}$ and $d_1, \dots, d_p \in k$ be such that D acts as multiplication by d_i on $\widehat{V}_{\nu_i}$. Then, writing $h = \tilde{\eta}_1(t_1) \cdots \tilde{\eta}_\ell(t_\ell)$, $D = \theta(h)$ is equivalent to

$$\prod_{j=1}^{\ell} t_j^{\langle \nu_i, \mu_j^\vee \rangle} = d_i \text{ for } 1 \le i \le p. \tag{5.14}$$

Let $M = (\langle \nu_i, \mu_j^\vee \rangle)_{1 \le i \le p, 1 \le j \le \ell}$. We compute an upper triangular form of M using elementary row operations (adding two rows, multiplying a row by -1; equivalently, we can multiply M by a scalar to ensure that it has integer entries, use the Hermite normal form algorithm (see [Sim94]), then divide by the scalar again). Analogous operations can be applied to (5.14) (multiplying two rows, taking the inverse of a row). So we may assume that the matrix M is upper triangular. But then it is straightforward to find a solution (we remark that it is not necessarily unique, as θ may have a kernel contained in $\mathcal{H}$). It follows that we can compute h, and therefore we have a word in $\mathcal{G}$ mapping to A under θ. Finally, we rewrite this word to normal form using the techniques of the previous section.

Example 5.7.8 Let the root datum be simply connected of type A_2. We let $k = \mathbb{C}$, and $\widehat{V}$ be the highest weight module with highest weight $2\lambda_1$. Then $\dim \widehat{V} = 6$, and the corresponding $\mathfrak{g}$-module is given in Example 5.2.5. Using the data in that example, it is straightforward to compute the matrix representing a given element of $\mathfrak{g}$ or of $\mathcal{G}$. Below we use the results of such computations, without going into detail.

The sets $J_0 = \{1, 2\}$, $J_1 = \{2\}$, $J_2 = \emptyset$, and weights $\bar{\mu}_0 = 2\lambda_1$, $\bar{\mu}_1 = -2\lambda_1 + 2\lambda_2$ are computed in Example 5.7.2. We start with a matrix A whose inverse is

$$A^{-1} = \begin{pmatrix} 9 & 12 & -48 & 4 & -32 & 64 \\ 6 & 7 & -28 & 2 & -16 & 32 \\ -3 & -2 & 11 & 0 & 2 & -8 \\ 4 & 4 & -16 & 1 & -8 & 16 \\ -2 & -1 & 6 & 0 & 1 & -4 \\ 1 & 0 & -2 & 0 & 0 & 1 \end{pmatrix}.$$

This matrix is taken with respect to the basis B given in Example 5.2.5. We denote its elements by $v_1, \dots, v_6$, which are weight vectors of weights $2\lambda_1$, λ_2, $\lambda_1 - \lambda_2$, $-2\lambda_1 + 2\lambda_2$, $-\lambda_1$, $-2\lambda_2$. For the contravariant form we have $(v_1, v_1) = (v_4, v_4) = (v_6, v_6) = 1$.

The J-decomposition of w_0 is $w_0 = w_{0,2}w_{0,1}$, where $w_{0,2} = s_{\alpha_2}$, $w_{0,1} = s_{\alpha_1}s_{\alpha_2}$. Furthermore, $\Phi_{w_{0,1}^{-1}} = \{\alpha_1, \alpha_3 = \alpha_1 + \alpha_2\}$, $\Phi_{w_{0,2}^{-1}} = \{\alpha_2\}$. We are looking for $g \in \mathcal{G}$, $g = u_1 u_2 h \dot{w}_2 u_2' \dot{w}_1 u_1'$, where $u_i \in \mathcal{U}_{w_{0,i}^{-1}}$, $u_i' \in \mathcal{U}_{w_i}$, $h \in \mathcal{H}$ such that $\theta(g^{-1}) = A^{-1}$.

In the first step we compute $v = A^{-1}v_{\bar{\mu}_0} = A^{-1}v_1 = 9v_1 + \cdots + v_6$. The lowest weight vector occurring in v is v_6, with weight $\nu = -2\lambda_2$. The shortest element of W mapping $\bar{\mu}_0 = 2\lambda_1$ to ν is $s_{\alpha_2}s_{\alpha_1}$. So that is w_1^{-1} and it follows that $w_1 = s_{\alpha_1}s_{\alpha_2}$.

Note that $\Phi_{w_1} = \{\alpha_2, \alpha_3\}$. Hence $(u_1')^{-1} = \xi_{\alpha_2}(t_2)\xi_{\alpha_3}(t_3)$. First we determine t_2. We have $\theta(n_{\alpha_2})v_\nu = v_4$ and $r = -\langle \nu, \alpha_2^\vee \rangle = 2$. Also a (notation as above) is the coefficient of v_ν in v so $a = 1$, and at_2^2 is the coefficient of $\theta(n_{\alpha_2})v_\nu$ in v so $at_2^2 = 4$. As $\theta(\xi_{\alpha_2}(-2))v = 9v_1 - 3v_3 + v_6$ (the coefficient of v_4 in this is 0), we conclude $t_2 = 2$. In order to determine t_3 we set $v' = 9v_1 - 3v_3 + v_6$. We have $\theta(n_{\alpha_3})v_\nu = v_1$. We find $a = 1$ and $at_3^2 = 9$. Because $\theta(\xi_{\alpha_3}(3))v' = v_6$ it follows that $t_3 = -3$.

We set $A_1 = A\theta(\xi_{\alpha_3}(-3)\xi_{\alpha_2}(2))$, so that

$$A_1^{-1} = \begin{pmatrix} 0 & 0 & 0 & 4 & -20 & 25 \\ 0 & 0 & 0 & 2 & -9 & 10 \\ 0 & -2 & 5 & 0 & 2 & -5 \\ 0 & 0 & 0 & 1 & -4 & 4 \\ 0 & -1 & 2 & 0 & 1 & -2 \\ 1 & 0 & -2 & 0 & 0 & 1 \end{pmatrix}.$$

We write $u_1 = \xi_{\alpha_1}(-t_1)\xi_{\alpha_3}(-t_3)$ (we use the convention that after a value of a t_i has been determined, the symbol t_i is freed, so that t_3 is different from the earlier one). We determine t_1. We have $\theta(n_{\alpha_1})v_{\bar{\mu}_0} = v_4$, and $r = \langle \bar{\mu}_0, \alpha_1^\vee \rangle = 2$. The value of a is the coefficient of $v_\nu = v_6$ in $A_1^{-1}v_1$, which is 1. The value of at_1^2 is the coefficient of v_6 in $A_1^{-1}v_4$ (times a quotient of values of the contravariant form), which is 0. It follows that $t_1 = 0$. For t_3 again we have $r = 2$ and $a = 1$. In this case, $\theta(n_{\alpha_3})v_1 = v_6$. The coefficient of $v_\nu = v_6$ in $A_1^{-1}v_6$ is 1, and the values of the contravariant form are all 1. Hence $t_3^2 = 1$. The coefficient of v_6 in $A_1^{-1}\theta(\xi_{\alpha_3}(-1))^{-1}v_6$ is 0 and $t_3 = -1$. We conclude that $u_1 = \xi_{\alpha_3}(1)$.

Set $A_2 = \theta(\xi_{\alpha_3}(-1))A_1\theta(n_{\alpha_1}n_{\alpha_2})^{-1}$, then

$$A_2^{-1} = \begin{pmatrix} 1 & 0 & 0 & 0 & 0 & 0 \\ 0 & 2 & -5 & 0 & 0 & 0 \\ 0 & -1 & 2 & 0 & 0 & 0 \\ 0 & 0 & 0 & 4 & -20 & 25 \\ 0 & 0 & 0 & -2 & 9 & -10 \\ 0 & 0 & 0 & 1 & -4 & 4 \end{pmatrix}.$$

Now $v_{\bar{\mu}_1} = v_4$ and $A_2^{-1}v_4 = 4v_4 - 2v_5 + v_6$. Denote this last vector by v. The lowest weight occurring in v is $\nu = -2\lambda_1$. The shortest element mapping $\bar{\mu}_1$ to ν is s_{α_2}, so that $w_2 = s_{\alpha_2}$. Furthermore, $\Phi_{w_2} = \{\alpha_2\}$, whence $(u_2')^{-1} = \xi_{\alpha_2}(t_2)$. Now $r = -\langle \bar{\mu}_1, \alpha_2^\vee \rangle = 2$. The coefficient of v_ν in v is $a = 1$. Also, $\theta(n_{\alpha_2}) = v_4$, and therefore, $at_2^2 = 4$. Since $\theta(\xi_{\alpha_2}(-2))v = v_6$ it follows that $t_2 = 2$.

Set $A_3 = A_2\theta(\xi_{\alpha_2}(2))$, then

$$A_3^{-1} = \begin{pmatrix} 1 & 0 & 0 & 0 & 0 & 0 \\ 0 & 0 & -1 & 0 & 0 & 0 \\ 0 & -1 & 2 & 0 & 0 & 0 \\ 0 & 0 & 0 & 0 & 0 & 1 \\ 0 & 0 & 0 & 0 & 1 & -2 \\ 0 & 0 & 0 & 1 & -4 & 4 \end{pmatrix}.$$

Write $u_2 = \xi_{\alpha_2}(-t_2)$. Again $r = \langle \bar{\mu}_1, \alpha_2^\vee \rangle = 2$, and a equaling the coefficient of v_6 in $A_3^{-1}v_4$ is 1. Furthermore, $\theta(n_{\alpha_2})v_4 = v_6$. The coefficient of v_6 in $A_3^{-1}v_6$ is 4 which equals t_2^2. The coefficient of v_6 in $A_3^{-1}\theta(\xi_{\alpha_2}(-2))v_6$ is 0 and it follows that $t_2 = 2$ and $u_2 = \xi_{\alpha_2}(-2)$.

Set $A_4 = \theta(x_{\alpha_2}(2))A_3\theta(n_{\alpha_2})^{-1}$, then $A_4 = \text{diag}(1, -1, 1, 1, -1, 1)$. Since the root datum is of simply connected type, we have $\lambda_1 = \mu_1$, $\lambda_2 = \mu_2$, so that $\tilde{\eta}_i(t) = \eta_{\alpha_i}(t)$, $i = 1, 2$. Write $h = \tilde{\eta}_1(t_1)\tilde{\eta}_2(t_2)$. Because A_4 acts as the identity on $\widehat{V}_{2\lambda_1}$ and $\widehat{V}_{\lambda_1-\lambda_2}$, this yields the equations $t_1^2 = 1$, $t_1t_2^{-1} = 1$. Furthermore, A_4 acts as multiplication by -1 on $\widehat{V}_{\lambda_2}$, so that $t_2 = -1$. Hence $t_1 = t_2 = -1$. Therefore, with

$$\begin{aligned} g &= \xi_{\alpha_2}(-2)\xi_{\alpha_3}(1)\tilde{\eta}_1(-1)\tilde{\eta}_2(-1)n_2\xi_{\alpha_2}(-2)n_1n_2\xi_{\alpha_2}(-2)\xi_{\alpha_3}(3) \\ &= \xi_{\alpha_2}(-2)\xi_{\alpha_3}(1)\tilde{\eta}_1(-1)\tilde{\eta}_2(-1)n_1n_2n_1\xi_{\alpha_1}(-2)\xi_{\alpha_2}(-2)\xi_{\alpha_3}(3), \end{aligned}$$

we have $\theta(g) = A$.

5.8 Reductive algebraic groups

In this section we collect material on reductive algebraic groups. We limit our discussion to base fields of characteristic 0 because the proofs that we indicate use Lie algebras, and therefore only work in that characteristic. However, with the exception of Theorem 5.8.1 the results of this section are also true in characteristic $p > 0$. Throughout this section K denotes an algebraically closed field of characteristic 0.

Theorem 5.8.1 *Let $G \subset \mathrm{GL}(n, K)$ be an algebraic group with Lie algebra $\mathfrak{g}$. The following are equivalent*

(i) *G is reductive,*

(ii) *$\mathfrak{g} = \mathfrak{s} \oplus \mathfrak{d}$ (direct sum of ideals), where $\mathfrak{s}$ is semisimple and $\mathfrak{d}$ is abelian consisting of semisimple elements (of $\mathfrak{gl}(n, K)$),*

(iii) *every finite-dimensional rational G-module is completely reducible.*

Proof. Theorem 4.3.22, Theorem 4.2.2 and Corollary 4.3.11 show that (i) and (ii) are equivalent.

We now show that (ii) and (iii) are equivalent as well. Let $\rho : G \to \mathrm{GL}(V)$ be a finite-dimensional rational representation. It can be shown ([Bor91], Section 4.4) that $\mathrm{d}\rho$ respects the Jordan decomposition, i.e., if $x \in \mathfrak{g}$ has Jordan decomposition $x = s + n$, then $\mathrm{d}\rho(x) = \mathrm{d}\rho(s) + \mathrm{d}\rho(n)$ is the Jordan decomposition of $\mathrm{d}\rho(x)$. Suppose (ii) holds. By Theorem 2.12.3, V is a completely reducible $\mathfrak{g}$-module. By Corollary 4.2.10, we see that V is a completely reducible G°-module. Now we use the following fact of group theory: let $\mathcal{N}$ be a normal subgroup of the group $\mathcal{G}$, of finite index, and let $\mathcal{W}$ be a finite-dimensional $\mathcal{G}$-module, then $\mathcal{W}$ is completely reducible as a $\mathcal{G}$-module if and only if it is completely reducible as an $\mathcal{N}$-module ([Mos56], Lemma 3.1). From that we conclude that V is a completely reducible G-module. Conversely, if (iii) holds, we let V be the natural n-dimensional G-module. Again Corollary 4.2.10 shows that V is a completely reducible $\mathfrak{g}$-module. Theorem 2.12.3 finishes the proof. □

Lemma 5.8.2 *Let $G \subset \mathrm{GL}(n, K)$ be a reductive algebraic group, and $\mathfrak{g} = \mathrm{Lie}(G)$. Let $\mathfrak{h}$ be a Cartan subalgebra of $\mathfrak{g}$.*

(i) *$\mathfrak{h}$ is abelian and consists of semisimple elements.*

(ii) *$\mathfrak{h}$ is algebraic, and the corresponding connected algebraic subgroup H of G is a maximal torus.*

(iii) *Let T be a maximal torus of G and $\mathfrak{t} = \mathrm{Lie}(T)$. Then $\mathfrak{t}$ is a Cartan subalgebra of $\mathfrak{g}$.*

(iv) *Two maximal tori of G are conjugate under G.*

Proof. Write $\mathfrak{g} = \mathfrak{s} \oplus \mathfrak{d}$, as in Theorem 5.8.1(ii). Then $\mathfrak{h}$ is the direct sum of a Cartan subalgebra of $\mathfrak{s}$ and $\mathfrak{d}$. Therefore, Proposition 2.11.5 implies that $\mathfrak{h}$ is abelian and consists of semisimple elements.

Let $\tilde{\mathfrak{h}}$ be the algebraic hull of $\mathfrak{h}$. The elements of $\mathfrak{h}$ are simultaneously diagonalizable. So using Lemma 4.3.3, Corollary 4.3.7 and Proposition 4.3.8 we see that $\tilde{\mathfrak{h}}$ is abelian and consists of semisimple elements. Therefore, $\tilde{\mathfrak{h}}$ normalizes $\mathfrak{h}$, whence $\tilde{\mathfrak{h}} = \mathfrak{h}$.

Now H is a torus (again use Lemma 4.3.3 and Proposition 4.3.8). If H is contained in a bigger torus H' then $\mathfrak{h}' = \mathrm{Lie}(H')$ normalizes $\mathfrak{h}$, whence $\mathfrak{h} = \mathfrak{h}'$ and $H = H'$ by Theorem 4.2.2(i).

For (iii), $\mathfrak{t}$ is contained in a Cartan subalgebra $\mathfrak{t}'$ of $\mathfrak{g}$. By the previous point, $\mathfrak{t}' = \mathrm{Lie}(T')$ and T' is a maximal torus of G. By Theorem 4.2.2(i), $T = T'$ so that $\mathfrak{t} = \mathfrak{t}'$.

Let T' be a second maximal torus of G, and $\mathfrak{t}' = \mathrm{Lie}(T')$. Then by (iii), and Proposition 4.3.2, there is a $g \in G$ such that $\mathrm{Ad}(g)(\mathfrak{t}) = \mathfrak{t}'$. Now gTg^{-1} is a connected algebraic subgroup of G with Lie algebra $\mathrm{Ad}(g)(\mathfrak{t}) = \mathfrak{t}'$ (Theorem 4.2.1). So $gTg^{-1} = T'$ (Theorem 4.2.2(i)). □

Let G be an algebraic group. A subgroup $B \subset G$ is said to be a *Borel subgroup* if it is algebraic, connnected, solvable and maximal with these properties. Such subgroups always exist because a connected solvable subgroup of maximal dimension is clearly a Borel subgroup.

Let $\mathfrak{g} = \mathrm{Lie}(G)$. A subalgebra $\mathfrak{b} \subset \mathfrak{g}$ is called a *Borel subalgebra* if it is solvable and maximal with that property. Note that a Borel subalgebra $\mathfrak{b} \subset \mathfrak{g}$ is algebraic. Indeed, its algebraic hull, $\mathfrak{b}'$, is solvable (Lemma 4.3.17), and hence $\mathfrak{b} = \mathfrak{b}'$. Furthermore, if $B \subset G$ is a Borel subgroup, then $\mathfrak{b} = \mathrm{Lie}(B)$ is a Borel subalgebra. (If not, then it is contained in a Borel subalgebra $\mathfrak{b}'$, which is algebraic, and we let B' be the corresponding connected subgroup of G. From Theorem 4.3.22 it follows that B' is solvable, and by Theorem 4.2.2(i) we have $B \subset B'$.)

Let $\mathfrak{b}_1, \mathfrak{b}_2$ be Borel subalgebras of $\mathfrak{g}$. Then in [Hum78], Theorem 16.4, it is shown that $\mathfrak{b}_1, \mathfrak{b}_2$ are conjugate under the group generated by all $\exp(\mathrm{ad} x)$ for $x \in \mathfrak{g}$ such that $\mathrm{ad} x$ is nilpotent. In the same way as in the proof of Proposition 4.3.2 it follows that there is a $g \in G$ such that $\mathrm{Ad}(g)(\mathfrak{b}_1) = \mathfrak{b}_2$. Since a Cartan subalgebra of $\mathfrak{g}$ lies in a Borel subalgebra, this implies that every Borel subalgebra contains a Cartan subalgebra of $\mathfrak{g}$. In the same way as in Lemma 5.8.2(iv) we see that two Borel subgroups of G are conjugate under G. Since a maximal torus of G is contained in a Borel subgroup, this implies that every Borel subgroup of G contains a maximal torus of G.

Now let G be a connected reductive algebraic group and $T \subset G$ a maximal torus. Write $\mathfrak{g} = \mathrm{Lie}(G)$, $\mathfrak{t} = \mathrm{Lie}(T)$, and $\mathfrak{g} = \mathfrak{s} \oplus \mathfrak{d}$ as in Theorem 5.8.1. Since $\mathfrak{t}$ is a Cartan subalgebra of $\mathfrak{g}$, $\mathfrak{t} \cap \mathfrak{s}$ is a Cartan subalgebra of $\mathfrak{s}$. Let Φ denote the corresponding root system and W the Weyl group of Φ. Then

$$\mathfrak{g} = \mathfrak{t} \oplus \bigoplus_{\alpha \in \Phi} \mathfrak{g}_\alpha$$

is the root space decomposition of $\mathfrak{g}$ where each $\mathfrak{g}_\alpha = \mathfrak{s}_\alpha$ is contained in $\mathfrak{s}$. Since W acts on $\mathfrak{t} \cap \mathfrak{s}$ we also get a W-action on $\mathfrak{t}$ by letting each $w \in W$ act as the identity on $\mathfrak{d}$. Note that this is compatible with the formula $s_\alpha(h) = h - \alpha(h)h_\alpha$, given in Remark 2.9.9, as $\alpha(d) = 0$ for all $d \in \mathfrak{d}$ (since $\mathfrak{d}$ is central).

The normalizer $N_G(T)$ of T in G, acts naturally on $\mathfrak{t}$. Indeed, let $g \in N_G(T)$. Then $\mathrm{Int}(g)(T) = T$ so that $\mathrm{Ad}(g)(\mathfrak{t}) = \mathfrak{t}$.

Theorem 5.8.3 *Let $G \subset \mathrm{GL}(n, K)$ be a connected reductive algebraic group. Let B be a Borel subgroup of G, containing a maximal torus T of G. Use the notation above. Then*

(i) *There is a surjective homomorphism $\eta : N_G(T) \to W$ with kernel T such that $\mathrm{Ad}(g)(x) = \eta(g) \cdot x$, for all $g \in N_G(T)$, $x \in \mathfrak{t}$.*

(ii) *For $w \in W$ fix a $\dot{w} \in N_G(T)$ such that $\eta(\dot{w}) = w$. Then*

$$G = \bigcup_{w \in W} B\dot{w}B.$$

Proof. First assume that G is semisimple. From $\mathfrak{g}$, $V = K^n$, K we can construct a Chevalley group $\widehat{G}$ as in Section 5.2. Since $\mathrm{Lie}(\widehat{G}) = \mathfrak{g}$ (Theorem 5.3.9(ii)), it follows that $G = \widehat{G}$ (Theorem 4.2.2(i)). Let N, H, U be as in Section 5.2.4. Then $\mathrm{Lie}(H) = \mathfrak{t}$ by Proposition 5.3.6, whence $H = T$ (again Theorem 4.2.2(i)). By Proposition 5.2.27, $N = N_G(T)$, so that the existence of η follows from Lemma 5.2.22. Lemma 5.2.13 now yields the second statement of (i). Let $B_0 = UH$. Then B_0 is a Borel subgroup of G (it is straightforward to verify that its Lie algebra is a Borel subalgebra of $\mathfrak{g}$, using Propositions 5.3.4 and 5.3.6). So there is a $g \in G$ with $gB_0g^{-1} = B$. The Lie algebras of maximal tori of G lying in B are Cartan subalgebras of $\mathfrak{b} = \mathrm{Lie}(B)$. So by Proposition 4.3.2 we may assume that $gTg^{-1} = T$, implying that $gNg^{-1} = N$. Set $\dot{w}' = g^{-1}\dot{w}g$. Then the $\dot{w}'$ also form a set of coset representatives of T in N. Hence by Theorem 5.2.23(i), G is the union of the sets $B_0\dot{w}'B_0$. But $gB_0\dot{w}'B_0g^{-1} = B\dot{w}B$, so we get the same for the sets $B\dot{w}B$.

For the general case write $\mathfrak{g} = \mathfrak{s} \oplus \mathfrak{d}$, as above. Let D and $\widetilde{G}$ be the connected algebraic subgroups of G with Lie algebras $\mathfrak{d}$, $\mathfrak{s}$ respectively. Then D is a central torus and by arguments similar to those used in the proof of Theorem 4.3.22, $G = \widetilde{G}D$. Set $\widetilde{T} = T \cap \widetilde{G}$. Then $\widetilde{T}$ is a maximal torus of $\widetilde{G}$ and $T = \widetilde{T}D$. A short verification shows that $N_G(T) = N_{\widetilde{G}}(\widetilde{T})D$. By the first part of the proof we have a homomorphism $\tilde{\eta} : N_{\widetilde{G}}(\widetilde{T}) \to W$ with kernel $\widetilde{T}$. Now for $g \in N_G(T)$, $g = \tilde{g}d$, where $\tilde{g} \in \widetilde{G}$ and $d \in D$, we set $\eta(g) = \tilde{\eta}(\tilde{g})$. This is well defined, for if $g \in N_G(T)$ can be written $g = \tilde{g}_1d_1 = \tilde{g}_2d_2$, then $\tilde{g}_1 = \tilde{g}_2d$ with $d \in D$. But then $\mathrm{Ad}(\tilde{g}_1) = \mathrm{Ad}(\tilde{g}_2)$ as $\mathrm{Ad}(d)$ is the identity, whence $\tilde{\eta}(\tilde{g}_1) = \tilde{\eta}(\tilde{g}_2)$. It now immediately follows that η is a homomorphism with the required properties.

Furthermore, $\widetilde{B} = B \cap \widetilde{G}$ is a Borel subgroup of $\widetilde{G}$. By the first part of the proof, $\widetilde{G}$ is the union of the sets $\widetilde{B}\dot{w}\widetilde{B}$ for $w \in W$. Hence G is the union of the sets $\widetilde{B}\dot{w}\widetilde{B}D = \widetilde{B}D\dot{w}\widetilde{B}D = B\dot{w}B$. □

Corollary 5.8.4 *Let $G \subset \mathrm{GL}(n, K)$ be a connected reductive algebraic group. Let B_1, B_2 be Borel subgroups of G. Then $B_1 \cap B_2$ contains a maximal torus of G.*

Proof. As seen above, there is a $g \in G$ with $gB_1g^{-1} = B_2$. By the previous theorem we can write $g = b_1\dot{w}b_2$ where $b_i \in B_1$ and $\dot{w} \in N_G(T)$, where T is a maximal torus of G contained in B_1. Then $b_2B_1b_2^{-1} = B_1$ and $\dot{w}B_1\dot{w}^{-1}$ contains T. Therefore, $b_1Tb_1^{-1}$ is a maximal torus of G contained in both B_1 and B_2. □

Corollary 5.8.5 *Let $G \subset \mathrm{GL}(n, K)$ be a connected reductive algebraic group. Let $\mathfrak{t}$ be a Cartan subalgebra of $\mathfrak{g} = \mathrm{Lie}(G)$, and W the corresponding Weyl group. Two elements of $\mathfrak{t}$ are conjugate under G if and only if they are conjugate under W.*

Proof. By Lemma 5.8.2, there is a maximal torus T of G with $\mathfrak{t} = \mathrm{Lie}(T)$. The "if" part is immediate from Theorem 5.8.3. Let $h_1, h_2 \in \mathfrak{t}$. Let $Z = \{g \in G \mid \mathrm{Ad}(g)(h_2) = h_2\}$; this is an algebraic subgroup of G with Lie algebra $\mathfrak{z} = \{x \in \mathfrak{g} \mid [x, h_2] = 0\}$ (Corollary 4.2.11). Suppose that there is a $g \in G$ with $\mathrm{Ad}(g)(h_1) = h_2$. Then $\mathfrak{t}$ and $\mathrm{Ad}(g)(\mathfrak{t})$ are two Cartan subalgebras of $\mathfrak{g}$ that contain h_2. Since Cartan subalgebras of $\mathfrak{g}$ are abelian (Lemma 5.8.2), they are Cartan subalgebras of $\mathfrak{z}$. Hence by Proposition 4.3.2 there is a $g' \in Z$ with $\mathrm{Ad}(g')\mathrm{Ad}(g)(\mathfrak{t}) = \mathfrak{t}$. Set $g'' = g'g$; then $\mathrm{Ad}(g'')(h_1) = h_2$ and $\mathrm{Ad}(g'')(\mathfrak{t}) = \mathfrak{t}$ so that $g'' \in N_G(T)$. Finally, by Theorem 5.8.3, g'' maps to an element of W having the same action on $\mathfrak{t}$. □

5.9 Regular subalgebras of semisimple Lie algebras

Let $\mathfrak{g}$ be a semisimple Lie algebra over the algebraically closed field K of characteristic 0, and let G be its adjoint group. This means that G is the Chevalley group over K constructed from the adjoint representation of G (see Section 5.3). So this group acts naturally on $\mathfrak{g}$. A subalgebra $\mathfrak{a}$ of $\mathfrak{g}$ is said to be *regular* if it is normalized by a Cartan subalgebra of $\mathfrak{g}$. Here we describe an algorithm due to Dynkin to classify the regular semisimple subalgebras of G up to conjugacy by G.

Let $\mathfrak{h}$ be a fixed Cartan subalgebra of $\mathfrak{g}$. A subalgebra of $\mathfrak{g}$ is said to be $\mathfrak{h}$-*regular* if it is normalized by $\mathfrak{h}$. Since all Cartan subalgebras of $\mathfrak{g}$ are G-conjugate (Proposition 4.3.2), we may restrict our discussion to classifying the $\mathfrak{h}$-regular semisimple subalgebras.

In the following we let Φ be the root system of $\mathfrak{g}$ with respect to $\mathfrak{h}$, W its Weyl group, and ℓ its rank. Also we use a fixed Chevalley basis of $\mathfrak{g}$ consisting of x_α, for $\alpha \in \Phi$, and $h_1, \dots, h_\ell \in \mathfrak{h}$.

Let $\Gamma = \{\beta_1, \dots, \beta_m\}$ be a set of roots. This set is called a *weak π-system* if $\beta_i - \beta_j \notin \Phi$ for all i, j. It is called a *π-system* if in addition its elements are linearly independent.

We note that for a weak π-system $\{\beta_1, \dots, \beta_m\}$, we have $(\beta_i, \beta_j) \leq 0$ by Lemma 2.8.8. The Dynkin diagram of the weak π-system $\Gamma = \{\beta_1, \dots, \beta_m\}$ is the Dynkin diagram of the matrix $(\langle \beta_i, \beta_j^\vee \rangle)_{1 \leq i,j \leq m}$, as defined in Section 2.8.1. The system is said to be *indecomposable* if this diagram is connected.

Proposition 5.9.1 *Let $\Gamma = \{\beta_1, \dots, \beta_m\}$ be a π-system. Let $\mathfrak{a}_\Gamma \subset \mathfrak{g}$ be the subalgebra generated by $x_{\beta_i}, x_{-\beta_i}$, $1 \leq i \leq m$. Then $\mathfrak{a}_\Gamma$ is an $\mathfrak{h}$-regular semisimple subalgebra of $\mathfrak{g}$. Its root system Ψ consists of all $\alpha \in \Phi$ that can be written as linear combinations of the β_i with integral coefficients. Γ is a basis of simple roots of Ψ. Every $\mathfrak{h}$-regular semisimple subalgebra of $\mathfrak{g}$ can be obtained in this way. Finally, two subalgebras $\mathfrak{a}_\Gamma$, $\mathfrak{a}_{\Gamma'}$ are G-conjugate if and only if Γ, Γ' are W-conjugate.*

Proof. Because $\mathfrak{a}_\Gamma$ is generated by subspaces that are stable under $\mathrm{ad}h$, for $h \in \mathfrak{h}$, it follows that $\mathfrak{a}_\Gamma$ is $\mathfrak{h}$-regular.

Let $C = (\langle \beta_i, \beta_j^\vee \rangle)_{1 \leq i,j \leq m}$. It is straightforward to see that C is a Cartan matrix (for condition 3 in Definition 2.8.2: we may assume that Γ is indecomposable, and for the matrix D we use a multiple of $\mathrm{diag}((\beta_1, \beta_1), \dots, (\beta_m, \beta_m))$).

The algebra $\mathfrak{a}_\Gamma$ also contains the elements $h_{\beta_i} = [x_{\beta_i}, x_{-\beta_i}]$ for $1 \leq i \leq m$. For $\alpha \in \Phi$ we have $[h_{\beta_i}, x_\alpha] = \langle \alpha, \beta_i^\vee \rangle x_\alpha$. For $i \neq j$ consider the element $y_{i,j} = (\mathrm{ad}x_{-\beta_i})^{-C(j,i)+1}(x_{-\beta_j})$. A short calculation shows that $[x_{\beta_i}, y_{i,j}] = 0$ and $[h_{\beta_i}, y_{i,j}] = (C(j,i) - 2)y_{i,j}$. But $C(j,i) - 2 < 0$. It follows that $y_{i,j}$ generates a finite-dimensional irreducible $\mathfrak{sl}_2$-module of negative highest weight. This is impossible, hence $y_{i,j} = 0$. Similarly we have $(\mathrm{ad}x_{\beta_i})^{-C(j,i)+1}(x_{\beta_j}) = 0$. Hence the x_{β_i}, $x_{-\beta_i}$, h_{β_i} satisfy the Serre relations (see [Ser66], Chapter VI, Section 4). This implies that the algebra they generate (i.e., $\mathfrak{a}_\Gamma$) is a quotient of the semisimple Lie algebra $\mathfrak{u}$ corresponding to the Cartan matrix C by an ideal. This ideal is the sum of some of the simple ideals of $\mathfrak{u}$. But since the x_{β_i}, $x_{-\beta_i}$, h_{β_i} are non-zero, this ideal must be zero. It follows that $\mathfrak{a}_\Gamma$ is semisimple and Γ is a basis of simple roots of its root system.

Let $\beta = \sum_{i=1}^m m_i \beta_i$ where all $m_i \in \mathbb{Z}$, lie in Φ. Suppose that there are positive as well as negative coefficients. Write $I_\gamma = \{i \mid m_i > 0\}$, $I_\delta = \{i \mid m_i < 0\}$, and $\gamma = \sum_{i \in I_\gamma} m_i \beta_i$, $\delta = \sum_{i \in I_\delta} -m_i \beta_i$. Then $\beta = \gamma - \delta$. Suppose $\langle \gamma, \beta_i^\vee \rangle \leq 0$ for all $i \in I_\gamma$. As we saw in the proof of Proposition 2.8.10, the set $\{\beta_i \mid i \in I_\gamma\} \cup \{\gamma\}$ is linearly independent, which clearly is not the case. Hence there is an $i_0 \in I_\gamma$ with $\langle \gamma, \beta_{i_0}^\vee \rangle > 0$. As $\langle \delta, \beta_{i_0}^\vee \rangle \leq 0$, it follows that $\langle \beta, \beta_{i_0}^\vee \rangle > 0$. Therefore, $s_{\beta_{i_0}}(\beta) = \beta - m\beta_{i_0}$, $m > 0$. So if $\gamma \neq \beta_{i_0}$, by Lemma 2.8.9, we see that $(\gamma - \beta_{i_0}) - \delta \in \Phi$. By a similar argument, if $\delta \notin \Gamma$, we can subtract a β_{j_0} from δ and stil obtain an element of Φ. Performing these operations a number of times we arrive at a root of the form $\beta_i - \beta_j$, which is excluded. We conclude that the m_i are either all non-negative or all non-positive. Suppose that they are all non-negative. As seen above, there is an i such that $m_i > 0$ and $\gamma = \beta - \beta_i \in \Phi$. By induction, $x_\gamma \in \mathfrak{a}_\Gamma$. Because $[x_{\beta_i}, x_\gamma]$ is a non-zero multiple of x_β (this can be seen from Theorem 2.9.13), the latter also lies in $\mathfrak{a}_\Gamma$. If all coefficients are non-positive, we use a similar reasoning. The conclusion is that Ψ is the root system of $\mathfrak{a}_\Gamma$.

Conversely, if $\mathfrak{b}$ is an $\mathfrak{h}$-regular semisimple subalgebra, it is spanned by $\mathfrak{h} \cap \mathfrak{b}$ along with the x_α such that $x_\alpha \in \mathfrak{b}$. Because it is semisimple, the set of all those α forms a root system, and by letting $\Gamma = \{\beta_1, \dots, \beta_m\}$ be equal to a basis of simple roots, we see that Γ is a π-systen, and $\mathfrak{b}$ is generated by x_{β_i}, $x_{-\beta_i}$, $1 \leq i \leq m$.

For the final statement, let $g_0 \in G$ be such that $g_0 \cdot \mathfrak{a}_\Gamma = \mathfrak{a}_{\Gamma'}$. Consider the group $N_G(\mathfrak{a}_{\Gamma'}) = \{g \in G \mid g \cdot \mathfrak{a}_{\Gamma'} = \mathfrak{a}_{\Gamma'}\}$. By Corollary 4.2.9 its Lie algebra is $\mathfrak{n}_\mathfrak{g}(\mathfrak{a}_{\Gamma'}) = \{x \in \mathfrak{g} \mid [x, \mathfrak{a}_{\Gamma'}] \subset \mathfrak{a}_{\Gamma'}\}$. We have that $\mathfrak{h}$, $g_0 \cdot \mathfrak{h}$ are two Cartan subalgebras of $\mathfrak{n}_\mathfrak{g}(\mathfrak{a}_{\Gamma'})$. So by Theorem 4.3.2, there is a $g_1 \in N_G(\mathfrak{a}_{\Gamma'})$ with $g_1 g_0 \cdot \mathfrak{h} = \mathfrak{h}$. Let $H \subset G$ be the maximal torus with Lie algebra $\mathfrak{h}$ (Lemma 5.8.2). Then $g = g_1 g_0$ lies in $N_G(H)$. Let η be as in Theorem 5.8.3 and $w = \eta(g)$. Then for $h \in \mathfrak{h}$ we see that $[h, g \cdot x_\beta] = g \cdot [g^{-1} \cdot h, x_\beta] = \beta(w^{-1}(h)) g \cdot x_\beta$. It is straightforward to see that $w(\beta)(h) = \beta(w^{-1}(h))$. It follows that $g \cdot x_\beta$ is a multiple of $x_{w\beta}$. So w maps the root system Ψ of $\mathfrak{a}_\Gamma$ to the root system Ψ' of $\mathfrak{a}_{\Gamma'}$. As seen in the proof of Proposition 2.8.12, two bases of simple roots of a root system are conjugate under the Weyl group of that root system. So by composing with an element of the Weyl group of Ψ' (which is a subgroup of W), we obtain a $w' \in W$ with $w'(\Gamma) = \Gamma'$. The reverse implication is shown by similar arguments. □

Proposition 5.9.2 *Let $\Gamma = \{\beta_0, \dots, \beta_m\}$ be a weak π-system of rank m (that is, the space spanned by its elements has dimension m). Suppose Γ is indecomposable. Then its Dynkin diagram is one of the following:*

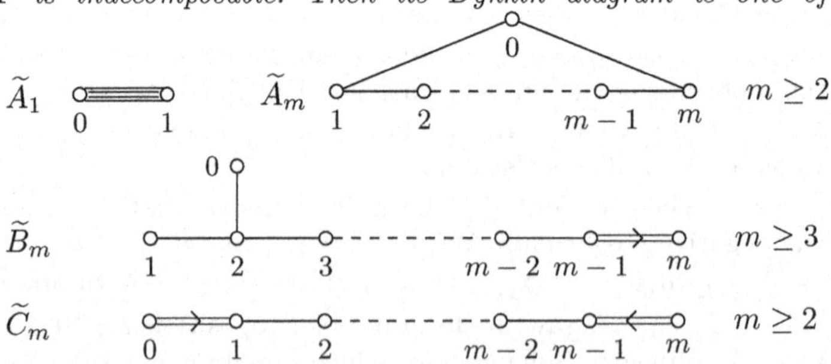

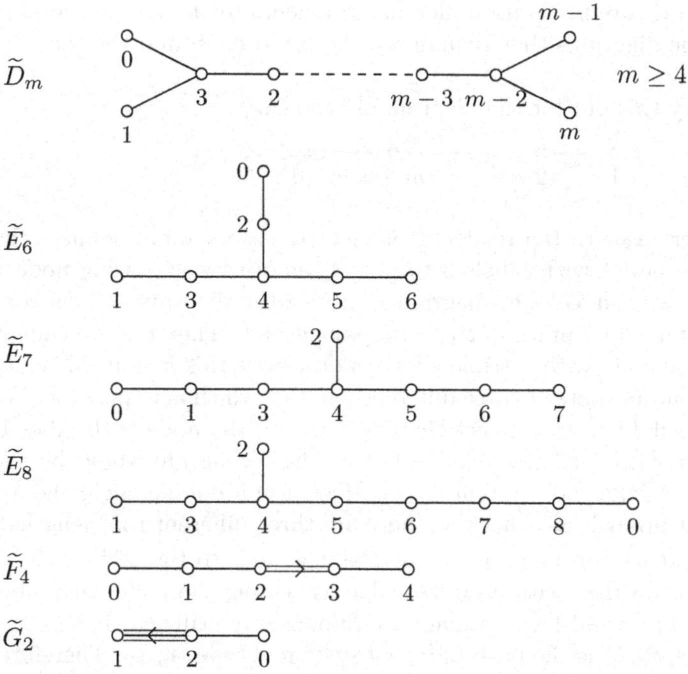

Proof. We may suppose that $\beta_1, \dots, \beta_m$ are linearly independent. Hence they form a π-system. Write $\beta_0 = \sum_{i=1}^{m} a_i\beta_i$, with $a_i \in \mathbb{R}$. We have $\langle \beta_0, \beta_i^\vee \rangle = 0, -1, -2, -3$, and at least one of them is non-zero. So $(a_1, \dots, a_m)$ is the solution of a non-homogeneous system of linear equations, whose matrix is the Cartan matrix of the β_i, $1 \leq i \leq m$. It is known that the coefficients of the inverse of a Cartan matrix (of an indecomposable root system) are all positive rational numbers (see [OV90], p. 295). Since β_0 is connected to every component of the diagram of the π-system $\{\beta_1, \dots, \beta_m\}$, all of the a_i are negative rational numbers. So, possibly after interchanging β_0, β_1, we may assume that $\beta_1, \dots, \beta_m$ is indecomposable and $\beta_0 = \sum_{i=1}^{m} a_i\beta_i$ where each a_i is a negative rational number. It follows that if we leave out exactly one β_i where $0 \leq i \leq m$, the remaining ones form a π-system. So if in the diagram of Γ we erase one node and all edges incident with it then what remains is the Dynkin diagram of a root system. This severely limits the possibilities for the diagram of Γ.

Let $\{\gamma_1, \dots, \gamma_r\}$, $\gamma_i \in \Phi$, be a weak π-system such that its diagram is the Dynkin diagram of a root system. Then the matrix $((\gamma_i, \gamma_j))$ is non-singular, so that the γ_i are linearly independent and in fact form a π-system. It follows that the diagram of Γ does not equal the Dynkin diagram of a root system.

Now we go through the list of the Dynkin diagrams of the root systems (as in Theorem 2.8.5), and try all possibilities to extend it with one node, to see what the diagram of Γ can be. The above principles eliminate most of the possibilities. For the diagrams which are not eliminated in this way, and are not listed in the theorem, we use different arguments to show that they are not possible. The diagrams that remain are the listed ones, and the theorem is proved.

Here we carry this out for the diagram of type C_m,

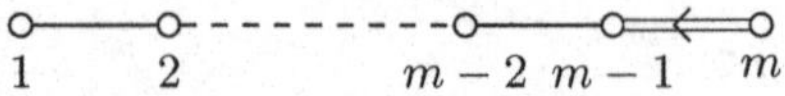

leaving the other cases to the reader. It is clear that we cannot connect the new node to the nodes with labels 3 to $m-1$, as otherwise erasing node 1 will not lead to a valid Dynkin diagram. Because the diagram of Γ is connected, Γ is contained in an irreducible component of Φ. Therefore we cannot connnect the new node with a triple edge, as otherwise this irreducible component would contain roots of three different lengths, which is impossible. We consider the possibilities to connect the new node to the node with label 1. We cannot do it with a single edge, as the resulting diagram would be the Dynkin diagram of the root system C_{m+1}. If we use a double edge, the arrow has to point inward (as otherwise there are three different root lengths), yielding $\widetilde{C}_m$. Next we consider connecting the new node to the node with label 2. We cannot do that with a double edge as erasing the node with label m would not yield a valid Dynkin diagram. Suppose we utilize a single edge. Note that $\beta_2, \ldots, \beta_{m-1}$ is the basis of a root system of type A_{m-2}. Therefore, $\gamma_2 = \sum_{i=2}^{m-1} \beta_i$ is a root, and $\Gamma' = \{\beta_0, \beta_1, \gamma_2, \beta_m\}$ is a weak π-system. It has to be linearly dependent because its diagram is not the Dynkin diagram of a root system. Furthermore, the diagram of $\{\beta_1, \gamma_2, \beta_m\}$ is of type C_3, so these roots form a basis of a root subsystem of that type. By solving the linear equations that are equivalent to $\langle \beta_0, \beta_1^\vee \rangle = \langle \beta_0, \beta_m^\vee \rangle = 0$, $\langle \beta_0, \gamma_2^\vee \rangle = -1$, we see that $\beta_0 = -\beta_1 - 2\gamma_2 - \beta_m$. By inspecting the root system of type C_3, β_0 is a root in it, but $\beta_0 - \beta_1$ as well, so that Γ' cannot be a weak π-system. Finally we have to consider the possibilities to attach the new node to the node with label m. It is only possible to do it with one edge when $m = 4$. However, in that case $\beta_0 = -\beta_1 - 2\beta_2 - 3\beta_3 - 2\beta_4$, and the C_4 subsystem spanned by $\beta_1, \ldots, \beta_4$ has no such root. It is possible to use a double edge only if $m = 2$, but in that case $\beta_0 = -2\beta_1 - 2\beta_2$, and the C_2 subsystem spanned by β_1, β_2 has no such root. □

A weak π-system corresponding to each diagram in Proposition 5.9.2 is constructed as follows. Let $\widetilde{X}_m$ be the type of a diagram and Ψ be the root system of type X_m, with basis of simple roots $\beta_1, \ldots, \beta_m$. There is a unique root $\gamma \in \Psi$ of maximal height (see, for example, [MT11], Appendix B), and we set $\beta_0 = -\gamma$. Then obviously, $\Gamma = \{\beta_0, \ldots, \beta_m\}$ is a weak π-system. By inspecting each root system, we see that the diagram of Γ is of type $\widetilde{X}_m$. By arguments as in the proof of Proposition 5.9.2, this also means that if we have an indecomposable weak π-system $\Gamma = \{\beta_0, \ldots, \beta_m\}$, whose diagram is of type $\widetilde{X}_m$ such that β_0 corresponds to the node with label 0 in the list of the mentioned proposition, then $\beta_1, \ldots, \beta_m$ form a basis of simple roots of a root subsystem of type X_m with β_0 as the lowest root.

Now let $\Gamma = \{\alpha_1, \dots, \alpha_\ell\}$ be a π-system in Φ. Choose an indecomposable subset $\Gamma_0 = \{\alpha_{i_1}, \dots, \alpha_{i_r}\}$. Let Ψ_0 be the root subsystem of Φ spanned by Γ_0. Let β_0 be minus the root of maximal height of Ψ_0 and Γ' be the set obtained from Γ by adding β_0 and erasing one of the α_{i_j}, $1 \leq j \leq r$. Then Γ' is a π-system of cardinality ℓ. We say that Γ' is obtained from Γ by an *elementary transformation.*

Let Ψ, Ψ' be the root subsystems of Φ spannned by respectively Γ and Γ'. Then Ψ contains all elements of Γ' so that $\Psi \supseteq \Psi'$.

Proposition 5.9.3 *Let Γ be a π-system of rank ℓ. There is a basis of simple roots Δ of Φ such that Γ can be obbtained from Δ by a sequence of elementary transformations.*

Proof. Let Ψ be the root system spanned by Γ. If $\Psi = \Phi$, then Ψ is a basis of Φ, and there is nothing to prove. Use the lexicographical order with respect to the basis Γ (Section 2.8.2). Relative to that order we take a lowest root β not contained in Ψ. Set $\Gamma_0 = \Gamma \cup \{\beta\}$. This is a weak π-system of rank ℓ and cardinality $\ell + 1$. It follows that its diagram is the union of a number of ordinary Dynkin diagrams (as listed in Theorem 2.8.5) and one diagram D of type $\widetilde{X}_m$ (Proposition 5.9.2). Then β corresponds to a node in D. We add β and remove the root from Γ corresponding to the node labeled 0 in D, obtaining the π-system Γ'. (Note that β cannot correspond to the node 0, as that would mean $\beta \in \Psi$.) It is clear that Γ can be obtained from Γ' by an elementary transformation. Moreover, letting Ψ' be the root subsystem spanned by Γ' yields $\Psi' \supsetneq \Psi$ (indeed, $\beta \in \Psi'$ and $\beta \notin \Psi$). It follows that after a finite number of steps of this kind we arrive at a basis of Φ. □

This yields the following procedure to obtain the π-systems in Φ of rank ℓ, up to W-conjugacy. We start with a basis of simple roots Δ of Φ. Then we apply all possible elementary transformations and add the resulting π-systems to the list. Of course, if Γ is a π-system, and Γ' is a π-system obtained from it by an elementary transformation and the diagrams of Γ and Γ' are the same, the root subsystems they span are equal, and therefore we do not add Γ'. Note that this must terminate, as the root subsystem spanned by a derived π-system is smaller than the root subsystem spanned by the original system. Once we are finished, we remove W-conjugate copies from the list.

Example 5.9.4 Consider the root system Φ of type B_4, with set of simple roots Δ. From the diagram $\widetilde{B}_4$ we see that with one elementary transformation it is possible to obtain the π-systems of types $2A_1 + B_2$, $A_1 + A_3$, D_4. Applying an elementary transformation to a π-system of type A_m only yields A_m, so nothing new is obtained that way. Therefore, the only useful operation on the system of type $2A_1 + B_2$ is to apply an elementary transformation to the piece of type B_2, leading to the new system $4A_1$. The system of type $A_1 + A_3$ has no elementary transformations leading to a new system. Finally, the system of type D_4 can be transformed to $4A_1$. We see that potentially we have two different π-systems of type $4A_1$. However, they are conjugate under the Weyl group. We conclude that there are five π-systems of rank 4, up to conjugacy by the Weyl group.

Lemma 5.9.5 *Every π-system in Φ is contained in a π-system of rank ℓ.*

Proof. Let Γ be a π-system of rank less than ℓ. Extend Γ to a basis of the space spanned by Φ, and use the lexicographical order with respect to that basis. Let $\beta \in \Phi$ be a lowest root (with respect to that order) not contained in the space spanned by Γ. Then $\Gamma \cup \{\beta\}$ is a π-system of higher rank. □

After we obtain the π-systems of rank ℓ, we add all their subsets, and again remove W-conjugate copies. The result is a list of all π-systems in Φ, up to W-conjugacy. By Proposition 5.9.2, this yields the list of all $\mathfrak{h}$-regular semisimple subalgebras of $\mathfrak{g}$, up to G-conjugacy.

Remark 5.9.6 In order to execute this algorithm we need a method to decide whether two sets of roots are W-conjugate. More generally, we can consider two sets of weights $\{\mu_1, \dots, \mu_r\}$, $\{\nu_1, \dots, \nu_r\}$ in in the weight lattice P (Section 2.8.3), and ask whether they are W-conjugate. Here we do not go into the details and refer to [Gra11]. The main idea is that it is straightforward to decide whether there is a $w_1 \in W$ with $w_1(\mu_1) = \nu_1$ (Remark 2.8.30) and find such a w_1. Then we replace W by the stabilizer W_{ν_1} of ν_1 in W, which is a reflection subgroup, and continue with W_{ν_1} and the sets $\{w_1(\mu_2), \dots, w_1(\mu_r)\}$, $\{\nu_2, \dots, \nu_m\}$.

5.10 Notes

The material of Sections 5.1 to 5.3 is mainly taken from Steinberg's famous lecture notes ([Ste67]). Here we stress the role played by the abstract group $\mathcal{G}$ of Section 5.1, as this group is useful for computations as well (Sections 5.6, 5.7). Also we give formulas for the constants $c_{i,j}^{\alpha,\beta}$ and $c_{\alpha,\beta}$, as they are fundamental for the algorithms in the last two sections.

Sections 5.6 and 5.7 are based on the paper by Cohen, Murray and Taylor ([CMT04]) in which algorithms are given for the wider class of reductive groups.

Dynkin's algorithm (Section 5.9) appeared in [Dyn52]. It is a step in his method to classify the semisimple subalgebras of a semisimple Lie algebra. The Borel-de Siebenthal theorem ([MT11], Theorem B.18) yields a different approach to the problem of classifying the regular semisimple subalgebras.

Chapter 6

Generators of Arithmetic Groups

Arithmetic groups appeared in the work of mathematicians like Gauss, Hermite and Minkowski who studied the action of $\mathrm{GL}(n, \mathbb{Z})$ on the set of quadratic forms in n variables. The motivating problem for this is to determine which set of integers is represented by a given form. This led to the reduction theory of quadratic forms, which describes a fundamental set for this action, yielding an algorithm for deciding whether two forms are in the same $\mathrm{GL}(n, \mathbb{Z})$-orbit. In the first half of the 20th century Siegel considered arithmetic subgroups of the classical groups, generalizing many aspects of the reduction theory. In the second half of that century, the work of Borel and Harish-Chandra extended the reduction theory to general algebraic groups over $\mathbb{Q}$.

Roughly speaking, an arithmetic group is the group of integral points of an algebraic group defined over the rational numbers. A consequence of the reduction theory, due to Borel and Harish-Chandra, is that such a group is finitely generated. So the problem is to devise algorithms that, given an algebraic group, find generators of an arithmetic subgroup. This problem is the subject of this chapter. After an introduction and a section on algorithms for lattices in $\mathbb{Z}^n$, we give algorithms for two special classes of algebraic groups: unipotent groups and tori.

No efficient algorithm exists that finds a generating set of a general arithmetic group. Such an algorithm would have a great number of applications. For example, it would become possible to compute the automorphism group of a finitely generated nilpotent group (see [GS80b]) and compute generators of the unit group of the integral group ring of a finite group.

6.1 Brief introduction to arithmetic groups

Let $G \subset \mathrm{GL}(n, \mathbb{C})$ be an algebraic group defined over $\mathbb{Q}$. For a ring $R \subseteq \mathbb{C}$ we set $G(R) = G \cap \mathrm{GL}(n, R)$. We let $V = \mathbb{C}^n$ be the natural G-module and set $V_{\mathbb{Q}} = \mathbb{Q}^n$.

A *lattice* in $V_{\mathbb{Q}}$ is a finitely generated subgroup of $V_{\mathbb{Q}}$ (note that this space is a group under addition). A lattice L in $V_{\mathbb{Q}}$ has a basis: a finite set of linearly independent vectors $v_1, \dots, v_r$ such that L consists of all $m_1 v_1 + \cdots + m_r v_r$

where $m_i \in \mathbb{Z}$ for all i. The number r is called the *dimension* of L, and if $r = n$ then we say that L is *full-dimensional.* In this section all lattices will be assumed to be full-dimensional. For a lattice L we denote its stabilizer in G by G_L, i.e, $G_L = \{g \in G \mid g(L) = L\}$. An example is $L = \mathbb{Z}^n$, in which case we have $G_L = G(\mathbb{Z})$.

We also need a piece of terminology from general group theory. Let H_1 and H_2 be subgroups of a group H. Then H_1 and H_2 are said to be *commensurable* if $H_1 \cap H_2$ has finite index in H_1 and in H_2. It is an exercise in group theory to show that commensurability is an equivalence relation on the subgroups of a group H.

Let L be a lattice in $V_\mathbb{Q}$ and fix a basis of L. For $g \in G$, we denote the matrix of g with respect to the fixed basis of L by (g_{ij}). Then $g \in G_L$ if and only if all g_{ij} lie in $\mathbb{Z}$. For $d \in \mathbb{Z}$, $d > 0$ set

$$G_L(d) = \{g \in G_L \mid (g_{ij}) = I_n \bmod d\}.$$

This is the kernel of the group homomorphism $G_L \to \mathrm{GL}(n, \mathbb{Z}) \to \mathrm{GL}(n, \mathbb{Z}/d\mathbb{Z})$, $g \mapsto (g_{ij}) \mapsto (g_{ij} \bmod d)$. So $G_L(d)$ is a normal subgroup of finite index. It is called a *congruence subgroup of level d.*

Proposition 6.1.1 *Let $G \subset \mathrm{GL}(n, \mathbb{C})$ and $H \subset \mathrm{GL}(m, \mathbb{C})$ be two algebraic groups defined over $\mathbb{Q}$. Let L and M be lattices in $V_\mathbb{Q} = \mathbb{Q}^n$ and $W_\mathbb{Q} = \mathbb{Q}^m$ respectively. Let $\varphi : G \to H$ be a morphism of algebraic groups defined over $\mathbb{Q}$. Then G_L has a subgroup of finite index whose image under φ is contained in H_M.*

Proof. We first prove this for the case where $L = \mathbb{Z}^n$ and $M = \mathbb{Z}^m$. We let $R = \mathbb{C}[x_{ij}, \frac{1}{\det(x_{ij})}]$ be the coordinate ring of $\mathrm{GL}(n, \mathbb{C})$ (see Example 3.1.3). There are polynomials $p_{k,l} \in R$ (for $1 \le k, l \le m$) having coefficients in $\mathbb{Q}$ such that $\varphi(g)_{k,l} = p_{k,l}(g)$. Write $p_{k,l} = \frac{1}{(\det(x_{ij}))^s} \tilde{p}_{k,l}$ where $\tilde{p}_{k,l} \in R_0 = \mathbb{Q}[x_{ij} \mid 1 \le i, j \le n]$. Define polynomials $q_{k,l} \in R_0$ by $\tilde{p}_{k,l} - \delta_{k,l} = q_{k,l}(x_{ij} - \delta_{i,j})$ (where $\delta_{k,l}$ denotes the Kronecker delta function). Since $\varphi(I_n) = I_m$ we have $\tilde{p}_{k,l}(I_n) = \delta_{k,l}$. We infer that the $q_{k,l}$ have no constant term. Let d be a common denominator of the coefficients of all $q_{k,l}$. Let $g \in G_L(d)$, then $g = I_n \bmod d$, whence $q_{k,l}(g - I_n) \in \mathbb{Z}$. As $g \in \mathrm{GL}(n, \mathbb{Z})$, $\det(g) = \pm 1$. Therefore $p_{k,l}(g) \in \mathbb{Z}$. We conclude that $\varphi(g)M \subset M$ for all $g \in G_L(d)$. Hence for a given $g \in G_L(d)$ we also have $\varphi(g)^{-1}M \subset M$, and therefore $\varphi(g)M = M$. So $G_L(d)$ is the finite index subgroup we are seeking.

For the general case we use maps $\sigma : G \to \mathrm{GL}(n, \mathbb{C})$, $\tau : H \to \mathrm{GL}(m, \mathbb{C})$, where $\sigma(g)$, $\tau(h)$ are the matrices of g and h with respect to fixed bases of L and M respectively. (So $\sigma(g) = AgA^{-1}$, $\tau(h) = BhB^{-1}$ for certain matrices A, B.) Set $\widetilde{G} = \sigma(G)$, $\widetilde{H} = \tau(H)$. Then $\sigma(G_L)$, $\tau(H_M)$ are the stabilizers of $\mathbb{Z}^n$ and $\mathbb{Z}^m$ in $\widetilde{G}$ and $\widetilde{H}$ respectively. Furthermore, we have the morphism $\tau\varphi\sigma^{-1} : \widetilde{G} \to \widetilde{H}$. By the first part of the proof, there is a finite-index subgroup

S of $\widetilde{G}_{\mathbb{Z}^n}$ such that $\tau\varphi\sigma^{-1}(S)$ is contained in $\widetilde{H}_{\mathbb{Z}^m}$. So $\sigma^{-1}(S)$ is the subgroup of G_L that we want. □

Corollary 6.1.2 *Let L and L' be two lattices in $V_{\mathbb{Q}}$. Then G_L and $G_{L'}$ are commensurable.*

Proof. This follows from Proposition 6.1.1 by considering the morphism $\varphi : G \to \mathrm{GL}(n, \mathbb{C})$ mapping $g \in G$ to its matrix with respect to a fixed basis of L'. □

Definition 6.1.3 *A subgroup $\Gamma \subset G(\mathbb{Q})$ is called* arithmetic *if it is commensurable with $G(\mathbb{Z})$.*

By Corollary 6.1.2 it follows that an arithmetic group is commensurable with G_L where L is any lattice in $V_{\mathbb{Q}}$. The next result shows that the arithmetic subgroups of $G(\mathbb{Q})$ are the G_L where L is a lattice in $V_{\mathbb{Q}}$ and their finite-index subgroups.

Proposition 6.1.4 *Let $\Gamma \subset G(\mathbb{Q})$ be an arithmetic subgroup. Then any lattice L in $V_{\mathbb{Q}}$ is contained in a Γ-invariant lattice M. Moreover Γ is of finite index in G_M.*

Proof. Because Γ is commensurable with G_L, $\Gamma \cap G_L$ is of finite index in Γ; in other words, Γ has a finite index subgroup stabilizing L. This implies that $\{g(L) \mid g \in \Gamma\}$ is a finite set of lattices. Their sum M is therefore a Γ-stable lattice containing L. Finally, Γ is commensurable with G_M, but also contained in G_M; therefore it is of finite index in G_M. □

Proposition 6.1.5 *Let L be a lattice in $V_{\mathbb{Q}}$. Let H and N be algebraic subgroups of G defined over $\mathbb{Q}$ such that $G = H \ltimes N$ where the isomorphism of closed sets $H \times N \to G$ (see Section 3.3) is defined over $\mathbb{Q}$. Then $H_L N_L$ is a finite-index subgroup of G_L (and hence an arithmetic subgroup of $G(\mathbb{Q})$).*

Proof. Define $\varphi : G \to H$ by $\varphi(hn) = h$, for $h \in H$, $n \in N$. By Proposition 6.1.1 there is a finite-index subgroup Γ of G_L such that $\varphi(\Gamma) \subset H_L$. Then for $g \in \Gamma$ we see that $g = \varphi(g)(\varphi(g)^{-1}g) \in H_L N_L$. It follows that $\Gamma \subset H_L N_L$. This implies that $H_L N_L$ has finite index in G_L. □

The next theorem is one of the main results in the theory of arithmetic groups and has been proved by Borel and Harish-Chandra ([BHC62]).

Theorem 6.1.6 *An arithmetic subgroup of $G(\mathbb{Q})$ is finitely presented.*

Some problems now come to mind. One is to compute a generating set of an arithmetic subgroup Γ of $G(\mathbb{Q})$ given G. A potentially more challenging task is to compute a finite presentation for Γ as well (that is, given generators $\gamma_1, \dots, \gamma_m$ of Γ, and letting F_m be the free group on m generators $f_1, \dots, f_m$, compute a finite set $R \subset F_m$ such that $f_i \mapsto \gamma_i$ induces an isomorphism $F_m/N_R \to \Gamma$ where N_R is the normal subgroup of F_m generated by R). Finally we can consider these problems for an arithmetic subgroup Γ that is specified in some way (for instance $\Gamma = G(\mathbb{Z})$). Grunewald and Segal ([GS80a]) devised a number of algorithms that solve the first problem for any arithmetic subgroup. However, many of their algorithms are not usable in practice. For an example, in Algorithm 5.1.1 of their paper, in order to find a set of polynomials in $\mathbb{C}[\mathrm{GL}(n, \mathbb{C})]$ having coefficients in $\mathbb{Q}$ and satisfying certain properties, all sequences of such polynomials of finite length are enumerated. It is clear that such a procedure will never terminate for examples that are not completely trivial. (This problem is acknowledged by the authors who wrote "... the content would not really satisfy a constructively-minded mathematician.") The question whether the algorithms in [GS80a] can be improved to such a degree that they become practical for at least some non-trivial inputs, has still not been resolved.

Here we shall be concerned with the first problem, that is, to compute an arithmetic subgroup of a $G(\mathbb{Q})$. This problem splits into two subcases: G unipotent, and G reductive. Indeed, by Theorem 4.3.22 we have $G = H \ltimes N$, where H is reductive and N is unipotent. Let Γ_H be an arithmetic subgroup of $H(\mathbb{Q})$ and L be a Γ_H-invariant lattice in $V_{\mathbb{Q}}$ such that Γ_H is of finite index in H_L (Proposition 6.1.4; it is straightforward to compute such an L; see Example 6.1.11). Section 6.3 contains an algorithm to compute generators of $\Gamma_N = N_L$. Then Γ_N is normalized by Γ_H and $\Gamma_N\Gamma_H$ is of finite index in $H_L N_L$, and hence of finite index in G_L by Proposition 6.1.5 (note that the map α in the proof of Theorem 4.3.22 is defined over $\mathbb{Q}$). As already mentioned, in Section 6.3 we deal with the unipotent case. The problem for reductive groups is generally much more difficult. Below we list some examples of cases that can be handled. In Section 6.4 we describe an algorithm to compute generators of $G(\mathbb{Z})$ for the case where G is a torus.

Remark 6.1.7 There are many other intriguing algorithmic problems for arithmetic groups, apart from the ones mentioned above. For example, given an algebraic group G and a finite set $S \subset G(\mathbb{Q})$, decide whether the group generated by S is an arithmetic subgroup of G. For an introduction to that and a solution for the case where G is solvable, we refer to [DFdG15]. For a second problem, we remark that an arithmetic group has a finite number of finite subgroups up to conjugacy ([PR94], Theorem 4.3); the problem is to find them. This appears to be extremely difficult; see [PP80] for results on $\mathrm{GL}(n, \mathbb{Z})$. A further obvious question is how a generating set of an arithmetic group Γ can be used to obtain information on Γ. A first step in this direction has been achieved in [DFH15], where algorithms are given for arithmetic subgroups Γ

of $\mathrm{SL}(n, \mathbb{Z})$, $n \geq 3$. They use a congruence subgroup of Γ to decide certain questions on Γ by doing computations in $\mathrm{SL}(n, \mathbb{Z}/d\mathbb{Z})$.

Example 6.1.8 Let A be an associative algebra over $\mathbb{Q}$ with an identity element which we denote by e. Let $\mathcal{O} \subset A$ be a $\mathbb{Z}$-order; this means that $\mathcal{O}$ consists of $k_1a_1 + \cdots + k_na_n$, where $k_i \in \mathbb{Z}$ and $\{a_1, \ldots, a_n\}$ is a basis of A such that $a_ia_j \in \mathcal{O}$ for all i, j. Suppose also that $e \in \mathcal{O}$. For $a \in A$ let $\rho(a)$ denote the matrix of the left multiplication by a with respect to a fixed basis of $\mathcal{O}$. We can extend ρ to an algebra homomorphism $\rho : \mathbb{C} \otimes A \to M_n(\mathbb{C})$. Let $(\mathbb{C} \otimes A)^*$ denote the unit group of $\mathbb{C} \otimes A$ (i.e., it consists of the invertible elements of $\mathbb{C}\otimes A$). Then $G = \rho((\mathbb{C}\otimes A)^*)$ is an algebraic subgroup of $\mathrm{GL}(n, \mathbb{C})$ (see Example 3.6.9). Let $\mathcal{O}^*$ denote the unit group of $\mathcal{O}$. Then $\rho(\mathcal{O}^*) = G(\mathbb{Z})$ (indeed, let $\alpha \in \mathbb{C} \otimes A$ be such that $\rho(\alpha) \in G(\mathbb{Z})$, then $\alpha = \alpha \cdot (1 \otimes e)$ lies in the $\mathbb{Z}$-span of the $1 \otimes a_i$).

Now we will briefly indicate some classes of algebras A for which the problem of finding generators of $\mathcal{O}^*$ has been resolved with some degree of success.

- Let A be an algebraic extension of $\mathbb{Q}$ of finite degree and $\mathcal{O}$ be the ring of integers of A. The classical theorem stating that $\mathcal{O}^*$ is finitely generated (and also giving a simple formula for the number of generators needed) is called Dirichlet's unit theorem. A widely used algorithm for obtaining generators of $\mathcal{O}^*$ is due to Buchmann ([Buc90]). In this approach the unit group is obtained as a by-product of an algorithm for determining the class group. The main idea of the latter is to enumerate the ideals of $\mathcal{O}$ of norm smaller than an appropriate bound, and subsequently generate enough relations between these ideals.

- Let $A = \mathbb{Q}\mathcal{G}$ be the group ring over $\mathbb{Q}$ of a finite group $\mathcal{G}$, and $\mathcal{O} = \mathbb{Z}\mathcal{G}$. There is a large body of work on finding explicit constructions of generators of subgroups of finite index of $\mathcal{O}^*$, starting with a construction of Bass for the case where $\mathcal{G}$ is abelian. In particular, these are arithmetic groups. We refer to [Seh93] for an overview.

- Let A be a quaternion algebra over $\mathbb{Q}$; that is, A has basis 1, i, j, k, subject to the relations $i^2 = a$, $j^2 = b$, $k = ij = -ji$, where $a, b \in \mathbb{Q}$ are two given parameters. Here $\mathcal{O}$ can be a maximal order or, for example, the order spanned by the basis of A in case $a, b \in \mathbb{Z}$. One method with roots going back at least to a 1925 paper of Ford ([For25]) considers the group of units of norm 1 denoted $\mathcal{O}_1^*$. This group is embedded in $\mathrm{SL}(2, \mathbb{R})$, which acts on the upper half plane in $\mathbb{C}$ by Moebius transformations. The main step of the method finds a fundamental domain for the action of $\mathcal{O}_1^*$ from which a set of generators is easily deduced. See [Kat92] Section 5.6 for a detailed example. A different algorithm proposed in [BCNS15] builds on the work of Voronoï, dating back to the beginning of the 20th century.

Example 6.1.9 Let d be a positive square-free integer. Let $G_d \subset \mathrm{GL}(2,\mathbb{C})$ be the group consisting of all matrices of the form $\begin{pmatrix} \alpha & \beta \\ d\beta & \alpha \end{pmatrix}$ (see Example 3.1.4). Then $G_d(\mathbb{Z})$ consists of all such matrices with $\alpha, \beta \in \mathbb{Z}$ and $\alpha^2 - d\beta^2 = \pm 1$. (This is known as Pell's equation.) Now $\mathbb{Z}[\sqrt{d}]^* = \{x + y\sqrt{d} \mid x, y \in \mathbb{Z}, x^2 - dy^2 = \pm 1\}$. So we get an isomorphism $G_d(\mathbb{Z}) \to \mathbb{Z}[\sqrt{d}]^*$ mapping a matrix as above to $\alpha + \beta\sqrt{d}$. By Dirichlet's unit theorem, $\mathbb{Z}[\sqrt{d}]^*$ is generated by -1 and a fundamental unit $u_0 = x_0 + y_0\sqrt{d}$. From this we immediately get two generators of $G_d(\mathbb{Z})$. For example, if $d = 109$, we can take $x_0 = 8890182$ and $y_0 = 851525$, so that $G_{109}(\mathbb{Z})$ is generated by

$$\begin{pmatrix} -1 & 0 \\ 0 & -1 \end{pmatrix}, \quad \begin{pmatrix} 8890182 & 851525 \\ 92816225 & 8890182 \end{pmatrix}.$$

Example 6.1.10 Let G be a simply connected Chevalley group over $\mathbb{C}$, defined in Section 5.2.3, generated by $x_\alpha(t)$, for $t \in \mathbb{C}$, $\alpha \in \Phi$. It can be shown that $G(\mathbb{Z})$ is generated by the elements $x_\alpha(1)$ ([Ste67], Theorem 18, Corollary 3). Using these generators, it is possible to devise a finite presentation for $G(\mathbb{Z})$ (see [Beh75], Satz 3.1(b)).

Example 6.1.11 Let $\mathfrak{g}$ be a 3-dimensional simple Lie algebra over $\mathbb{C}$ and suppose $\mathfrak{g}$ has a basis whose structure constants lie in $\mathbb{Q}$. Then $\mathrm{ad}\mathfrak{g}$ is algebraic (see the argument used for $\mathfrak{s}$ in Theorem 4.3.20). Let G denote the connected algebraic subgroup of $\mathrm{GL}(\mathfrak{g})$ with Lie algebra $\mathrm{ad}\mathfrak{g}$. Here we show that we can compute an arithmetic subgroup of G.

There is a basis x, y, z of $\mathfrak{g}$ such that $[x, y] = z$, $[x, z] = \lambda y$, $[y, z] = \mu x$, where $\lambda, \mu \in \mathbb{Q}$ ([Jac79], Section I.4). Let A be the quaternion algebra over $\mathbb{Q}$ with parameters λ, $-\mu$, that is, A has basis 1, i, j, k and the multiplication is determined by $i^2 = \lambda$, $j^2 = -\mu$, $ij = -ji = k$. Let A_{Lie} denote the Lie algebra with the same underlying space as A, but with the commutator as the algebra operation. The elements $\hat{x} = \frac{1}{2}i$, $\hat{y} = \frac{1}{2}j$, $\hat{z} = \frac{1}{2}k$ satisfy the commutation relations of x, y, z. Hence they span a subalgebra $\hat{\mathfrak{g}}$ of A_{Lie} isomorphic to $\mathfrak{g}$. Let $\rho : A \to \mathrm{End}(A)$ be the left regular representation of A. Then

$$\rho(a + bi + cj + dk) = \begin{pmatrix} a & \lambda b & -\mu c & \lambda\mu d \\ b & a & -\mu d & \mu c \\ c & -\lambda d & a & \lambda b \\ d & -c & b & a \end{pmatrix}. \tag{6.1}$$

Let $H \subset \mathrm{GL}(4,\mathbb{C})$ be the algebraic group consisting of the matrices (6.1) with determinant 1. Using Theorem 4.2.2, Example 3.6.9 we see that $\mathrm{Lie}(H) = \rho(\hat{\mathfrak{g}})$. The *reduced norm* of A is defined to be $N(a + bi + cj + dk) = a^2 - \lambda b^2 + \mu c^2 - \lambda\mu d^2$. A small calculation shows that $\det \rho(a + bi + cj + dk) = N(a + bi + cj + dk)^2$. Let $\mathcal{O}$ be an order in A. Let $t_1, \dots, t_m$ be generators of the group $\mathcal{O}_1^*$ (see Example 6.1.8) and $t_0 \in \mathcal{O}$ be an element of reduced norm -1. Then the group $H_{\mathcal{O}}$, consisting of $h \in H$ with $h\mathcal{O} = \mathcal{O}$ is generated by the $\rho(t_i)$, $0 \le i \le m$. Identifying $\mathfrak{g}$ and $\hat{\mathfrak{g}}$, we obtain a surjective morphism of algebraic groups $\varphi : H \to G$, where $\varphi(h)$ is the conjugation by h on $\rho(\hat{\mathfrak{g}})$. It is known that the image of an arithmetic group under a surjective morphism of algebraic groups is an arithmetic subgroup of the image group ([PR94], Theorem 4.1). We conclude that the $\varphi(t_i)$ generate an arithmetic subgroup of G.

For example, set $\lambda = 3$, $\mu = -5$. The quaternion algebra A with these parameters is considered in [Kat92], Example G of Section 5.6. There an order $\mathcal{O}$ with basis $1, i, \frac{1+j}{2}, \frac{j+k}{2}$ is used. The cited example has the generators of $\mathcal{O}_1^*$, $t_1 = \frac{3}{2} - \frac{1}{2}j$, $t_2 = 2 - i$, $t_3 = 2 - \frac{3}{2}i + \frac{1}{2}k$, $t_4 = 2 + \frac{3}{2}i + \frac{1}{2}k$, to which we add $t_0 = \frac{1}{2} + \frac{1}{2}j$. The element of G corresponding to t_3 is

$$\begin{pmatrix} -\frac{13}{2} & 10 & -\frac{45}{2} \\ -6 & 7 & -18 \\ \frac{9}{2} & -6 & \frac{29}{2} \end{pmatrix}.$$

We leave it to the reader to write the other matrices. We see that the t_i do not leave $L = \mathbb{Z}^3$ invariant. By Proposition 6.1.4, there exists a lattice L' left invariant by the t_i. It is straightforward to compute it using the proof of that proposition. Initially we set $L' = L$. Let t_j be a generator such that t_jL' is not contained in L', then replace L' by $L' + t_jL'$. In this example we only need to extend the lattice once, and denoting the basis elements of $\mathbb{Z}^3$ by e_1, e_2, e_3, we have that L' is spanned by $\frac{1}{2}(e_1 + e_3)$, e_2, e_3.

6.2 Algorithms for lattices

In this section we describe some algorithms for lattices in $\mathbb{Z}^n$. We used some of these in previous chapters, but they will play a most prominent role in this chapter.

By definition, a lattice in $\mathbb{Z}^n$ is a finitely generated subgroup of $\mathbb{Z}^n$. It is not difficult to see that a lattice L has a basis, which means that there are $v_1, \ldots, v_l \in L$ such that each $v \in L$ can be written uniquely as $v = m_1v_1 + \cdots + m_lv_l$, $m_i \in \mathbb{Z}$. We will frequently write the elements of a basis as the rows of a matrix.

One of our basic workhorses is the *Smith normal form* algorithm. Given an $m \times n$ matrix A with coefficients in $\mathbb{Z}$, this algorithm finds an $m \times n$ integral matrix S and integral unimodular matrices P and Q such that S is in Smith normal form (this means that there is an $r \leq m$ such that $d_i = S(i,i)$ is positive for $1 \leq i \leq r$, d_i divides d_{i+1} for $1 \leq i < r$ and S has no other non-zero entries), and $S = PAQ$. One important property of the Smith normal form is the following:

Lemma 6.2.1 *For integers $d > 0$, s let $[s]_d$ denote the class of s in $\mathbb{Z}/d\mathbb{Z}$. Let $L \subset \mathbb{Z}^n$ be a lattice and A be a matrix whose rows form a basis of L. Let $S = PAQ$ and d_i be as above. Define a map $\phi : \mathbb{Z}^n \to M = \mathbb{Z}/d_1\mathbb{Z} \oplus \cdots \mathbb{Z}/d_r\mathbb{Z} \oplus$*

$\mathbb{Z}^{n-r}$ *by* $\phi(v) = ([x_1]_{d_1}, \ldots, [x_r]_{d_r}, x_{r+1}, \ldots, x_n)$, *where* $(x_1, \ldots, x_n) = vQ$. *Then* ϕ *is a surjective group homomorphism with kernel* L.

Here we do not go into algorithms for computing the Smith normal form or the proof of the lemma; for both we refer to [Sim94], Chapter 8.

Corollary 6.2.2 *Let the notation be as in Lemma 6.2.1 and* $q_1, \ldots, q_n$ *denote the rows of* Q^{-1}. *They form a basis of* $\mathbb{Z}^n$ *and* $d_1q_1, \ldots, d_rq_r$ *are a basis of* L.

Proof. The first statement follows from the fact that Q is unimodular. Let $v \in \mathbb{Z}^n$. By Lemma 6.2.1, $v \in L$ if and only if $vQ = (a_1, \ldots, a_r, 0, \ldots, 0)$ where d_i divides a_i for $1 \leq i \leq r$. Writing $v = \sum_i \alpha_i q_i$ we see that $vQ = (\alpha_1, \ldots, \alpha_n)$. This yields the second statement. □

Let $L \subset \mathbb{Z}^n$ be a lattice. It is called *pure* if $\mathbb{Z}^n/L$ is torsion-free. This is the same as saying that $mv \in L$ for $m \in \mathbb{Z}$ and $v \in \mathbb{Z}^n$, implies that $v \in L$.

Let $v_1, \ldots, v_r \in \mathbb{Z}^n$ be a basis of a lattice $L \subset \mathbb{Z}^n$. We form the $r \times m$ matrix A with the v_i as rows. By computing the Smith normal form of A we can effectively compute the homomorphism $\phi : \mathbb{Z}^n \to M$ with kernel L, as shown in Lemma 6.2.1. Let T denote the torsion submodule of M (i.e., $T = \mathbb{Z}/d_1\mathbb{Z} \oplus \cdots \oplus \mathbb{Z}/d_r\mathbb{Z}$). Then $\phi^{-1}(T)$ is the smallest pure lattice containing L. We call this the *purification* of L. We conclude that the Smith normal form algorithm yields an algorithm to compute the purification of a lattice.

Now we give two algorithms that use the purification algorithm. In these we use the following observation. Let W be a subspace of $\mathbb{Q}^n$; then $W \cap \mathbb{Z}^n$ is a pure lattice. Moreover, let B be an integral matrix whose rows span W; then the purification of the lattice spanned by the rows of B is $W \cap \mathbb{Z}^n$.

Algorithm 6.2.3 (Intersection of lattice and subspace) *Input: an* $n \times n$ *integral matrix* A *whose rows span an* n*-dimensional lattice* L *in* $\mathbb{Z}^n$, *and an* $m \times n$ *matrix* B *whose rows span an* m*-dimensional* $\mathbb{Q}$*-subspace* W *of* $\mathbb{Q}^n$. *Output: an* $n \times n$ *integral matrix whose rows span* L *and whose first* m *rows span the lattice* $W \cap L$.

1. *Let* $e_1, \ldots, e_n$ *and* $b_1, \ldots, b_m$ *denote the rows of* A *and* B *respectively. Write* $b_i = \sum_{j=1}^n \beta_{ij} e_j$ *and let* $B' = (\beta_{ij})$; *if necessary multiply the rows of* B' *by integers to get integral entries.*

2. *Let* C *be a matrix whose rows form a basis of the purification of the lattice spanned by the rows of* B'.

3. *Let* S, P, Q *be the output of the Smith normal form algorithm with input* C.

4. *Return* $Q^{-1}A$.

Lemma 6.2.4 *Algorithm 6.2.3 is correct.*

Proof. The idea is to use the given basis of L as a basis of $\mathbb{Q}^n$. Define $\psi : \mathbb{Q}^n \to \mathbb{Q}^n$ by letting $\psi(v)$ be the vector containing the coefficients of v with respect to the basis of L. So after the first step the rows of B' form a basis of $\psi(W)$. Of course $\psi(L) = \mathbb{Z}^n \subset \mathbb{Q}^n$.

The rows of C form a basis of $\psi(W) \cap \psi(L) = \psi(W \cap L)$. Furthermore, as $\psi(W \cap L)$ is pure, the Smith normal form S of C has diagonal entries equal to 1. Therefore the rows of Q^{-1} form a basis of $\mathbb{Z}^n$ and the first m rows form a $\mathbb{Z}$-basis of $\psi(W \cap L)$ (this follows from Lemma 6.2.1). Note that for $v \in \mathbb{Q}^n$ we have $\psi^{-1}(v) = vA$. Therefore the rows of $Q^{-1}A$ form a basis of L and the first m are a $\mathbb{Z}$-basis of $W \cap L$. □

Algorithm 6.2.5 (integral relations) *Input: An $m \times n$ matrix A with rational coefficients.*
Output: an $m \times n$ integral matrix whose rows are a basis of the lattice

$$\Lambda = \{(e_1, \ldots, e_m) \in \mathbb{Z}^m \mid \sum_{i=1}^{m} e_i a_i \in \mathbb{Z}^n\},$$

where $a_1, \ldots, a_m$ are the rows of A.

1. *Let M be the matrix obtained by appending the $n \times n$ identity matrix at the bottom of A.*
2. *Let $v_1, \ldots, v_m$ be a basis of the space $\{v \in \mathbb{Q}^{m+n} \mid vM = 0\}$. If necessary multiply each v_i by an integer to ensure that it has integral coefficients.*
3. *Let B be a matrix whose rows span the purification of the lattice spanned by the v_i.*
4. *Output the rows of B with the last n coefficients deleted.*

Lemma 6.2.6 *Algorithm 6.2.5 is correct.*

Proof. Note that the matrix M has rank n; therefore in Step 2 we find m linearly independent basis vectors. Set

$$\Lambda' = \{e = (e_1, \ldots, e_{m+n}) \in \mathbb{Z}^{m+n} \mid eM = 0\}.$$

Then $(e_1, \ldots, e_{m+n}) \mapsto (e_1, \ldots, e_m)$ is a bijection $\Lambda' \to \Lambda$. Now after Step 3 B is a basis of Λ'. We conclude that the output is a basis of Λ. □

Remark 6.2.7 On some occasions (in this chapter and in Chapter 4) we need an algorithm for the following task. Let $\alpha_1, \ldots, \alpha_n$ be elements of a number field F. Find a basis of the relation lattice Λ consisting of $(e_1, \ldots, e_n) \in \mathbb{Z}^n$

such that $\alpha_1^{e_1} \cdots \alpha_n^{e_n} = 1$. In [Ge93], Guoqiang Ge developed an attractive algorithm for this. It uses approximations of the logarithms of the $\tau(\alpha_i)$ where τ runs over the embeddings of F in $\mathbb{C}$, along with the celebrated LLL algorithm. We will not go further into detail on that.

6.3 Arithmetic subgroups of unipotent groups

In a 1949 paper ([Mal49]), Malcev studied lattices in nilpotent Lie groups, which is a concept similar to arithmetic groups. He showed that a dense discrete subgroup Γ of a nilpotent Lie group N is finitely generated. Roughly the proof goes as follows. If N is abelian, N is a vector space and Γ is generated by a basis of that vector space. Let $\pi : N \to N/[N,N]$ be the projection. Since $\Gamma \cap [N,N]$ is a lattice in the latter we can conclude by induction that $\Gamma \cap [N,N]$ is finitely generated. The next step is to show that $\pi(\Gamma)$ is a lattice in $N/[N,N]$. So, by the abelian case, $\pi(\Gamma)$ is finitely generated. By putting these together it is immediately seen that Γ is finitely generated. By also keeping track of the number of generators, the same induction shows that Γ has a generating set of $\dim N$ elements.

Now let $G \subset \mathrm{GL}(n, \mathbb{C})$ be a unipotent algebraic group and $L \subset \mathbb{Z}^n$ be a full-dimensional lattice. We consider the problem of finding a generating set of G_L. We could proceed as in the above outlined proof of Malcev. However, the problem is to construct the quotient $G/[G,G]$ and exactly describe $\pi(G_L)$. Here we map to a "simpler" group, but in such a way that we can characterize the image of G_L (Theorem 6.3.10).

This section is divided into three subsections. In the first subsection we collect some material on $\mathfrak{T}$-groups (the arithmetic subgroups of unipotent algebraic groups are examples of $\mathfrak{T}$-groups). In Section 6.3.2 we describe how the simpler group we mentioned above is obtained. The final subsection contains the algorithm based on this simpler group. We also recover Malcev's theorem on the number of generators (Proposition 6.3.12).

6.3.1 $\mathfrak{T}$-groups

Arithmetic subgroups of unipotent algebraic groups are finitely generated, nilpotent and torsion free. In the literature such groups are called $\mathfrak{T}$-groups. Here we summarize some properties of these groups.

A $\mathfrak{T}$-group $\mathcal{G}$ admits a series of normal subgroups

$$\mathcal{G} = \mathcal{G}_1 \supset \mathcal{G}_2 \supset \cdots \supset \mathcal{G}_{r+1} = \{1\}$$

such that $\mathcal{G}_i/\mathcal{G}_{i+1}$ is inifinite cyclic and central in $\mathcal{G}/\mathcal{G}_{i+1}$, for $1 \leq i \leq r$. Let $g_i \in \mathcal{G}_i$ be such that $g_i\mathcal{G}_{i+1}$ generates $\mathcal{G}_i/\mathcal{G}_{i+1}$. Then by induction every element of $\mathcal{G}_i$ can be written uniquely as a normal word $g_i^{e_i} \cdots g_r^{e_r}$ where $e_i \in \mathbb{Z}$. In particular, every element of $\mathcal{G}$ can be written uniquely as $g_1^{e_1} \cdots g_n^{e_n}$. We call the sequence $g_1, \ldots, g_r$ a *$\mathfrak{T}$-sequence* for $\mathcal{G}$. Such a sequence yields the following presentation of $\mathcal{G}$:

$$\mathcal{G} = \langle g_1, \ldots, g_r \mid g_i^{-1} g_j g_i = g_j w_{ij} \text{ for } 1 \leq i < j \leq r \rangle,$$

where $w_{ij} \in \mathcal{G}_{j+1}$, because $\mathcal{G}_j / \mathcal{G}_{j+1}$ is central in $\mathcal{G}/\mathcal{G}_{j+1}$. Such a presentation is called a *polycyclic presentation* of $\mathcal{G}$. The number r depends only on $\mathcal{G}$, not on the chosen series. It is called the *Hirsch length* of $\mathcal{G}$.

Now we consider the following problem. Let $\mathcal{G}$ be a $\mathfrak{T}$-group, with $\mathfrak{T}$-sequence as above. Let $\varphi : \mathcal{G} \to A$ be a homomorphism of groups, where A is an abelian group. Then $\mathcal{K} = \ker \varphi$ is also a $\mathfrak{T}$-group, and the problem is to find a $\mathfrak{T}$-sequence for $\mathcal{K}$. For this we set $a_i = \varphi(g_i)$, $1 \leq i \leq r$ and set

$$R = \{(e_1, \ldots, e_r) \in \mathbb{Z}^r \mid a_1^{e_1} \cdots a_r^{e_r} = 1\}.$$

This is a lattice in $\mathbb{Z}^r$ and therefore it has a basis $e(1), \ldots, e(m)$. By $e(l)_j$ we denote the j-th coordinate of $e(l)$. Let E denote the matrix with rows $e(1), \ldots, e(m)$. We assume that E is in echelon form, i.e., there is a sequence of integers $1 \leq i_1 < i_2 < \cdots < i_m \leq r$ such that $e(l)_j = 0$ for $1 \leq j < i_l$ and $e(l)_{i_l} \neq 0$, where $1 \leq l \leq m$. The next lemma shows that our problem is solved by defining $k_l \in \mathcal{K}$ as

$$k_l = g_1^{e(l)_1} \cdots g_r^{e(l)_r}.$$

Lemma 6.3.1 *$k_1, \ldots, k_m$ is a $\mathfrak{T}$-sequence for $\mathcal{K}$.*

Proof. Let $\mathcal{K}_l$ be the subgroup of $\mathcal{K}$ generated by $k_l, \ldots, k_m$. Also we let $\mathcal{K}_{m+1}$ be the trivial subgroup. The indices i_l are defined as above, and we also set $i_{m+1} = r + 1$, $i_0 = 0$. We claim that $\mathcal{K} \cap \mathcal{G}_j = \mathcal{K}_{l+1}$ for $i_l < j \leq i_{l+1}$ and $0 \leq l \leq m$. This is shown by descending induction on l. We start with $l = m$ for which the statement is trivial since R does not have elements $(f_1, \ldots, f_r)$ with $f_i = 0$ for $1 \leq i \leq i_m$. Now suppose the claim holds for an $l \leq m$. We show that $\mathcal{K} \cap \mathcal{G}_{i_l} = \mathcal{K}_l$. Let $g = g_{i_l}^{e_{i_l}} \cdots g_r^{e_r} \in \mathcal{K} \cap \mathcal{G}_{i_l}$. Set $e_i = 0$ for $1 \leq i < i_l$. Then $(e_1, \ldots, e_r) \in R$ and by the echelon property e_{i_l} must be a multiple of $t = e(l)_{i_l}$, say $e_{i_l} = st$. Then $k_l^{-s} g$ lies in $\mathcal{K} \cap \mathcal{G}_{i_l+1} = \mathcal{K}_{l+1}$ so that $g \in \mathcal{K}_l$. The other inclusion is trivial. Finally, for j such that $i_{l-1} < j < i_l$ we have $\mathcal{K} \cap \mathcal{G}_j = \mathcal{K} \cap \mathcal{G}_{i_l}$ as R has no elements $(f_1, \ldots, f_r)$ with $f_i = 0$ for $i < j$ and $f_j \neq 0$.

The claim implies that $\mathcal{K}_1 = \mathcal{K}$, so that $\mathcal{K}$ is generated by $k_1, \ldots, k_m$. It also implies that $\mathcal{K}_l$ is a normal subgroup of $\mathcal{K}$ (in fact, of $\mathcal{G}$) for all l. Moreover, for $g \in \mathcal{K}$, $h \in \mathcal{K}_l = \mathcal{K} \cap \mathcal{G}_{i_l}$ we have $g^{-1} h g \in \mathcal{K} \cap \mathcal{G}_{i_l+1} = \mathcal{K}_{l+1}$, so that $\mathcal{K}_l / \mathcal{K}_{l+1}$ is central in $\mathcal{K}/\mathcal{K}_{l+1}$. Obviously, $k_l \mathcal{K}_{l+1}$ generates $\mathcal{K}_l / \mathcal{K}_{l+1}$. □

6.3.2 The derived representation

In this section we introduce the main concepts and derive the main results used in the final algorithm, which will be described in the next subsection. The Lie algebra $\mathfrak{g}$ of G will play an important role: although all constructions can be defined using G only, our computations will use $\mathfrak{g}$. We recall from Section 4.3.2 that we have the maps $\log : G \to \mathfrak{g}$ and $\exp : \mathfrak{g} \to G$ which are each other's inverses. We will illustrate our constructions by a simple example.

Example 6.3.2 Let

$$G = \left\{ \begin{pmatrix} 1 & 0 & a & b \\ 0 & 1 & c & \frac{1}{2}c^2 \\ 0 & 0 & 1 & c \\ 0 & 0 & 0 & 1 \end{pmatrix} \text{ with } a, b, c \in \mathbb{C} \right\}.$$

This is a unipotent algebraic subgroup of $\mathrm{GL}(4, \mathbb{C})$. Its Lie algebra $\mathfrak{g}$ is spanned by

$$\begin{pmatrix} 0 & 0 & 1 & 0 \\ 0 & 0 & 0 & 0 \\ 0 & 0 & 0 & 0 \\ 0 & 0 & 0 & 0 \end{pmatrix}, \begin{pmatrix} 0 & 0 & 0 & 1 \\ 0 & 0 & 0 & 0 \\ 0 & 0 & 0 & 0 \\ 0 & 0 & 0 & 0 \end{pmatrix}, \begin{pmatrix} 0 & 0 & 0 & 0 \\ 0 & 0 & 1 & 0 \\ 0 & 0 & 0 & 1 \\ 0 & 0 & 0 & 0 \end{pmatrix}.$$

Let $V = \mathbb{C}^n$ be the natural G-module. There is a series of submodules $0 = V_0 \subset V_1 \subset \cdots \subset V_s = V$ of V such that for $g \in G$ and $v \in V_i$ we have $gv = v \bmod V_{i-1}$. These are also $\mathfrak{g}$-submodules, and we have $\mathfrak{g} \cdot V_i \subset V_{i-1}$, $1 \leq i < s$. Here we take a series that is as short as possible, i.e., we set $V_1 = \{v \in V \mid xv = 0 \text{ for all } x \in \mathfrak{g}\}$, $V_2 = \{v \in V \mid xv \in V_1 \text{ for all } x \in \mathfrak{g}\}$ and so on. Note that by Proposition 2.4.1 the V_i are non-zero, and by Corollary 4.2.8 the correct series is found this way. Furthermore, if $s = 1$ it is easy to find generators of G_L as G is trivial in that case.

We set $V^* = V_{s-1} \oplus V/V_1$, and note that this is a G-module (and a $\mathfrak{g}$-module) in a natural way. By $\rho : G \to \mathrm{GL}(V^*)$ we denote the corresponding representation. The corresponding representation of $\mathfrak{g}$ is the differential of ρ, $d\rho : \mathfrak{g} \to \mathfrak{gl}(V^*)$. It is straightforward to see that for $x \in \mathfrak{g}$ we have $\rho(\exp(x)) = \exp(d\rho(x))$. We call ρ the *derived representation.*

In V^* we have a full-dimensional lattice $L^* = L \cap V_{s-1} + (L + V_1)/V_1$. Furthermore, setting $V_i^* = V_i \oplus V_{i+1}/V_1$, for $0 \leq i < s$, we get a series of subspaces $0 = V_0^* \subset V_1^* \subset \cdots \subset V_{s-1}^* = V^*$ such that for $g \in G$ and $w \in V_i^*$ we have $\rho(g)w = w \bmod V_{i-1}^*$. We note that the length of this series is less than the length of the original series for V.

Example 6.3.3 Let the notation be as in Example 6.3.2 and $e_1, \ldots, e_4$ denote the standard basis vectors of $V = \mathbb{C}^4$. Then $V_1 = \langle e_1, e_2 \rangle$, $V_2 = \langle e_1, e_2, e_3 \rangle$, $V_3 = V$. Set $\bar{e}_i = e_i + V_1$ for $i = 3, 4$. Then V^* has basis $(e_1, 0)$, $(e_2, 0)$, $(e_3, 0)$, $(0, \bar{e}_3)$, $(0, \bar{e}_4)$. Expressing elements of $\mathrm{GL}(V^*)$ by their matrices relative to this basis, the derived representation looks like

$$\rho : G \to \mathrm{GL}(5, \mathbb{C}), \quad \begin{pmatrix} 1 & 0 & a & b \\ 0 & 1 & c & \frac{1}{2}c^2 \\ 0 & 0 & 1 & c \\ 0 & 0 & 0 & 1 \end{pmatrix} \mapsto \left(\begin{array}{ccc|cc} 1 & 0 & a & 0 & 0 \\ 0 & 1 & c & 0 & 0 \\ 0 & 0 & 1 & 0 & 0 \\ \hline 0 & 0 & 0 & 1 & c \\ 0 & 0 & 0 & 0 & 1 \end{array}\right).$$

We take $L = \mathbb{Z}^4$, then L^* is the $\mathbb{Z}$-span of the basis elements above, i.e., $L^* = \mathbb{Z}^5$.

Now we consider the chain of subspaces $V_1 \subset V_{s-1} \subset V_s = V$. By applying Algorithm 6.2.3 we get a basis of L such that the first l basis elements are a basis of $V_{s-1} \cap L$. By applying the algorithm again, this time to V_1 and $V_{s-1} \cap L$, we get a basis of the latter such that the first k basis elements are a basis of $V_1 \cap L$ (here $k \leq l$). So we get a basis $u_1, \dots, u_n$ of L such that $u_1, \dots, u_k$ form a basis of $V_1 \cap L$ and $u_1, \dots, u_l$ form a basis of $V_{s-1} \cap L$. Let W_1 be the subspace of V spanned by $u_{k+1}, \dots, u_n$ and W_{s-1} the subspace spanned by $u_{l+1}, \dots, u_n$. Then

$$\begin{aligned} &W_{s-1} \subset W_1 \\ &V_1 \oplus W_1 = V = V_{s-1} \oplus W_{s-1} \\ &V_1 \cap L + W_1 \cap L = L = V_{s-1} \cap L + W_{s-1} \cap L. \end{aligned} \tag{6.2}$$

Let $\xi : V \to V_1$ be the projection corresponding to the decomposition $V = V_1 \oplus W_1$. Let $\mathrm{Hom}(W_{s-1}, V_1)$ be the space of all linear maps $W_{s-1} \to V_1$. Define a linear map $\varepsilon : \mathrm{End}(V) \to \mathrm{Hom}(W_{s-1}, V_1)$, by $\varphi \mapsto \xi \circ \varphi|_{W_{s-1}}$. This map will play an important role in the sequel; in a sense it picks the piece of an $n \times n$ matrix forgotten by ρ.

Also set

$$\Lambda = \{\gamma \in \mathrm{Hom}(W_{s-1}, V_1) \mid \gamma(L \cap W_{s-1}) \subset L \cap V_1\}.$$

This is a full-dimensional lattice in $\mathrm{Hom}(W_{s-1}, V_1)$; if we represent the elements of $\mathrm{Hom}(W_{s-1}, V_1)$ by matrices with respect to bases of $L \cap W_{s-1}$ and $L \cap V_1$, Λ is the set of elements represented by integral matrices.

Example 6.3.4 Let the notation be as in Example 6.3.3. Then $W_{s-1} = \langle e_4 \rangle$ and $W_1 = \langle e_3, e_4 \rangle$. Let $a \in \mathrm{End}(V)$ and (a_{ij}) be its matrix with respect to the standard basis. Then $\varepsilon(a)$ is represented by the matrix $\left(\begin{smallmatrix} a_{14} \\ a_{24} \end{smallmatrix}\right)$ with respect to the given bases of W_{s-1} and V_1.

Let $N = \ker \rho$, which is an algebraic subgroup of G with Lie algebra $\mathfrak{n} = \ker \mathrm{d}\rho$ (Theorem 4.2.4).

Proposition 6.3.5 (i) *ε maps $\mathfrak{n}$ injectively into* $\mathrm{Hom}(W_{s-1}, V_1)$.

(ii) *For $x \in \mathfrak{n}$ we have* $\exp(x) = I_n + x$ *and* $\varepsilon(\exp(x)) = \varepsilon(x)$.

Proof. Let $x \in \mathfrak{n}$ be such that $\varepsilon(x) = 0$. Since $x \in \mathfrak{n}$, $xV \subset V_1$. So for $v \in V$ we have $\xi \circ x(v) = xv$. So $\varepsilon(x) = 0$ implies that $x(W_{s-1}) = 0$. Since also $x(V_{s-1}) = 0$ as $x \in \mathfrak{n}$, we conclude that $x = 0$.

For $x \in \mathfrak{n}$ we have $x(V) \subset V_1$ and $x(V_{s-1}) = 0$. So because $V_1 \subset V_{s-1}$ it follows that $x^2 = 0$, whence $\exp(x) = I_n + x$. The final statement follows from $\varepsilon(I_n) = 0$. □

Set $M = \varepsilon(N)$. By the previous proposition (also using Proposition 4.3.15) we see that $M = \varepsilon(\mathfrak{n})$, so that M is a subspace of $\mathrm{Hom}(W_{s-1}, V_1)$ of the same dimension as $\mathfrak{n}$.

Example 6.3.6 Let the notation be as in Example 6.3.4. Then $\mathfrak{n}$ is spanned by $\begin{pmatrix} 0&0&0&1\\ 0&0&0&0\\ 0&0&0&0\\ 0&0&0&0 \end{pmatrix}$. Hence M is spanned by $\left(\begin{smallmatrix}1\\0\end{smallmatrix}\right)$.

Lemma 6.3.7 *For $g \in G$ and $h \in N$ we have $\varepsilon(gh) = \varepsilon(g) + \varepsilon(h)$.*

Proof. Let $v \in W_{s-1}$. Since h is the identity on V/V_1 we have $h(v) - v \in V_1$. Furthermore, g is the identity on V_1, so $g(h(v) - v) = h(v) - v$. To this we apply ξ and use $\xi(v) = 0$ (since $v \in W_{s-1} \subset W_1$) to obtain $\xi \circ gh(v) - \xi \circ g(v) = \xi \circ h(v)$. □

Lemma 6.3.8 *Let G_{L^*} be the subgroup of G consisting of all $g \in G$ such that $\rho(g)L^* = L^*$. Let $g \in G_{L^*}$. Then $g \in G_L$ if and only if $\varepsilon(g) \in \Lambda$.*

Proof. Suppose $g \in G_L$. Then $g(L \cap W_{s-1}) \subset L$, so $\xi \circ g(L \cap W_{s-1}) \subset \xi(L) = L \cap V_1$, implying $\varepsilon(g) \in \Lambda$.

For the converse, suppose $\varepsilon(g) \in \Lambda$. From $g \in G_{L^*}$ we get $g(L \cap V_{s-1}) = L \cap V_{s-1}$ and $g(L) \subset L + V_1$. Furthermore, $\varepsilon(g) \in \Lambda$ amounts to $\xi \circ g(L \cap W_{s-1}) \subset L \cap V_1$. So since $\xi^{-1}(L \cap V_1) = L \cap V_1 + W_1$ we obtain $g(L \cap W_{s-1}) \subset L \cap V_1 + W_1$. Since also $g(L \cap W_{s-1}) \subset g(L) \subset L + V_1$ we have $g(L \cap W_{s-1}) \subset (L + V_1) \cap (L \cap V_1 + W_1)$. But the latter is equal to L. Therefore, since $g(L \cap V_{s-1}) \subset L \cap V_{s-1}$ we conclude that $g(L) \subset L$. Now represent g by its matrix with respect to a basis of L. This matrix has integral entries. But its determinant is 1, so its inverse is integral as well. Hence $g^{-1}(L) \subset L$ and we get $g(L) = L$. □

Proposition 6.3.9 *Let G_{L^*} be as in Lemma 6.3.8. Define $\psi : G_{L^*} \to \mathrm{Hom}(W_{s-1}, V_1)/\Lambda$ (the latter is a quotient of abelian groups) by $\psi(g) = \varepsilon(g) + \Lambda$. Then ψ is a homomorphism of groups with kernel G_L.*

Proof. Let $g_1, g_2 \in G_{L^*}$ and $l \in L$. Then $g_2(L) \subset L + V_1$ (as $g_2 \in G_{L^*}$) and $g_2(l) = l \bmod V_{s-1}$ (as $g_2 \in G$). So $g_2(l) - l \in V_{s-1} \cap (L + V_1)$. But the latter is equal to $V_1 + (L \cap V_{s-1})$. Hence $g_2(l) - l = v + l'$ where $v \in V_1$, $l' \in L \cap V_{s-1}$. So $g_1(g_2(l) - l) = v + g_1(l')$ and $g_1(l') \in L \cap V_{s-1}$, as $g_1 \in G_{L^*}$. It follows that $g_1g_2(l) - g_1(l) - g_2(l) = g_1(l') - l' - l$. Composing with ξ and using $\xi(L) = L \cap V_1$ we obtain $(\xi \circ g_1g_2 - \xi \circ g_1 - \xi \circ g_2)(l) \in L \cap V_1$. This implies that $\varepsilon(g_1g_2) - \varepsilon(g_1) - \varepsilon(g_2) \in \Lambda$, so that ψ is a group homomorphism. Lemma 6.3.8 shows that its kernel is G_L. □

Now we set $Q = \rho(G)$, which is a unipotent subgroup of $\mathrm{GL}(V^*)$. We consider its arithmetic subgroup Q_{L^*}. Note that $M+\Lambda$ is a subgroup of the abelian group $\mathrm{Hom}(W_{s-1}, V_1)$. Define the map $\Psi : Q_{L^*} \to \mathrm{Hom}(W_{s-1}, V_1)/(M + \Lambda)$ by $\Psi(q) = \varepsilon(g) + M + \Lambda$ where $g \in G$ is any preimage of $q \in Q$. We claim that this is well-defined. In order to see that, let $g, g' \in G$ be such that $\rho(g) = \rho(g') = q$. Then $g^{-1}g' \in N$ and by Lemma 6.3.7 we infer that $\varepsilon(g') = \varepsilon(gg^{-1}g') = \varepsilon(g) + \varepsilon(g^{-1}g')$. Hence $\varepsilon(g') = \varepsilon(g) \bmod M$.

Theorem 6.3.10 *The map Ψ is a homomorphism of groups with kernel $\rho(G_L)$.*

Proof. The fact that Ψ is a homomorphism of groups follows from Proposition 6.3.9. Let $q \in \rho(G_L)$ and $g \in G_L$ be such that $\rho(g) = q$. Then by Proposition 6.3.9, $\varepsilon(g) \in \Lambda$, whence $q \in \ker \Psi$. Conversely, let $q \in \ker \Psi$ and let $g \in G$ be such that $\rho(g) = q$. Then $g \in G_{L^*}$ (notation as in Lemma 6.3.8). As $q \in \ker \Psi$ there is an $m \in M$ such that $\varepsilon(g) + m \in \Lambda$. Let $h \in N$ be such that $\varepsilon(h) = m$. Then $\varepsilon(gh) \in \Lambda$ by Lemma 6.3.7. Since $N \subset G_{L^*}$ we have $gh \in G_{L^*}$. By Lemma 6.3.8 we conclude that $gh \in G_L$. But $\rho(gh) = \rho(g) = q$, so $q \in \rho(G_L)$. □

Example 6.3.11 Let the notation be as in Examples 6.3.2, 6.3.3 and 6.3.4. Then Q consists of

$$\left(\begin{array}{ccc|cc} 1 & 0 & a & 0 & 0 \\ 0 & 1 & c & 0 & 0 \\ 0 & 0 & 1 & 0 & 0 \\ \hline 0 & 0 & 0 & 1 & c \\ 0 & 0 & 0 & 0 & 1 \end{array}\right),$$

where $a, c \in \mathbb{C}$. Here $L^* = \mathbb{Z}^5$, so Q_{L^*} is a free abelian group of rank 2 generated by

$$q_1 = \left(\begin{array}{ccc|cc} 1 & 0 & 1 & 0 & 0 \\ 0 & 1 & 0 & 0 & 0 \\ 0 & 0 & 1 & 0 & 0 \\ \hline 0 & 0 & 0 & 1 & 0 \\ 0 & 0 & 0 & 0 & 1 \end{array}\right) \quad q_2 = \left(\begin{array}{ccc|cc} 1 & 0 & 0 & 0 & 0 \\ 0 & 1 & 1 & 0 & 0 \\ 0 & 0 & 1 & 0 & 0 \\ \hline 0 & 0 & 0 & 1 & 1 \\ 0 & 0 & 0 & 0 & 1 \end{array}\right).$$

In order to compute the map Ψ, we need preimages of q_1, q_2 under ρ. For example we can take

$$g_1 = \begin{pmatrix} 1 & 0 & 1 & \frac{1}{2} \\ 0 & 1 & 0 & 0 \\ 0 & 0 & 1 & 0 \\ 0 & 0 & 0 & 1 \end{pmatrix} \quad g_2 = \begin{pmatrix} 1 & 0 & 0 & 0 \\ 0 & 1 & 1 & \frac{1}{2} \\ 0 & 0 & 1 & 1 \\ 0 & 0 & 0 & 1 \end{pmatrix},$$

then $\rho(g_i) = q_i$. As $\varepsilon(g_1) = \begin{pmatrix} \frac{1}{2} \\ 0 \end{pmatrix}$ and $\varepsilon(g_2) = \begin{pmatrix} 0 \\ \frac{1}{2} \end{pmatrix}$, it follows that

$$\Psi(q_1) = \left(\begin{smallmatrix} 0 \\ 0 \end{smallmatrix}\right) + M + \Lambda, \quad \Psi(q_2) = \begin{pmatrix} 0 \\ \frac{1}{2} \end{pmatrix} + M + \Lambda.$$

Hence $\ker \Psi$ is generated by q_1 and q_2^2. Therefore, by Theorem 6.3.10, $\rho(G_L)$ is also generated by these elements.

Next we require elements of G_L that map to q_1, q_1^2. Note that $g_1 N$ is exactly the set of elements mapping to g_1. Let $h \in N$ then by Lemma 6.3.7, $\varepsilon(g_1 h) = \varepsilon(g_1) + \varepsilon(h)$. Moreover, $g_1 h \in G_L$ if and only if $\varepsilon(g_1 h) \in \Lambda$ by Lemma 6.3.8. Furthermore, there is an $x \in \mathfrak{n}$ such that $\varepsilon(x) = \varepsilon(h)$ (Proposition 6.3.5). Let

$$x = \begin{pmatrix} 0 & 0 & 0 & -\frac{1}{2} \\ 0 & 0 & 0 & 0 \\ 0 & 0 & 0 & 0 \\ 0 & 0 & 0 & 0 \end{pmatrix}.$$

Then $\varepsilon(g_1) + \varepsilon(x) = \left(\begin{smallmatrix} 0 \\ 0 \end{smallmatrix}\right)$. Using Proposition 6.3.5 again we see that by taking $h = \exp(x)$ we have $\hat{g}_1 = g_1 h \in G_L$. Furthermore, setting $\hat{g}_2 = g_2$ we see that $\hat{g}_2^2$ already lies in G_L.

Finally, we observe that the kernel of the restriction of ρ to G_L is N_L. Hence $\hat{g}_1 N_L$ and $\hat{g}_2^2 N_L$ generate G_L / N_L. Moreover, N_L is generated by

$$n = \begin{pmatrix} 1 & 0 & 0 & 1 \\ 0 & 1 & 0 & 0 \\ 0 & 0 & 1 & 0 \\ 0 & 0 & 0 & 1 \end{pmatrix}.$$

It follows that G_L is generated by $\hat{g}_1$, $\hat{g}_2^2$, $\exp(n)$ (something that in this example could also have been observed directly).

6.3.3 The algorithm

The algorithm proceeds along the lines of Example 6.3.11: first generators of Q_{L^*} are computed, followed by generators of $\ker \Psi$, which are pulled back to G and modified so that the resulting elements lie in G_L, and finally generators of N_L are added. We refrain from formulating the algorithm as a list of steps to be executed. Instead, we describe how the main computational problems in the outlined procedure can be solved. First of all, we assume that the input to the algorithm is the Lie algebra $\mathfrak{g}$ of G, given by a basis consisting of elements of $\mathfrak{gl}(n, \mathbb{Q})$. Since G is connected (Corollary 4.3.11), this determines G completely (Theorem 4.2.2). We set $V = \mathbb{C}^n$, $V_{\mathbb{Q}} = \mathbb{Q}^n$ and let $L \subset \mathbb{Z}^n$

be a full-dimensional lattice in $V_{\mathbb{Q}}$. We use the notation introduced in Section 6.3.2.

1. The first step is to compute a chain of subspaces $0 = V_0 \subset V_1 \subset \cdots \subset V_s = V$ such that $gv = v \bmod V_{i-1}$ for all $v \in V_i$, $g \in G$, and where s is as small as possible. As noted in the previous section, we can obtain this from $\mathfrak{g}$ by computing a basis of $V_i = \{v \in V \mid xv \in V_{i-1} \text{ for all } x \in \mathfrak{g}\}$ for $i > 0$. Subsequently we compute bases of V^* and L^*, and of $\mathfrak{q} = \mathrm{d}\rho(\mathfrak{g})$ (which is the Lie algebra of Q by Theorem 4.2.1).

2. The group Q, acting on V^*, admits a chain of subspaces with length strictly less than s, so by a recursive call we can compute a set of generators of Q_{L^*}. Let $q_1, \dots, q_t$ be the result.

3. As described in Section 6.3.2, by using Algorithm 6.2.3, we can compute bases of subspaces W_1, W_{s-1} with (6.2). This method also produces bases of $L \cap V_{s-1}$ and $L \cap V_1$. We use these bases to represent elements of $\mathrm{Hom}(W_{s-1}, V_1)$ as matrices. Then Λ consists of the elements represented by an integral matrix. So it is straightforward to obtain a $\mathbb{Z}$-basis of Λ. Computing $\varepsilon(a)$, for an $a \in \mathrm{End}(V)$, can be done by standard linear algebra: for $w \in W_{s-1}$ write $aw = v + w$, where $v \in V_1$, $w \in W_1$, then $\varepsilon(a)w = v_1$.

4. Next we consider computing generators of N_L. We note that $N_L = \{\exp(x) \mid x \in \mathfrak{n},\ \varepsilon(x) \in \Lambda\}$. (Indeed, if $g \in N_L$, then $g = \exp(x)$ for a certain $x \in \mathfrak{n}$. By Proposition 6.3.5, $g = I_n + x$ and $\varepsilon(g) = \varepsilon(x)$, implying $\varepsilon(x) \in \Lambda$. Conversely, if $x \in \mathfrak{n}$ is such that $\varepsilon(x) \in \Lambda$, then $x \cdot W_{s-1} \cap L \subset L$, and also $x \cdot V_{s-1} = 0$ as $x \in \ker \mathrm{d}\rho$, so that $\exp(x)L = L$.) So we compute a basis of $M = \varepsilon(\mathfrak{n})$. Subsequently we compute a basis $m_1, \dots, m_l$ and of $M \cap \Lambda$ using Algorithm 6.2.3. Then we let $x_i \in \mathfrak{n}$ be such that $\varepsilon(x_i) = m_i$ and set $n_i = \exp(x_i)$, which are the required generators of N_L.

5. For $1 \leq i \leq t$ let $y_i \in \mathfrak{g}$ be a preimage of $\log q_i$ under $\mathrm{d}\rho$, and set $g_i = \exp(y_i)$. Then $\rho(g_i) = q_i$. As remarked in Section 6.3.2, for $x \in \mathfrak{g}$ we have $\rho(\exp(x)) = \exp(\mathrm{d}\rho(x))$, so this follows from Lemma 4.3.12.

6. Let A denote the additive group $\mathrm{Hom}(W_{s-1}, V_1)/(M + \Lambda)$. Set $a_i = \Psi(g_i)$, and $R = \{(e_1, \dots, e_t) \in \mathbb{Z}^t \mid e_1 a_1 + \cdots + e_t a_t = 0\}$. Previously, we obtained a basis $\gamma_1, \dots, \gamma_k$ of Λ such that $\gamma_1, \dots, \gamma_j$ form a basis of $M \cap \Lambda$. Set $p = k - j$ and define a map $\sigma : \mathrm{Hom}(W_{s-1}, V_1) \to \mathbb{C}^p$ by $\sigma(\gamma) = (c_{j+1}, \dots, c_k)$ where the c_i are defined by $\gamma = \sum_i c_i \gamma_i$. Then σ is a linear map, and $\gamma \in M + \Lambda$ if and only if $\sigma(\gamma) \in \mathbb{Z}^p$. Set $u_i = \varepsilon(g_i)$, then $(e_1, \dots, e_t) \in R$ if and only if $\sum_i e_i u_i \in M + \Lambda$ if and only if $\sum_i e_i \sigma(u_i) \in \mathbb{Z}^p$. So we get a basis of R by applying Algorithm 6.2.5 with input the matrix whose rows are $\sigma(g_i)$. Subsequently we echelonize this basis; let $b_1, \dots, b_r$ denote the result. If $b_i = (e_1, \dots, e_t)$ then set $q_{b_1} = q_1^{e_1} \cdots q_t^{e_t}$. By Lemma 6.3.1 along with Theorem 6.3.10, the q_{b_i} form a $\mathcal{T}$-sequence for $\ker \Psi$.

7. Now we want to pull back the q_{b_i} to G_L. First, if $b_i = (e_1, \dots, e_t)$, we set $h_i = g_1^{e_1} \cdots g_t^{e_t}$. Then $\rho(h_i) = q_{b_i}$. We compute $m_i \in M$ and $\lambda_i \in \Lambda$ such that $\varepsilon(h_i) = m_i + \lambda_i$. Next, $z_i \in \mathfrak{n}$ will be such that $\varepsilon(z_i) = m_i$. Finally we set $\hat{h}_i = h_i \exp(-z_i)$. Then $\rho(\hat{h}_i) = q_{b_i}$ and $\varepsilon(\hat{h}_i) = \varepsilon(h_i) - m_i \in \Lambda$ (Lemma 6.3.7), so that $\hat{h}_i \in G_L$ (Lemma 6.3.8).

8. The elements $\hat{h}_1, \dots, \hat{h}_r$ and $n_1, \dots, n_l$ generate G_L. This is shown by the argument used in Example 6.3.11.

Proposition 6.3.12 *The output of the algorithm above is a $\mathcal{T}$-sequence for G_L. Moreover, the number of elements in this sequence equals* $\dim G$ *(so the Hirsch length of G_L is* $\dim G$*).*

Proof. First we claim that $\mathfrak{n}$ is central in $\mathfrak{g}$. In order to see that, let $x \in \mathfrak{g}$, $y \in \mathfrak{n}$ and $v \in V$. Since y acts trivially on V/V_1 we have $y \cdot v \in V_1$, whence $xy \cdot v = 0$. Furthermore, $x \cdot v \in V_{s-1}$, on which y acts trivially as well, so that $yx \cdot v = 0$. It follows that $[x, y] \cdot v = 0$. This holds for all $v \in V$, hence $[x, y] = 0$.

It follows that N is central in G (indeed, every $\exp(y)$, $y \in \mathfrak{n}$ commutes with every $\exp(x)$, $x \in \mathfrak{g}$) and N_L is central in G_L. By induction the $\hat{h}_i N_L$, $1 \leq i \leq r$ form a $\mathcal{T}$-sequence of G_L/N_L and $r = \dim Q$, which is equal to $\dim G - \dim N$. It follows that $\hat{h}_1, \dots, \hat{h}_r, n_1, \dots, n_l$ form a $\mathcal{T}$-sequence of G_L. Moreover, since $l = \dim \mathfrak{n} = \dim N$ we obtain $l + r = \dim G$. □

Remark 6.3.13 The dimension of V^* in general is greater than the dimension of V. In the worst case it can be about twice the dimension of V. It is not difficult to construct examples of groups G where this happens repeatedly: in each step of the recursion the dimension of the underlying vector space roughly doubles. So for those kinds of groups the algorithm has an exponential complexity.

6.4 Arithmetic subgroups of tori

Let $G \subset \mathrm{GL}(n, \mathbb{C})$ be a torus, that is, a connected diagonalizable algebraic group defined over $\mathbb{Q}$. Here we outline an algorithm to obtain generators of $G(\mathbb{Z})$. We assume that G is given by its Lie algebra $\mathfrak{g} \subset \mathfrak{gl}(n, \mathbb{C})$, which in turn is given by a basis consisting of matrices with coefficients in $\mathbb{Q}$.

This section is divided into two subsections. In the first we show how to obtain generators of the unit group of an order in a semisimple commutative matrix algebra. This algorithm is of interest in its own right (for example, it can be applied to obtain generators of the unit group of an integral abelian group ring). In the second subsection we describe the algorithm for tori: the first step is to embed a torus in a larger torus that is the unit group of a commutative associative algebra, the second step is to obtain generators of the arithmetic group corresponding to this larger algebraic group, using the algorithm of the first subsection, and the final step is to describe the original torus as an intersection of kernels of characters, which can then be used to obtain generators of the arithmetic group we started with.

6.4.1 Unit groups of orders in toral matrix algebras

We say that an associative algebra over $\mathbb{Q}$ is *toral* if it is semisimple, abelian and has an identity element. Here we fix such an algebra A, and denote its unit element by e. Let $\mathcal{O}$ be an order in A (see Example 6.1.8). By $\mathcal{O}^*$ we denote the unit group of $\mathcal{O}$, that is $\mathcal{O}^* = \{a \in \mathcal{O} \mid ab = 1 \text{ for some } b \in \mathcal{O}\}$. In this section we describe an algorithm for computing a set of generators of $\mathcal{O}^*$.

By the theorem of Wedderburn-Artin (see [Hun80], Section IX.3) a toral algebra A is a direct sum $A = A_1 \oplus \cdots \oplus A_s$ where the A_i are ideals that are simple associative algebras over $\mathbb{Q}$, implying that they are isomorphic to field extensions of $\mathbb{Q}$ of finite degree.

A non-zero $\epsilon \in A$ is said to be an *idempotent* if $\epsilon^2 = \epsilon$. Two idempotents ϵ_1, ϵ_2 are *orthogonal* if $\epsilon_1\epsilon_2 = 0$. An idempotent is said to be *primitive* if it is not the sum of orthogonal idempotents.

The decomposition of A as a direct sum of simple ideals, as above, corresponds to a decomposition of e as sum of primitive orthogonal idempotents $e = e_1 + \cdots + e_s$. Here $A_i = e_iA$, and conversely, e_i is the identity element of A_i. Given a basis of A we can compute the e_i as follows. Let $a_0 \in A$ be a splitting element (see Section 4.5.2) and $p \in \mathbb{Q}[x]$ be its minimal polynomial. Factorize $p = p_1 \cdots p_s$, where the p_i are irreducible and pairwise distinct. (As A is toral it has no nilpotent elements, and therefore p is square-free.) Let $q_i = p_1 \cdots p_{i-1}p_{i+1} \cdots p_s$. Then $\gcd(q_1, \ldots, q_s) = 1$, hence there are polynomials f_i such that $f_1q_1 + \cdots + f_rq_r = 1$. Setting $e_i = h_iq_i(a_0)$ yields the required idempotents. We refer to [Gra00], Lemma A.2.1 for a detailed proof of this.

As remarked above, $A_i = e_iA$ is isomorphic to a field extension of $\mathbb{Q}$ of finite degree. Furthermore, $e_i\mathcal{O}$ is an order in A_i. So by using the known algorithms for computing generators of the unit group of an order in a number field (see Example 6.1.8), we can compute generators of $(e_i\mathcal{O})^*$.

The main part of the algorithm is a procedure to combine the generators of $(e_i\mathcal{O})^*$ into generators of $\mathcal{O}^*$. To describe this procedure we consider a subalgebra B of A, an order $\mathcal{P}$ of B and two orthogonal (but not necessarily primitive) idempotents $\epsilon_1, \epsilon_2 \in B$ such that $\epsilon_1 + \epsilon_2 = \epsilon$, where ϵ is the identity element of B. We assume that we have generators of the unit groups of the orders $\epsilon_i\mathcal{P}$, and we seek generators of $\mathcal{P}^*$. In the course of our description we occasionally give a construction focusing on $\epsilon_1\mathcal{P}$: in those cases the same construction works for $\epsilon_2\mathcal{P}$.

We set $J = (\epsilon_1\mathcal{P} \cap \mathcal{P}) + (\epsilon_2\mathcal{P} \cap \mathcal{P})$, which is an ideal of $\mathcal{P}$. So we can form the quotient ring $R = \mathcal{P}/J$. Observe that there is an integer $m > 0$ such that $m\epsilon_i \in \mathcal{P}$. This implies $m\mathcal{P} \subset J \subset \mathcal{P}$; thus R is a finite ring.

Define maps $\varphi_i : \epsilon_i\mathcal{P} \to R$ by $\varphi_i(\epsilon_i a) = a + J$. They are well defined: if $\epsilon_1 a = \epsilon_1 b$ for some $a, b \in \mathcal{P}$, then $a - b = \epsilon_2(a - b)$, whence $a - b \in J$. The φ_i are surjective ring homomorphisms with respective kernels $\epsilon_i\mathcal{P} \cap \mathcal{P}$.

Proposition 6.4.1 $\mathcal{P} = \{a_1 + a_2 \mid a_i \in \epsilon_i\mathcal{P}$ *and* $\varphi_1(a_1) = \varphi_2(a_2)\}$.

Proof. Let $a \in \mathcal{P}$. Set $a_i = \epsilon_i a$; then $a = a_1 + a_2$, $a_i \in \epsilon_i\mathcal{P}$ and $\varphi_1(a_1) = a+J = \varphi_2(a_2)$. Conversely, let $a_i \in \epsilon_i\mathcal{P}$ be such that $\varphi_1(a_1) = \varphi_2(a_2)$ and $a, b \in \mathcal{P}$ be such that $a_1 = \epsilon_1 a$, $a_2 = \epsilon_2 b$. Then $a+J = b+J$, and therefore $a-b \in J$. So $a = b+u_1+u_2$, where $u_i \in \epsilon_i\mathcal{P}\cap\mathcal{P}$. Multiplying by ϵ_1 we obtain $\epsilon_1 a = \epsilon_1 b+u_1$, and $a_1+a_2 = \epsilon_1 b+u_1+\epsilon_2 b = \epsilon b+u_1 = b+u_1$, which lies in $\mathcal{P}$. □

Corollary 6.4.2 $\mathcal{P}^* = \{a_1 + a_2 \mid a_i \in (\epsilon_i\mathcal{P})^*$ *and* $\varphi_1(a_1) = \varphi_2(a_2)\}$.

The procedure is directly based on this corollary. We first list its main steps, and then comment in detail how they can be carried out.

1. Compute a basis of J, and construct $R = \mathcal{P}/J$.
2. Compute generators of the groups $H_i = \varphi_i((\epsilon_i\mathcal{P})^*) \subset R^*$ and $H = H_1 \cap H_2$.
3. Compute generators of the groups $\overline{H}_i = \varphi_i^{-1}(H) \subset (\epsilon_i\mathcal{P})^*$.
4. Compute generators of $\mathcal{P}^* = \{a_1 + a_2 \mid a_i \in \overline{H}_i$ and $\varphi_1(a_1) = \varphi_2(a_2)\}$.

For the first step we note that $\mathcal{P}$ is a free $\mathbb{Z}$-module of finite rank, and hence so is J. Moreover, the union of bases of $\epsilon_i\mathcal{P}\cap\mathcal{P}$ is a basis of J. We indicate how a basis of $\epsilon_1\mathcal{P} \cap \mathcal{P}$ can be obtained. Let $a_1, \dots, a_r$ be a basis of $\mathcal{P}$. Set $b_i = \epsilon_1 a_i$. The b_i form a generating set of the $\mathbb{Z}$-module $\epsilon_1\mathcal{P}$. Write $b_i = \sum_{j=1}^r \alpha_{ij} a_j$ where $\alpha_{ij} \in \mathbb{Q}$ and set $\bar{\alpha}_i = (\alpha_{i1}, \dots, \alpha_{ir})$. Then a $v = \sum_{i=1}^r \lambda_i b_i$ with $\lambda_i \in \mathbb{Z}$ lies in $\epsilon_1\mathcal{P} \cap \mathcal{P}$ if and only if $\sum_i \lambda_i\bar{\alpha}_i \in \mathbb{Z}^r$. Define

$$\Lambda = \{(\lambda_1, \dots, \lambda_r) \in \mathbb{Z}^r \mid \sum_{i=1}^r \lambda_i\alpha_i \in \mathbb{Z}^r\}.$$

Then $\lambda = (\lambda_1, \dots, \lambda_r) \mapsto v_\lambda = \sum_i \lambda_i b_i$ is a surjective $\mathbb{Z}$-linear map $\Lambda \to \epsilon_1\mathcal{P} \cap \mathcal{P}$. Using Algorithm 6.2.5 we compute a basis of Λ. The images of the basis elements form a generating set of $\epsilon_1\mathcal{P}\cap\mathcal{P}$. From that it is straightforward to obtain a basis of $\epsilon_1\mathcal{P} \cap \mathcal{P}$.

Using a Smith normal form computation and Lemma 6.2.1 we can find positive integers $d_1, \dots, d_s$, and an explicit $\mathbb{Z}$-linear map $\phi : \mathcal{P} \to M = \mathbb{Z}/d_1\mathbb{Z} \oplus \cdots \oplus \mathbb{Z}/d_s\mathbb{Z}$ with kernel J. We use ϕ to define a multiplication on M, making M into a finite ring, isomorphic to R.

In the second step we use the given generators of the $(\epsilon_i\mathcal{P})^*$. So we have generators of the groups H_i. Since R is a finite ring we can list the elements of the groups H_i and compute the intersection H. However, for a subsequent step we need a *normal set* of generators of H, that is, a set $h_1, \dots, h_l \in H$ such that the order of h_i is $m_i > 0$ and $h_1^{k_1} \cdots h_l^{k_l} \mapsto ([k_1]_{m_1}, \dots, [k_l]_{m_l})$ (notation as in Lemma 6.2.1) is an isomorphism $H \to \mathbb{Z}/m_1\mathbb{Z} \oplus \cdots \oplus \mathbb{Z}/m_l\mathbb{Z}$. Since H is a finite group, there are several ("brute-force") methods to obtain such a set. For example, let $\widetilde{H}$ be a subgroup with normal generating set $\tilde{h}_1, \dots, \tilde{h}_t$ and $\tilde{h} \in H \setminus \widetilde{H}$. By finding the smallest exponent m such that $\tilde{h}^m \in \widetilde{H}$ we find a basis of the lattice of multiplicative relations between the elements $\tilde{h}_1, \dots, \tilde{h}_t, \tilde{h}$. Then by using a Smith normal form computation and Lemma 6.2.1, we find a set of normal generators of the subgroup generated by $\widetilde{H}$ and $\tilde{h}$. We replace $\widetilde{H}$ by that group and continue.

In the sequel we let $\{h_1, \dots, h_l\}$ denote a fixed set of normal generators of H. Also we compute sets of normal generators of H_1 and H_2. We denote the order of a group element g by $|g|$.

In order to compute generators of the groups $\overline{H}_i$ in the third step, we first construct generators of the kernel of φ_i (seen as group homomorphism $(\epsilon_i\mathcal{P})^* \to R^*$), to which we add inverse images of the h_i. We outline how to obtain generators of the kernel of φ_1. Let $x_1, \dots, x_k$ be generators of $(\epsilon_1\mathcal{P})^*$. Let $\{y_1, \dots, y_p\}$ be the set of normal generators of H_1, as computed previously. Define integers γ_{ij} by $\varphi_1(x_i) = \prod_j y_j^{\gamma_{ij}}$. Then $x_1^{n_1} \cdots x_k^{n_k} \in \ker \varphi_1$ if and only if $\sum_{i=1}^k n_i\gamma_{ij} = 0 \bmod |y_j|$ for $1 \le j \le p$. We form the matrix C with columns $(\gamma_{1j}, \dots, \gamma_{kj}, |y_j|)$, $1 \le j \le p$. Let $\mathcal{L}$ be the lattice of all $v \in \mathbb{Z}^{k+1}$ such that $vC = 0$. A basis of $\mathcal{L}$ can, for example, be computed with Algorithm 6.2.3. A basis element $(n_1, \dots, n_k, z)$ is then mapped to $x_1^{n_1} \cdots x_k^{n_k}$. All elements so obtained form a generating set of $\ker \varphi_1$.

For the final step we let $u_1, \dots, u_p, v_1, \dots, v_q$ be generating sets of $\overline{H}_1$, $\overline{H}_2$ respectively. Set

$$L = \{(\alpha_1, \dots, \alpha_p, \beta_1, \dots, \beta_q) \in \mathbb{Z}^{p+q} \mid \varphi_1(u_1^{\alpha_1} \cdots u_p^{\alpha_p}) = \varphi_2(v_1^{\beta_1} \cdots v_q^{\beta_q})\}.$$

Then by Corollary 6.4.2,

$$\mathcal{P}^* = \{u_1^{\alpha_1} \cdots u_p^{\alpha_p} + v_1^{\beta_1} \cdots v_q^{\beta_q} \mid (\alpha_1, \dots, \alpha_p, \beta_1, \dots, \beta_q) \in L\}.$$

Moreover, L is a lattice, hence has a finite basis. Furthermore, the elements of $\mathcal{P}^*$ corresponding to the elements of a basis of L generate $\mathcal{P}^*$. So the problem of finding a generating set of $\mathcal{P}^*$ is reduced to finding a basis of L. This can be done in similar fashion to the construction of generators of $\ker \varphi_1$ above. Define $\mu_{ij}, \nu_{ij} \in \mathbb{Z}$ by $\varphi_1(u_i) = \prod_{j=1}^l h_j^{\mu_{ij}}$, $\varphi_2(v_i) = \prod_{j=1}^l h_j^{\nu_{ij}}$. Then $(\alpha_1, \dots, \alpha_p, \beta_1, \dots, \beta_q) \in L$ if and only if

$$\sum_{i=1}^{p} \mu_{ij}\alpha_i - \sum_{k=1}^{q} \nu_{kj}\beta_k = 0 \bmod |h_j| \text{ for } 1 \le j \le l.$$

Let S be the integral matrix with columns $(\mu_{1j},\dots,\mu_{pj},-\nu_{1j},\dots,-\nu_{qj},|h_j|)$ for $1\leq j\leq l$. Using Algorithm 6.2.3 we obtain a basis of the lattice consisting of all $v\in\mathbb{Z}^{p+q+1}$ such that $vS=0$. For each v in this basis we take the vector consisting of the first $p+q$ coordinates. This way we obtain a basis of L.

This ends the intermezzo on finding generators of $\mathcal{P}^*$, and we return to our original setting. For $1\leq j<s$ set $E_j=e_1+\cdots+e_j$ and suppose we have generators of $(E_j\mathcal{O})^*$ (this is certainly true for $j=1$). Setting $B=E_{j+1}A$, $\mathcal{P}=E_{j+1}\mathcal{O}$, $\epsilon_1=E_j$, $\epsilon_2=e_{j+1}$ and using the procedure described above we find generators of $(E_{j+1}\mathcal{O})^*$. After $s-1$ iterations we find generators of $(E_s\mathcal{O})^*=\mathcal{O}^*$.

Example 6.4.3 Let $A\subset M_4(\mathbb{Q})$ be the associative algebra with one generated by

$$a=\begin{pmatrix}0&1&0&0\\0&0&1&0\\0&0&0&1\\-16&0&10&0\end{pmatrix}.$$

The minimal polynomial of a is $f=x^4-10x^2+16$ (a is the companion matrix of that polynomial). So $\dim A=4$. Let $\mathcal{O}=A\cap M_4(\mathbb{Z})$, then $\mathcal{O}$ is spanned by I_4,a,a^2,a^3. In the sequel we also write 1 instead of I_4 to ease the notation a bit. We have $f=(x^2-2)(x^2-8)$ and $\frac{1}{6}(x^2-2)-\frac{1}{6}(x^2-8)=1$. So we have two idempotents, $e_1=\frac{1}{6}(a^2-2)$ and $e_2=-\frac{1}{6}(a^2-8)$.

Set $b_i=e_ia$, $i=1,2$. Then $e_i\mathcal{O}$ is spanned (over $\mathbb{Z}$) by e_i, b_i. Furthermore, $b_1^2-8e_1=0$, so $e_1A\cong\mathbb{Q}(\sqrt{2})$ and $e_1\mathcal{O}$ is isomorphic to the order spanned by $1,2\sqrt{2}$. The unit group of this order is generated by $-1,-2\sqrt{2}+3$ (this is established by a short calculation in MAGMA). So $(e_1\mathcal{O})^*$ is generated by $-e_1$ and $-b1+3e_1$. Similarly, $(e_2\mathcal{O})^*$ is generated by $-e_2$ and $-b_2+e_2$.

Obviously, $e_1\mathcal{O}\cap\mathcal{O}$ and $e_2\mathcal{O}\cap\mathcal{O}$ are spanned by $6e_1=a^2-2$, $6b_1=a^3-2a$ and $6e_2=a^2-8$, $6b_2=a^3-8a$ respectively. So we immediately have a basis of J. By a Smith normal form computation we find the following basis of $\mathcal{O}$: $v_1=-2+a^2$, $v_2=-2a+a^3$, $v_3=1-2a+a^3$, $v_4=-a$. This basis has the property that sending $v=\sum_i m_iv_i\in\mathcal{O}$ to $([m_3]_6,[m_4]_6)$ (notation as in Lemma 6.2.1) defines a homomorphism of abelian groups $\mathcal{O}\to\mathbb{Z}/6\mathbb{Z}\oplus\mathbb{Z}/6\mathbb{Z}$ with kernel J. So as abelian groups, $R\cong\mathbb{Z}/6\mathbb{Z}\oplus\mathbb{Z}/6\mathbb{Z}$.

Next we define a multiplication on $\mathbb{Z}/6\mathbb{Z}\oplus\mathbb{Z}/6\mathbb{Z}$ so that the map above defines a ring homomorphism. Write $\omega_1=([1]_6,[0]_6)$, $\omega_2=([0]_6,[1]_6)$. Then ω_1 and ω_2 are the images of v_3 and v_4 respectively. Computing products in $\mathcal{O}$ and mapping back to $\mathbb{Z}/6\mathbb{Z}\oplus\mathbb{Z}/6\mathbb{Z}$, we obtain $\omega_1^2=\omega_1$, $\omega_1\omega_2=\omega_2$, $\omega_2^2=2\omega_1$.

Now $\varphi_1(-e_1)=-\omega_1$, $\varphi_1(-b_1+3e_1)=3\omega_1+\omega_2$. Let x denote this last element and H_1 be the group generated by $-\omega_1$, x. Then $H_1=\{\omega_1,-\omega_1,x,-x\}$ and $x^2=-\omega_1$. So H_1 is cyclic with generator x. Similarly, $\varphi_2(-e_2)=-\omega_1$ and $\varphi_2(-b_2+e_2)=\omega_1+\omega_2$, which we denote by y. Let H_2 be the group generated by $-\omega_1$, y. Then H_2 is cyclic of order 8 with generator y (and $y^4=-\omega_1$). By explicitly computing all elements of the H_i we obtain $H=H_1\cap H_2=\{\omega_1,-\omega_1\}$.

Now we compute generators of $\ker\varphi_i$. First, $\varphi_1(-e_1) = x^2$, $\varphi_1(-b_1+3e_1) = x$. So we consider the matrix with single column $(2,1,4)$ (as the order of x is 4). A basis of the kernel of this matrix consists of $(1,2,-1)$, $(0,4,-1)$. Hence $\ker\varphi_1$ is generated by $(-e_1)(-b_1+3e_1)^2$ and $(-b_1+3e_1)^4$. Observe that the second element is the square of the first, so the kernel is generated by $(-e_1)(-b_1+3e_1)^2$. Setting $\overline{H}_1 = \varphi_1^{-1}(H)$ we see that $\overline{H}_1$ is generated by $-e_1$, $(-b_1+3e_1)^2$.

By analogous computations we see that the kernel of φ_2 is generated by $(-e_2)(-b_2+e_2)^4$ and $\overline{H}_2 = \varphi_2^{-1}(H)$ by $-e_2$, $(-b_2+e_2)^4$.

Finally we compute generators of $\mathcal{O}^*$. The images of the generators of $\overline{H}_1$ under φ_1 are $-\omega_1$, $-\omega_1$. Exactly the same holds for the generators of $\overline{H}_2$. So we consider the matrix with single column $(1,1,-1,-1,2)$ (the order of $-\omega_1$ is 2). A basis of the integral kernel of this matrix consists of $(1,0,0,1,0)$, $(0,1,0,1,0)$, $(0,0,1,1,1)$, $(0,0,0,2,1)$. This yields the following generators of $\mathcal{O}^*$: $c_1 = -e_1+(-b_2+e_2)^4$, $c_2 = (-b_1+3e_1)^2+(-b_2+e_2)^4$, $c_3 = e_1 - e_2(-b_2+e_2)^4$, $c_4 = (-b_2+e_2)^8$. It is straightforward to see that $c_3 = -c_1$ and $c_4 = c_3^2$. So $\mathcal{O}^*$ is generated by $-I_4$, c_1, c_2. We have

$$c_1 = \begin{pmatrix} 23 & -16 & -3 & 2 \\ -32 & 23 & 4 & -3 \\ 48 & -32 & -7 & 4 \\ -64 & 48 & 8 & -7 \end{pmatrix}, \quad c_2 = \begin{pmatrix} 17 & -14 & 0 & 1 \\ -16 & 17 & -4 & 0 \\ 0 & -16 & 17 & -4 \\ 64 & 0 & -56 & 17 \end{pmatrix}.$$

Remark 6.4.4 Note that in the application of the main step of the algorithm, $\epsilon_2 B$ is always a number field and $\epsilon_2\mathcal{P}$ is an order in it. Moreover, $R \cong \epsilon_2\mathcal{P}/(\epsilon_2\mathcal{P}\cap\mathcal{P})$.

In [HPP03] and [KP05] algorithms are given to compute the structure of the groups $(\mathcal{Q}/I)^*$ where $\mathcal{Q}$ is an order in a number field and I is an ideal in $\mathcal{Q}$. So we can use these to compute a normal set of generators of R^*. Using such a set makes the calculation of sets of normal generators of the groups H_1, H_2 and H much easier (in particular, we do not need to use the brute force methods outlined above).

Remark 6.4.5 Let $\mathcal{G}$ be a finite abelian group. Then $\mathbb{Q}\mathcal{G}$ is a toral algebra. The algorithm of this section can be used to obtain generators of the unit group $(\mathbb{Z}\mathcal{G})^*$. In [FGP13] practical experiences with these computations are reported and the indices of the group of Hoechsmann units in $(\mathbb{Z}\mathcal{G})^*$ are obtained for $\mathcal{G}$ up to order 110.

6.4.2 Generators of arithmetic subgroups of tori

As before we let $G \subset \mathrm{GL}(n,\mathbb{C})$ be a torus defined over $\mathbb{Q}$ with Lie algebra $\mathfrak{g} \subset \mathfrak{gl}(n,\mathbb{C})$. We assume that we have a basis of $\mathfrak{g}$ consisting of elements in $\mathfrak{gl}(n,\mathbb{Q})$. Let A be the associative algebra with one over $\mathbb{Q}$ generated by $\mathfrak{g}$. Since $\mathfrak{g}$ consists of commuting semisimple matrices, A is toral.

We let A_c be the associative algebra with one over $\mathbb{C}$ generated by $\mathfrak{g}$. Then

A is a $\mathbb{Q}$-form of A_c. Furthermore, the set of invertible elements denoted A_c^* is a diagonalizable algebraic group defined over $\mathbb{Q}$ whose Lie algebra contains $\mathfrak{g}$ (see Example 3.6.9). Therefore, $G \subset A_c^*$ (Theorem 4.2.2). Note that A_c^* is an open set in a vector space, and therefore is connected.

Let $\mathfrak{a} = \mathrm{Lie}(A_c)$; as a vector space this is the same as A_c. Section 4.5.2 shows how to obtain an extension field $k \supset \mathbb{Q}$, and $X \in \mathrm{GL}(n, k)$ such that $X\mathfrak{a}X^{-1}$ consists of diagonal matrices. This implies the same for $XA_c^*X^{-1}$ and hence for XGX^{-1} and $X\mathfrak{g}X^{-1}$ as well. The same section has a procedure for finding a basis of a pure lattice $\Lambda \subset \mathbb{Z}^n$ such that $XA_c^*X^{-1} = \mathrm{D}(\Lambda)$ (notation as in Example 3.9.6). We use the same procedure to find a basis of a pure lattice L such that $XGX^{-1} = \mathrm{D}(L)$. As $G \subset A_c^*$ we see that $\Lambda \subset L$. By applying Algorithm 6.2.3 twice we find a basis $v_1, \dots, v_n$ of $\mathbb{Z}^n$ such that the first r basis elements form a basis of Λ, and the first s basis elements are a basis of L (so $r \leq s$).

Write $v_i = (d_1, \dots, d_n)$ and define $\chi_i : \mathrm{D}(n, \mathbb{C}) \to \mathbb{C}^*$ by $\chi_i(\mathrm{diag}(\alpha_1, \dots, \alpha_n)) = \alpha_1^{d_1} \cdots \alpha_n^{d_n}$. For $1 \leq i \leq s$ define $\bar{\chi}_i : A_c^* \to \mathbb{C}^*$ by $\bar{\chi}_i(g) = \chi_i(XgX^{-1})$. Then the $\bar{\chi}_i$ are characters of A_c^*. Moreover, as $XGX^{-1} = \{g \in XA_c^*X^{-1} \mid \chi_i(g) = 1 \text{ for } r+1 \leq i \leq s\}$ we have

$$G = \{g \in A_c^* \mid \bar{\chi}_i(g) = 1 \text{ for } r+1 \leq i \leq s\}.$$

Set $t = s - r$. We have constructed an extension k of $\mathbb{Q}$ of finite degree and characters $\psi_1, \dots, \psi_t$ of A_c^* such that G is the intersection of the kernels of the ψ_i (i.e., $\psi_i = \bar{\chi}_{r+i}$).

Set $\mathcal{O} = A \cap M_n(\mathbb{Z})$. The purification algorithm (Section 6.2) can be used to construct a basis of $\mathcal{O}$. Then by the algorithm of Section 6.4.1 we can compute a generating set $\{a_1, \dots, a_m\}$ of $\mathcal{O}^*$. For $1 \leq i \leq m$ we use the algorithm indicated in Remark 6.2.7 to compute a basis of the lattice $\mathcal{L}_i = \{(e_1, \dots, e_m) \in \mathbb{Z}^m \mid \psi_i(a_i)^{e_1} \cdots \psi_i(a_m)^{e_m} = 1\}$. Next we compute a basis of the intersection of all $\mathcal{L}_i$ and let a basis element $(e_1, \dots, e_m)$ of that correspond to the element $a_1^{e_1} \cdots a_m^{e_m}$. By $g_1, \dots, g_l$ we denote all elements of $\mathcal{O}^*$ so obtained.

We note that $G(\mathbb{Z}) = \{a \in \mathcal{O}^* \mid \psi_i(a) = 1 \text{ for } 1 \leq i \leq t\}$. In view of the above constructions, this immediately implies that $G(\mathbb{Z})$ is generated by $g_1, \dots, g_l$.

Example 6.4.6 Let a be as in Example 6.4.3. As in Section 4.3.1 we let $\mathfrak{g}(a)$ denote the algebraic hull of a. The eigenvalues of a are $\sqrt{2}, -\sqrt{2}, 2\sqrt{2}, -2\sqrt{2}$. Hence $\Lambda_{\mathbb{Q}}$ (notation as in Section 4.4) is spanned by $(1,1,0,0)$, $(0,2,1,0)$, $(0,0,1,1)$. This implies that $\dim \mathfrak{g}(a) = 1$, and therefore $\mathfrak{g}(a)$ is spanned by a. $G(a) \subset \mathrm{GL}(4, \mathbb{C})$ denotes the connected algebraic group with Lie algebra $\mathfrak{g}(a)$. We compute generators of $G(a)(\mathbb{Z})$.

We first compute characters that define $G(a)$ as a subgroup of A_c^* (where A, as in Example 6.4.3, is the associative algebra with one generated by a).

Set

$$X = \begin{pmatrix} 16 & 8\sqrt{2} & -2 & -\sqrt{2} \\ 16 & -8\sqrt{2} & -2 & \sqrt{2} \\ 16 & 4\sqrt{2} & -8 & -2\sqrt{2} \\ 16 & -4\sqrt{2} & -8 & 2\sqrt{2} \end{pmatrix}.$$

Then $XaX^{-1} = \mathrm{diag}(\sqrt{2}, -\sqrt{2}, 2\sqrt{2}, -2\sqrt{2})$. Set $\tilde{\mathfrak{g}} = X\mathfrak{g}(a)X^{-1}$ and $\widetilde{G} = XG(a)X^{-1}$. Then

$$\tilde{\mathfrak{g}} = \{\mathrm{diag}(d_1, d_2, d_3, d_4) \mid d_i \in \mathbb{C} \text{ and } d_1 + d_2 = 2d_2 + d_3 = d_3 + d_4 = 0\}.$$

In other words, $\tilde{\mathfrak{g}}$ is equal to $\mathfrak{d}(\Lambda)$ where $\Lambda \subset \mathbb{Z}^4$ is the lattice spanned by the same basis vectors as $\Lambda_{\mathbb{Q}}$. Since $\dim A = 4$, we have $\widetilde{A}_c^* = XA_c^*X^{-1}$ equal to the group $\mathrm{D}(4, \mathbb{C})$. Define $\chi_i : \mathrm{D}(4, \mathbb{C}) \to \mathbb{C}^*$ $(i = 1, 2, 3)$ by $\chi_1(\bar{\beta}) = \beta_1\beta_2$, $\chi_2(\bar{\beta}) = \beta_2^2\beta_3$, $\chi_3(\bar{\beta}) = \beta_3\beta_4$, where $\bar{\beta} = \mathrm{diag}(\beta_1, \beta_2, \beta_3, \beta_4) \in \mathrm{D}(4, \mathbb{C})$. Then $\widetilde{G} = \{\bar{\beta} \in \mathrm{D}(4, \mathbb{C}) \mid \chi_i(\bar{\beta}) = 1\}$. Define $\psi_i : A_c^* \to \mathbb{C}^*$ by $\psi_i(c) = \chi_i(XcX^{-1})$. The ψ_i, for $i = 1, 2, 3$ are characters of A_c^* and $G(a)$ is the intersection of their kernels.

As seen in Example 6.4.3, $A_c^*(\mathbb{Z})$ is generated by $-I_4$, c_1, c_2. Now $\psi_1(c_1) = 1$, $\psi_2(c_1) = -408\sqrt{2} - 577$, $\psi_3(c_1) = 1$, $\psi_1(c_2) = 1$, $\psi_2(c_2) = 12\sqrt{2} + 17$, $\psi_3(c_2) = 1$. Furthermore, $\psi_1(-I_4) = \psi_3(-I_4) = 1$ and $\psi_2(-I_4) = -1$. So an element $(-I_4)^k c_1^l c_2^m$ lies in $G(a)(\mathbb{Z})$ if and only if $(-1)^k(-408\sqrt{2}-577)^l(12\sqrt{2}+17)^m = 1$. Let $\mathcal{L} \subset \mathbb{Z}^3$ be the lattice consisting of all (k, l, m) satisfying this condition. A basis of $\mathcal{L}$ consists of $(-1, 1, -2)$, $(-2, 0, 0)$. Hence $G(a)(\mathbb{Z})$ is generated by $(-I_4)^{-1}c_1c_2^{-2}$ and $(-I_4)^{-2}$. However, the second element is the identity. Therefore $G(a)(\mathbb{Z})$ is generated by a single element:

$$(-I_4)^{-1}c_1c_2^{-2} = \begin{pmatrix} -215 & -84 & 99 & 36 \\ -576 & -215 & 276 & 99 \\ -1584 & -576 & 775 & 276 \\ -4416 & -1584 & 2184 & 775 \end{pmatrix}.$$

6.5 Notes

For more in-depth treatments of the theory behind Theorem 6.1.6 we refer to [Bor69] and [PR94].

The account of Malcev's proof in Section 6.3 is borrowed from Raghunathan [Rag72].

Section 6.3 is based on [GP09], which contains some of the main results of the Ph.D. thesis of Andrea Pavan [Pav09]. The latter thesis also contains an algorithm for the case where G is a torus. Our treatment of that case in Section 6.4 is based on [FGP13].

Chapter 7

Invariants of Algebraic Groups

The roots of invariant theory can be traced back at least to Gauss who, in his *Disquisitiones Arithmeticae*, showed that the discriminant of a binary form in two indeterminates upon a linear change of variables changes with the square of the determinant of the change of variables (so that the discriminant is an invariant under the group $\mathrm{SL}(2, \mathbb{C})$). In the 19th century the subject saw tremendous progress made by some of the most famous mathematicians of the time and culminating in the work of Hilbert, who proved, among other results, that the invariant ring of a reductive group is finitely generated. From the outset, there has been a strong computational side to the theory of invariants. As an example we mention the determination of an explicit set of generating invariants of the group $\mathrm{SL}(2, \mathbb{C})$ acting on the space of homogeneous polynomials of degree n (see Example 7.1.3). This is a complex problem, and only recently has the case $n = 10$ been solved ([BP10]).

In this chapter we describe algorithms that arise in the invariant theory of algebraic groups. We are mostly concerned with (linearly) reductive groups, the exception being the algorithm in Section 7.2 for computing generators of the invariant field (which works for any algebraic group).

An algebraic group G is called *linearly reductive* if every finite-dimensional rational representation of G is completely reducible. If the base field K is of characteristic 0, a connected algebraic group is linearly reductive if and only if it is reductive (Theorem 5.8.1).

Let an algebraic group G act on a vector space V. Let $K[V]$ denote the ring of polynomial functions on V. Then G acts on $K[V]$ by $g \cdot f(v) = f(g^{-1}v)$. By $K[V]^G$ we denote the *invariant ring* consisting of $f \in K[V]$ such that $g \cdot f = f$ for all $g \in G$. If G is linearly reductive then by Hilbert's theorem, the ring $K[V]^G$ is finitely generated. This poses the problem of computing a set of generating invariants. In Section 7.3 an algorithm for this purpose due to Derksen [Der99] is described.

The nullcone of V is the closed set where all homogeneous invariants of positive degree vanish. It was introduced by Hilbert and extensively studied since then. In Section 7.4 we describe a stratification of the nullcone, focusing on the case where the base field has characteristic 0, and give an algorithm due to Popov [Pop03] to compute elements in the Lie algebra that determine this stratification. This yields information on the nullcone. Using this algorithm it is possible, for example, to determine the dimension of the highest dimensional irreducible component of the nullcone.

7.1 The Derksen ideal

In this section we look at the *Derksen ideal*, which plays a main role in Derksen's algorithm for computing generators of the invariant ring of a reductive algebraic group. It also appears in an algorithm for computing generators of the invariant field corresponding to a rational action of an algebraic group. For this reason we treat it separately.

Let G be an algebraic group over the algebraically closed field K, and let $\rho : G \to \mathrm{GL}(V)$ be a rational representation of G. Let $n = \dim V$, and write $K[x, y]$ for $K[x_1, \dots, x_n, y_1, \dots, y_n]$. Then

$$D_\rho = \{f \in K[x, y] \mid f(v, \rho(g)v) = 0 \text{ for all } v \in V, g \in G\},$$

is the Derksen ideal corresponding to ρ. (Note that in this definition we have tacitly chosen a basis of V with respect to which we represent every element of V as a coefficient vector.) It is the vanishing ideal of the graph of ρ, $\Gamma_\rho = \{(v, \rho(g)v) \mid v \in V, g \in G\} \subset V \times V$. By the Hilbert basis theorem (Theorem 1.1.3), D_ρ is finitely generated. Below is an algorithm to find a generating set for it.

Algorithm 7.1.1 *Input: polynomials $p_1, \dots, p_t \in K[z] = K[z_1, \dots, z_s]$, which generate the vanishing ideal of an algebraic group $G \subset K^s$ and a representation $\rho : G \to \mathrm{GL}(n, K)$ given by $g \mapsto (\rho_{ij}(g))$ with $\rho_{ij} \in K[z]$.*
Output: generators of D_ρ.

1. *Let I be the ideal of $K[z, x, y]$ generated by $p_1, \dots, p_t$ along with $y_i - \sum_{j=1}^n \rho_{i,j} x_j$ for $1 \le i \le n$.*

2. *Compute generators of the ideal $I \cap K[x, y]$ (Section 1.6), and return those.*

Proposition 7.1.2 *Algorithm 7.1.1 is correct.*

Proof. Consider the set $U = \{(g, v, \rho(g)v) \mid v \in V, g \in G\} \subset K^s \times V \times V$. We show that $I = \mathcal{I}(U)$. Let $\psi \in \mathcal{I}(U)$ then substituting $y_i \mapsto \sum_{j=1}^n \rho_{i,j} x_j$ we obtain a polynomial $\tilde{\psi}$ with the properties that $\tilde{\psi} \in \mathcal{I}(U)$ and $\psi \in I$ if and only if $\tilde{\psi} \in I$. Furthermore, $\tilde{\psi}$ does not depend on the y_i, so we can write $\tilde{\psi} = \sum_r q_r(z) a_r(x)$, where we may assume that the $a_r \in K[x]$ are linearly independent over K. Thus, for $g \in G$ we have $\tilde{\psi}(g, x) = \sum_r q_r(g) a_r(x)$. As $\tilde{\psi}(g, v) = 0$ for all $v \in V$, we infer that for each $g \in G$, the polynomial $\tilde{\psi}(g, x)$ is the zero polynomial. It follows that for each r with $q_r(g) \neq 0$, we have $a_r = 0$. Furthermore, if $q_r(g) = 0$ for all $g \in G$, then q_r lies in the ideal generated by the p_i. It follows that $\tilde{\psi}$, and hence also ψ, lies in I. The other inclusion is trivial.

Now set $J_0 = I \cap K[x,y]$. By Lemma 1.6.2, $\mathcal{V}(J_0) = \overline{\Gamma}_\rho$. Since $I = \mathcal{I}(U)$ we see that I is radical, hence so is J_0, and therefore, using the Nullstellensatz, $J_0 = \mathcal{I}(\Gamma_\rho) = D_\rho$. □

It is also possible to formulate an algorithm for computing generators of D_ρ using a parametrized dense subset of G (as in Section 4.8) instead of generators of the vanishing ideal of G. We do not explicitly formulate it, but illustrate it in an example.

Example 7.1.3 Let $G = \mathrm{SL}(2,K)$ where K is of characteristic 0 and x, y be two indeterminates. Then G acts on the polynomials in x, y in the following way. Let $g = \begin{pmatrix} a & b \\ c & d \end{pmatrix} \in G$. Then $g \cdot x = ax + cy$, $g \cdot y = bx + dy$. Furthermore, for $p \in K[x,y]$ we have $g \cdot p = p(g \cdot x, g \cdot y)$. We see that G stabilizes the subspace of homogeneous polynomials of degree n.

Note that as a G-module the space spanned by x, y is isomorphic to K^2. Moreover, the G-module spanned by the homogeneous polynomials of degree n is isomorphic to $\mathrm{Sym}^n(K^2)$ (see Example 3.7.3). This leads to the following description of the action of the Lie algebra of G.

Let $\mathfrak{g} = \mathfrak{sl}(2,K)$ be the Lie algebra of G, spanned by h, e, f as in Example 2.1.4. Then $\mathfrak{g}$ acts on $K[x,y]$ by derivations: $e \cdot y = x$, $f \cdot x = y$, $h \cdot x = x$, $h \cdot y = -y$ and $e \cdot x = f \cdot y = 0$, and $u \cdot pq = (u \cdot p)q + p(u \cdot q)$ for $u \in \mathfrak{g}$ and $p, q \in K[x,y]$.

Consider the space V_2 of homogeneous polynomials of degree 2. Let the corresponding representation of G be denoted ρ_2. We use the basis x^2, xy, y^2 of V_2. Then

$$\mathrm{d}\rho_2(e) = \begin{pmatrix} 0 & 1 & 0 \\ 0 & 0 & 2 \\ 0 & 0 & 0 \end{pmatrix}, \quad \exp(\xi_1 \mathrm{d}\rho_2(e)) = \begin{pmatrix} 1 & \xi_1 & \xi_1^2 \\ 0 & 1 & 2\xi_1 \\ 0 & 0 & 1 \end{pmatrix},$$

$$\mathrm{d}\rho_2(f) = \begin{pmatrix} 0 & 0 & 0 \\ 2 & 0 & 0 \\ 0 & 1 & 0 \end{pmatrix}, \quad \exp(\xi_2 \mathrm{d}\rho_2(f)) = \begin{pmatrix} 1 & 0 & 0 \\ 2\xi_2 & 1 & 0 \\ \xi_2^2 & \xi_2 & 1 \end{pmatrix},$$

and letting $T(s,t)$ denote the isomorphism $\mathbb{G}_m = \mathcal{V}(st-1) \subset K^2 \to H$ where H is the algebraic subgroup of $\mathrm{GL}(V_2)$ corresponding to $\mathrm{d}\rho_2(h)$, we have

$$\mathrm{d}\rho_2(h) = \begin{pmatrix} 2 & 0 & 0 \\ 0 & 0 & 0 \\ 0 & 0 & -2 \end{pmatrix}, \quad T(s,t) = \begin{pmatrix} s & 0 & 0 \\ 0 & 1 & 0 \\ 0 & 0 & t \end{pmatrix}.$$

Then $(\xi_1, \xi_2, s, t) \mapsto \exp(\xi_2 \mathrm{d}\rho_2(f)) T(s,t) \exp(\xi_1 \mathrm{d}\rho_2(e))$ is a dominant map $\mathbb{A}^2 \times \mathbb{G}_m \to \rho_2(G)$.

In order to abbreviate the notation a bit we set $\theta = sx_1 + s\xi_1 x_2 + s\xi_1^2 x_3$. Then a short calculation shows that

$$\exp(\xi_2 \mathrm{d}\rho_2(f)) T(s,t) \exp(\xi_1 \mathrm{d}\rho_2(e)) \begin{pmatrix} x_1 \\ x_2 \\ x_3 \end{pmatrix} = \begin{pmatrix} \theta \\ 2\xi_2\theta + x_2 + 2\xi_1 x_3 \\ \xi_2^2\theta + \xi_2 x_2 + 2\xi_1\xi_2 x_3 + t x_3 \end{pmatrix}.$$

Let u_1, u_2, u_3 denote the coordinates of the last vector and I be the ideal of $K[\xi_1, \xi_2, s, t, x_1, x_2, x_3, y_1, y_2, y_3]$ generated by $\{st-1, y_1-u_1, y_2-u_2, y_3-u_3\}$. Then $D_{\rho_2} = I \cap K[x_1, x_2, x_3, y_1, y_2, y_3]$. This intersection is computed in 0.0 seconds, and is generated by the single polynomial $x_1x_3 - \frac{1}{4}x_2^2 - y_1y_3 + \frac{1}{4}y_2^2$, which consequently generates D_{ρ_2}.

Remark 7.1.4 As with most algorithms relying on Gröbner basis computations, the running time of the algorithms in this section can increase sharply for larger input. For the group G from Example 7.1.3 and the module V_n of homogeneous polynomials of degree n, the Derksen ideal is computed within 1 second for $n = 2, 3, 4$, but I have not been able to terminate the computation for $n = 5$ (for memory reasons).

7.2 The invariant field

Let G be an algebraic group over the algebraically closed field K and $\rho : G \to \mathrm{GL}(V)$ be a rational representation. Let F be the field of quotients of $K[V]$. Since G acts on $K[V]$ by $g \cdot f(v) = f(\rho(g)^{-1}v)$, it also acts on F by

$$g \cdot \frac{f_1}{f_2} = \frac{g \cdot f_1}{g \cdot f_2}.$$

It is the objective of this section to describe an algorithm to find generators of the *invariant field*,

$$F^G = \{q \in F \mid g \cdot q = q \text{ for all } g \in G\}.$$

By choosing a basis of V, we identify $K[V]$ with $K[x_1, \dots, x_n]$ and F with $K(x_1, \dots, x_n)$. Let $D_\rho \subset K[x, y]$ (notation as in the previous section) be the Derksen ideal corresponding to ρ. Let $f_1, \dots, f_s \in K[x, y]$ generate D_ρ and $\overline{D}_\rho$ be the ideal of $F[y] = F[y_1, \dots, y_n]$ generated by $f_1, \dots, f_s$. Also by letting an element of G act on the coefficients of a polynomial in $F[y]$, we obtain a G-action on $F[y]$.

Lemma 7.2.1 *$\overline{D}_\rho$ is G-invariant.*

Proof. For $f \in K[x, y] \subset F[y]$ and $g \in G$ we have $g \cdot f(v, w) = f(\rho(g)^{-1}v, w)$, for all $v, w \in V$. Let $g, g' \in G$, $v \in V$ and $f \in D_\rho$, then, setting $w = \rho(g)^{-1}v$,

$$g \cdot f(v, \rho(g')v) = f(\rho(g)^{-1}v, \rho(g')v) = f(w, \rho(g'g)w) = 0.$$

Hence $g \cdot f \in D_\rho$. Since $\overline{D}_\rho$ is generated by elements of D_ρ, this implies that $\overline{D}_\rho$ is invariant under G. □

Let B be a reduced Gröbner basis of $\overline{D}_\rho$ (with respect to any monomial order of $F[y]$). Let $q_1, \ldots, q_m \in F$ be the totality of coefficients of the elements of B. The next proposition yields an immediate algorithm to compute generators of F^G.

Proposition 7.2.2 $F^G = K(q_1, \ldots, q_m)$.

Proof. Let $g \in G$; then Lemma 7.2.1 implies that $g(B)$ is also a reduced Gröbner basis of $\overline{D}_\rho$. By the uniqueness of reduced Gröbner bases (Theorem 1.5.6), it follows that $g \cdot f = f$ for all $f \in B$. Hence the q_i lie in F^G.

Let $q \in F^G$ and write $q = \frac{f_1}{f_2}$, where $f_i \in K[x_1, \ldots, x_n]$. Let $\bar{f}_i$ be the image of f_i under the isomorphism $K[x] \to K[y]$, mapping $x_j \mapsto y_j$, $1 \leq j \leq n$. Let $\psi = \bar{f}_1 f_2 - f_1 \bar{f}_2 \in K[x, y]$ and $g \in G$. Since $q \in F^G$ we have $(g^{-1} \cdot f_1) f_2 = f_1 (g^{-1} \cdot f_2)$ and this implies that

$$\psi(v, \rho(g)v) = f_1(\rho(g)v) f_2(v) - f_1(v) f_2(\rho(g)v) = 0.$$

It follows that $\psi \in D_\rho$, and hence $\bar{f}_1 - q\bar{f}_2 = \frac{1}{f_2}\psi \in \overline{D}_\rho$. Let $\hat{f}_i$ be the normal form of $\bar{f}_i$ modulo B (computed with Algorithm 1.5.2). Then $\hat{f}_1 - q\hat{f}_2$ is also in normal form modulo B. But as $\bar{f}_1 - q\bar{f}_2$ lies in $\overline{D}_\rho$, its normal form is 0. Therefore, $q = \frac{\hat{f}_1}{\hat{f}_2}$. Set $L = K(q_1, \ldots, q_m)$. As B, $\bar{f}_1$, $\bar{f}_2$ are all contained in $L[y]$, the same holds for $\hat{f}_1$ and $\hat{f}_2$. The conclusion is that $q \in L$. □

Example 7.2.3 Let K be of characteristic $p > 0$ and G be the algebraic group of Example 3.6.3,

$$G = \left\{ \begin{pmatrix} a & 0 & 0 \\ 0 & a^p & b \\ 0 & 0 & 1 \end{pmatrix} \mid a, b \in K, a \neq 0 \right\}.$$

Let $V = M_3(K)$, the space of 3×3 matrices. Then G acts on V by $g \cdot A = gA$. We denote the corresponding representation by ρ. By e_{ij} denote the 3×3 matrix with a 1 on position (i, j) and zeros elsewhere. We use the basis $e_{11}, e_{12}, e_{13}, e_{21}, \ldots, e_{33}$ of V. With Algorithm 7.1.1 we compute the ideal D_ρ. In Step 1 the ideal I is generated by the polynomials $da-1$, $y_i - ax_i$, $1 \leq i \leq 3$, $y_i - a^p x_i - bx_{i+3}$, $4 \leq i \leq 6$, $y_i - x_i$, $7 \leq i \leq 9$. Taking $p = 7$, the elimination ideal $I \cap K[x, y]$ is computed by MAGMA in 61 seconds. The output is too bulky to report here (it consists of 115 polynomials). A Gröbner basis of the ideal generated by D_ρ in $K(x_1,\dots,x_9)[y_1,\dots,y_9]$ computed in 0.04 seconds, is

$$
\begin{aligned}
&y_1 - \tfrac{x_1}{x_3} y_3,\\
&y_2 - \tfrac{x_2}{x_3} y_3,\\
&y_3^7 - \tfrac{x_3^7 x_9}{x_5x_9 - x_6x_8} y_5 + \tfrac{x_3^7 x_8}{x_5x_9-x_6x_8} y_6,\\
&y_4 - \tfrac{x_4x_9-x_6x_7}{x_5x_9-x_6x_8} y_5 + \tfrac{x_4x_8-x_5x_7}{x_5x_9-x_6x_8} y_6,\\
&y_7 - x_7, y_8 - x_8, y_9 - x_9.
\end{aligned}
$$

We conclude that $K(x_1,\dots,x_9)^G$ is equal to

$$K\left(\tfrac{x_1}{x_3}, \tfrac{x_2}{x_3}, \tfrac{x_3^7}{x_5x_9-x_6x_8}, \tfrac{x_3^7}{x_5x_9-x_6x_8}, \tfrac{x_4x_9-x_6x_7}{x_5x_9-x_6x_8}, \tfrac{x_4x_8-x_5x_7}{x_5x_9-x_6x_8}, x_7, x_8, x_9\right).$$

Let $g \in G$ and $f = x_5x_9 - x_6x_8$. Then $g \cdot f = a^{-7} f$. This is expressed by saying that f is a semi-invariant. Furthermore, according to a theorem of Chevalley ([Che51], Section II.2, Théorème 1), G is exactly the subgroup of $\mathrm{GL}(3, K)$ leaving the above rational functions invariant.

7.3 Computing invariants of reductive groups

Let G be a linearly reductive group and V a rational G-module. Here we look at the invariant ring $K[V]^G$. The next subsection shows a theorem by Hilbert that $K[V]^G$ is finitely generated. In the subsequent subsection an algorithm is given to compute generators of $K[V]^G$.

7.3.1 Reynolds operator

We start with a general lemma from group theory.

Lemma 7.3.1 *Let $\mathcal{G}$ be a group and M a finite-dimensional completely reducible $\mathcal{G}$-module over a field k. Let $M^{\mathcal{G}} \subset M$ be the subspace consisting of all $m \in M$ with $g \cdot m = m$ for all $g \in \mathcal{G}$. Let $N \subset M$ be the subspace spanned by all $g \cdot m - m$ for $m \in M$ and $g \in \mathcal{G}$. Then $M^{\mathcal{G}}$ and N are $\mathcal{G}$-submodules of M and $M = M^{\mathcal{G}} \oplus N$.*

Proof. It is clear that $M^{\mathcal{G}}$ is stable under $\mathcal{G}$. Let $g, g' \in \mathcal{G}$ and $m \in M$. Then $g'(gm - m) = g'g(g')^{-1}(g'm) - (g'm)$, which lies in N. Hence N is $\mathcal{G}$-stable as well. Since M is completely reducible, there are $\mathcal{G}$-submodules U and N' such that $M = M^{\mathcal{G}} \oplus N'$ and $M = U \oplus N$. Let $m \in U$ and $g \in G$. Then $gm - m \in U \cap N = 0$. So $U \subset M^{\mathcal{G}}$. Let $m \in M$ and write $m = m_1 + m_2$ where $m_1 \in M^{\mathcal{G}}$ and $m_2 \in N'$. Then for $g \in \mathcal{G}$ we have $gm - m = gm_2 - m_2 \in N'$. This implies that $N \subset N'$. Now $\dim(M) = \dim(M^{\mathcal{G}}) + \dim(N') = \dim(U) + \dim(N)$. Hence the inclusions $U \subset M^{\mathcal{G}}$ and $N \subset N'$ imply that $M^{\mathcal{G}} = U$ and $N = N'$. □

For $\mathcal{G}$ and M as in the lemma we let $\mathcal{P}_M : M \to M^{\mathcal{G}}$ be the projection of M onto $M^{\mathcal{G}}$, along N.

Now we return to the situation where G is a linearly reductive group and V a rational G-module. The next lemma follows immediately from the fact that G stabilizes the space consisting of the homogeneous polynomials in $K[V]$ of a fixed degree.

Lemma 7.3.2 *Every $f \in K[V]$ is contained in a finite-dimensional G-submodule of $K[V]$.*

Let $f \in K[V]$ and $M \subset K[V]$ be a finite-dimensional G-submodule containing f (Lemma 7.3.2). Define $R_V(f) = \mathcal{P}_M(f)$, and note that this does not depend on the choice of M. Then we see that $R_V(f_1 + f_2) = R_V(f_1) + R_V(f_2)$ by choosing an M containing both f_1 and f_2. So $R_M : K[M] \to K[M]^G$ is a linear map. The map R_V is called the *Reynolds operator* of the G-module V.

Proposition 7.3.3 (i) *$R_V(f) = f$ for $f \in K[V]^G$,*

(ii) *$R_V(g \cdot f) = R_V(f)$ for $f \in K[V]$ and $g \in G$,*

(iii) *$R_V(fh) = fR_V(h)$ for $f \in K[V]^G$ and $h \in K[V]$,*

(iv) *$R_V(M) = M \cap K[V]^G$ for all G-submodules $M \subset K[V]$.*

Proof. The first two statements are immediate from the definition of R_V.

For the third statement, let M be a finite-dimensional G-submodule of $K[V]$ containing f and h (Lemma 7.3.2). Write $M = M^G \oplus N$ where N is spanned by all $gm - m$ for $g \in G$ and $m \in M$ (Lemma 7.3.1). By definition of R_V, there are $m_i \in M$, $g_i \in G$ such that $h = R_V(h) + \sum_i g_i \cdot m_i - m_i$. Hence $fh = fR_V(h) + \sum_i f(g_i \cdot m_i) - fm_i = fR_V(h) + \sum_i g_i \cdot (fm_i) - fm_i$ (observe that $g_i \cdot (fm_i) = (g_i \cdot f)(g_i \cdot m_i) = f(g_i \cdot m_i)$ as $f \in K[V]^G$). Since $fR_V(h) \in K[V]^G$, we see that $R_V(fh) = fR_V(h)$.

By the construction of R_V, we see that $R_V(M) \subset M$, whence the last statement. □

Theorem 7.3.4 (Hilbert) *$K[V]^G$ is a finitely generated algebra.*

Proof. Let J be the ideal of $K[V]$ generated by all elements of $K[V]^G$ with constant term 0. By Hilbert's basis theorem (Theorem 1.1.3), there are $f_1, \ldots, f_s \in K[V]^G$ that generate J as an ideal of $K[V]$. We claim that $K[V]^G$ is generated by $f_1, \ldots, f_s$. Since the homogeneous parts of invariants are also invariant, we may assume that the f_i are homogeneous. Let $\psi \in K[V]^G$ be homogeneous. Then $\psi = \sum_i a_i f_i$ where the $a_i \in K[V]$ are homogeneous. After applying the Reynolds operator R_V (and using Proposition 7.3.3(iii)), we may assume that the $a_i \in K[V]^G$. By induction on the degree, the a_i lie in the algebra generated by $f_1, \dots, f_s$. It follows that ψ lies in that algebra. □

7.3.2 Computing generators of the invariant ring

In this section we describe Derksen's algorithm for computing generators of the invariant ring $K[V]^G$.

Fixing a basis of V, we write $K[x] = K[x_1, \dots, x_n] = K[V]$ and let $J \subset K[x]$ be the ideal generated by all homogeneous invariants of degree > 0. From the proof of Theorem 7.3.4 it follows that if the homogeneous invariants $f_1, \dots, f_s$ generate J as an ideal, they generate $K[V]^G$ as K-algebra. One problem is that ideal generators of J are not necessarily invariant. However, we have the following.

Proposition 7.3.5 *Let $f_1, \dots, f_r$ be ideal generators of J. Then $R_V(f_i)$ $1 \leq i \leq r$ generate $K[V]^G$.*

Proof. Let $h_1, \dots, h_t$ be homogeneous generators of $K[V]^G$ and $U \subset K[x]$ be the maximal ideal generated by $x_1, \dots, x_n$. We consider the space UJ and the quotient J/UJ. Since the h_i are also ideal generators of J, the elements $h_i \bmod UJ$ span J/UJ.

Because J and UJ are G-stable, $R_V(J) \subset J$ and $R_V(UJ) \subset UJ$ (Proposition 7.3.3(iv)). Hence R_V induces a linear map $R_V : J/UJ \to J/UJ$, which is the identity because the $h_i \bmod UJ$ span J/UJ. Therefore $f_i \bmod UJ = R_V(f_i) \bmod UJ$. So J/UJ is spanned by the $R_V(f_i) \bmod UJ$. We claim that the $\tilde{f}_i = R_V(f_i)$ generate J as an ideal. Indeed, let $h \in J$, then we can write $h = \sum_i c_i \tilde{f}_i + \sum_j a_j g_j$ where $c_j \in K$, $a_j \in U$ and $g_j \in J$. Here the a_j and g_j can be assumed to be homogeneous. Hence $\deg(g_j) < \deg(h)$. So by induction the g_j are contained in the ideal generated by the $\tilde{f}_i$ and so is h. □

Now we let $K[x, y]$ be as in Section 7.1 and $\rho : G \to \mathrm{GL}(V)$ be the representation afforded by V.

Proposition 7.3.6 *Let $D_\rho \subset K[x, y]$ be the Derksen ideal corresponding to ρ and $Y = \langle y_1, \dots, y_n \rangle$ denote the ideal of $K[x, y]$ generated by the y_i. Then $J + Y = D_\rho + Y$.*

Proof. Let $f \in K[x]$ be a homogeneous invariant of degree > 0. Then $f(x) = (f(x) - f(y)) + f(y)$ and $f(y) \in Y$. Moreover, $f(x) - f(y) \in D_\rho$ since, for $v \in V$, $f(v) - f(\rho(g)v) = f(v) - f(v) = 0$ (as f is invariant).

Now let $f(x, y) \in D_\rho$. Then we can write

$$f(x, y) = g(x) + \sum_i h_i(x) p_i(y), \tag{7.1}$$

where the $p_i(y) \in K[y]$ are homogeneous. Note that G acts on $V \times V$ by $g \cdot (v, w) = (v, \rho(g)w)$. Let $R : K[x, y] \to K[x, y]^G$ be the corresponding Reynolds operator. All elements of $K[x]$ are invariant, so by applying R to (7.1) we obtain

$$R(f(x, y)) = g(x) + \sum_i h_i(x) R(p_i(y))$$

(using Proposition 7.3.3.(i)). Let $\varphi : K[x, y] \to K[x]$ be the homomorphism defined by $\varphi(q(x, y)) = q(x, x)$. Since the identity lies in G, we have $q(v, v) = 0$ for all $v \in V$, $q \in D_\rho$. Hence $\varphi(D_\rho) = \{0\}$. Because D_ρ is stable under G we have $R(f(x, y)) \in D_\rho$ and thus

$$0 = \varphi(R(f(x, y))) = g(x) + \sum_i h_i(x) \varphi(R(p_i(y))).$$

Now $\varphi(R(p_i(y)))$ are homogeneous invariants of degree > 0. Hence $g(x) \in J$ and $f(x, y) \in J + Y$. It follows that $D_\rho \subset J + Y$ and the proposition is proved. □

The algorithm for computing a set of generators of the invariant ring is as follows:

1. Compute generators $f_1, \ldots, f_r$ of D_ρ (Section 7.1).
2. Return the polynomials $R_V(f_i(x, 0))$ (where $f_i(x, 0)$ is obtained from f_i by substituting 0 for the y_i).

Note that this works correctly. Indeed, by Proposition 7.3.6 the $f_i(x, 0)$ generate J. Therefore by Proposition 7.3.5, $K[V]^G$ is generated by the $R_V(f_i(x, 0))$.

It is, however, not immediately clear how to execute the second step. Here we describe an alternative to applying the Reynolds operator to find generating invariants from the polynomials $\tilde{f}_i = f_i(x, 0)$. First, note that we may assume that the $\tilde{f}_i$ are homogeneous, as otherwise we replace them by their homogeneous components (which we can do as J is a homogeneous ideal). For $d \geq 0$ let $K[V]_d$ be the space spanned by the homogeneous polynomials of degree d. Write $d_i = \deg \tilde{f}_i$. Then $R_V(\tilde{f}_i)$ is also of degree d_i so that $K[V]^G$ is generated by $K[V]^G_{d_i}$, $1 \leq i \leq r$. As J is homogeneous we have, setting $J_d = J_d \cap K[V]_d$, that $K[V]^G_{d_i} \subset J_{d_i}$. Now fix an i. Because we have homogeneous ideal generators of J, we can compute a basis $h_1, \ldots, h_m$ of J_{d_i}. Subsequently, we set $F = a_1 h_1 + \cdots + a_m h_m$, where the a_i are unknown coefficients. Then we compute $(\rho_{ij}(z)) \cdot F - F$ (where ρ_{ij} is as in Algorithm

7.1.1). This gives us a linear combination of $h_1, \ldots, h_m$, with coefficients that are polynomial in the z_i and linear in the a_i. Then we compute the normal form of these coefficients with respect to the ideal generated by the defining polynomials of G (note that this may require another Gröbner basis computation). Now F is invariant if and only if the resulting coefficients are all zero. This gives linear equations for the a_i, which we can solve. The solution space gives us the homogeneous generators of $K[V]^G_{d_i}$.

This approach can be made more efficient. Suppose we have already computed bases of $K[V]^G_{d_j}$ for all j such that $d_j < d_i$. Let $\widehat{A}$ denote the subalgebra generated by the elements of those bases. Then we can compute a basis of $\widehat{A} \cap K[V]_{d_i}$ and let $h_1, \ldots, h_s$ in the above computation be a basis of that space. These h_i are already invariant, so we can set $F = a_{s+1}h_{s+1} + \cdots + a_m h_m$, and proceed as above.

If G is connected, the base field is of characteristic 0, and $\mathfrak{g} = \mathrm{Lie}(G)$ is available, we can also set up a set of linear equations directly. Indeed, $f \in K[V]$ is G-invariant if and only if $x \cdot f = 0$ for $x \in \mathfrak{g}$ (Corollary 4.2.11). So by letting x vary through a basis of $\mathfrak{g}$ and using the basis $h_1, \ldots, h_m$ as above, we directly get a set of linear equations whose solution yields a basis of $K[V]^G_{d_i}$. Here, in order to define the action of $\mathfrak{g}$ some care is needed. Let $v_1, \ldots, v_n$ be the basis of V, implicit in the identification $K[V] = K[x_1, \ldots, x_n]$. The indeterminates $x_1, \ldots, x_n$ can be seen as a basis of the dual V^* by $x_i(v_j) = \delta_{ij}$. Now G acts on V^* by $g \cdot \psi(v) = \psi(g^{-1}v)$, and on $\mathrm{Sym}^r(V^*)$ as in Example 3.7.3. This leads to the action of G on $K[V]$ that we use here. We have that $\mathfrak{g}$ acts on V^* by $x \cdot \psi(v) = \psi(-xv)$. The action of $\mathfrak{g}$ on $K[V]$ is described in Example 3.7.3.

Example 7.3.7 Let G and ρ_2 be as in Example 7.1.3 in which it is seen that D_{ρ_2} is generated by $f_1 = x_1x_3 - \frac{1}{4}x_2^2 - y_1y_3 + \frac{1}{4}y_2^2$. So $\tilde{f}_1 = f_1(x, 0) = x_1x_3 - \frac{1}{4}x_2^2$. We see that the invariant ring is generated by $K[V]^G_2$. A basis of J_2 consists of f_1. Therefore, no computation is necessary and we conclude that $K[V_2]^G = K[x_1x_3 - \frac{1}{4}x_2^2]$.

Example 7.3.8 Here we illustrate the use of linear equations instead of evaluating the Reynolds operator. Let G be as in Example 7.1.3 and consider the space V_4 of homogeneous polynomials of degree 5. We use the notation of the cited example and the basis of V_4 consisting of the following elements:

$$x^4,\ x^3y,\ x^2y^2,\ xy^3,\ y^4.$$

Then $e \cdot x^{4-s}y^s = sx^{5-s}y^{s-1}$, $f \cdot x^{4-s}y^s = (4-s)x^{3-s}y^{s+1}$ for $0 \leq s \leq 4$. The Derksen ideal is generated by three elements of degrees 2, 3 and 4. This leads to

$$\tilde{f}_1 = 12x_1x_5 - 3x_2x_4 + x_3^2$$
$$\tilde{f}_2 = 96x_1x_3x_5 - 27x_1x_4^2 - 27x_2^2x_5 + 3x_2x_3x_4$$

(and we omit $\tilde{f}_3$). This implies that J_2 is spanned by $\tilde{f}_1$; hence this polynomial has to be invariant. Furthermore, J_3 is spanned by $x_i\tilde{f}_1$, $1 \leq i \leq 5$, and $\tilde{f}_2$. (And these are linearly independent.) We have $e \cdot x_i = -ix_{i+1}$, $f \cdot x_i = -(6-i)x_{i-1}$, where we set $x_0 = x_6 = 0$. Therefore, $e \cdot x_i\tilde{f}_1 = -ix_{i+1}\tilde{f}_1$, $f \cdot x_i\tilde{f}_1 = -(6-i)x_{i-1}\tilde{f}_1$. Furthermore, $e \cdot \tilde{f}_2 = -6x_4\tilde{f}_1$, $f \cdot \tilde{f}_2 = -6x_2\tilde{f}_1$. We see that, up to scalar multiples, there is one element in J_3 annihilated by e and f, namely

$$-2x_3\tilde{f}_1 + \tilde{f}_2 = 72x_1x_3x_5 - 27x_1x_4^2 - 27x_2^2x_5 + 9x_2x_3x_4 - 2x_3^3.$$

Performing this procedure also for J_4, we again obtain one element, but it is equal to $\tilde{f}_1^2$. In conclusion, $K[V_4]^G$ is generated by the two invariants given above.

7.4 The nullcone

Let $G \subset \mathrm{GL}(m, K)$ be a linearly reductive algebraic group, and $\rho : G \to \mathrm{GL}(V)$ a rational representation. Let $\mathcal{N}_G(V)$ be the closed set defined by all homogeneous invariants in $K[V]^G$ of positive degree. This set is called the *nullcone.* It is obviously stable under G so that $\mathcal{N}_G(V)$ is a union of G-orbits. We would like to obtain a list of these orbits, but in general that seems too difficult. Instead we describe a stratification of the nullcone (which in many cases coincides with the subdivision into orbits). This makes it possible to study certain geometrical aspects of the nullcone, such as its dimension (more precisely, the dimension of its irreducible component of highest dimension).

We start with general remarks on the nullcone. Then we have a subsection devoted to the concept of *characteristic* of an element of the nullcone. These characteristics are up to conjugacy finite in number and yield the stratification mentioned earlier. Some properties of this stratification are given in Section 7.4.2. In Section 7.4.3 we describe an algorithm for listing the characteristics up to conjugacy.

Remark 7.4.1 If $K[V]^G$ is generated by 1 (in other words, G does not have homogeneous invariants of positive degree), then $\mathcal{N}_G(V) = V$.

Definition 7.4.2 *A $v \in V$ is said to be* unstable *with respect to the action of G if $0 \in \overline{Gv}$. If v is not unstable with respect to G, it is called* semistable *with respect to the action of G.*

By the next proposition, the nullcone is exactly the set of unstable elements of V. In the proof we use the construction of a categorical quotient, which we sketch; for more details we refer to [VP89], Section 4.4 and [DK02], Section 2.3.

Proposition 7.4.3 *A $v \in V$ lies in $\mathcal{N}_G(V)$ if and only if $0 \in \overline{G \cdot v}$.*

Proof. Let $f_1, \dots, f_r$ be homogeneous invariants of positive degree, generating $K[V]^G$ (Theorem 7.3.4). Let I be the ideal consisting of $\psi \in K[y_1, \dots, y_r]$ such that $\psi(f_1, \dots, f_r) = 0$. Let $X = \mathcal{V}(I) \subset K^r$ and $\pi : V \to X$ be the regular map defined by $\pi(v) = (f_1(v), \dots, f_r(v)$ (the closed set X is said to be a *categorical quotient* of V by the action of G). We have $\pi^*(\bar{y}_i) = f_i$ (where $\bar{y}_i$ is the image of y_i in $K[X]$), so π^* is an isomorphism $K[X] \to K[V]^G$.

Let $Y \subset V$ be a G-invariant closed set, $J_Y = \mathcal{I}(Y)$. Then $\mathcal{I}(\pi(Y))$ is mapped by π^* onto $J_Y^G = J_Y \cap K[V]^G$. Let Z be a second G-invariant closed set, $J_Z = \mathcal{I}(Z)$. Then π^* maps $\mathcal{I}(\pi(Y \cap Z))$ to $(J_Y + J_Z)^G$. By applying the Reynolds operator R_V and using Proposition 7.3.3(iv), we see that this is equal to $J_Y^G + J_Z^G$. So $\pi(Y \cap Z) = \overline{\pi(Y)} \cap \overline{\pi(Z)}$. Now let $v \in \mathcal{N}_G(V)$ and $Z = \overline{G \cdot v}$. As observed in Section 3.12, Z is stable under G. Let $Y = \{0\}$. Then $\pi(Y) \cap \pi(Z) = \{0\}$, whence $Y \cap Z$ is non-empty.

For the converse, let $f \in K[V]^G$ be homogeneous of positive degree. Write $\alpha = f(v)$ and set $\tilde{f} = f - \alpha$. Then $\tilde{f}$ vanishes on $G \cdot v$ and hence on $\overline{G \cdot v}$. As $0 \in \overline{G \cdot v}$, it follows that $\alpha = 0$. □

Example 7.4.4 Here we let K be of characteristic 0. Let $\mathfrak{g} = \mathrm{Lie}(G)$ and assume that $\mathfrak{g}$ is semisimple. Consider the adjoint representation $\mathrm{Ad} : G \to \mathrm{GL}(\mathfrak{g})$. Then we claim that

$$\mathcal{N}_G(\mathfrak{g}) = \{x \in \mathfrak{g} \mid \mathrm{ad}x \text{ is nilpotent}\}.$$

Let $x \in \mathfrak{g}$; then the coefficients of the characteristic polynomial of $\mathrm{ad}x$ are polynomial functions on $\mathfrak{g}$. Moreover, a short calculation shows that $\mathrm{Ad}(g)\mathrm{ad}x\mathrm{Ad}(g)^{-1} = \mathrm{ad}\mathrm{Ad}(g)(x)$, for $g \in G$, $x \in \mathfrak{g}$. Therefore, the coefficients of the characteristic polynomial are homogeneous invariants under the action of G. So if $x \in \mathcal{N}_G(\mathfrak{g})$, all coefficients of the characteristic polynomial of $\mathrm{ad}x$ are 0. In other words, $\mathrm{ad}x$ is nilpotent.

Conversely, let $x \in \mathfrak{g}$ be such that $\mathrm{ad}x$ is nilpotent. Then there are $h, y \in \mathfrak{g}$ (Theorem 2.13.2) such that $[h, x] = 2x$, $[h, y] = -2y$, $[x, y] = h$. (So that h, x, y span a subalgebra isomorphic to $\mathfrak{sl}(2, K)$, see Example 2.1.4.) The differential of Ad is $\mathrm{ad} : \mathfrak{g} \to \mathrm{End}(\mathfrak{g})$ (Theorem 3.8.1), so the Lie algebra of $\mathrm{Ad}(G)$ is $\mathrm{ad}\mathfrak{g}$ (Theorem 4.2.1). We have that $\mathrm{ad}h$ is diagonalizable with integral eigenvalues (Section 2.9.1) and use a basis of eigenvectors of $\mathrm{ad}h$ to represent elements of $\mathrm{End}(\mathfrak{g})$ by matrices. Then $\mathrm{ad}h = \mathrm{diag}(k_1, \dots, k_n)$ with $k_i \in \mathbb{Z}$. Set

$$\Lambda = \{(e_1, \dots, e_n) \in \mathbb{Z}^n \mid \sum_{i=1}^{n} k_i e_i = 0\},$$

which is a pure lattice of rank $n - 1$. By Lemma 4.3.3 the smallest algebraic subgroup H of $\mathrm{GL}(\mathfrak{g})$ whose Lie algebra contains $\mathrm{ad}h$ is a torus with defining lattice Λ (see Section 3.9). Therefore H has dimension 1 and its Lie algebra is spanned by ad$h$. Furthermore, $\mathrm{diag}(k_1, \dots, k_n) \in \Lambda^\perp$ (Section 3.9). Let $\lambda : \mathbb{G}_m \to H$ be the corresponding cocharacter. Then $\lambda(t) \cdot x = t^2 x$. It follows that $\mathrm{Ad}(G)x$ contains all non-zero multiples of x. Therefore, the closure of $\mathrm{ad}(G)x$ contains 0 and by Proposition 7.4.3 we conclude that x lies in the nullcone.

The following is an important criterion in the theory of the nullcone. It was first proved by Hilbert for a particular case and later by Mumford in general. For a proof we refer to [Kra84], Section III.2.

Theorem 7.4.5 (Hilbert Mumford criterion) *Let $v \in V$. Then $v \in \mathcal{N}_G(V)$ if and only if there is a morphism of algebraic groups $\lambda : \mathbb{G}_m \to G$ such that* $\lim_{t\to 0} \lambda(t) \cdot v = 0$.

Remark 7.4.6 Concerning the condition $\lim_{t\to 0} \lambda(t) \cdot v = 0$ we note the following. Let $\lambda : \mathbb{G}_m \to G$ be a morphism of algebraic groups. Then $\mathbb{G}_m$ acts on V via the composition of λ and ρ. Proposition 3.9.3(iii) and the fact that every character of $\mathbb{G}_m$ is of the form $t \mapsto t^m$ for some $m \in \mathbb{Z}$ imply that there is a basis $\{v_1, \dots, v_n\}$ of V with respect to which $\rho(\lambda(t)) = \mathrm{diag}(t^{m_1}, \dots, t^{m_n})$. Let $v \in V$ and write $v = \sum_i \alpha_i v_i$. Then $\lim_{t\to 0} \lambda(t) \cdot v = 0$ if and only if for all i such that $\alpha_i \neq 0$ we have $m_i > 0$.

Remark 7.4.7 Let $v \in \mathcal{N}_G(V)$ and λ be as in Theorem 7.4.5. Let T be a fixed maximal torus of G. Then $\lambda(\mathbb{G}_m)$ is a torus, and hence contained in a maximal torus. Since maximal tori are conjugate under G (see Lemma 5.8.2 for a proof in characteristic 0), it follows from Theorem 7.4.5 that v is conjugate to an element that is unstable with respect to the fixed maximal torus T.

Example 7.4.8 Theorem 7.4.5 makes it possible to discuss the elements of the nullcone without computing any invariants. In this example we see that occasionally it even yields a characterization of the the elements of the nullcone. Let $G = \mathrm{SL}(2, K)$ and consider its action on the space V_d of homogeneous polynomials in variables x, y of degree d (see Example 7.1.3). Let

$$T = \left\{ \sigma_t = \begin{pmatrix} t & 0 \\ 0 & t^{-1} \end{pmatrix} \mid t \in K^* \right\}$$

be the standard maximal torus of G. Let $f = \sum_{i=0}^d a_i x^i y^{d-i} \in V_d$, then $\sigma_t \cdot f = \sum_{i=0}^d t^{2i-d} a_i x^i y^{d-i}$. It follows that $\lim_{t\to 0} \sigma_t \cdot f$ exists and is 0 if and only if $a_i = 0$ for all i such that $2i - d \leq 0$. This is equivalent to f being divisible by x^m where $m > \lfloor \frac{d}{2} \rfloor$. Also $\lim_{t\to 0} \sigma_{t^{-1}} \cdot f$ exists and is 0 if and only if f is divisible by y^m where $m > \lfloor \frac{d}{2} \rfloor$. In view of Remark 7.4.7, we conclude that f is unstable with respect to G if and only if f has a linear factor of multiplicity $m > \lfloor \frac{d}{2} \rfloor$.

7.4.1 Characteristics

Let $v \in \mathcal{N}_G(V)$. In general, there is more than one cocharacter λ as in Theorem 7.4.5. In the first edition of his book ([Mum65]), Mumford posed the problem to find a canonical class of λ for a given v. Kempf ([Kem78]) and Rousseau ([Rou78]) independently found a solution to this problem. Their theory was used by Hesselink ([Hes79]) to define a stratification of the nullcone. As an example of an application of this stratification, Hesselink was able to show that $\mathcal{N}_G(V)$, with $G = \mathrm{SL}(3, K)$ and V the space of homogeneous polynomials of degree 5 in three indeterminates, has two irreducible components, whereas Hilbert mistakenly suggested three.

In this section we assume that K is of characteristic 0 and G is connected (it is not difficult to show that $\mathcal{N}_G(V) = \mathcal{N}_{G^\circ}(V)$, so this last condition is not restrictive). Here we describe a version of the Kempf-Rousseau theory that works with elements of the Lie algebra (the characteristics), rather than with cocharacters $\lambda : \mathbb{G}_\mathrm{m} \to G$ (very roughly, a characteristic h corresponds to $\mathrm{d}\lambda(1)$).

Let $\mathfrak{t}$ be a Cartan subalgebra of $\mathfrak{g}$. By Lemma 5.8.2, $\mathfrak{t}$ consists of commuting semisimple elements and is algebraic. Let T be the corresponding connected algebraic subgroup of G. By the same lemma, T is a maximal torus of G.

Let $A \in \mathrm{GL}(m, K)$ be such that $\mathfrak{t}_A = A\mathfrak{t}A^{-1}$ consists of diagonal matrices. Then the same holds for $T_A = ATA^{-1}$. Let $\Lambda \subset \mathbb{Z}^m$ be the defining lattice of T_A (Section 3.9). As seen in Example 4.2.5, $\mathfrak{t}_A = \mathfrak{d}(\Lambda)$. So as a subspace of the space of all diagonal matrices, $\mathfrak{t}_A$ is given by linear equations with coefficients in $\mathbb{Z}$. Let $\mathfrak{t}_{A,\mathbb{Q}}$ denote the set of all elements of $\mathfrak{t}_A$ with coefficients in $\mathbb{Q}$. Then $\mathfrak{t}_{A,\mathbb{Q}}$ is a vector space over $\mathbb{Q}$ of dimension $\dim \mathfrak{t}$. Set $\mathfrak{t}_\mathbb{Q} = A^{-1}\mathfrak{t}_{A,\mathbb{Q}}A$. Then $\mathfrak{t}_\mathbb{Q}$ is the subspace of $\mathfrak{t}$ consisting of the elements with rational eigenvalues. It is a vector space over $\mathbb{Q}$ of dimension $\dim \mathfrak{t}$.

A semisimple element h of $\mathfrak{g}$ is said to be *rational* if it has rational eigenvalues. This is equivalent to saying that h lies in $\mathfrak{t}'_\mathbb{Q}$ for some Cartan subalgebra $\mathfrak{t}'$.

Mapping $g \mapsto AgA^{-1}$ yields an isomorphism of algebraic groups $T \to T_A$. Therefore, its comorphism is an isomorphism $K[T_A] \to K[T]$ mapping a $\psi \in K[T_A]$ to ψ_A, defined by $\psi_A(g) = \psi(AgA^{-1})$. It follows that

$$X^*(T) = \{\psi_A \mid \psi \in X^*(T_A)\}. \tag{7.2}$$

Let $\chi \in X^*(T)$; then its differential, $\mathrm{d}\chi : \mathfrak{t} \to K$ lies in the dual space $\mathfrak{t}^*$. Because of (7.2) we have

$$\mathfrak{t}_\mathbb{Q} = \{x \in \mathfrak{t} \mid \mathrm{d}\chi(x) \in \mathbb{Q} \text{ for all } \chi \in X^*(T)\}.$$

We define a bilinear form $(\ ,\ ) : \mathfrak{g} \times \mathfrak{g} \to K$ by $(x, y) = \mathrm{Tr}(xy)$. This form is G-invariant, and its restriction to $\mathfrak{t}$ is non-degenerate. Moreover, if $x, y \in \mathfrak{t}_\mathbb{Q}$ then $(x, y) \in \mathbb{Q}$.

As seen in Section 3.9, $X^*(T)$ is (as an abelian group) isomorphic to a lattice. Let $\chi_1, \dots, \chi_r$ be a basis of $X^*(T)$. By $\mathfrak{t}^*_\mathbb{Q}$ we denote the $\mathbb{Q}$-vector space spanned by the differentials $d\chi_i$, $1 \leq i \leq r$. This vector space contains the differential of every character in $X^*(T)$. We have a linear map $\nu : \mathfrak{t}_\mathbb{Q} \to \mathfrak{t}_\mathbb{Q}^*$, defined by $\nu(h)(h') = (h, h')$. (Note that for $h \in \mathfrak{t}_\mathbb{Q}$ there is an integer q such that qh has integral eigenvalues. Then $h' \mapsto (qh, h')$ is the differential of a character in $X^*(T)$.) Since $(\, , \,)$ is non-degenerate, ν is bijective.

Set $\mathfrak{t}_\mathbb{R} = \mathbb{R} \otimes \mathfrak{t}_\mathbb{Q}$ and extend the form $(\, , \,)$ to $\mathfrak{t}_\mathbb{R}$ by linearity. This yields a norm: $\|h\| = \sqrt{(h,h)}$ for $h \in \mathfrak{t}_\mathbb{R}$. Also we set $\mathfrak{t}_\mathbb{R}^* = \mathbb{R} \otimes \mathfrak{t}_\mathbb{Q}^*$. By linearity we extend ν to a map $\mathfrak{t}_\mathbb{R} \to \mathfrak{t}_\mathbb{R}^*$.

Let $\rho : G \to \mathrm{GL}(V)$ be a rational representation of G. For $\chi \in X^*(T)$ set

$$V_\chi = \{v \in V \mid t \cdot v = \chi(t)v \text{ for all } t \in T\}.$$

By Proposition 3.9.3(iii), V is the direct sum of spaces of this form. The Lie algebra $\mathfrak{g}$ acts on V via the differential of ρ. Therefore for $v \in V_\chi$ and $x \in \mathfrak{t}$ we have $x \cdot v = d\chi(x)v$.

Let $v \in V$; then we can uniquely write $v = v_{\chi_1} + \cdots + v_{\chi_s}$ where $v_{\chi_i} \in V_{\chi_i}$ are non-zero and all χ_i are distinct. Define $P(T, v) = \{\chi_1, \ldots, \chi_s\}$ and let $S(T, v)$ be the convex hull in $\mathfrak{t}_\mathbb{R}$ of the $\nu^{-1}(d\chi_i)$, $1 \leq i \leq s$. This is a convex polytope in the euclidean space $\mathfrak{t}_\mathbb{R}$. Furthermore, for $h \in \mathfrak{t}_\mathbb{R}$ we define

$$m(v, h) = \min_{1 \leq i \leq s} d\chi_i(h).$$

Lemma 7.4.9

$$m(v, h) = \min_{h' \in S(T,v)} (h, h').$$

Proof. $h' \mapsto (h, h')$ is a linear function $\mathfrak{t}_\mathbb{R} \to \mathbb{R}$. So on $S(T, v)$ it takes its minimum at one of the vertices $\nu^{-1}(d\chi_i)$. But $(h, \nu^{-1}(d\chi_i)) = d\chi_i(h)$. □

Lemma 7.4.10 *There is a unique point of $S(T, v)$ of minimum norm denoted $\tilde{h}_{T,v}$. We have $\tilde{h}_{T,v} \in \mathfrak{t}_\mathbb{Q}$.*

Proof. The first statement follows from the convexity of $S(T, v)$. We may suppose that $\tilde{h}_{T,v}$ is not a vertex of $S(T, v)$ and that it is non-zero. Then $0 \notin S(T, v)$ and $\tilde{h}_{T,v}$ lies in the interior of an i-dimensional face of $S(T, v)$, where $i > 0$. As all $\nu^{-1}(d\chi_i) \in \mathfrak{t}_\mathbb{Q}$, these faces are defined by linear equations and inequalities with coefficients in $\mathbb{Q}$. Furthermore, $\| \, \|^2 : \mathfrak{t}_\mathbb{R} \to \mathbb{R}$ is a quadratic form with rational coefficients. We consider its restriction to the face containing $\tilde{h}_{T,v}$. Using the criterion of differential calculus, its minima are given by linear equations with rational coefficients. Therefore $\tilde{h}_{T,v} \in \mathfrak{t}_\mathbb{Q}$. □

Lemma 7.4.11 *Suppose $\tilde{h}_{T,v} \neq 0$; then $(\tilde{h}_{T,v}, \tilde{h}_{T,v}) = m(v, \tilde{h}_{T,v})$.*

Proof. Write $\tilde{h}$ instead of $\tilde{h}_{T,v}$. By Lemma 7.4.9 we see that $m(v, \tilde{h}) \leq (\tilde{h}, \tilde{h})$. Suppose there is an $h' \in S(T, v)$ with $(\tilde{h}, h') < (\tilde{h}, \tilde{h})$. Then we consider the line from $\tilde{h}$ to h' given by $h(t) = \tilde{h} + t(h' - \tilde{h})$, $0 \leq t \leq 1$. Since $S(T, v)$ is convex, it contains this line. Set $f(t) = (h(t), h(t))$; then $f(t) = a + bt + ct^2$ with $b < 0$. So $f'(0) < 0$, whence there are points on the line with smaller norms than $\tilde{h}$, contrary to the construction of $\tilde{h}$. In view of Lemma 7.4.9 we conclude that $(\tilde{h}, \tilde{h}) = m(v, \tilde{h})$. □

Lemma 7.4.12 *A $v \in V$ is unstable with respect to T if and only if $\tilde{h}_{T,v} \neq 0$.*

Proof. As in the previous proof we write $\tilde{h} = \tilde{h}_{T,v}$. Suppose $\tilde{h} \neq 0$. There is a positive integer q such that $q\mathrm{d}\rho(\tilde{h})$ has integral eigenvalues. As in Example 7.4.4 we obtain a corresponding cocharacter $\lambda : \mathbb{G}_{\mathrm{m}} \to \rho(T)$. For $\chi \in P(T, v)$ and $v_\chi \in V_\chi$, we have $\lambda(t)v_\chi = t^{m_\chi} v_\chi$ where $m_\chi = \mathrm{d}\chi(q\tilde{h})$. So by Lemma 7.4.11, all m_χ are positive, and hence v is unstable with respect to T.

Suppose v is unstable with respect to T. By the Hilbert-Mumford criterion (Theorem 7.4.5), there exists a cocharacter $\lambda : \mathbb{G}_{\mathrm{m}} \to T$ such that $\lim_{t \to 0} \lambda(t)v = 0$. Set $h_0 = \mathrm{d}\lambda(1)$. Let the matrix A be defined as above. Then $A\lambda(t)A^{-1} = \mathrm{diag}(t^{l_1}, \ldots, t^{l_m})$ where $l_i \in \mathbb{Z}$. Let $\chi \in X^*(T)$; by (7.2) we have $\chi = \psi_A$ where $\psi \in X^*(T_A)$. Write $\psi(\mathrm{diag}(a_1, \ldots, a_m)) = a_1^{k_1} \cdots a_m^{k_m}$. Then $\chi(\lambda(t)) = t^N$ with $N = \sum_i k_i l_i$. This yields $\mathrm{d}\chi(h_0) = \mathrm{d}\chi(\mathrm{d}\lambda(1)) = \mathrm{d}(\chi \circ \lambda)(1) = N$, so that $h_0 \in \mathfrak{t}_{\mathbb{Q}}$. If $\chi \in P(T, v)$, the N above has to be positive (see Remark 7.4.6). So $\mathrm{d}\chi(h_0) > 0$ as well. But $\mathrm{d}\chi(h_0) = (\nu^{-1}(\mathrm{d}\chi), h_0)$. It follows that the vertices of $S(T, v)$ lie in the half space defined by $(h, h_0) > 0$. Therefore $S(T, v)$ is contained in that half space, whence $\tilde{h} \neq 0$. □

Example 7.4.13 Let $G = \mathrm{SL}(3, \mathbb{C})$ and consider the action of G on the 10-dimensional space V_3 of homogeneous polynomials of degree 3 in the indeterminates x, y, z. The action is defined in an analogous way to Example 7.1.3. The Lie algebra $\mathfrak{g}$ acts by derivations on V_3. Let $\mathfrak{t}$ be the standard Cartan subalgebra of $\mathfrak{g}$ consisting of the diagonal matrices of trace 0. Let $\lambda_1, \lambda_2 \in \mathfrak{t}^*$ denote the fundamental weights. Then the weight of a basis element $x^i y^j z^k$ is $(i - j)\lambda_1 + (j - k)\lambda_2$. The inner product on $\mathfrak{t}^*$ can be normalized so that $(\lambda_i, \lambda_i) = 2$, $i = 1, 2$ and $(\lambda_1, \lambda_2) = 1$. Now setting

$$u_1 = \tfrac{1}{\sqrt{2}}\lambda_1, \quad u_2 = \tfrac{1}{\sqrt{6}}(\lambda_1 - 2\lambda_2),$$

we obtain an orthonormal basis of $\mathfrak{t}^*$. Transforming the weights of V_3 to this basis and letting $au_1 + bu_2$ correspond to the point (a, b), we have the following picture:

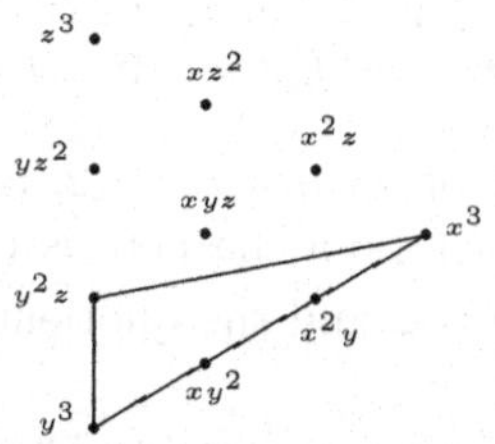

The Weyl group is isomorphic to the symmetric group on three points and acts on the weights in the picture by permuting x, y and z. So the maximal convex region with weights as vertices not containing 0 (i.e., the weight of xyz), up to the action of the Weyl group, is the triangle displayed in the picture. Therefore, by Lemma 7.4.12, in view also of Remark 7.4.7, we see that any unstable vector in V_3 with respect to the action of G is G-conjugate to an element of the space spanned by the vectors in the triangle.

Lemma 7.4.14 *Let $v \in V$ be unstable with respect to T. Then for all non-zero $h \in \mathfrak{t}_{\mathbb{R}}$ we have*

$$\frac{m(v, \tilde{h}_{T,v})}{\|\tilde{h}_{T,v}\|} \geq \frac{m(v,h)}{\|h\|},$$

and we have equality if and only if $h = \tau \tilde{h}_{T,v}$, with $\tau \in \mathbb{R}_{>0}$.

Proof. Again we write $\tilde{h}$ for $\tilde{h}_{T,v}$. Suppose there is no positive real τ with $h = \tau\tilde{h}$. Then

$$\begin{aligned}
\frac{m(v,h)}{\|h\|} &= \min_{h' \in S(T,v)} (h', \frac{h}{\|h\|}) \text{ (Lemma 7.4.9)}\\
&\leq (\tilde{h}, \frac{h}{\|h\|}) \text{ (as } \tilde{h} \in S(T,v))\\
&< (\tilde{h}, \frac{\tilde{h}}{\|\tilde{h}\|}) = \frac{m(v,\tilde{h})}{\|\tilde{h}\|} \text{ (Lemma 7.4.11).}
\end{aligned}$$

□

Suppose $v \in V$ is unstable with respect to T. Then $\tilde{h}_{T,v} \neq 0$ by Lemma 7.4.12. As $\tilde{h}_{T,v}$ lies in $\mathfrak{t}_{\mathbb{Q}}$ (Lemma 7.4.10), $m(v, \tilde{h}_{T,v}) > 0$ (Lemma 7.4.11), and $m(v, \tau\tilde{h}_{T,v}) = \tau m(v, \tilde{h}_{T,v})$ for all positive τ (Lemma 7.4.9), there is a unique $\tau_0 \in \mathbb{Q}_{>0}$ such that $m(v, \tau_0\tilde{h}_{T,v}) = 2$. We define $h_{T,v} = \tau_0\tilde{h}_{T,v}$.

Let $h \in \mathfrak{g}$ be a rational semisimple element. For $\tau \in \mathbb{Q}$ set

$$V_\tau(h) = \{v \in V \mid hv = \tau v\} \text{ and } V_{\geq\tau}(h) = \bigoplus_{\tau' \geq \tau} V_{\tau'}(h).$$

Definition 7.4.15 *Let $v \in V$. A* characteristic *of v is any shortest rational semisimple element h of $\mathfrak{g}$ such that $v \in V_{\geq 2}(h)$.*

If h is a characteristic of v, $m(v,h) = 2$, as otherwise there is a $q \in \mathbb{Q}$, $q < 1$ such that $v \in V_{\geq 2}(qh)$. So a characteristic of v is a rational semisimple element such that $m(v,h) = 2$ and the quotient $\frac{m(v,h)}{\|h\|}$ is maximal among all such h.

Note that a $v \in V$ having a characteristic is necessarily unstable with respect to G (this is seen in the same way as in Example 7.4.4). Conversely, suppose that v is instable with respect to G. By the Hilbert-Mumford criterion, there is a maximal torus S of G such that v is unstable with respect to S. Then the set of rational semisimple elements h with $v \in V_{\geq 2}(h)$, contains $h_{S,v}$ and is therefore non-empty. It follows that v has a characteristic. It is, however, not clear how to find a characteristic of a given $v \in \mathcal{N}_G(V)$. The next lemma, which immediately follows from Lemma 7.4.14, shows that a Cartan subalgebra has precisely one candidate for this.

Lemma 7.4.16 *If $\mathfrak{t}_{\mathbb{Q}}$ has a characteristic of v, it is equal to $h_{T,v}$.*

Next we show a criterion (Theorem 7.4.22) for deciding whether a given rational semisimple element h is a characteristic of a $v \in \mathcal{N}_G(V)$. For that theorem we need some preparation.

Let $v \in V$ be unstable with respect to G. Then we set

$$\mathcal{C}(v) = \{h \in \mathfrak{g} \mid h \text{ is a characteristic of } v\}.$$

Let $h \in \mathfrak{g}$ be a rational semisimple element. Set $\mathfrak{z}(h) = \mathfrak{g}_0(h) = \{x \in \mathfrak{g} \mid [h,x] = 0\}$. Then $\mathfrak{z}(h)$ is an algebraic subalgebra of $\mathfrak{g}$, since, by Corollary 4.2.8 and Theorem 3.8.1, it is the Lie algebra of

$$Z(h) = \{g \in G \mid \mathrm{Ad}(g)(h) = h\}^{\circ}.$$

In the following we write $W = K^m$, which is the standard G-module.

Lemma 7.4.17 *$\mathfrak{z}(h) = \mathfrak{a} \oplus \mathfrak{b}$ (direct sum of ideals) where $\mathfrak{a}$ is semisimple and $\mathfrak{b}$ is abelian and consists of semisimple elements (of* End(W)*). In particular, $\mathfrak{z}(h)$ is reductive.*

Proof. As G is reductive, we can write $\mathfrak{g} = \mathfrak{s} \oplus \mathfrak{d}$ (direct sum of ideals) where $\mathfrak{s}$ is semisimple and $\mathfrak{d}$ is the centre consisting of semisimple elements (Theorem 5.8.1). So $h = h_1 + h_2$ with $h_1 \in \mathfrak{s}$, $h_2 \in \mathfrak{d}$, and $\mathfrak{z}(h) = \mathfrak{c}_{\mathfrak{s}}(h_1) \oplus \mathfrak{d}$. Let $\kappa_{\mathfrak{s}}$ be the Killing form of $\mathfrak{s}$. By an argument like the one used for Proposition 2.9.2(iii) we see that the restriction of $\kappa_{\mathfrak{s}}$ to $\mathfrak{c}_{\mathfrak{s}}(h_1)$ is non-degenerate. Let $x \in \mathfrak{c}_{\mathfrak{s}}(h_1)$ and write $x = s + n$ with $\mathrm{ad}_{\mathfrak{s}} s$ semisimple and $\mathrm{ad}_{\mathfrak{s}} n$ nilpotent (Theorem 2.7.9). As $\mathrm{ad}_{\mathfrak{s}}(x)(h_1) = 0$ we also have $\mathrm{ad}_{\mathfrak{s}}(s)(h_1) = \mathrm{ad}_{\mathfrak{s}}(n)(h_1) = 0$ (Proposition 2.2.3) so that $s, n \in \mathfrak{c}_{\mathfrak{s}}(h_1)$. As in the proof of Proposition 2.12.5 we infer that $\mathfrak{c}_{\mathfrak{s}}(h_1) = \mathfrak{a} \oplus \mathfrak{d}'$ where $\mathfrak{a}$ is semisimple $\mathfrak{d}'$ is abelian and consists of elements

y such that $\mathrm{ad}_{\mathfrak{s}} y$ is semisimple. From Corollary 2.11.6 it follows that all such y are semisimple (as elements of $\mathrm{End}(W)$) We set $\mathfrak{b} = \mathfrak{d}' \oplus \mathfrak{d}$, and obtain the statement of the lemma. □

Let $h \in \mathfrak{g}$ be rational semisimple; then for $\tau \in \mathbb{Q}$ we define

$$\mathfrak{g}_\tau(h) = \{x \in \mathfrak{g} \mid [h, x] = \tau x\}$$

and the subalgebra

$$\mathfrak{p}(h) = \bigoplus_{\tau \geq 0} \mathfrak{g}_\tau(h).$$

Note that $\mathfrak{p}(h) = \mathfrak{z}(h) \oplus \mathfrak{n}$, where $\mathfrak{n} = \oplus_{\tau > 0} \mathfrak{g}_\tau(h)$. For $\eta \in \mathbb{Q}$ we write $W_\eta(h) = \{w \in W \mid hw = \eta w\}$. Then, for $v \in W_\eta(h)$ and $x \in \mathfrak{g}_\tau(h)$ we have $x \cdot v \in W_{\eta+\tau}(h)$. Hence all elements of $\mathfrak{n}$ are nilpotent, and by Theorem 4.3.14, $\mathfrak{n}$ is algebraic. As seen above, $\mathfrak{z}(h)$ is algebraic. So by Corollary 4.3.7, $\mathfrak{p}(h)$ is algebraic. By $P(h)$ we denote the unique connected algebraic subgroup of G with Lie algebra $\mathfrak{p}(h)$.

Now $\mathfrak{p}(h)$ leaves $V_{\geq \tau}(h)$ invariant, and hence the same holds for $P(h)$ (Corollary 4.2.10). Also observe that $m(v, h) = \max\{\tau \in \mathbb{Q} \mid v \in V_{\geq \tau}(h)\}$. This implies that

$$m(v, h) = m(pv, h) \text{ for all } p \in P(h). \tag{7.3}$$

We also have

$$m(v, h) = m(gv, \mathrm{Ad}(g)(h)) \text{ for all } g \in G. \tag{7.4}$$

Indeed, let $u \in V_\tau(h)$, for some $\tau \in \mathbb{Q}$. Then $\mathrm{Ad}(g)(h)gu = ghg^{-1}gu = ghu = \tau gu$ so that $gu \in V_\tau(\mathrm{Ad}(g)(h))$. Therefore, if we decompose $v = u_1 + \cdots + u_t$ as a sum of h-eigenvectors, then $gv = gu_1 + \cdots + gu_t$ is the decomposition of gv as sum of $\mathrm{Ad}(g)(h)$-eigenvectors and both decompositions have the same eigenvalues. In particular, (7.4) entails the following lemma.

Lemma 7.4.18 *Let h be a characteristic of $v \in V$ and $g \in G$. Then* $\mathrm{Ad}(g)(h)$ *is a characteristic of gv.*

Lemma 7.4.19 *Let $v \in V$ be unstable with respect to G. Let $h \in \mathcal{C}(v)$, and $p \in P(h)$. Then* $\mathrm{Ad}(p)(h) \in \mathcal{C}(v)$.

Proof. As $(\ ,\ )$ is G-invariant we have $\|\mathrm{Ad}(p)(h)\|^2 = \|h\|^2$. Using (7.4) and (7.3), we see that $m(v, \mathrm{Ad}(p)(h)) = m(p^{-1}v, h) = m(v, h) = 2$. It follows that $\mathrm{Ad}(p)(h) \in \mathcal{C}(h)$. □

For the next lemma, let $h \in \mathfrak{g}$ be a rational semisimple element. Then it lies in a Cartan subalgebra of $\mathfrak{g}$ (Corollary 2.5.10). This Cartan subalgebra is contained in $\mathfrak{z}(h)$ and therefore in $\mathfrak{p}(h)$. So it is a Cartan subalgebra of $\mathfrak{p}(h)$ as well. Since the Cartan subalgebras of $\mathfrak{p}(h)$ are conjugate under $P(h)$ (Proposition 4.3.2), it follows that any Cartan subalgebra of $\mathfrak{p}(h)$ is also a Cartan subalgebra of $\mathfrak{g}$.

Lemma 7.4.20 *Let $v \in V$ be unstable with respect to G, $h \in \mathcal{C}(v)$ and $\mathfrak{t}$ be a Cartan subalgebra of $\mathfrak{p}(h)$. Then $\mathfrak{t}_{\mathbb{Q}} \cap \mathcal{C}(v)$ consists of one element $h_{T,v}$. Moreover, h and $h_{T,v}$ are conjugate under $P(h)$.*

Proof. Let $\hat{\mathfrak{t}}$ be a Cartan subalgebra of $\mathfrak{p}(h)$ such that $h \in \hat{\mathfrak{t}}_{\mathbb{Q}}$. Then there is a $p \in P(h)$ such that $\mathfrak{t} = \mathrm{Ad}(p)(\hat{\mathfrak{t}})$ (Proposition 4.3.2). It follows that $\mathrm{Ad}(p)(h) \in \mathfrak{t}$. But also $\mathrm{Ad}(p)(h) \in \mathcal{C}(v)$ (Lemma 7.4.19). So $\mathfrak{t}_{\mathbb{Q}} \cap \mathcal{C}(v)$ is non-empty, and by Lemma 7.4.16, $\mathfrak{t}_{\mathbb{Q}} \cap \mathcal{C}(v) = \{h_{T,v}\}$. We also see that $h_{T,v} = \mathrm{Ad}(p)(h)$. □

Lemma 7.4.21 *Let $v \in V$ be unstable with respect to G and $h_1, h_2 \in \mathcal{C}(v)$. Then $P(h_1) = P(h_2)$ and there is a $p \in P(h_1)$ such that $h_2 = \mathrm{Ad}(p)(h_1)$.*

Proof. Note that the $\mathfrak{p}(h_i)$ contain a Borel subalgebra of $\mathfrak{g}$. Using Corollary 5.8.4 we see that $\mathfrak{p}(h_1) \cap \mathfrak{p}(h_2)$ contains a Cartan subalgebra of $\mathfrak{g}$; denote it $\mathfrak{t}$. By Lemma 7.4.20, $\mathcal{C}(v) \cap \mathfrak{t}_{\mathbb{Q}}$ consists of a unique element we denote h. By the same lemma there are $p_1 \in P(h_1)$, $p_2 \in P(h_2)$ such that $\mathrm{Ad}(p_1)(h_1) = h = \mathrm{Ad}(p_2)(h_2)$.

Now $\mathfrak{g}_\tau(\mathrm{Ad}(p_1)(h_1)) = \mathrm{Ad}(p_1)(\mathfrak{g}_\tau(h_1))$, implying $\mathfrak{p}(\mathrm{Ad}(p_1)(h_1)) = \oplus_{\tau \geq 0} \mathfrak{g}_\tau(\mathrm{Ad}(p_1)(h_1)) = \mathrm{Ad}(p_1)(\mathfrak{p}(h_1)) = \mathfrak{p}(h_1)$. So by Theorem 4.2.2(i) we obtain $P(h) = P(\mathrm{Ad}(p_1)(h_1)) = P(h_1)$. In the same way we see that $P(h) = P(h_2)$. Furthermore, by setting $p = p_2^{-1}p_1$ we have $h_2 = \mathrm{Ad}(p)(h_1)$. □

Let $h \in \mathfrak{g}$ be a rational semisimple element. As in Lemma 7.4.17 we write $\mathfrak{z}(h) = \mathfrak{a} \oplus \mathfrak{b}$. Here $\mathfrak{b}$ is the centre of $\mathfrak{z}(h)$. By setting $W = \mathfrak{z}(h)$ and $U = 0$ in Corollary 4.2.8, we see that the centre is algebraic and by B we denote the corresponding connected subgroup of $Z(h)$. As h is rational, the map $h' \mapsto (h, h')$ is, up to a rational scalar factor, the differential of a character χ of B. So $\mathfrak{b}' = \{h' \in \mathfrak{b} \mid (h, h') = 0\}$ is an algebraic subalgebra of $\mathfrak{b}$ (it is the Lie algebra of the kernel of χ; see Theorem 4.2.4). Since $\mathfrak{a}$ is semisimple, it is algebraic as well. Furthermore, as $[\mathfrak{z}(h), \mathfrak{z}(h)] = \mathfrak{a}$, it follows that $\mathfrak{a}$ is orthogonal to h with respect to the form $(\, , \,)$ (use Lemma 2.1.2). So, setting

$$\tilde{\mathfrak{z}}(h) = \{x \in \mathfrak{z}(h) \mid (h, x) = 0\}$$

we have $\tilde{\mathfrak{z}}(h) = \mathfrak{a} \oplus \mathfrak{b}'$ and $\tilde{\mathfrak{z}}(h)$ is algebraic. By $\widetilde{Z}(h)$ we denote the connected algebraic subgroup of $Z(h)$ with Lie algebra $\tilde{\mathfrak{z}}(h)$.

Theorem 7.4.22 (Kirwan-Ness) *Let $v \in V$ be unstable with respect to G. Let $h \in \mathfrak{g}$ be a rational semisimple element such that $v \in V_{\geq 2}(h)$ and v_2 denote the projection of v on $V_2(h)$. Then h is a characteristic of v if and only if v_2 is semistable with respect to the action of $\widetilde{Z}(h)$.*

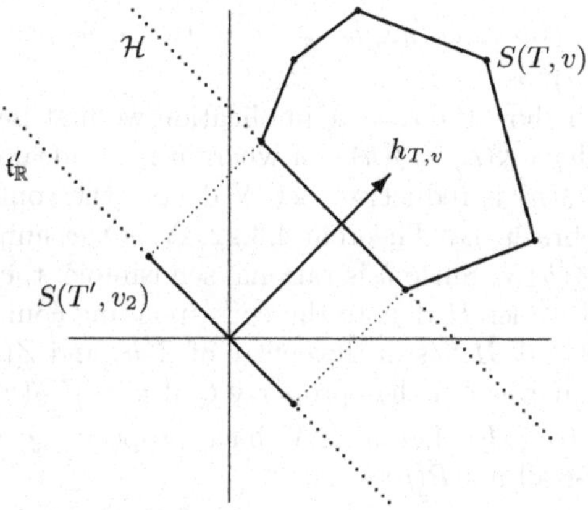

Proof. (The sketch above illustrates some of the constructions used in this proof.) Let $\mathfrak{t}$ be a Cartan subalgebra of $\mathfrak{z}(h)$. Then $\mathfrak{t}$ contains h and is also a Cartan subalgebra of $\mathfrak{g}$. Furthermore, $\mathfrak{t} = \mathfrak{t}' \oplus \langle h \rangle$, where $\mathfrak{t}' = \{h' \in \mathfrak{t} \mid (h, h') = 0\}$ is a Cartan subalgebra of $\tilde{\mathfrak{z}}(h)$. In the same way as for the subalgebra $\mathfrak{b}'$ above, we see that $\mathfrak{t}'$ is algebraic and let T' be the corresponding connected algebraic group. Then $T' \subset \widetilde{Z}(h)$.

Observe that $\mathfrak{t}_{\mathbb{R}} = \mathfrak{t}'_{\mathbb{R}} \oplus \langle h \rangle$ and let $\pi : \mathfrak{t}_{\mathbb{R}} \to \mathfrak{t}'_{\mathbb{R}}$ be the corresponding projection. Also define $\sigma : \mathfrak{t}^*_{\mathbb{R}} \to (\mathfrak{t}'_{\mathbb{R}})^*$ by restriction: $\sigma(\mu)(h') = \mu(h')$ for $h' \in \mathfrak{t}'_{\mathbb{R}}$. Note that a character of T' is the restriction of a character of T, and therefore the same holds for their differentials. The image of σ is indeed $(\mathfrak{t}'_{\mathbb{R}})^*$. Also we have a map $\theta : \mathfrak{t}'_{\mathbb{R}} \to (\mathfrak{t}'_{\mathbb{R}})^*$ defined similarly to ν so that θ is the restriction of ν to $\mathfrak{t}'_{\mathbb{R}}$.

Let $\mu \in \mathfrak{t}^*_{\mathbb{R}}$. Then for $h' \in \mathfrak{t}'_{\mathbb{R}}$ we have $(\theta^{-1}(\sigma(\mu)), h') = \sigma(\mu)(h') = \mu(h')$, and, writing $\nu^{-1}(\mu) = h_1 + ch$, with $h_1 \in \mathfrak{t}'_{\mathbb{R}}$, we see that $(\pi(\nu^{-1}(\mu)), h') = (h_1, h') = (\nu^{-1}(\mu), h') = \mu(h')$. It follows that

$$\theta^{-1}(\sigma(\mu)) = \pi(\nu^{-1}(\mu)). \tag{7.5}$$

Set $\mathcal{H} = \{h' \in \mathfrak{t}_{\mathbb{R}} \mid (h, h') = 2\}$, which is a hyperplane orthogonal to h. Let $\mathcal{H}^+$ denote the set of all h' with $(h, h') > 2$. Let $\chi \in P(T, v)$ be such that $d\chi(h) = 2$, then $\nu^{-1}(d\chi) \in \mathcal{H}$. On the other hand, if $d\chi(h) > 2$, then $\nu^{-1}(d\chi) \in \mathcal{H}^+$.

We have that π maps $\mathcal{H}$ bijectively onto $\mathfrak{t}'_{\mathbb{R}}$ and h to 0. Furthermore, by (7.5), $\pi(\mathcal{H} \cap S(T, v)) = S(T', v_2)$.

Suppose h is a characteristic of v and v_2 is unstable with respect to $\widetilde{Z}(h)$. By the Hilbert-Mumford criterion and the conjugacy of maximal tori (Lemma 5.8.2(iv)), there is a $g \in \widetilde{Z}(h)$ such that gv_2 is unstable with respect to T'. Since $g \in Z(h)$ we have $\mathrm{Ad}(g)(h) = h$. By Lemma 7.4.18, h is also a characteristic of gv. So after replacing v by gv, we have $\tilde{h}_{T',v_2} \neq 0$ (Lemma 7.4.12).

By Lemma 7.4.20, $h = h_{T,v}$, which is a positive scalar multiple of $\tilde{h}_{T,v}$. Since $\pi(\mathcal{H} \cap S(T, v)) = S(T', v_2)$, $\tilde{h}_{T,v} \in S(T, v)$ and $\pi(h) = 0$, we see that

$0 \in S(T', v_2)$, contradicting $\tilde{h}_{T',v_2} \neq 0$. Hence v_2 has to be semistable with respect to $\widetilde{Z}(h)$.

In order to show the reverse implication we first investigate the structure of $P(h)$. We have $\mathfrak{p}(h) = \mathfrak{z}(h) \oplus \mathfrak{n}$ where $\mathfrak{n}$ is an ideal consisting of nilpotent elements and $\mathfrak{z}(h)$ is reductive. Let N denote the connected algebraic group with Lie algebra $\mathfrak{n}$. By Theorem 4.3.22, N is the unipotent radical of $P(h)$ and $P(h) = Z(h)N$. Since h is rational semisimple, the subalgebra it spans is algebraic and we let H denote the corresponding connected algebraic group. Since $[h, \mathfrak{z}(h)] = 0$, H lies in the centre of $Z(h)$ and $Z(h) = \widetilde{Z}(h)H$.

We say that a $w \in V$ has property Q if $w \in V_{\geq 2}(h)$ and w_2 is semistable with respect to $\widetilde{Z}(h)$. Let $w \in V$ have property Q; we claim that pw has property Q for all $p \in P(h)$.

Let $g \in \widetilde{Z}(h)$. As seen in the proof of (7.4), $(gw)_2 = gw_2$. In particular, also $(gw)_2$ is semistable with respect to $\widetilde{Z}(h)$. Furthermore, as $\mathrm{Ad}(g)(h) = h$, $gw \in V_{\geq 2}(h)$.

Let $g \in H$. Then $g \in T$ so that $P(T, w) = P(T, gw)$. Hence $gw \in V_{\geq 2}(h)$. Furthermore, $(gw)_2 = gw_2$. As g centralizes $\widetilde{Z}(h)$ we have $\widetilde{Z}(h) \cdot (gw_2) = g(\widetilde{Z}(h) \cdot w_2)$. Since 0 does not lie in the closure of $\widetilde{Z}(h) \cdot w_2$, it does not lie in the closure of $\widetilde{Z}(h) \cdot (gw_2)$ either. In other words, $(gw)_2$ is semistable with respect to $\widetilde{Z}(h)$.

Let $g \in N$. Then for $u \in V_\tau(h)$ we have $gu = u + u'$ where $u' \in \oplus_{\tau' > \tau} V_{\tau'}(h)$. Hence $(gw)_2 = w_2$, so that $(gw)_2$ is semistable with respect to $\widetilde{Z}(h)$. Also $gw \in V_{\geq 2}(h)$.

In all cases we see that gw has property Q as well. Therefore, pw has property Q for all $p \in P(h)$. The claim above is proved.

Let $\hat{h}$ be a positive scalar multiple of h such that $(\hat{h}, h) = 2$ (i.e., $\hat{h} \in \mathcal{H}$). Let $w \in V$ have property Q; then we claim that $\tilde{h}_{T,w} = \hat{h}$. Indeed, in the same way as for v we have $\pi(\mathcal{H} \cap S(T, w)) = S(T', w_2)$. So if $\hat{h} \notin S(T, w)$, then $S(T', w_2)$ does not contain 0, and by Lemma 7.4.12 w_2 is unstable with respect to $\widetilde{Z}(h)$, which is impossible. Therefore we must have $\hat{h} \in S(T, w)$, whence $\hat{h} = \tilde{h}_{T,w}$.

Now suppose that v_2 is semistable with respect to $\widetilde{Z}(h)$. Hence v has property Q. By the first claim above, pv has property Q for all $p \in P(h)$. By the second claim, $\tilde{h}_{T,pv} = \hat{h}$ for all $p \in P(h)$.

Let h' be a characteristic of v. As seen in the proof of Lemma 7.4.21, $\mathfrak{p}(h) \cap \mathfrak{p}(h')$ contains a Cartan subalgebra of $\mathfrak{g}$. Denote it $\hat{\mathfrak{t}}$. By Lemma 7.4.20, $\hat{\mathfrak{t}}_{\mathbb{Q}} \cap \mathcal{C}(v) \neq \emptyset$. Since $\mathfrak{t}$ and $\hat{\mathfrak{t}}$ are conjugate under $P(h)$, there is a $p \in P(h)$ such that $\mathfrak{t}_{\mathbb{Q}} \cap \mathcal{C}(pv) \neq \emptyset$. As shown above, $\tilde{h}_{T,pv} = \tilde{h}_{T,v}$ which implies that $\mathfrak{t}_{\mathbb{Q}} \cap \mathcal{C}(pv) = \{h\}$. Now Lemma 7.4.19 entails $\mathrm{Ad}(p)(h) \in \mathcal{C}(pv)$ and that yields $h \in \mathcal{C}(v)$. □

Example 7.4.23 Let $\mathfrak{g}$ be a semisimple Lie algebra over $\mathbb{C}$, and let G be its adjoint group. This is the Chevalley group whose construction starts with the adjoint representation (Section 5.2.3) and the field $\mathbb{C}$. So by Theorem 5.3.9(ii), $\mathrm{Lie}(G) = \mathrm{ad}\mathfrak{g}$. Potentially, G acts in two ways on $\mathfrak{g} \cong \mathrm{ad}\mathfrak{g}$: by the adjoint representation it acts on $\mathrm{ad}\mathfrak{g}$ and by construction it acts on $\mathfrak{g}$. However, these two actions coincide: for $g \in G$ and $x \in \mathfrak{g}$ we have $\mathrm{Ad}(g)(\mathrm{ad}x) = g(\mathrm{ad}x)g^{-1} = \mathrm{ad}g(x)$. Therefore, to fix the notation, we will only consider the action of G on $\mathfrak{g}$ and write $g \cdot x$ (or gx) instead of $g(x)$.

As seen in Example 7.4.4, the nullcone in this case consists of the $e \in \mathfrak{g}$ such that $\mathrm{ad}e$ is nilpotent. Let e be such an element; then by the Jacobson-Morozov theorem (Section 2.13), there are $h, f \in \mathfrak{g}$ such that (h, e, f) is an $\mathfrak{sl}_2$-triple. Then h (or, more precisely, $\mathrm{ad}h$) is a rational semisimple element of $\mathfrak{g}$. We claim that h is a characteristic of e. This is shown using the Kempf-Ness theorem. To explain how that works we need theory not covered in this book, so we content ourselves with a brief sketch.

First we describe the content of the Kempf-Ness theorem. Let $\mathcal{G}$ be a reductive algebraic group defined over $\mathbb{R}$ and V a rational $\mathcal{G}$-module. Let $\mathcal{K} \subset \mathcal{G}$ be a compact form of $\mathcal{G}$ ([OV90], Section 5.2.3). On V it is possible to choose a Hermitian scalar product $\langle\ ,\ \rangle$, invariant under $\mathcal{K}$. Let $v \in V$. Then the Kempf-Ness theorem states that the orbit $\mathcal{G}v$ is closed if and only if there is a $w \in \mathcal{G}v$ such that $\langle \xi w, w\rangle = 0$ for all $\xi \in \mathrm{Lie}(\mathcal{K})$ (see [KN79], [VP89] and [Los06]).

Let $\mathfrak{a} \subset \mathfrak{g}$ be the subalgebra spanned by h, e, f. Let $\mathfrak{a}_0$ be the real span of the elements $\imath h$, $e - f$, $\imath(e + f)$ (where $\imath$ denotes the imaginary unit). Then $\mathfrak{a}_0$ is a compact real form of $\mathfrak{a}$. By [Oni04] Section 6, Proposition 3, there is a compact real form $\mathfrak{u}$ of $\mathfrak{g}$ such that $\mathfrak{a}_0 \subset \mathfrak{u}$. Then $\mathfrak{g} = \mathfrak{u} + \imath\mathfrak{u}$, and we let $\tau : \mathfrak{g} \to \mathfrak{g}$ be the corresponding conjugation. Then $\tau(h) = -h$, $\tau(e) = -f$, $\tau(f) = -e$. Let κ denote the Killing form of $\mathfrak{g}$; then the form $\langle\ ,\ \rangle$ defined by

$$\langle x, y\rangle = -\langle x, \tau(y)\rangle$$

is a Hermitian scalar product on $\mathfrak{g}$. Let $K \subset G$ be the subgroup consisting of $g \in G$ such that $g \cdot \mathfrak{u} = \mathfrak{u}$. Then K is a compact form of G, with Lie algebra $\mathfrak{u}$ (or, more precisely, $\mathrm{ad}\mathfrak{u}$). Moreover, $\langle\ ,\ \rangle$ is K-invariant (cf., [OV90], Chapter 5, Problem 17).

From $\tau(h) = -h$ it follows that $\mathfrak{z}(h)$ is stable under τ. As κ is invariant under τ, the same follows for $\tilde{\mathfrak{z}}(h)$. So $\tilde{\mathfrak{z}}(h) \cap \mathfrak{u}$ is a compact form of $\tilde{\mathfrak{z}}(h)$. Now $\widetilde{K} = K \cap \widetilde{Z}(h)$ has Lie algebra $\tilde{\mathfrak{z}}(h) \cap \mathfrak{u}$, so $\widetilde{K}$ is a compact form of $\widetilde{Z}(h)$. For $x \in \tilde{\mathfrak{z}}(h)$ we have, using Lemma 2.1.2,

$$\langle x \cdot e, e\rangle = \langle [x, e], e\rangle = -\kappa([x, e], \tau(e)) = \kappa(x, [e, f]) = \kappa(x, h) = 0.$$

Hence by the Kempf-Ness theorem, the orbit $\widetilde{Z}(h) \cdot e$ is closed. The Kirwan-Ness theorem therefore shows that h is a characteristic of e.

7.4.2 Stratification of the nullcone

Let $\mathfrak{t}$ be a fixed Cartan subalgebra of $\mathfrak{g}$. We say that an $h \in \mathfrak{t}_{\mathbb{Q}}$ is a characteristic if it is the characteristic of a $v \in \mathcal{N}_G(V)$. The weights of V are finite in number, hence the collection of subsets of the weights is finite and so is the number of characteristics contained in $\mathfrak{t}_{\mathbb{Q}}$ (see Lemma 7.4.16). Let $h_1, \ldots, h_t$ denote the characteristics in $\mathfrak{t}_{\mathbb{Q}}$, up to G-conjugacy.

Let $h \in \mathfrak{t}_{\mathbb{Q}}$ be a characteristic; then define $N(h)$ to be the set of $v \in \mathcal{N}_G(V)$ whose characteristic is G-conjugate to h (or, equivalently, such that there is a $g \in G$ such that gv has characteristic h by Lemma 7.4.18).

As the Cartan subalgebras of $\mathfrak{g}$ are G-conjugate (Proposition 4.3.2), a characteristic of a $v \in \mathcal{N}_G(V)$ is G-conjugate to an element of $\mathfrak{t}_{\mathbb{Q}}$. So $\mathcal{N}_G(V)$ is the disjoint union of the $N(h_i)$ and $\{0\}$. The sets $N(h_i)$ are called the *strata* of $\mathcal{N}_G(V)$.

Let $h \in \mathfrak{t}_{\mathbb{Q}}$ be a characteristic and $V^0_{\geq 2}(h)$ be the open subset of $V_{\geq 2}(h)$ consisting of all $v \in V_{\geq 2}(h)$ whose projection on $V_2(h)$ is semistable with respect to $\widetilde{Z}(h)$. (So $V^0_{\geq 2}(h)$ is the direct sum of the complement of $\mathcal{N}_{\widetilde{Z}(h)}(V_2(h))$ and $\oplus_{\tau>2} V_\tau(h)$.) By Theorem 7.4.22, $V^0_{\geq 2}(h)$ is the set of all elements of $V_{\geq 2}(h)$ having characteristic h.

Proposition 7.4.24 *Let $h \in \mathfrak{t}_{\mathbb{Q}}$ be a characteristic. We have*

(i) $\overline{N(h)} = GV_{\geq 2}(h)$,

(ii) $\dim \overline{N(h)} = \dim V_{\geq 2}(h) + \dim \mathfrak{g} - \dim \mathfrak{p}(h)$.

Proof. Let $g \in G$ be such that $\mathrm{Ad}(g)(h) = h$. Then we claim that $g \in P(h)$. Indeed, $\mathrm{Ad}(g)$ stabilizes the subspaces $\mathfrak{g}_\tau(h)$, so that $\mathrm{Ad}(g)(\mathfrak{p}(h)) = \mathfrak{p}(h)$. But $\mathrm{Int}(g)(P(h))$ is an algebraic subgroup of G with Lie algebra $\mathrm{Ad}(g)(\mathfrak{p}(h)) = \mathfrak{p}(h)$. We see that $\mathrm{Int}(g)(P(h)) = P(h)$ (as both groups are connected, using Theorem 4.2.2). Now $P(h)$ is a parabolic subgroup of G (it contains a Borel subgroup), so we can apply [Bor91], Theorem 11.16, stating that a parabolic subgroup of G is equal to its own normalizer.

Now we show that

$$\mathfrak{p}(h) = \{x \in \mathfrak{g} \mid x \cdot V_{\geq 2}(h) \subset V_{\geq 2}(h)\}.$$

It is obvious that $\mathfrak{p}(h)$ is contained in the right-hand side. For the other inclusion, let $x \in \mathfrak{g}$ map $V_{\geq 2}$ to itself and suppose that $x \notin \mathfrak{p}(h)$. Write $x = \sum_{\tau \in \mathbb{Q}} x_\tau$ where $x_\tau \in \mathfrak{g}_\tau(h)$, and let τ_0 be minimal such that $x_{\tau_0} \neq 0$. Then $\tau_0 < 0$. Let $v \in V^0_{\geq 2}(h)$, and let v_2 be its projection on $V_2(h)$. Then $x_{\tau_0} \cdot v_2 \in V_{2+\tau_0}(h)$, whence $x_{\tau_0} \cdot v_2 = 0$. Furthermore, x_{τ_0} is nilpotent so that $H_0 = \{\exp(tx_{\tau_0}) \mid t \in K\}$ is the smallest algebraic subgroup of G whose Lie algebra contains x_{τ_0} (Lemma 4.3.1). Let $g \in H_0$ then $gv_2 = v_2$. Note that h is a characteristic of v_2 by Theorem 7.4.22. So $\mathrm{Ad}(g)(h)$ also is a characteristic of v_2 by (7.4). Therefore, by Lemma 7.4.21, there is a $p \in P(h)$

with $\mathrm{Ad}(p)(h) = \mathrm{Ad}(g)(h)$. By the claim above it follows that $g \in P(h)$. Therefore, $x_{\tau_0} \in \mathfrak{p}(h)$, a contradiction.

Let R be the algebraic subgroup of G consisting of those g with $g(V_{\geq 2}(h)) = V_{\geq 2}(h)$. By Corollary 4.2.9, $\mathrm{Lie}(R) = \mathfrak{p}(h)$. So $P(h) = R^\circ$. Moreover, using [Bor91], Theorem 11.16, stating that parabolic subgroups are connected, we infer that $R = P(h)$.

Next we show that $GV_{\geq 2}(h)$ is closed. The quotient $G/P(h)$ can be made into an algebraic variety ([Bor91], Theorem 6.8), and, since $P(h)$ is parabolic, this variety is complete ([Bor91], Corollary 11.2). By definition that means that for any variety X the image of the projection $G/P(h) \times X \to X$ maps closed subsets to closed subsets. Let $X = V$ and consider

$$\Gamma = \{(gP(h), v) \mid v \in g(V_{\geq 2}(h))\},$$

which is a closed subset of $G/P(h) \times V$. The image of Γ under the projection $G/P(h) \times V \to V$ is $GV_{\geq 2}(h)$, which therefore is closed.

For (ii), let $\sigma : G \times V_{\geq 2} \to V$, $\sigma(g, v) = gv$. By (i) the image of σ is $\overline{N(h)}$. Let $w \in \sigma(G \times V^0_{\geq 2}(h))$ and consider the fibre $\sigma^{-1}(w) = \{(g, v) \mid \sigma(g, v) = w\}$. Let (g_0, v_0), (g_1, v_1) be two elements of $\sigma^{-1}(w)$. Then $v_0, v_1 \in V^0_{\geq 2}(h)$ and with $g = g_0^{-1} g_1$ we have $gv_1 = v_0$. So h and $\mathrm{Ad}(g)(h)$ are characteristics of v_0 (Lemma 7.4.18). As seen in this proof, this implies that $g \in P(h)$. So $g_1 = g_0 p$ for a certain $p \in P(h)$, and hence $v_1 = p^{-1} v_0$. It follows that $\sigma^{-1}(w) = \{(g_0 p, p^{-1} v_0) \mid p \in P(h)\}$. Moreover, the map $P(h) \to \sigma^{-1}(w)$, $p \mapsto (g_0 p, p^{-1} v_0)$ is injective. Hence $\dim \sigma^{-1}(w) = \dim P(h)$.

Now [Sha94], Section I.6.3, Theorem 7, states the following: let $f : X \to Y$ be a surjective regular map between irreducible varieties; then there is a non-empty open subset U of Y such that $\dim f^{-1}(y) = \dim X - \dim Y$ for all $y \in U$. In our situation we obtain a non-empty open subset U_1 of $\sigma(G \times V_{\geq 2}(h))$ such that $\dim \sigma^{-1}(u) = \dim G + \dim V_{\geq 2}(h) - \dim \overline{N(h)}$ for all $u \in U_1$. By Theorem 1.4.4, $\sigma(G \times V^0_{\geq 2}(h))$ contains an open set U_2 of $\overline{N(h)}$. If we take a $u \in U_1 \cap U_2$, we obtain (ii). □

Remark 7.4.25 It is possible to start the development using a different scalar product, satisfying some properties (we will not specify which these should be) so that all proofs go through. Unfortunately the stratification can depend on that (for an example see [Hes79], Remark 4.10). If $\mathfrak{g}$ is simple, up to non-zero scalar multiples, there is only one G-invariant bilinear form. (Indeed, each such form induces a G-equivariant linear map $\mathfrak{g} \mapsto \mathfrak{g}^*$ so if we have two forms, we get two such maps, and composing one with the inverse of the other yields a G-equivariant map $\mathfrak{g} \to \mathfrak{g}$, which must be a scalar times the identity as $\mathfrak{g}$ is an irreducible G-module.) In that case, the stratification is uniquely determined.

7.4.3 Computing characteristics of the strata

Let G be a connected reductive algebraic group over the algebraically closed field K of characteristic 0. Let V be a rational G-module. Here we describe an algorithm to compute the characteristics of the elements of $\mathcal{N}_G(V)$, up to G-conjugacy. A second output of the algorithm is the list of the dimensions of the closures of the corresponding strata.

The algorithm recursively calls itself. The base case, where the output is returned without performing a recursion, occurs when the representation is trivial. This means that every element of G acts as the identity. In that case 0 is the only unstable element and there are no characteristics.

As input to the algorithm we assume that we have the Lie algebra $\mathfrak{g}$ (given by a basis and multiplication table), containing a fixed Cartan subalgebra $\mathfrak{t}$, and the $\mathfrak{g}$-module V. We let $\mu_1, \dots, \mu_m$ be the weights of V (so these are elements of $\mathfrak{t}^*$). By W we will denote the Weyl group of $\mathfrak{g}$ with respect to $\mathfrak{t}$. Since two elements of $\mathfrak{t}$ are G-conjugate if and only if they are W-conjugate (Corollary 5.8.5), we seek the list of characteristics in $\mathfrak{t}$ up to W-conjugacy.

The algorithm consists of two basic steps. First a finite list of candidate characteristics is produced. In the second step we verify which of the candidates really are characteristics.

The first step of the algorithm is based on the following observation. Let $h \in \mathfrak{t}_{\mathbb{Q}}$ be a characteristic of a $v \in \mathcal{N}_G(V)$ and v_2 be the projection of v on $V_2(h)$. By Theorem 7.4.22, h is also a characteristic of v_2. Write $v_2 = \sum_i v_{\lambda_i}$, where the λ_i are weights of $\mathfrak{t}$, and v_{λ_i} are non-zero weight vectors of weight λ_i. Let $\mathcal{H} \subset \mathfrak{t}_{\mathbb{R}}$ denote the hyperplane consisting of all h' with $(h, h') = 2$. Let S denote the convex hull of the $\nu^{-1}(\lambda_i)$ (which is contained in $\mathcal{H}$). Let $\tilde{h}$ be a positive scalar multiple of h lying in $\mathcal{H}$. Then $\tilde{h} \in S$ (Lemma 7.4.16).

It follows that we can find all characteristics in $\mathfrak{t}$ by looping over all subsets of the set of weights of V and for each subset check whether its convex hull S (or more precisely, the convex hull of its image under ν^{-1}) can come from a vector v with characteristic h as above. Specifically, we set $\mathbf{H} = \{\nu^{-1}(\mu_1), \dots, \nu^{-1}(\mu_m)\}$. Since W preserves the weights of V (this follows from Lemma 2.11.3), it also stabilizes $\mathbf{H}$. (Indeed: for $w \in W$, $\lambda \in \mathfrak{t}^*$, we have, by Remark 2.9.9, $w(\lambda) = \lambda \circ w^{-1}$ implying that $w(\nu^{-1}(\lambda)) = \nu^{-1}(w(\lambda))$ because the bilinear form $(\ ,\ )$ is G-invariant, and therefore also W-invariant, by Theorem 5.8.3(i).) The algorithm first computes a list of representatives of the orbits of W on the set of all subsets of $\mathbf{H}$ of size up to the rank of $\mathfrak{g}$. Then for each subset M the following steps are performed:

1. Let A be the smallest affine space containing M (let $h_0 \in M$, then $A = h_0 + \mathcal{V}$, where $\mathcal{V}$ is the subspace of $\mathfrak{t}_{\mathbb{R}}$ spanned by all elements of the form $h - h_0$, $h \in M$). If A contains 0, we discard M. Otherwise we perform the next two steps.
2. Let $M' = A \cap \mathbf{H}$ and $\tilde{h} \in \mathfrak{t}_{\mathbb{R}}$ be the point on A closest to 0 (it is the intersection point of A with the subspace of $\mathfrak{t}_{\mathbb{R}}$ consisting of all vectors perpendicular to $\mathcal{V}$).

3. If $\tilde{h}$ lies in the convex hull of M', we let h be the positive scalar multiple of $\tilde{h}$ such that $(h, \tilde{h}) = 2$, and add h to our list. Otherwise we discard M.

Finally, from the list obtained, we remove all W-conjugate copies. The observation above implies that the resulting list contains all characteristics in $\mathfrak{t}_{\mathbb{Q}}$ up to W-conjugacy.

For the second step, let h be an element of the list obtained in the first step. It is our objective to establish whether h is a characteristic. For this, compute the subalgebras $\mathfrak{z}(h) = \mathfrak{c}_{\mathfrak{g}}(h)$ and

$$\tilde{\mathfrak{z}}(h) = \{x \in \mathfrak{z}(h) \mid (h, x) = 0\},$$

as well as a basis of $V_2(h)$. We now recursively call the algorithm with inputs $\tilde{\mathfrak{z}}(h)$ and $V_2(h)$. The result is a list of characteristics with the dimensions of the closures of the corresponding strata. If no closure has dimension equal to the dimension of $V_2(h)$, h is a characteristic and otherwise it is not by Theorem 7.4.22.

Finally we compute the dimensions of the closures of the strata using Proposition 7.4.24.

Remark 7.4.26 It is possible to reformulate the algorithm so that it operates with weights and roots only. The input to the algorithm consists of a basis B of the fixed Cartan subalgebra $\mathfrak{t}$ (this can be a basis of a vector space of dimension equal to the rank of $\mathfrak{g}$; it does not need to be realized explicitly as a subspace of $\mathfrak{g}$), the restriction of the bilinear form $(\ ,\ )$ to $\mathfrak{t}$, the weights P of the G-module V together with their multiplicities, and the positive roots Φ^+ of the root system of $\mathfrak{g}$. A weight or root λ can be represented by the vector containing the values $\lambda(x)$ for x in the basis B of $\mathfrak{t}$. Instead of working in the space $\mathfrak{t}$, we work in the space $\mathfrak{t}^*$ and take convex hulls of sets of weights, instead of the sets formed by their images under ν^{-1}.

If $\mathfrak{g}$ is simple, this input can be computed knowing the root system, and the highest weight only. Indeed, for the bilinear form we can take the Killing form, since any other G-invariant bilinear form on $\mathfrak{g}$ is a scalar multiple of that. For the weights and their multiplicities we can use the methods of Section 2.11.3.

Then instead of the elements $\tilde{h}$, we obtain $\tilde{\lambda} \in \mathfrak{t}^*$. We let λ be a scalar multiple of $\tilde{\lambda}$ such that $(\lambda, \tilde{\lambda}) = 2$, $\mathfrak{t}' = \{x \in \mathfrak{t} \mid \lambda(x) = 0\}$, and B' be a basis of it. Furthermore, P' consists of the restrictions to $\mathfrak{t}'$ of those $\mu \in P$ such that $(\mu, \lambda) = 2$. Similarly, Φ'^+ is the set of the restrictions to $\mathfrak{t}'$ of those $\alpha \in \Phi^+$ such that $(\lambda, \alpha) = 0$. Then the input to the recursive call consists of B', the same form $(\ ,\ )$, P', and Φ'^+.

In order to compute the dimension of the closure of the stratum with characteristic h, we need $\dim \mathfrak{g}$, $\dim \mathfrak{p}(h)$ and $\dim V_{\geq 2}(h)$. Knowing only $\lambda = \nu(h)$ we compute these as follows. Firstly, $\dim \mathfrak{g} = |B| + 2|\Phi^+|$. Secondly,

$$\dim \mathfrak{p}(h) = |B| + 2|\{\alpha \in \Phi^+ \mid (\alpha, \lambda) = 0\}| + |\{\alpha \in \Phi^+ \mid (\alpha, \lambda) > 0\}|.$$

Thirdly, $\dim V_{\geq 2}(h)$ is equal to to the sum of the multiplicities of the weights μ such that $(\lambda, \mu) \geq 2$.

Remark 7.4.27 1. In the recursive step, before deciding whether a candidate element of $\tilde{\mathfrak{z}}(h)$ is a characteristic, we can compute the dimension of the hypothetical stratum corresponding to it and proceed only with the decision procedure (that is, with further recursive calls) if the dimension is equal to $\dim V_2(h)$.

2. In one step of the algorithm it is necessary to decide whether a given point in a real euclidean space lies in the convex hull of a set of other points. Kalantari ([Kal15]) devised a very elegant solution for this problem. Here we will not go into the details but make a few observations. This algorithm uses a parameter $\varepsilon > 0$. It outputs "yes" or "no". If the output is "no" then the given point is certainly outside the convex hull. If the output is "yes", then depending on the value of ε, it can still be the case that the given point is outside the convex hull. The smaller ε, the closer the point has to be to the convex hull in order to get the wrong answer. However, a smaller ε leads also to a longer run time. For our application this is not so important: if we incidentally get a wrong answer, we include a candidate characteristic too many, which will be discovered not to be a characteristic using the recursive step.

3. Since the algorithm computes representatives of the W-orbits in the set of all r-element subsets of $\mathbf{H}$ for r up to the rank of $\mathfrak{g}$, the complexity of the algorithm increases sharply with the rank. For an example, consider the simple Lie algebra of type B_n and V irreducible of highest weight equal to the fundamental weight λ_n. Then for $n = 4, 5, 6$, the author's implementation of the algorithm in GAP4 takes 0.1, 2.3 and 114.2 seconds.

4. Using Proposition 7.4.24(i) it is straightforward to formulate an algorithm for computing polynomials defining the closure of a stratum given its characteristic. This is similar to computing the closure of an orbit (Section 4.8); we omit the details.

Example 7.4.28 Let $G = \mathrm{SL}(2, \mathbb{C})$ and V the irreducible representation of dimension n. For simplicity we assume that $n = 2r$ is even (the odd case goes in the same way). Here we only consider 1-element sets of weights. Let h, e, f be the standard basis of $\mathfrak{sl}(2, \mathbb{C})$ (Example 2.1.4). Let λ_1 denote the fundamental weight, i.e., $\lambda_1(h) = 1$. The Weyl group is of order 2 and the nontrivial element maps λ_1 to $-\lambda_1$. Therefore, up to the action of the Weyl group, the 1-element sets of weights are $M_i = \{(2i-1)\lambda_1\}$ for $1 \leq i \leq r$. Starting with M_i we find $\tilde{\lambda} = (2i-1)\lambda_1$, $B' = \emptyset$, $\Phi'^{+} = \emptyset$ so that $\widetilde{Z}(h)$ is the trivial group, having no unstable elements. Therefore this yields a characteristic. The dimension of the closure of the corresponding stratum is $r - i + 2$.

Example 7.4.29 Consider $G = \mathrm{SL}(3, \mathbb{C})$ acting on the space V_3 of Example 7.4.13. We consider 1-element and 2-element sets of weights. Up to the action of the Weyl group these are contained in the triangle of the figure in Example 7.4.13. Again using the Weyl group we see that we can limit ourselves to considering the sets

$$\{x^3\},\ \{x^2y\},\ \{x^3, y^3\},\ \{y^3, y^2z\},\ \{x^3, y^2z\},\ \{y^2z, xy^2\},\ \{y^2z, x^2y\}.$$

(Here we indicate a weight by its weight vector. Furthermore, the line from x^3 to y^3 contains a few more 2-element sets which, however, yield the same affine space.) It is immediately seen that we can discard $\{y^3, y^2z\}$ since the closest point to 0 on the line through these points does not lie in their convex hull.

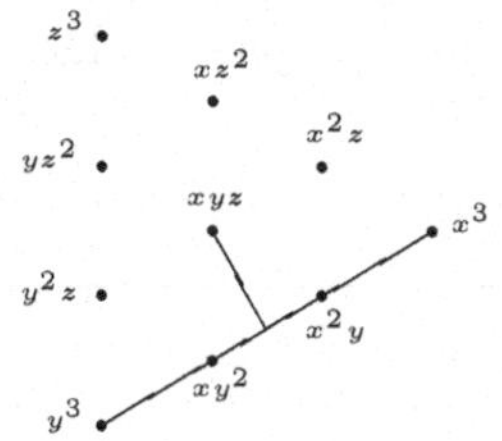

Here we use the set-up described in Remark 7.4.26. The fundamental weights are denoted λ_1, λ_2, as in Example 7.4.13. A basis of $\mathfrak{t}$ is $B = \{h_1, h_2\}$. Here $\lambda_i(h_j) = \delta_{ij}$. The form $(\ ,\ )$, restricted to $\mathfrak{t}^*$ is the one given in Example 7.4.13. The positive roots are $\Phi^+ = \{\alpha_1, \alpha_2, \alpha_1 + \alpha_2\}$. The weights are given in the figure above (see also Example 7.4.13); all have multiplicity 1.

Consider the set $M = \{x^3, y^3\}$; then A is the line from x^3 to y^3 drawn in the picture. So $M' = \{x^3, x^2y, xy^2, y^3\}$. Using the basis u_1, u_2 described in Example 7.4.13, we have that x^3 corresponds to $3\sqrt{2}u_1$ and y^3 to $-\frac{3}{2}\sqrt{2}u_1 - \frac{3}{2}\sqrt{6}u_2$. The picture also displays the line from 0 perpendicular to A. The point of A closest to 0 is $\frac{3}{4}(\sqrt{2}u_1 - \sqrt{6}u_2) = \frac{3}{2}\lambda_2$. So $\tilde{\lambda} = \frac{3}{2}\lambda_2$ and we find $\lambda = \frac{2}{3}\lambda_2$, and $B' = \{h_1\}$. Also we have $(\lambda, \alpha_1) = 0$, $(\lambda, \alpha_2) = 2$. Hence $\Phi'^+ = \{\alpha_1\}$. Furthermore, the weights of $V_2(h)$ are the ones on A, i.e., $3\lambda_1$, $\lambda_1 + \lambda_2$, $-\lambda_1 + 2\lambda_2$, $-3\lambda_1 + 3\lambda_2$. Thus $P' = \{(3), (1), (-1), (-3)\}$. We see that $V_2(h)$ is the irreducible 4-dimensional $\mathfrak{sl}(2, \mathbb{C})$-module. This module has two strata, whose closures have dimensions 2 and 3 (Example 7.4.28). In particular, there is no closure with dimension 4. Therefore, $\nu^{-1}(\lambda)$ is a characteristic. The dimension of the closure of the corresponding stratum is 6.

Now let $M = \{y^2z, xy^2\}$. Then we find $\tilde{\lambda} = \frac{3}{2}(\lambda_1 - \lambda_2)$. Hence $B' = \{h_1 + h_2\}$, $\Phi'^+ = \{\alpha_1 + \alpha_2\}$, $P' = \{(1), (-1)\}$. Now $V_2(h)$ is the 2-dimensional $\mathfrak{sl}(2, \mathbb{C})$-module. It has one stratum of dimension 2 (Example 7.4.28). In other words, all elements of $V_2(h)$ are unstable with respect to $\widetilde{Z}(h)$. Therefore, we do not find a characteristic in this case.

In the end we find five characteristics. The corresponding weights are $-\frac{2}{3}\lambda_1 + 2\lambda_2$, $-\frac{2}{3}\lambda_1 + \frac{4}{3}\lambda_2$, $\frac{2}{3}\lambda_2$, $\frac{1}{3}(\lambda_1 + \lambda_2)$, $\frac{1}{3}\lambda_1$. The dimensions of their closures are, respectively, 8, 7, 6, 5, 3. In particular we see that the irreducible component of the nullcone of highest dimension has dimension 8. It is possible to show that in this case the nullcone is irreducible. Moreover, the strata are orbits in this case. Furthermore, for the representation in this example, all orbits are known ([Kra84], Section I.7).

Remark 7.4.30 We say that the G-module V has the closure property if the closures of all strata are unions of strata. If V has the closure property, it is of interest to determine which strata are contained in the closure of a given one. This would make it possible, for example, to determine the number of connected components of the nullcone. In [Hes79], Section 6, some examples are given that have the closure property, along with the relative closure diagrams. The same section contains an example not having the closure property.

7.5 Notes

For the algorithm of Section 7.3.2 we followed [DK02]. Originally this algorithm appeared in [Der99]. In [DK08] algorithms are developed for computing the invariant ring in other situations (for base fields of characteristic $p > 0$ and for unipotent groups).

The first algorithm for computing generators of the invariant field (Section 7.2) appeared in [MQB99]. It has been simplified and generalized in [Kem07], on which our treatment is based. We remark that the algorithm from [Kem07] works more generally for a regular G-action on a variety.

A version of Theorem 7.4.22 can be found in the works of Kirwan ([Kir84]) and Ness ([Nes84]). The treatment given here is based on the Kempf-Rousseau theory ([Kem78] and [Rou78]), and follows [Slo89] and [VP89].

The algorithm of Section 7.4.3 is taken from [Pop03]. In that paper a construction involving trees is used and here replaced by a series of recursive calls. This leads to the same result; the difference is that the operation of going down the tree has been moved "behind the scenes".

Chapter 8

Nilpotent Orbits

In Chapter 7 we looked at a stratification of the nullcone of a representation of a reductive algebraic group. Here we consider two classes of representations of reductive groups where the strata coincide with orbits of the group. The first class concerns the action of a semisimple algebraic group on its Lie algebra. For this action the nullcone was studied widely since the 1960's, with roots going back to Dynkin's work ([Dyn52]). The second class is formed by the so-called θ-groups, introduced by Vinberg in the 1970's. They are obtained from a grading of a semisimple Lie algebra and share many properties of the groups in the first class. In fact, it is possible to view both classes as one by allowing the trivial grading. However, the adjoint groups also have a number of special properties (as an example we mention the weighted Dynkin diagram uniquely identifying a nilpotent orbit), motivating a separate treatment.

Throughout this chapter we let K be an algebraically closed field of characteristic 0. Let $\mathfrak{g}$ be a semisimple Lie algebra over K, and G the adjoint group of $\mathfrak{g}$. This means that G is constructed from the adjoint representation of $\mathfrak{g}$ with the methods of Chapter 5. We have $\mathrm{Lie}(G) = \mathrm{ad}_{\mathfrak{g}}\mathfrak{g}$, which we identify with $\mathfrak{g}$ and therefore we will say that $\mathfrak{g}$ is the Lie algebra of G. For an algebraic subgroup H of G, we view its Lie algebra $\mathfrak{h}$ as a subalgebra of $\mathfrak{g}$. (Although it really is a subalgebra of $\mathrm{ad}_{\mathfrak{g}}\mathfrak{g}$; so we identify the subalgebra $\mathfrak{h} \subset \mathfrak{g}$ with $\mathrm{ad}_{\mathfrak{g}}\mathfrak{h} \subset \mathrm{ad}_{\mathfrak{g}}\mathfrak{g}$.) As noted in Example 7.4.23, G acts in two ways on $\mathfrak{g}$: via the adjoint representation on $\mathrm{ad}_{\mathfrak{g}}\mathfrak{g}$, and by construction as a subgroup of $\mathrm{Aut}(\mathfrak{g})$. However, as seen in the mentioned example, these actions coincide. Therefore, for $g \in G$, $x \in \mathfrak{g}$ we will write $g \cdot x$ for the result of letting g act on x.

In this chapter we study the *nilpotent* orbits of G on $\mathfrak{g}$. We recall that an element $e \in \mathfrak{g}$ is said to be nilpotent if $\mathrm{ad} e$ is nilpotent. Furthermore, a G-orbit in $\mathfrak{g}$ is called nilpotent if it consists of nilpotent elements. The θ-groups are certain reductive algebraic subgroups G_0 of G, acting on a subspace $\mathfrak{g}_1$ of $\mathfrak{g}$. We also look at the nilpotent orbits of these groups. As noted earlier, in both cases the set of nilpotent elements is equal to the nullcone whose strata are orbits.

We could also take a more general semisimple algebraic group $\widehat{G}$ constructed from $\mathfrak{g}$ as in Chapter 5 and consider the adjoint action on its Lie algebra $\hat{\mathfrak{g}}$ (which is given by Theorem 5.3.9). However, the image of $\widehat{G}$ in $\mathrm{GL}(\hat{\mathfrak{g}})$ is isomorphic to G, and the $\widehat{G}$-orbits in $\hat{\mathfrak{g}}$ are in one-to-one correspondence with the G-orbits in $\mathfrak{g}$.

In this chapter we use the notation introduced elsewhere in this book. In particular, $\mathfrak{h}$ is a fixed Cartan subalgebra of $\mathfrak{g}$ and Φ is the root system of $\mathfrak{g}$ with respect to $\mathfrak{h}$. We denote a fixed basis of simple roots of Φ by $\Delta = \{\alpha_1, \dots, \alpha_\ell\}$. Occasionally we also use a Chevalley basis of $\mathfrak{g}$, whose elements we denote by $h_1, \dots, h_\ell, x_\alpha, \alpha \in \Phi$. The Weyl group of Φ will be denoted W.

8.1 Weighted Dynkin diagrams

In this section we show that the nilpotent G-orbits in $\mathfrak{g}$ are parametrized by a set of so-called weighted Dynkin diagrams. We also briefly indicate how to list these diagrams for G of classical type.

From Section 2.13 we recall that a triple (h, e, f) of elements of a Lie algebra is called an $\mathfrak{sl}_2$-triple if $[h, e] = 2e$, $[h, f] = -2f$ and $[e, f] = h$.

Lemma 8.1.1 *Let (h, e, f) and (h, e, f') be $\mathfrak{sl}_2$-triples in a Lie algebra $\mathfrak{a}$ of characteristic 0. Then $f = f'$.*

Proof. Let $v_0 = f - f'$. Then $[e, v_0] = 0$ and $[h, v_0] = -2v_0$. Let V be the span of all $(\mathrm{ad} f)^i v_0$ for $i \geq 0$. Then V is an irreducible $\mathfrak{sl}_2$-module with highest weight -2 but that is not possible so $V = 0$ and $v_0 = 0$. □

Lemma 8.1.2 *Let $\mathfrak{a}$ be a Lie algebra of characteristic 0. Let $h \in \mathfrak{a}$ and $\mathfrak{n}$ be a subalgebra of $\mathfrak{a}$ such that $[h, \mathfrak{n}] = \mathfrak{n}$, and $\mathrm{ad}_{\mathfrak{a}} z$ is nilpotent for all $z \in \mathfrak{n}$. Then*

$$\{\exp(\mathrm{ad} z)(h) \mid z \in \mathfrak{n}\} = h + \mathfrak{n}.$$

Proof. Let A denote the set on the left-hand side. Since $[h, \mathfrak{n}] = \mathfrak{n}$ it is obvious that $A \subset h + \mathfrak{n}$. For the reverse inclusion, let $u \in \mathfrak{n}$. We claim that $h + u \in A + \mathfrak{n}^k$ for all $k \geq 1$. The lemma follows from that.

Since $A \subset h + \mathfrak{n}$, the claim is obvious for $k = 1$. Suppose $h + u = \exp(\mathrm{ad} z)(h) + y$ for certain $z \in \mathfrak{n}$ and $y \in \mathfrak{n}^k$. Since $\mathrm{ad} h : \mathfrak{n} \to \mathfrak{n}$ is bijective, its restriction to $\mathfrak{n}^k$ has the same property. Hence $y = [y', h]$ for a certain $y' \in \mathfrak{n}^k$. Then

$$\exp(\mathrm{ad}(z + y'))(h) - \exp(\mathrm{ad} z)(h) \in [y', h] + \mathfrak{n}^{k+1}.$$

So $\exp(\mathrm{ad}(z + y'))(h) \in h + u - y + [y', h] + \mathfrak{n}^{k+1} = h + u + \mathfrak{n}^{k+1}$. □

Proposition 8.1.3 *Let $\mathfrak{a}$ be a Lie algebra of characteristic 0. Let (h, e, f) be an $\mathfrak{sl}_2$-triple in $\mathfrak{a}$ and (h', e, f') be a second $\mathfrak{sl}_2$-triple. Then there is a nilpotent $z \in \mathfrak{a}$ such that, setting $\sigma = \exp(\mathrm{ad} z)$, we have $\sigma(h) = h'$, $\sigma(e) = e$ and $\sigma(f) = f'$.*

Proof. Let $\mathfrak{n} = \text{im}(\text{ad}e) \cap \ker(\text{ad}e)$, which is a subalgebra of $\mathfrak{a}$. Write $\mathfrak{a}$ as a direct sum of $\mathfrak{sl}_2$-modules. Then $\ker(\text{ad}e)$ is spanned by the highest weight vectors of all summands. Furthermore, $\text{im}(\text{ad}e)$ is spanned by all weight vectors except the lowest ones. It follows that $\mathfrak{n}$ has a basis cconsisting of $\text{ad}h$-eigenvectors with eigenvalues ≥ 1. So $[h, \mathfrak{n}] = \mathfrak{n}$. Furthermore, $\mathfrak{a}$ is the direct sum of the spaces $\mathfrak{a}_k = \{x \in \mathfrak{g} \mid [h, x] = kx\}$. Let $z \in \mathfrak{n}$, then $\text{ad}z(\mathfrak{a}_k) \subset \oplus_{l>k}\mathfrak{a}_l$ and $\text{ad}_{\mathfrak{a}}z$ is nilpotent.

We have $[e, h - h'] = 0$, $[e, f - f'] = h - h'$. Hence $h' \in h + \mathfrak{n}$. Therefore, by Lemma 8.1.2, there is a $z \in \mathfrak{n}$ with, setting $\sigma = \exp(\text{ad}z)$, $\sigma(h) = h'$. Since $[z, e] = 0$ we get $\sigma(e) = e$. Lemma 8.1.1 finally implies that $\sigma(f) = f'$. □

Theorem 8.1.4 *Let (h, e, f) and (h', e', f') be two $\mathfrak{sl}_2$-triples in $\mathfrak{g}$. Then the following are equivalent:*

(i) *e and e' are conjugate under G.*

(ii) *(h, e, f) and (h', e', f') are conjugate under G.*

(iii) *h and h' are conjugate under G.*

Proof. Proposition 8.1.3 implies that (i) implies (ii). It is obvious that (ii) implies (iii). So it remains to show that (iii) implies (i). For this we may assume that $h = h'$. Let $\mathfrak{g}(k) = \{x \in \mathfrak{g} \mid [h, x] = kx\}$ and $G_h = \{g \in G \mid g(h) = h\}$. Then we have a regular map $\psi : G_h \to \mathfrak{g}(2)$ by $\psi(g) = g(e)$. The differential of ψ is $\text{ad}e : \mathfrak{g}(0) \to \mathfrak{g}(2)$ (Theorem 3.8.1). Writing $\mathfrak{g}$ as a direct sum of $\mathfrak{sl}_2$-modules, we see that $\text{ad}e$ is surjective. Hence ψ is dominant (Proposition 1.4.2), and therefore, $\psi(G_h)$ contains an open set of $\mathfrak{g}(2)$ (Theorem 1.4.4). Similarly, defining $\psi' : G_h \to \mathfrak{g}(2)$ by $\psi'(g) = g(e')$, $\psi'(G_h)$ contains an open set of $\mathfrak{g}(2)$. Two open sets have a non-empty intersection, so (i) follows. □

As seen in Remark 2.9.9, the Weyl group W acts on $\mathfrak{h}$. Let $\mathfrak{h}_{\mathbb{Q}}$ be the vector space over $\mathbb{Q}$ spanned by $h_1, \ldots, h_\ell$. Then every $h \in \mathfrak{h}$ such that $\text{ad}_{\mathfrak{g}}h$ has rational eigenvalues lies in $\mathfrak{h}_{\mathbb{Q}}$ and the action of W preserves $\mathfrak{h}_{\mathbb{Q}}$. Furthermore, the set $C = \{h \in \mathfrak{h}_{\mathbb{Q}} \mid \alpha(h) \geq 0 \text{ for all } \alpha \in \Delta\}$ is a fundamental domain for the action of W. (This is shown in the same way as Theorem 2.8.29.)

Corollary 8.1.5 *Let $e' \in \mathfrak{g}$ be nilpotent; then e' is G-conjugate to a nilpotent $e \in \mathfrak{g}$ lying in an $\mathfrak{sl}_2$-triple (h, e, f) such that $h \in C$. Moreover, $h \in C$ is uniquely determined.*

Proof. By the Jacobson-Morozov theorem (Theorem 2.13.2), there is an $\mathfrak{sl}_2$-triple (h', e', f'). As $\text{ad}_{\mathfrak{g}}h'$ is semisimple, h' lies in a Cartan subalgebra of $\mathfrak{g}$ (Corollary 2.5.10). Since all Cartan subalgebras of $\mathfrak{g}$ are conjugate under G (Proposition 4.3.2), we see that (h', e', f') is conjugate to a triple (h'', e'', f'') with $h'' \in \mathfrak{h}$. Furthermore, h'' is W-conjugate and hence G-conjugate (Corollary 5.8.5) to an element $h \in C$. The uniqueness of h follows from the fact that C is a fundamental domain for the action of W and Corollary 5.8.5. □

Proposition 8.1.6 *Let (h, e, f) be as in Corollary 8.1.5. Then $\alpha(h) \in \{0, 1, 2\}$ for $\alpha \in \Delta$.*

Proof. Write $\mathfrak{g}(l) = \{x \in \mathfrak{g} \mid [h, x] = lx\}$. Set $R = \{\alpha \in \Phi \mid \alpha(h) = 2\}$. Note that $\alpha(h) \geq 0$ when $\alpha > 0$ and $\alpha(h) \leq 0$ when $\alpha < 0$ so $R \subset \Phi^+$. Furthermore, $e = \sum_{\alpha \in R} c_\alpha x_\alpha$. Let $\beta \in \Delta$ and $\alpha \in R$. Then for $[x_{-\beta}, x_\alpha]$ we have three possibilities: it is 0, it lies in $\mathfrak{h}$, or it is a multiple of x_γ for some $\gamma \in \Phi^+$. In particular, $[x_{-\beta}, e]$ lies in $\oplus_{l \geq 0} \mathfrak{g}(l)$ but it also lies in $\mathfrak{g}(2 - \beta(h))$. So if $[x_{-\beta}, e] \neq 0$, then $2 - \beta(h) \geq 0$, whence $\beta(h) \in \{0, 1, 2\}$. If, on the other hand, $[x_{-\beta}, e] = 0$ and $\beta(h) > 0$, the irreducible $\mathfrak{sl}_2$-module generated by $x_{-\beta}$ has negative highest weight, which is impossible. Hence in this case $\beta(h) = 0$. □

Remark 8.1.7 We remark that an $h \in \mathfrak{h}$ is uniquely determined by the values $\alpha(h)$, $\alpha \in \Delta$. Furthermore, given $a_i \in K$, $1 \leq i \leq \ell$, it is straightforward to find an $h \in \mathfrak{h}$ such that $\alpha_i(h) = a_i$ for all i. Indeed, let $C = (\langle \alpha_i, \alpha_j^\vee \rangle)$ be the Cartan matrix of Φ and a be the vector with coordinates a_i. Let $b = C^{-1}a$. Then $h = b_1 h_1 + \cdots + b_\ell h_\ell$ has the required property.

By putting together Theorem 8.1.4, Corollary 8.1.5, and Proposition 8.1.6 we see that every nilpotent orbit Ge is determined uniquely by a set of values $\alpha(h) \in \{0, 1, 2\}$, $\alpha \in \Delta$. We now take the Dynkin diagram of Δ and label the node corresponding to α by $\alpha(h)$. This is the *weighted Dynkin diagram* of the orbit. In particular we see that the number of nilpotent orbits is finite, bounded as it is by 3^ℓ, where ℓ is the rank of Φ.

In general not all 3^ℓ labelings yield a weighted Dynkin diagram of a nilpotent orbit. So the question is which ones do. This question is answered differently for the classical (A, B, C, D) and the exceptional (E, F, G) types. First we briefly describe the answer for the classical types, sketching the proof for type A and for the other types referring to [CM93]. For the classical types we need the notion of a *partition* of a positive integer n: this is a non-increasing sequence of positive integers $(d_1, \ldots, d_r)$, with $d_i \geq d_{i+1}$, such that $n = \sum_i d_i$. So, for example, the partitions of $n = 5$ are $(1, 1, 1, 1, 1)$, $(2, 1, 1, 1)$, $(2, 2, 1)$, $(3, 1, 1)$, $(3, 2)$, $(4, 1)$, (5). We see that this notation can be a bit cumbersome. In order to shorten it, an exponential notation is also used, which means that the first three of the above partitions are written (1^5), $(2, 1^3)$, $(2^2, 1)$. The integers d_i are called the *parts* of the partition and the number of i such that d_i equals a given m is said to be the *multiplicity* of m in the partition.

In type A_ℓ the nilpotent orbits are in bijection with the partitions of $\ell + 1$. This is easily seen by considering $\mathrm{SL}(\ell+1, K)$ instead of G and $\mathfrak{g} = \mathfrak{sl}(\ell+1, K)$. The point is that $\mathrm{SL}(\ell+1, K)$ has exactly the same orbits on $\mathfrak{g}$ as $\mathrm{GL}(\ell+1, K)$

(which also acts on $\mathfrak{g}$ by conjugation). Therefore, a nilpotent element of $\mathfrak{g}$ can be conjugated under $\mathrm{SL}(\ell+1, K)$ to an element in Jordan normal form. The sizes of the Jordan blocks then show the partition corresponding to the given orbit. This immediately gives an algorithm to determine a representative of the orbit corresponding to a given partition. It is also straightforward to indicate a corresponding $\mathfrak{sl}_2$-triple (h, e, f). If the partition is $(d_1, \dots, d_r)$, the element h of this triple is block diagonal, with blocks $D(d_i) = \mathrm{diag}(d_i-1, d_i-3, \dots, -d_i+3, -d_i+1)$, $1 \leq i \leq r$ ([CM93], Section 3.6). The Weyl group W is isomorphic to $S_{\ell+1}$, the symmetric group of degree $\ell+1$. It acts on the Cartan subalgebra consisting of the diagonal matrices by permuting the diagonal entries. Here the fundamental domain C consists of all matrices $\mathrm{diag}(a_1, \dots, a_{\ell+1})$ of trace 0 such that $a_1 \geq a_2 \geq \dots \geq a_{\ell+1}$. In order to obtain the element of C that is W-conjugate to h, we form the sequence containing all entries of the $D(d_i)$. Writing this sequence in non-increasing order, we obtain $b_1, \dots, b_{\ell+1}$. The weighted Dynkin diagram is then

For example, we compute the weighted Dynkin diagram corresponding to the partition $(2^2, 1)$. The sequence of the b_i's is $1, 1, 0, -1, -1$, leading to

In type B_ℓ, the nilpotent orbits are in bijection with the partitions of $2\ell+1$, in which even integers occur with even multiplicity. For example, in type B_3 the nilpotent orbits are parametrized by the partitions (1^5), $(2^2, 1^3)$, $(3, 1^4)$, $(5, 1^2)$, $(3^2, 1)$, $(3, 2^2)$, (7). There is an explicit method to attach a representative to a partition using as in type A_ℓ a realization of the Lie algebra of type B_ℓ as a matrix Lie algebra ([CM93], Recipe 5.2.4). Let $\tau = (d_1, \dots, d_r)$ be a partition as above. For each d_i we form the sequence $d_i-1, d_i-3, \dots, -d_i+1$. We take the elements of these sequences together to form a single sequence $b_0, b_1 \dots, b_{2\ell}$ where $b_0 = 0$, $b_1, \dots, b_\ell$ are non-negative, in non-increasing order, and $b_{\ell+i} = -b_i$ for $1 \leq i \leq \ell$. (Let, for example, $\tau = (2, 2, 1, 1, 1)$. Then $b_1 = b_2 = 1$, $b_3 = 0$.) Then the weighted Dynkin diagram corresponding to τ is

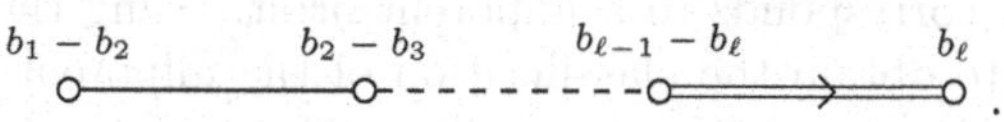

In type C_ℓ, the nilpotent orbits are in bijection with the partitions of 2ℓ, in which odd integers occur with even multiplicity. For example, in type C_3 the nilpotent orbits are parametrized by the partitions (1^6), $(2, 1^4)$, $(2^2, 1^2)$, (2^3), (3^2), $(4, 1^2)$, $(4, 2)$, (6). Also in this case, starting with a partition $\tau = (d_1, \dots, d_r)$, we take together all integers in the sequences $d_i - 1, \dots, -d_i + 1$, and reorder them to arrive at a sequence $b_1, \dots, b_{2\ell}$ such that $b_1, \dots, b_\ell$ are non-negative and in non-increasing order and $b_{\ell+i} = -b_i$ for $1 \leq i \leq \ell$. For example, when $\tau = (4, 1, 1)$ we arrive at $b_1 = 3$, $b_2 = 1$ and $b_3 = 0$. The weighted Dynkin diagram corresponding to τ is

In type D_ℓ the procedure is a bit trickier. Here the nilpotent orbits are parametrized by the partitions of 2ℓ in which even integers have even multiplicity, with the addition that *very even* partitions (that is, those with only even parts) correspond to two orbits (whereas the other partitions correspond to just one orbit). For example, in type D_4 the relevant partitions are (1^8), $(2^2, 1^4)$, (2^4), $(3, 1^5)$, $(3, 2^2, 1)$, $(3^2, 1^2)$, (4^2), $(5, 1^3)$, $(5, 3)$, $(7, 1)$. Here (2^4) and (4^2) correspond to two orbits, the other partitions to only one. In order to obtain the weighted Dynkin diagram corresponding to the partition $\tau = (d_1, \dots, d_r)$, we first form the sequence $b_1, \dots, b_\ell$ exactly as in type C_ℓ. The weighted Dynkin diagram is then

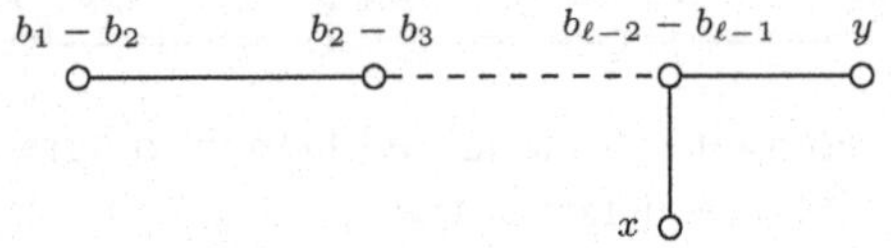

If τ is not very even, then $x = y = b_{\ell-1}$. On the other hand, if τ is very even, we have two weighted Dynkin diagrams: one with $x = 0$, $y = 2$, the other with $x = 2$, $y = 0$. In both cases $y = b_{\ell-1} - b_\ell$ and $x = b_{\ell-1} + b_\ell$. If τ is not very even, these are equal, as $b_\ell = 0$. In the very even case, $b_{\ell-1} = b_\ell = 1$, so that $y = 0$, $x = 2$, and the orbit with these values inverted exists as well.

The story is very different when dealing with the exceptional types. Here the nilpotent orbits are not parametrized by sets of partitions. In view of Theorem 8.1.4, Dynkin's classification of the subalgebras of type A_1 of the Lie algebras of exceptional type ([Dyn52]) yields the classification of the nilpotent orbits. However, his lists are not entirely correct. Many corrections were published in [Ela75]. A uniform method to classify nilpotent orbits has been devised by Bala and Carter ([BC76a] and [BC76b]). The explicit lists of weighted Dynkin diagrams are published widely, for example in [CM93]. In the next section we will describe a simple algorithm for deciding whether a given weighted Dynkin diagram corresponds to a nilpotent orbit. Using that algorithm it is straightforward to obtain the classification of the nilpotent orbits in the exceptional cases.

8.2 Computing representatives and weighted Dynkin diagrams

We discuss two algorithmic problems suggested by the results of the previous section: computing a representative of a nilpotent orbit given by a weighted Dynkin diagram and deciding whether two given nilpotent elements are conjugate under G.

We start with a lemma whose proof is entirely analogous to the one of Theorem 4.2.1; therefore we omit it.

Lemma 8.2.1 *Let H be a connected algebraic group over K and $\rho : H \to \mathrm{GL}(V)$ be a rational representation. Let $v \in V$, $O = \rho(H)v$ its orbit and $X = \overline{O}$. Define the regular map $\varphi : H \to X$ by $\varphi(g) = \rho(g)v$. Then $\mathrm{d}_e\varphi : \mathrm{Lie}(H) \to T_v(X)$ is surjective.*

Proposition 8.2.2 *Let $a_i \in \{0, 1, 2\}$ be given for $1 \leq i \leq \ell$. Let $h \in \mathfrak{h}$ be such that $\alpha_i(h) = a_i$ for all i. For $l \in \mathbb{Z}$, set $\mathfrak{g}_l = \{x \in \mathfrak{g} \mid [h, x] = lx\}$. Define $O = \{x \in \mathfrak{g}_2 \mid [\mathfrak{g}_0, x] = \mathfrak{g}_2\}$.*

(i) *O is non-empty and open in $\mathfrak{g}_2$.*

(ii) *Let $e \in O$. Then either there exists $f \in \mathfrak{g}_{-2}$ such that (h, e, f) is an $\mathfrak{sl}_2$-triple or h does not lie in an $\mathfrak{sl}_2$-triple.*

Proof. Let $H = \{g \in G \mid g \cdot h = h\}$. This is an algebraic subgroup of G with Lie algebra $\mathfrak{g}_0$ (Corollary 4.2.8). Set $G_0 = H^\circ$. Since $[\mathfrak{g}_0, \mathfrak{g}_2] \subset \mathfrak{g}_2$, also $g \cdot \mathfrak{g}_2 = \mathfrak{g}_2$ for all $g \in G_0$ (Corollary 4.2.10). It can be shown that G_0 has a finite number of orbits in $\mathfrak{g}_2$ (Corollary 8.3.8). So there is a unique orbit O' of dimension equal to $\dim \mathfrak{g}_2$ and by Lemma 3.12.1 it is open in $\mathfrak{g}_2$. Let $x \in O'$, then $[\mathfrak{g}_0, x]$ is the tangent space to (the closure of) O' at x (Lemma 8.2.1). Hence $O' \subset O$. On the other hand, if $x \in O$, by the same argument, the dimension of its orbit is equal to $\dim \mathfrak{g}_2$, whence the G_0-orbit of x is equal to O'. We conclude that $O' = O$.

Suppose h lies in the $\mathfrak{sl}_2$-triple (h, e', f'). Let $\mathfrak{a}$ be the subalgebra of $\mathfrak{g}$ spanned by these elements. Write $\mathfrak{g}$ as a direct sum of irreducible $\mathfrak{a}$-modules. Let M denote such a module and M_r, $r \in \mathbb{Z}$ be its 1-dimensional weight spaces. If 2 occurs as a weight of M, then 0 is a weight of M (Theorem 2.9.1) and $\mathrm{ad}e' : M_0 \to M_2$ is surjective. Because $\mathfrak{g}_2$ is the sum of such M_2's, and $\mathfrak{g}_0$ contains all M_0's, we see that $\mathrm{ad}e' : \mathfrak{g}_0 \to \mathfrak{g}_2$ is surjective. In particular, $e' \in O$. Above we have seen that O is a single G_0-orbit, so there is a $g \in G_0$ with $g \cdot e' = e$. Setting $f = g \cdot f'$ we obtain the required $\mathfrak{sl}_2$-triple. □

This proposition immediately gives an algorithm, given a weighted Dynkin diagram, for finding an $\mathfrak{sl}_2$-triple (h, e, f) such that e lies in the corresponding orbit. Indeed, first we find $h \in \mathfrak{h}$ such that $\alpha_i(h) = a_i$ where a_i is the i-th label of the diagram (Remark 8.1.7). Second, we compute bases of the spaces $\mathfrak{g}_0$ and $\mathfrak{g}_2$. Third, by repeatedly trying random elements we quickly find an $e \in \mathfrak{g}_2$ such that $[\mathfrak{g}_0, e] = \mathfrak{g}_2$. Finally, by solving a system of linear equations we find the f of the triple (h, e, f).

This algorithm is easily modified into an algorithm for testing whether a given labeling of the Dynkin diagram corresponds to a nilpotent orbit. The only difference is the last step: the linear equations may not have a solution, in which case the labeling does not correspond to a nilpotent orbit (Proposition 8.2.2(ii)).

Example 8.2.3 Let $\mathfrak{g}$ be the Lie algebra of type A_2, with multiplication table as in Example 2.9.14. Consider the diagram with labels 2, 0. Then $h = \frac{4}{3}h_1 + \frac{2}{3}h_2$, $\mathfrak{g}_0 = \langle h_1, h_2, x_{\alpha_2}, x_{-\alpha_2}\rangle$, $\mathfrak{g}_2 = \langle x_{\alpha_1}, x_{\alpha_3}\rangle$, $\mathfrak{g}_{-2} = \langle x_{-\alpha_1}, x_{-\alpha_3}\rangle$. It is immediate that $e = x_{\alpha_1}$ lies in O. Let $f = ax_{-\alpha_1} + bx_{-\alpha_3}$. Then $[e, f] = ah_1 - bx_{-\alpha_2}$, which cannot be equal to h. We conclude that the given labeling does not correspond to a nilpotent orbit.

Example 8.2.4 Executing the latter procedure for $\mathfrak{g}$ of type E_8 took about 65 seconds, resulting in 69 weighted Dynkin diagrams of non-zero nilpotent orbits.

Now we look at the converse problem: given a nilpotent $e \in \mathfrak{g}$, find the weighted Dynkin diagram of its orbit. This is equivalent to deciding whether two nilpotent elements are G-conjugate. The first step is to compute an $\mathfrak{sl}_2$-triple (h, e, f) (Section 2.13). Now the most immediate approach is to compute a Cartan subalgebra of $\mathfrak{g}$ containing h, the corresponding root system; find a Weyl group conjugate of h lying in the fundamental domain C (Corollary 8.1.5); and finally to compute the values of the simple roots on that conjugate of h. However, finding the root system of the computed Cartan subalgebra requires computing the eigenvalues of some of its elements, and these may lie in a high-degree algebraic extension of $\mathbb{Q}$. Therefore this algorithm is very difficult to use in practice. A much more efficient method is based on the following lemma.

Lemma 8.2.5 *Let $\rho : \mathfrak{g} \to \mathfrak{gl}(V)$ be a representation of $\mathfrak{g}$. If $h, h' \in \mathfrak{g}$ are G-conjugate, $\rho(h)$, $\rho(h')$ are conjugate under* $\mathrm{GL}(V)$.

Proof. The group G is generated by elements of the form $\exp(\mathrm{ad}x)$, where $x \in \mathfrak{g}$ is nilpotent. So we may assume that h and h' are conjugate under one such $\exp(\mathrm{ad}x)$. By the Jacobson-Morozov theorem, x lies in an $\mathfrak{sl}_2$-triple, so $\rho(x)$ is nilpotent as well. Using Lemma 2.3.1, $\rho(\exp(\mathrm{ad}x)(h)) = \exp(\rho(x))\rho(h)\exp(-\rho(x))$, proving the lemma. □

Let ρ be as in the lemma. Let (h, e, f) be an $\mathfrak{sl}_2$-triple in $\mathfrak{g}$. Combining Theorems 2.7.6 and 2.9.1, we see that $\rho(h)$ has integral eigenvalues. We let

$s_\rho(h)$ denote the sequence of the eigenvalues, with multiplicities of $\rho(h)$ in non-decreasing order. Let (h', e', f') be a second $\mathfrak{sl}_2$-triple. By Lemma 8.2.5, if h and h' are G-conjugate, $s_\rho(h) = s_\rho(h')$. Now we show a result allowing the converse conclusion.

Proposition 8.2.6 *Let $\mathfrak{g}$ be simple. Number the Dynkin diagram of the root system of $\mathfrak{g}$ as in Theorem 2.8.5. The fundamental weights are denoted $\lambda_1, \dots, \lambda_\ell$ (Section 2.8.3). Let $h, h' \in \mathfrak{g}$ be as above.*

(i) *Set $\lambda = \lambda_1$ if $\mathfrak{g}$ is of type A_ℓ, B_ℓ, C_ℓ, E_6, G_2, and $\lambda = \lambda_\ell$ if $\mathfrak{g}$ is of type E_7, E_8. Let ρ be the irreducible highest weight representation of $\mathfrak{g}$ of highest weight λ. If $s_\rho(h) = s_\rho(h')$, then h and h' are G-conjugate.*

(ii) *If $\mathfrak{g}$ is of type D_ℓ, set $\lambda = \lambda_1$, $\mu = \lambda_\ell$. Let ρ and φ be irreducible highest weight representations of $\mathfrak{g}$ of highest weights λ and μ respectively. If $s_\rho(h) = s_\rho(h')$ and $s_\varphi(h) = s_\varphi(h')$ then h and h' are G-conjugate.*

Proof. This is shown by case-by-case considerations. If $\mathfrak{g}$ is of type A_ℓ, then ρ is the standard $(\ell+1)$-dimensional representation of $\mathfrak{sl}(\ell+1, K)$. The weights of this representation are $\mu_i = \lambda_1 - \alpha_1 - \cdots - \alpha_i$ for $0 \leq i \leq \ell$. As noticed above, replacing h and h' by G-conjugates, does not change $s_\rho(h)$ and $s_\rho(h')$. So we may assume that h and h' lie in the fixed Cartan subalgebra $\mathfrak{h}$, and that $\alpha_i(h)$ and $\alpha_i(h')$ are non-negative for $1 \leq i \leq \ell$. But then $\mu_i(h) \geq \mu_{i+1}(h)$, $0 \leq i < \ell$. It follows that from $s_\rho(h)$ we can recover $\mu_i(h)$, and hence $\alpha_i(h)$ for $1 \leq i \leq \ell$. Similarly, $s_\rho(h')$ determines $\alpha_i(h')$. It follows that $h = h'$.

The proof in the cases B_ℓ and C_ℓ is similar. It is a bit more work, but not difficult, to calculate the weights of ρ in these cases. One way of doing that is by inspecting the explicit constructions of these representations given in [Jac79] Section IV.6 or [Hum78] Section 1.2. Then the same line of argument as for A_ℓ works here.

The exceptional cases are best settled by direct (computer) calculations. Going through the list of nilpotent orbits, we compute an $\mathfrak{sl}_2$-triple (h, e, f) for each such orbit, and then compute $s_\rho(h)$, and check that all resulting sequences are different.

Now we treat D_ℓ. In this case, ρ is again the standard representation of $\mathfrak{g}$. The weights of smallest height (see Section 2.11.4 for the definition of this concept) of ρ are $\mu_i = \lambda - \alpha_1 - \cdots - \alpha_i$ for $0 \leq i \leq \ell - 2$. These are followed by two weights of height $\ell - 1$, namely $\mu_{\ell-2} - \alpha_{\ell-1}$ and $\mu_{\ell-2} - \alpha_\ell$. Again, after replacing h and h' by suitable G-conjugates, we may assume that $\alpha_i(h)$ and $\alpha_i(h')$ are non-negative. Then the first $\ell - 1$ values of $s_\rho(h)$ are $\mu_i(h)$ for $0 \leq i \leq \ell - 2$. So $s_\rho(h)$ determines the values of $\alpha_i(h)$, $1 \leq i \leq \ell - 2$ and $\lambda(h)$. Furthermore, $s_\varphi(h)$ determines $\mu(h)$. Since λ and μ along with the α_i, $1 \leq i \leq \ell - 2$, span $\mathfrak{h}^*$, we infer that $s_\rho(h)$ and $s_\varphi(h)$ determine h. The same is true for h', so again it follows that $h = h'$. □

Now the algorithm for deciding whether two nilpotent elements e and e' in $\mathfrak{g}$ are G-conjugate is straightforward. First assume that $\mathfrak{g}$ is simple and not of type D_ℓ. Then we construct the representation ρ of the previous proposition using the method of Section 2.11.4. Next, we compute two $\mathfrak{sl}_2$-triples (h, e, f) and (h', e', f') (Section 2.13). Then e and e' are G-conjugate if and only if $s_\rho(h) = s_\rho(h')$ (combine Theorem 8.1.4, Proposition 8.2.6(i)). If $\mathfrak{g}$ is of type D_ℓ, we do the same, except that we consider the representations ρ and φ, as in Proposition 8.2.6(ii). If $\mathfrak{g}$ is semisimple, we decompose $\mathfrak{g}$ as a direct sum of simple ideals, decompose e and e' accordingly, and treat each simple piece separately.

If we are interested in the weighted Dynkin diagram of the orbit Ge, we precompute a list of $s_\rho(h')$ (in case $\mathfrak{g}$ is simple and not of type D_ℓ), where (h', e', f') runs through the $\mathfrak{sl}_2$-triples corresponding to the known list of weighted Dynkin diagrams of nilpotent orbits in $\mathfrak{g}$. Subsequently, we compute $s_\rho(h)$ and look it up in the list. If $\mathfrak{g}$ is of type D_ℓ or semisimple, we make the appropriate modifications to this procedure.

Example 8.2.7 Let $\mathfrak{g}$ be the simple Lie algebra of type A_2. There are two non-zero nilpotent orbits, the weighted Dynkin diagrams having labels $(1, 1)$ and $(2, 2)$. The first elements of corresponding $\mathfrak{sl}_2$-triples are, respectively, $h_1 + h_2$ and $2h_1 + 2h_2$. Letting ρ be the 3-dimensional representation with highest weight λ_1, we have $s_\rho(h_1 + h_2) = (1, 0, -1)$ and $s_\rho(2h_1 + 2h_2) = (2, 0, -2)$. Let e be as in Example 2.13.3 in which an $\mathfrak{sl}_2$-triple (h, e, f) is computed with $h = 2x_{\alpha_1} + 2h_1 + 2h_2$. We have

$$\rho(h) = \begin{pmatrix} 2 & 2 & 0 \\ 0 & 0 & 0 \\ 0 & 0 & -2 \end{pmatrix}.$$

We see that $s_\rho(h) = (2, 0, -2)$. We conclude that the weighted Dynkin diagram of the orbit of e has labels $(2, 2)$.

8.3 θ-groups

This section briefly introduces the theory of θ-groups, which were introduced by Vinberg in the 1970's ([Vin76] and [Vin79]). They form a class of representations of reductive algebraic groups, sharing many properties of the adjoint representation of a semisimple algebraic group. In particular, the strata of the nullcone are orbits that are consequently finite in number.

8.3.1 Cyclic gradings of $\mathfrak{g}$

Let $\mathcal{C}_m = \mathbb{Z}/m\mathbb{Z}$ be the cyclic group of order m so that $\mathcal{C}_\infty = \mathbb{Z}$. We consider $\mathcal{C}_m$-gradings of $\mathfrak{g}$:

$$\mathfrak{g} = \bigoplus_{i \in \mathcal{C}_m} \mathfrak{g}_i.$$

Here the $\mathfrak{g}_i$ are subspaces of $\mathfrak{g}$ such that $[\mathfrak{g}_i, \mathfrak{g}_j] \subset \mathfrak{g}_{i+j}$.

If $m \neq \infty$, a $\mathcal{C}_m$-grading yields an automorphism θ of $\mathfrak{g}$ of order m by setting $\theta(x) = \omega^i x$ for $x \in \mathfrak{g}_i$ where $\omega \in K$ is a primitive m-th root of unity. Conversely, if $\theta \in \mathrm{Aut}(\mathfrak{g})$ has order m, the eigenvalues of θ are ω^i, and the decomposition of $\mathfrak{g}$ into eigenspaces of θ gives a $\mathcal{C}_m$-grading. We see that specifying a $\mathcal{C}_m$-grading is the same as specifying an automorphism of order m. On the other hand, a $\mathbb{Z}$-grading of $\mathfrak{g}$ corresponds to a subgroup $\{\theta_t \mid t \in K^*\} \subset \mathrm{Aut}(\mathfrak{g})$ where $\theta_t(x) = t^i x$ for $x \in \mathfrak{g}_i$. In this case we let the grading correspond to an automorphism $\theta = \theta_t$ where $t \in K$ has infinite order (note that since we assume that the characteristic is 0, such t exist). This has the drawback that θ is not defined uniquely, but for our purposes that does not matter.

We say that two $\mathcal{C}_m$-gradings $\mathfrak{g} = \oplus_{i \in \mathcal{C}_m} \mathfrak{g}_i$ and $\mathfrak{g} = \oplus_{i \in \mathcal{C}_m} \mathfrak{g}'_i$, corresponding to automorphisms θ and θ', are equivalent if there is an automorphism φ of $\mathfrak{g}$ such that $\varphi(\mathfrak{g}_i) = \mathfrak{g}'_i$. This is the same as saying that $\varphi\theta\varphi^{-1} = \theta'$. Hence, if $m \neq \infty$, the $\mathcal{C}_m$-gradings are classified by the conjugacy classes in $\mathrm{Aut}(\mathfrak{g})$ of automorphisms of $\mathfrak{g}$ of order m. The latter classification has been carried out by a number of authors, for example, by Kac ([Kac69]; see also the account in [Hel78], Chapter X, Section 5) and Reeder ([Ree10]). Now consider the case where $m = \infty$. We can define a $\mathbb{Z}$-grading of $\mathfrak{g}$ by assigning a non negative degree to each simple root. Moreover, by stipulating $\deg(\alpha+\beta) = \deg(\alpha)+\deg(\beta)$ and $\deg(-\alpha) = -\deg(\alpha)$, the degree of every root is defined. We let $\mathfrak{g}_0$ be spanned by $\mathfrak{h}$, along with all $\mathfrak{g}_\alpha$ with $\deg(\alpha) = 0$. Furthermore, $\mathfrak{g}_i$ for $i \neq 0$ is spanned by all $\mathfrak{g}_\alpha$ with $\deg(\alpha) = i$. Then $\mathfrak{g} = \oplus_{i \in \mathbb{Z}} \mathfrak{g}_i$ is a $\mathbb{Z}$-grading of $\mathfrak{g}$. We call this a *standard* $\mathbb{Z}$-grading. An arbitrary $\mathbb{Z}$-grading is equivalent to a standard one. To see this, let $\mathfrak{g} = \oplus_{i \in \mathbb{Z}} \mathfrak{g}'_i$ be a $\mathbb{Z}$-grading of $\mathfrak{g}$. Define $d(x) = ix$ for $x \in \mathfrak{g}'_i$ and extend d to a map $\mathfrak{g} \to \mathfrak{g}$ by linearity. Then d is a derivation of $\mathfrak{g}$ and because all derivations of $\mathfrak{g}$ are inner (Proposition 2.7.7), there is a unique $h' \in \mathfrak{g}$ with $[h', x] = ix$ for $x \in \mathfrak{g}'_i$. The element h' is called the *defining element* of the given $\mathbb{Z}$-grading of $\mathfrak{g}$ (or also the defining element of $\mathfrak{g}$, if the grading is understood). Now h' is semisimple and thus lies in a Cartan subalgebra of $\mathfrak{g}$. Hence h' is G-conjugate to an element h'' of $\mathfrak{h}$ (Proposition 4.3.2). Because h'' has integral eigenvalues, it lies in $\mathfrak{h}_{\mathbb{Q}}$ and is therefore G-conjugate to an element h of C (notation as in Section 8.1). It follows that the $\mathbb{Z}$-grading we started with is equivalent to the $\mathbb{Z}$-grading with defining element h. But the latter is standard.

Theorem 8.3.1 *Consider a C_m-grading $\mathfrak{g} = \bigoplus_{i \in C_m} \mathfrak{g}_i$. Let κ denote the Killing form of $\mathfrak{g}$.*

(i) *$\mathfrak{g}_i$ contains the semisimple and nilpotent parts (Theorem 2.7.9) of its elements.*

(ii) *The restriction of κ to $\mathfrak{g}_0$ is non-degenerate.*

(iii) *Let $\mathfrak{h}_0$ be a Cartan subalgebra of $\mathfrak{g}_0$. Then the centralizer of $\mathfrak{h}_0$ in $\mathfrak{g}$ is a Cartan subalgebra of $\mathfrak{g}$. In particular, $\mathfrak{g}_0 \neq 0$.*

(iv) *$\mathfrak{g}_0$ is reductive in $\mathfrak{g}$.*

Proof. Let θ be the automorphism associated to the grading. Then θ maps nilpotent (respectively semisimple) elements of $\mathfrak{g}$ to nilpotent (respectively semisimple) elements of $\mathfrak{g}$. (Indeed, the minimal polynomial of $\mathrm{ad}x$ is the same as the minimal polynomial of $\mathrm{ad}\theta(x)$.) Let $x \in \mathfrak{g}_i$ and $x = x_s + x_n$ be its Jordan decomposition (Theorem 2.7.9). Then $\theta(x) = \theta(x_s) + \theta(x_n)$ is the Jordan decomposition of $\theta(x)$. But $\theta(x) = t^i x$ (for some $t \in K$), and $t^i x = t^i x_s + t^i x_n$ is the Jordan decomposition of $t^i x$. So $\theta(x_s) = t^i x_s$ and $\theta(x_n) = t^i x_n$, implying that $x_s, x_n \in \mathfrak{g}_i$.

Let $x \in \mathfrak{g}_i$, $y \in \mathfrak{g}_j$, then $(\mathrm{ad}x)(\mathrm{ad}y)$ maps $\mathfrak{g}_r$ to $\mathfrak{g}_{r+i+j}$. So if $i+j \neq 0$ then $\kappa(x,y) = 0$. Since κ is non-degenerate (Proposition 2.7.2), its restrictions to $\mathfrak{g}_i \times \mathfrak{g}_{-i}$ have to be non-degenerate as well. In particular, its restriction to $\mathfrak{g}_0$ is non-degenerate.

By Proposition 2.12.5, it follows that $\mathfrak{g}_0$ is either 0 or reductive in $\mathfrak{g}$. Now we prove (iii), thereby also proving (iv).

If $\mathfrak{g}_0 \neq 0$, then $\mathfrak{h}_0$ is a Cartan subalgebra of a subalgebra which is reductive in $\mathfrak{g}$, and hence consists of commuting semisimple elements. Therefore, $\mathfrak{h}_0$ is contained in a Cartan subalgebra $\mathfrak{h}'$ of $\mathfrak{g}$. If $\mathfrak{g}_0 = 0$, we immediately obtain the same conclusion. Let Φ' be the root system of $\mathfrak{g}$ with respect to $\mathfrak{h}'$. Then a linear combination of elements of $\mathfrak{h}'$ and root vectors y_α (where $\alpha \in \Phi'$) lies in the centralizer of $\mathfrak{h}_0$ if and only if for the y_α that appear with non-zero coefficient we have $\alpha(h) = 0$ for all $h \in \mathfrak{h}_0$. The set of all those roots forms a root subsystem of Φ'. Let $\mathfrak{s}$ be the subalgebra generated by all these y_α; then $\mathfrak{s}$ is semisimple. Note that since $\mathfrak{h}_0 \subset \mathfrak{g}_0$, an element lies in the centralizer of $\mathfrak{h}_0$ if and only if all of its homogeneous components do. Therefore, $\mathfrak{s}$ inherits the grading from $\mathfrak{g}$. Moreover, $\mathfrak{s}_0 = 0$. (Indeed, an $x \in \mathfrak{s}_0$ normalizes $\mathfrak{h}_0$ and therefore lies in $\mathfrak{h}_0$, but then it centralises $\mathfrak{s}$, and has to be zero as $\mathfrak{s}$ is semisimple.)

Now let $k > 0$ be such that $\mathfrak{s}_0, \dots, \mathfrak{s}_{k-1} = 0$. Let $x \in \mathfrak{s}_k$. Then $(\mathrm{ad}x)^r(\mathfrak{s}_i) \subset \mathfrak{s}_{rk+i}$ for all i. If $m = \infty$, it immediately follows that $\mathrm{ad}_{\mathfrak{s}}x$ is nilpotent. If $m \neq \infty$, let $r > 0$ be such that $k(r-1) < m - i \leq kr$. Then $kr + i = m + j$ where $0 \leq j < k$. So $(\mathrm{ad}x)^r(\mathfrak{s}_i) \subset \mathfrak{s}_j = 0$. Therefore, $\mathrm{ad}_{\mathfrak{s}}x$ is nilpotent. In the same way, $\mathrm{ad}_{\mathfrak{s}}y$ is nilpotent for $y \in \mathfrak{s}_{-k}$. Moreover, $[\mathfrak{s}_k, \mathfrak{s}_{-k}] \subset \mathfrak{s}_0 = 0$. So x and y commute and hence $(\mathrm{ad}_{\mathfrak{s}}x)(\mathrm{ad}_{\mathfrak{s}}y)$ is nilpotent.

Denoting the Killing form of $\mathfrak{s}$ by $\kappa_{\mathfrak{s}}$, we see that $\kappa_{\mathfrak{s}}(x, y) = 0$. But the restriction of $\kappa_{\mathfrak{s}}$ to $\mathfrak{s}_k \times \mathfrak{s}_{-k}$ is non-degenerate. It follows that $\mathfrak{s} = 0$. Hence the centralizer of $\mathfrak{h}_0$ is $\mathfrak{h}'$. □

Let G be the adjoint group of $\mathfrak{g}$ having Lie algebra $\mathrm{ad}_{\mathfrak{g}}\mathfrak{g}$. Consider again a $\mathcal{C}_m$-grading $\mathfrak{g} = \oplus_{i\in\mathcal{C}_m}\mathfrak{g}_i$ with corresponding automorphism θ. Define the algebraic group $H = \{g \in \mathrm{GL}(\mathfrak{g}) \mid \theta g = g\theta\}$. Since the equations that define H are linear, $\mathrm{Lie}(H) = \{x \in \mathfrak{gl}(\mathfrak{g}) \mid \theta x = x\theta\}$ (this follows from Lemma 3.6.8). Let G_0 be the identity component of $H \cap G$. Then by Theorem 4.2.2(ii), $\mathrm{Lie}(G_0) = \mathrm{Lie}(H) \cap \mathrm{Lie}(G) = \{\mathrm{ad}_{\mathfrak{g}}x \mid x \in \mathfrak{g} \text{ and } \mathrm{ad}_{\mathfrak{g}}\theta(x) = \mathrm{ad}_{\mathfrak{g}}x\}$ (as $\theta(\mathrm{ad}x)\theta^{-1} = \mathrm{ad}\theta(x)$). But the latter is identified with $\mathfrak{g}_0$. It follows that $\mathrm{ad}_{\mathfrak{g}}\mathfrak{g}_0$ is algebraic and G_0 is the unique connected subgroup of G with Lie algebra $\mathrm{ad}\mathfrak{g}_0$. Moreover, $\mathrm{ad}\mathfrak{g}_0(\mathfrak{g}_1) \subset \mathfrak{g}_1$. Therefore, by Corollary 4.2.10, $G_0 \cdot \mathfrak{g}_1 = \mathfrak{g}_1$.

The image of G_0 in $\mathrm{GL}(\mathfrak{g}_1)$ is called a θ-group. Usually we also say that G_0 is a θ-group, its action on $\mathfrak{g}_1$ being understood. By Theorems 8.3.1(iv), 5.8.1, G_0 is reductive.

Remark 8.3.2 Instead of looking at the action of G_0 on $\mathfrak{g}_1$ we can also consider its action on $\mathfrak{g}_r$ for any r. Everything then works the same way, so for simplicity, we restrict to the case $r = 1$. Alternatively, we may define a different graded Lie algebra $\bar{\mathfrak{g}}$, which is the direct sum of the spaces $\bar{\mathfrak{g}}_i = \mathfrak{g}_{ri}$. Then $\bar{\mathfrak{g}}_0 = \mathfrak{g}_0$ and $\bar{\mathfrak{g}}_1 = \mathfrak{g}_r$.

Let $\mathfrak{h}_0 \subset \mathfrak{g}_0$ be a Cartan subalgebra. We consider pairs (i, α) where $i \in \mathcal{C}_m$, $\alpha \in \mathfrak{h}_0^*$. These are added component wise: $(i, \alpha) + (j, \beta) = (i + j, \alpha + \beta)$. For such a pair, $\bar{\alpha} = (i, \alpha)$, set $\mathfrak{g}_{\bar{\alpha}} = \{x \in \mathfrak{g}_i \mid [h, x] = \alpha(h)x \text{ for all } h \in \mathfrak{h}_0\}$. We let $\overline{\Phi}$ be the set of all $\bar{\alpha} = (i, \alpha)$ for which $\mathfrak{g}_{\bar{\alpha}} \neq 0$. Moreover, the spaces $\mathfrak{g}_{\bar{\alpha}}$ are stable under $\mathfrak{c}_{\mathfrak{g}}(\mathfrak{h}_0)$, which is a Cartan subalgebra of $\mathfrak{g}$ (Theorem 8.3.1(iii)). In the same way as in the proof of Lemma 2.9.5 we see that $\dim \mathfrak{g}_{\bar{\alpha}} = 1$ for all $\bar{\alpha} \in \overline{\Phi}$. Hence

$$\mathfrak{g} = \mathfrak{c}_{\mathfrak{g}}(\mathfrak{h}_0) \oplus \bigoplus_{\bar{\alpha}\in\overline{\Phi}} \mathfrak{g}_{\bar{\alpha}}$$

is the root space decomposition of $\mathfrak{g}$ with respect to $\mathfrak{c}_{\mathfrak{g}}(\mathfrak{h}_0)$.

Let W_0 denote the Weyl group of the root system of $\mathfrak{g}_0$ with respect to $\mathfrak{h}_0$. Then W_0 acts on $\overline{\Phi}$ by $w \cdot (i, \alpha) = (i, w(\alpha))$. We show that this again lies in $\overline{\Phi}$. We note that $\mathfrak{h}_0$ is algebraic, and the corresponding connected subgroup H_0 of G_0 is a maximal torus of G_0 (Lemma 5.8.2(ii)). Set

$$N_{G_0}(\mathfrak{h}_0) = \{g \in G_0 \mid g(\mathfrak{h}_0) = \mathfrak{h}_0\}.$$

Let $g \in N_{G_0}(\mathfrak{h}_0)$. Then gH_0g^{-1} is a connected algebraic subgroup of G_0 with Lie algebra $g \cdot \mathfrak{h}_0 = \mathfrak{h}_0$. So by Theorem 4.2.2(i) it follows that $gH_0g^{-1} = H_0$ or, in other words, $g \in N_{G_0}(H_0)$. We conclude that $N_{G_0}(\mathfrak{h}_0) = N_{G_0}(H_0)$. Hence by Theorem 5.8.3, there is a surjective homomorphism $\eta : N_{G_0}(\mathfrak{h}_0) \to W_0$. Let $\bar{\alpha} = (i, \alpha) \in \overline{\Phi}$. For $g \in N_{G_0}(\mathfrak{h}_0)$, define α^g by $\alpha^g(h) = \alpha(g^{-1}(h))$ for $h \in \mathfrak{h}_0$. Then $g(\mathfrak{g}_{\bar{\alpha}}) = \mathfrak{g}_{(i,\alpha^g)}$. Moreover, if g is such that $\eta(g) = w$, then $\alpha^g = w(\alpha)$

(this is straightforward to see if $w = s_{\alpha_i}$ and the general case follows from that).

Example 8.3.3 Let $\mathfrak{g}$ be the simple Lie algebra of type D_4. The Dynkin diagram of type D_4 is labeled as follows

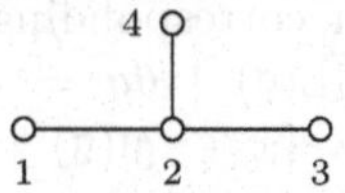

By $\Delta = \{\alpha_1, \dots, \alpha_4\}$ we denote a fixed set of simple roots. Let x_α for $\alpha \in \Phi$ and $h_1, \dots, h_4$ denote the elements of a fixed Chevalley basis of $\mathfrak{g}$. The h_i, x_{α_i}, $x_{-\alpha_i}$ form a canonical generating set of $\mathfrak{g}$. Replacing $x_{\pm\alpha_2}$ by $-x_{\pm\alpha_2}$ again yields a canonical generating set. So mapping $h_i \mapsto h_i$, $1 \leq i \leq 4$, $x_{\pm\alpha_i} \mapsto x_{\pm\alpha_i}$ for $i \neq 2$, and $x_{\pm\alpha_2} \mapsto -x_{\pm\alpha_2}$ extends to an automorphism θ of $\mathfrak{g}$ of order 2 (Theorem 2.9.12). We number the positive roots, other than the simple roots, of Φ as follows

$$\begin{aligned}
&\alpha_5 = \alpha_1 + \alpha_2, \quad \alpha_6 = \alpha_2 + \alpha_3, \quad \alpha_7 = \alpha_2 + \alpha_4, \quad \alpha_8 = \alpha_1 + \alpha_2 + \alpha_3 \\
&\alpha_9 = \alpha_1 + \alpha_2 + \alpha_4, \quad \alpha_{10} = \alpha_2 + \alpha_3 + \alpha_4, \quad \alpha_{11} = \alpha_1 + \alpha_2 + \alpha_3 + \alpha_4, \\
&\alpha_{12} = \alpha_1 + 2\alpha_2 + \alpha_3 + \alpha_4.
\end{aligned}$$

Then $\mathfrak{g} = \mathfrak{g}_0 \oplus \mathfrak{g}_1$ is the $\mathcal{C}_2$-grading of $\mathfrak{g}$ corresponding to θ where $\mathfrak{g}_0$ is spanned by $h_1, \dots, h_4$, along with $x_{\pm\alpha_i}$ for $i = 1, 3, 4, 12$ and $\mathfrak{g}_1$ is spanned by $x_{\pm\alpha_i}$ for $i \neq 1, 3, 4, 12$.

We see that the $x_{\pm\alpha_i}$ for $i = 1, 3, 4, 12$ pairwise commute so that $\mathfrak{g}_0$ is isomorphic to the direct sum of four copies of $\mathfrak{sl}(2, K)$. Corollary 5.3.10 yields a surjective morphism of algebraic groups $\phi : \mathrm{SL}(2, K)^4 \to G_0$. Let $\Phi_0 = \pm\{\alpha_i \mid i = 1, 3, 4, 12\}$, which is the root system of $\mathfrak{g}_0$. Denote the corresponding fundamental weights by λ_i, $i = 1, 3, 4, 12$ (so that $\lambda_i = \frac{1}{2}\alpha_i$). We have that $\mathfrak{g}_1$ is an irreducible $\mathfrak{g}_0$-module with highest weight vector $x_{\alpha_{11}}$ of weight $\lambda = \lambda_1 + \lambda_2 + \lambda_3 + \lambda_4$. So by composing ϕ and the morphism $G_0 \to \mathrm{GL}(\mathfrak{g}_1)$ we obtain an irreducible representation $\rho : \mathrm{SL}(2, 4)^4 \to \mathrm{GL}(\mathfrak{g}_1)$.

Let $V = K^2 \otimes K^2 \otimes K^2 \otimes K^2$, and define a representation $\sigma : \mathrm{SL}(2, K)^4 \to \mathrm{GL}(V)$ by $\sigma(g_1, \dots, g_4)(v_1 \otimes \cdots \otimes v_4) = (g_1 v_1) \otimes \cdots \otimes (g_4 v_4)$. This representation is irreducible with the same highest weight as ρ. It follows that ρ and σ are equivalent (Proposition 5.3.14). This representation is of interest in quantum information theory (see [BDD+10]).

Remark 8.3.4 Many more interesting representations of reductive algebraic groups arise as θ-groups. For some examples we refer to [VE78] and [GT99]. A detailed list is contained in [Kac80].

8.3.2 The nilpotent orbits of a θ-group

Let $\mathfrak{g} = \oplus_{i \in \mathcal{C}_m} \mathfrak{g}_i$ be a $\mathcal{C}_m$-graded semisimple Lie algebra and $G_0 \subset G$ be the corresponding θ-group. We say that an $e \in \mathfrak{g}_1$ is *nilpotent* if $\mathrm{ad}_{\mathfrak{g}} e$

is nilpotent. Here we study the G_0-orbits of nilpotent elements, also called nilpotent G_0-orbits.

Lemma 8.3.5 *Let $e \in \mathfrak{g}_1$ be a nilpotent element. Then there are $f \in \mathfrak{g}_{-1}$ and $h \in \mathfrak{g}_0$ such that (h, e, f) is an $\mathfrak{sl}_2$-triple.*

Proof. By Theorem 2.13.2, there is an $\mathfrak{sl}_2$-triple (h', e, f') with $h', f' \in \mathfrak{g}$. Write $h' = \sum_{i \in C_m} h_i$, with $h_i \in \mathfrak{g}_i$. Then $2e = [h', e] = \sum_i [h_i, e]$. Now $2e \in \mathfrak{g}_1$ and $[h_i, e] \in \mathfrak{g}_{i+1}$. So $[h_0, e] = 2e$ and $[h_i, e] = 0$ for $i \neq 0$. Set $h = h_0$. Also write $f' = \sum_{i \in C_m} f_i$, with $f_i \in \mathfrak{g}_i$. Then $h' = [e, f'] = \sum_i [e, f_i]$. We find that $[e, f_i] = h_{i+1}$, and in particular, $[e, f_{-1}] = h$. It follows that $[h, e] = 2e$ and $h \in \mathrm{ade}(\mathfrak{g})$. By Lemma 2.13.1 we see that e lies in an $\mathfrak{sl}_2$-triple (h, e, f). But then $f \in \mathfrak{g}_{-1}$. Indeed, let f'' be the $\mathfrak{g}_{-1}$-component of f, then (h, e, f'') is an $\mathfrak{sl}_2$-triple, so by Lemma 8.1.1, $f = f''$. □

An $\mathfrak{sl}_2$-triple as in the lemma is said to be *homogeneous*. We have the following analogue of Theorem 8.1.4.

Theorem 8.3.6 *Let (h, e, f) and (h', e', f') be two homogeneous $\mathfrak{sl}_2$-triples in $\mathfrak{g}$. Then the following are equivalent:*

(i) *e and e' are conjugate under G_0.*

(ii) *(h, e, f) and (h', e', f') are conjugate under G_0.*

(iii) *h and h' are conjugate under G_0.*

Proof. In order to see that (i) implies (ii), we may assume that $e = e'$. Set $\mathfrak{n} = \ker(\mathrm{ade}) \cap \mathrm{im}(\mathrm{ade}) \cap \mathfrak{g}_0$. As in the proof of Proposition 8.1.3, $\mathfrak{n}$ has a basis consisting of $\mathrm{ad}h$-eigenvectors with positive eigenvalues. So $[h, \mathfrak{n}] = \mathfrak{n}$. It also follows that $\mathrm{ad}_{\mathfrak{g}} z$ is nilpotent for all $z \in \mathfrak{n}$. Therefore, as in the proof of Proposition 8.1.3, we conclude that the triples (h, e, f) and (h', e, f') are G_0-conjugate.

It is obvious that (ii) implies (iii).

For the final implication we use an argument analogous to the one used in the proof of Theorem 8.1.4. Again we may assume that $h = h'$. Now we set $\mathfrak{g}_i(r) = \mathfrak{g}_i \cap \mathfrak{g}(r)$ and $G_{0,h} = \{g \in G_0 \mid g(h) = h\}$. Then we have the regular map $\psi : G_{0,h} \to \mathfrak{g}_1(2)$ by $\psi(g) = g(e)$. Again we infer that the image of ψ contains an open set. The desired conclusion is obtained as in the mentioned proof. □

Proposition 8.3.7 *With the notation of Section 7.4 we have $\mathcal{N}_{G_0}(\mathfrak{g}_1) = \{e \in \mathfrak{g}_1 \mid \mathrm{ad}_{\mathfrak{g}} e$ is nilpotent$\}$. The strata in $\mathcal{N}_{G_0}(\mathfrak{g}_1)$ equal the nilpotent G_0-orbits. Moreover, letting (h, e, f) be a homogeneous $\mathfrak{sl}_2$-triple, h is a characteristic of e.*

Proof. If $e \in \mathfrak{g}_1$ is such that $\mathrm{ad}_{\mathfrak{g}} e$ is nilpotent, then by Lemma 8.3.5, it lies in a homogeneous $\mathfrak{sl}_2$-triple. Now in the same way as in Example 7.4.4, it follows from the Hilbert Mumford criterion that $e \in \mathcal{N}_{G_0}(\mathfrak{g}_1)$. As seen in the same example, the coefficients of the characteristic polynomial of $\mathrm{ad}_{\mathfrak{g}} x$ for $x \in \mathfrak{g}_1$ are G-invariant, and hence G_0-invariant. Therefore, if $e \in \mathcal{N}_{G_0}(\mathfrak{g}_1)$, then $\mathrm{ad}_{\mathfrak{g}} e$ is nilpotent.

Let (h, e, f) be a homogeneous $\mathfrak{sl}_2$-triple. We show that h is a characteristic of e. Since $\mathfrak{g}_0 \subset \mathfrak{g}$, the form $(\ ,\ )$ that we use to define the strata for G_0 is the restriction of the corresponding form for G (which in this case is the Killing form). Let $\tilde{\mathfrak{z}}_0(h) = \{x \in \mathfrak{g}_0 \mid [x, h] = 0 \text{ and } (h, x) = 0\}$, and let $\widetilde{Z}_0(h) \subset G_0$ be the corresponding connected subgroup. Then by Theorem 4.2.2(i), $\widetilde{Z}_0(h) \subset \widetilde{Z}(h)$. Since the closure of the $\widetilde{Z}(h)$-orbit of e does not contain 0 (Example 7.4.23), neither does the closure of its $\widetilde{Z}_0(h)$-orbit. Hence h is a characteristic of e. By Theorem 8.3.6 we conclude that two different G_0-orbits in $\mathcal{N}_{G_0}(\mathfrak{g}_1)$ cannot lie in the same stratum. □

Corollary 8.3.8 *$\mathfrak{g}_1$ has a finite number of nilpotent G_0-orbits.*

Proof. This follows immediately from the previous proposition. For an alternative argument, let $\mathfrak{h}_0$ be a fixed Cartan subalgebra of $\mathfrak{g}_0$ contained in the Cartan subalgebra $\mathfrak{h}$ of $\mathfrak{g}$. Because two Cartan subalgebras of $\mathfrak{g}_0$ are G_0-conjugate (Proposition 4.3.2), and in view of Lemma 8.3.5, every nilpotent element in $\mathfrak{g}_1$ is G_0-conjugate to an e lying in a homogeneous $\mathfrak{sl}_2$-triple (h, e, f) with $h \in \mathfrak{h}_0$. Moreover, by Corollary 8.1.5, Proposition 8.1.6, a finite number of $h \in \mathfrak{h}$ lie in an $\mathfrak{sl}_2$-triple. □

8.3.3 Carrier algebras

In this section we attach a subalgebra, called carrier algebra, to a nilpotent element and show that the classification of the nilpotent orbits of a θ-group G_0 is equivalent to the classification of the carrier algebras up to G_0-conjugacy. The notation is as in the previous section.

We consider semisimple $\mathbb{Z}$-graded subalgebras $\mathfrak{s} \subset \mathfrak{g}$. This means that $\mathfrak{s} = \oplus_{i \in \mathbb{Z}} \mathfrak{s}_i$ with $\mathfrak{s}_i \subset \mathfrak{g}_i$ for all $i \in \mathbb{Z}$. This implies that $\mathfrak{s}$ is θ-stable. Note that here we deal with two gradings (the $\mathbb{Z}$-grading of $\mathfrak{s}$ and the C_m-grading of $\mathfrak{g}$), and that we use the same index i to denote potentially different objects, $i \in \mathbb{Z}$, $i \in C_m$. This will occur on more occasions in this section.

Because $\mathfrak{s}$ is semisimple, it is algebraic (see the proof of Theorem 4.3.20(v)). Let $S \subset G$ denote the corresponding connected group. In the same way as for $\mathfrak{g}_0$ we can see that $\mathfrak{s}_0$ is algebraic as well (let φ be the automorphism of $\mathfrak{s}$ corresponding to its grading and S_0 be the identity component of the subgroup of the elements of S that commute with φ; then $\mathrm{Lie}(S_0) = \mathfrak{s}_0$).

Since $\mathfrak{s}$ is $\mathbb{Z}$-graded, $\mathfrak{s}_1$ consists of nilpotent elements. So by Corollary

8.3.8, $\mathfrak{s}_1$ has a finite number of S_0-orbits. Hence there is a unique orbit of maximal dimension (which is $\dim \mathfrak{s}_1$). By Lemma 3.12.1, this orbit is open in $\mathfrak{s}_1$. By Lemma 8.2.1 we see that an $e \in \mathfrak{s}_1$ lies in the open orbit if and only if $[\mathfrak{s}_0, e] = \mathfrak{s}_1$.

A $\mathbb{Z}$-graded subalgebra $\mathfrak{s}$ is called *regular* if it is normalized by a Cartan subalgebra of $\mathfrak{g}_0$. (This definition is slightly different from the one in Section 5.9.)

We say that two $\mathbb{Z}$-graded subalgebras, $\mathfrak{s}$, $\mathfrak{s}'$, are G_0-conjugate if there is a $g \in G_0$ with $g \cdot \mathfrak{s}_i = \mathfrak{s}_i'$ for all i.

For a subspace $\mathfrak{v} \subset \mathfrak{g}$ we set $\mathfrak{c}_0(\mathfrak{v}) = \mathfrak{c}_\mathfrak{g}(\mathfrak{v}) \cap \mathfrak{g}_0$ and $\mathfrak{n}_0(\mathfrak{v}) = \mathfrak{n}_\mathfrak{g}(\mathfrak{v}) \cap \mathfrak{g}_0$ (see Section 2.1.4 for the definitions of $\mathfrak{c}_\mathfrak{g}$ and $\mathfrak{n}_\mathfrak{g}$). For $x \in \mathfrak{g}$ we write $\mathfrak{c}_0(x)$ and $\mathfrak{n}_0(x)$ for $\mathfrak{c}_0(\mathfrak{v})$ and $\mathfrak{n}_0(\mathfrak{v})$, where $\mathfrak{v}$ is the space spanned by x.

Lemma 8.3.9 *Let $\mathfrak{s} \subset \mathfrak{g}$ be a semisimple and θ-stable subalgebra. Then $\mathfrak{c}_0(\mathfrak{s})$ is reductive in $\mathfrak{g}$.*

Proof. Set $C = \mathfrak{c}_\mathfrak{g}(\mathfrak{s})$. Using Proposition 2.2.3 it is immediate that C contains the semisimple and nilpotent parts of its elements. Via the adjoint representation, $\mathfrak{g}$ is an $\mathfrak{s}$-module. Since C is a submodule, by Weyl's theorem (Theorem 2.7.6) there is a submodule U such that $\mathfrak{g} = C \oplus U$. Furthermore, $[\mathfrak{s}, U]$ is a submodule of U and is complemented in U by another submodule U'. Then $[\mathfrak{s}, U'] = 0$ so that $U' \subset C \cap U = 0$, and it follows that $U = [\mathfrak{s}, U]$. Now let $y \in C$, $x \in \mathfrak{s}$ and $u' \in U$. Then $\kappa(y, [x, u']) = \kappa([y, x], u') = 0$. Therefore the restriction of κ to C has to be non-degenerate.

Since $\mathfrak{s}$ is θ-stable, the same holds for C. So $C = \oplus_{i \in C_m} C_i$, with $C_i = C \cap \mathfrak{g}_i$. As seen above, the restriction of κ to C is non-degenerate. Therefore, its restriction to C_0 has to be non-degenerate as well. We have that $\mathfrak{g}_0$ contains the nilpotent and semisimple parts of its elements (Theorem 8.3.1(i)), and we saw above that the same holds for C. So we also get this property for C_0. We conclude that $\mathfrak{c}_0(\mathfrak{s}) = C_0$ is reductive in $\mathfrak{g}$ by Proposition 2.12.5. □

We say that a subalgebra $\mathfrak{t} \subset \mathfrak{g}$ is *toral* if all $\mathrm{ad}_\mathfrak{g} x$, for $x \in \mathfrak{t}$, commute and are semisimple. Let $\mathfrak{u} \subset \mathfrak{g}$ be a subalgebra which is reductive in $\mathfrak{g}$. Combining Theorem 2.12.3 and Proposition 2.11.5, $\mathfrak{h} \subset \mathfrak{u}$ is a Cartan subalgebra of $\mathfrak{u}$ if and only if it is a maximal toral subalgebra. The *rank* of $\mathfrak{u}$, denoted $\mathrm{rank}(\mathfrak{u})$, is the dimension of a Cartan subalgebra of $\mathfrak{u}$.

Let $\mathfrak{s}$ be a semisimple θ-stable subalgebra of $\mathfrak{g}$. Since all derivations of $\mathfrak{s}$ are inner (Proposition 2.7.7), we have $\mathfrak{n}_\mathfrak{g}(\mathfrak{s}) = \mathfrak{s} \oplus \mathfrak{c}_\mathfrak{g}(\mathfrak{s})$. Since also $\mathfrak{c}_\mathfrak{g}(\mathfrak{s})$ is θ-stable we get

$$\mathfrak{n}_0(\mathfrak{s}) = \mathfrak{s}_0 \oplus \mathfrak{c}_0(\mathfrak{s}). \tag{8.1}$$

Lemma 8.3.10 *Let $\mathfrak{s}$ be a semisimple regular $\mathbb{Z}$-graded subalgebra of $\mathfrak{g}$. Then* $\mathrm{rank}(\mathfrak{s}) = \mathrm{rank}(\mathfrak{g}_0) - \mathrm{rank}(\mathfrak{c}_0(\mathfrak{s}))$.

Proof. Since $\mathfrak{s}$ is regular $\mathbb{Z}$-graded, $\mathfrak{n}_0(\mathfrak{s})$ contains a Cartan subalgebra $\mathfrak{h}_0$ of $\mathfrak{g}_0$. Let $h \in \mathfrak{h}_0$. Again because all derivations of $\mathfrak{s}$ are inner, there is an

$h' \in \mathfrak{s}$ such that $[h', x] = [h, x]$ for all $x \in \mathfrak{s}$. Set $h'' = h - h'$, then $h'' \in \mathfrak{c}_{\mathfrak{g}}(\mathfrak{s})$. By (8.1) we have $h = \hat{h}' + \hat{h}''$ with $\hat{h}' \in \mathfrak{s}_0$ and $\hat{h}'' \in \mathfrak{c}_0(\mathfrak{s})$. For all $x \in \mathfrak{s}$, $[h, x] = [\hat{h}', x] = [h', x]$ so that $h' = \hat{h}'$ and $h'' = \hat{h}''$. The conclusion is that $\mathfrak{h}_0 = \mathfrak{h}' \oplus \mathfrak{h}''$ where $\mathfrak{h}'$ and $\mathfrak{h}''$ are maximal toral in $\mathfrak{s}_0$ and $\mathfrak{c}_0(\mathfrak{s})$ respectively. Moreover, since $\mathfrak{s}$ is $\mathbb{Z}$-graded, we have $\text{rank}(\mathfrak{s}) = \text{rank}(\mathfrak{s}_0)$. The lemma follows. □

Let $\mathfrak{t} \subset \mathfrak{g}_0$ be a toral subalgebra and $\lambda : \mathfrak{t} \to K$ a non-zero linear function. For $i \in \mathbb{Z}$ set

$$\mathfrak{g}_i(\mathfrak{t}, \lambda) = \{x \in \mathfrak{g}_i \mid [t, x] = i\lambda(t)x \text{ for all } t \in \mathfrak{t}\} \text{ and } \mathfrak{g}(\mathfrak{t}, \lambda) = \bigoplus_{i \in \mathbb{Z}} \mathfrak{g}_i(\mathfrak{t}, \lambda).$$

Lemma 8.3.11 $\mathfrak{g}(\mathfrak{t}, \lambda)$ *is reductive in* $\mathfrak{g}$.

Proof. Let Σ be the set of weights of $\mathfrak{t}$ when acting on $\mathfrak{g}$. Since $\mathfrak{t} \subset \mathfrak{g}_0$, $[\mathfrak{t}, \mathfrak{g}_i] \subset \mathfrak{g}_i$. Hence $\mathfrak{g}_i$ is a sum of $\mathfrak{t}$-weight spaces. For $\sigma \in \Sigma$, let $\mathfrak{g}_{i,\sigma}$ be the set of $x \in \mathfrak{g}_i$ such that $[t, x] = \sigma(t)x$ for all $t \in \mathfrak{t}$. Then $\mathfrak{g}$ is the direct sum of the non-zero $\mathfrak{g}_{i,\sigma}$. Moreover, $[\mathfrak{g}_{i,\sigma}, \mathfrak{g}_{j,\eta}] \subset \mathfrak{g}_{i+j,\sigma+\eta}$. Therefore, the restriction of κ to $\mathfrak{g}_{i,\sigma} \times \mathfrak{g}_{-i,-\sigma}$ is non-degenerate. Now $\mathfrak{g}_i(\mathfrak{t}, \lambda) = \mathfrak{g}_{i,i\lambda}$. Therefore, the restriction of κ to $\mathfrak{g}(\mathfrak{t}, \lambda)$ is non-degenerate. By Proposition 2.12.2(iii), $\mathfrak{g}(\mathfrak{t}, \lambda)$ is reductive. The centre of this algebra inherits its $\mathbb{Z}$-grading. Since λ is assumed to be non-zero, the spaces $\mathfrak{g}_i(\mathfrak{t}, \lambda)$ for $i \neq 0$ cannot contain central elements (note that $\mathfrak{t} \subset \mathfrak{g}_0(\mathfrak{t}, \lambda)$). Hence the centre of $\mathfrak{g}(\mathfrak{t}, \lambda)$ is contained in $\mathfrak{g}_0(\mathfrak{t}, \lambda)$. The restriction of κ to $\mathfrak{g}_0(\mathfrak{t}, \lambda)$ is non-degenerate. Furthermore, this subalgebra is equal to the centralizer of $\mathfrak{t}$ in $\mathfrak{g}_0$, and therefore contains the semisimple and nilpotent parts of its elements by Proposition 2.2.3. By Proposition 2.12.5, $\mathfrak{g}_0(\mathfrak{t}, \lambda)$ is reductive in $\mathfrak{g}$. By Theorem 2.12.3, its centre is a toral subalgebra of $\mathfrak{g}$. Hence the same follows for the centre of $\mathfrak{g}(\mathfrak{t}, \lambda)$. Again using Theorem 2.12.3, we conclude that $\mathfrak{g}(\mathfrak{t}, \lambda)$ is reductive in $\mathfrak{g}$. □

By $\mathfrak{s}(\mathfrak{t}, \lambda)$ we denote the derived algebra of $\mathfrak{g}(\mathfrak{t}, \lambda)$, which consequently is semisimple.

A regular reductive $\mathbb{Z}$-graded subalgebra of $\mathfrak{g}$ is called *complete* if it is not a proper $\mathbb{Z}$-graded subalgebra of a reductive regular $\mathbb{Z}$-graded subalgebra of the same rank.

Lemma 8.3.12 $\mathfrak{s}(\mathfrak{t}, \lambda)$ *is a semisimple and complete regular* $\mathbb{Z}$*-graded subalgebra of* $\mathfrak{g}$.

Proof. Write $\mathfrak{s} = \mathfrak{s}(\mathfrak{t}, \lambda)$. Let $\hat{\mathfrak{h}}$ be a Cartan subalgebra of $\mathfrak{g}_0$ containing $\mathfrak{t}$ and $h \in \hat{\mathfrak{h}}$. Then $[h, \mathfrak{t}] = 0$, and hence $[h, \mathfrak{g}_k(\mathfrak{t}, \lambda)] \subset \mathfrak{g}_k(\mathfrak{t}, \lambda)$. Hence $\mathfrak{g}(\mathfrak{t}, \lambda)$ and also $\mathfrak{s}$ are normalized by $\hat{\mathfrak{h}}$. Therefore, $\mathfrak{s}$ is regular $\mathbb{Z}$-graded.

Let $\mathfrak{s}'$ be a reductive regular $\mathbb{Z}$-graded subalgebra of $\mathfrak{g}$ containing $\mathfrak{s}$ and of the same rank. Then $\mathfrak{s}'$ cannot have a non-trivial centre, and hence is semisimple. So by Lemma 8.3.10, the ranks of $\mathfrak{c}_0(\mathfrak{s})$ and $\mathfrak{c}_0(\mathfrak{s}')$ are equal. As seen in the proof of Lemma 8.3.10, $\mathfrak{t} = \mathfrak{t}' \oplus \mathfrak{t}''$ where $\mathfrak{t}' \subset \mathfrak{s}_0$, $\mathfrak{t}'' \subset \mathfrak{c}_0(\mathfrak{s})$. Let $\mathfrak{h}_0$ be a Cartan subalgebra of $\mathfrak{c}_0(\mathfrak{s})$ containing $\mathfrak{t}''$. Also, let $\mathfrak{h}_0'$ be a Cartan subalgebra of $\mathfrak{c}_0(\mathfrak{s}')$. Since $\mathfrak{c}_0(\mathfrak{s}') \subset \mathfrak{c}_0(\mathfrak{s})$ and these algebras are of the same rank, $\mathfrak{h}_0'$ is also a Cartan subalgebra of $\mathfrak{c}_0(\mathfrak{s})$. Set $Z_{G_0}(\mathfrak{s}) = \{g \in G_0 \mid g(x) = x \text{ for all } x \in \mathfrak{s}\}$. By Corollary 4.2.8 we see that the Lie algebra of this group is $\mathfrak{c}_0(\mathfrak{s})$. So by Proposition 4.3.2, there is a $g \in Z_{G_0}(\mathfrak{s})$ with $g \cdot \mathfrak{h}_0' = \mathfrak{h}_0$, whence $g^{-1} \cdot \mathfrak{h}_0 \subset \mathfrak{c}_0(\mathfrak{s}')$. But that implies that $g \cdot \mathfrak{s}' \subset \mathfrak{c}_\mathfrak{g}(\mathfrak{h}_0)$ (indeed, let $x \in \mathfrak{s}'$, $h \in \mathfrak{h}_0$ then $[g \cdot x, h] = g[x, g^{-1} \cdot h] = 0$). From $g \cdot \mathfrak{s} = \mathfrak{s}$ it follows that $\mathfrak{s} \subset g \cdot \mathfrak{s}'$. Hence, after replacing $\mathfrak{s}'$ by $g \cdot \mathfrak{s}'$ we may assume that $\mathfrak{s}' \subset \mathfrak{c}_\mathfrak{g}(\mathfrak{h}_0)$.

Let $h_0 \in \mathfrak{s}_0$ and $h_0' \in \mathfrak{s}_0'$ be the defining elements of $\mathfrak{s}$ and $\mathfrak{s}'$ respectively. Since $\mathfrak{s}_i \subset \mathfrak{s}_i'$ we get $h_0 - h_0' \in \mathfrak{c}_\mathfrak{g}(\mathfrak{s})$. The ranks of $\mathfrak{s}$ and $\mathfrak{s}'$ are equal so that a Cartan subalgebra of $\mathfrak{s}$ is also a Cartan subalgebra of $\mathfrak{s}'$, whence $\mathfrak{c}_\mathfrak{g}(\mathfrak{s}) \cap \mathfrak{s}' = 0$. It follows that $h_0 = h_0'$. In particular, $[h_0, x] = ix$ for all $x \in \mathfrak{s}_i'$.

Let $t \in \mathfrak{t}$. Then $[t, x] = i\lambda(t)x$ for all $x \in \mathfrak{s}_i$. Write $t = t' + t''$ where $t' \in \mathfrak{t}' \subset \mathfrak{s}_0$, $t'' \in \mathfrak{t}'' \subset \mathfrak{c}_0(\mathfrak{s})$. Then for $x \in \mathfrak{s}_i$, $[t, x] = [t', x] = i\lambda(t')x$, and on the other hand, $[t, x] = i\lambda(t' + t'')x$, from which it follows that $\lambda(t'') = 0$. Also, $[t' - \lambda(t')h_0, x] = 0$ for all $x \in \mathfrak{s}_i$ so that $t' - \lambda(t')h_0$ lies in the centre of $\mathfrak{s}$ which is zero. So $t' = \lambda(t')h_0 = \lambda(t)h_0$. It follows that $\mathfrak{t}$ is contained in $\mathfrak{t}'' \oplus \langle h_0 \rangle$.

Let $x \in \mathfrak{s}_i'$, then because of the assumption on $\mathfrak{s}'$, $x \in \mathfrak{c}_\mathfrak{g}(\mathfrak{t}'')$. Also $[h_0, x] = ix$. Let $t \in \mathfrak{t}$; as seen above we can write $t = t'' + \lambda(t)h_0$ with $t'' \in \mathfrak{t}''$. Then $[t, x] = i\lambda(t)x$, as $[t'', x] = 0$. So $\mathfrak{s}_i' \subset \mathfrak{g}_i(\mathfrak{t}, \lambda)$ implying $\mathfrak{s}' \subset \mathfrak{g}(\mathfrak{t}, \lambda)$ and $\mathfrak{s}' \subset \mathfrak{s}$. We conclude $\mathfrak{s}' = \mathfrak{s}$. □

Let $e \in \mathfrak{g}_1$ be nilpotent, lying in the homogeneous $\mathfrak{sl}_2$-triple (h, e, f). Let $\mathfrak{a}$ be the subalgebra of $\mathfrak{g}$ spanned by h, e, f. Since $\mathfrak{a}$ is θ-stable, $\mathfrak{c}_0(\mathfrak{a})$ is reductive in $\mathfrak{g}$ (Lemma 8.3.9). Let $\mathfrak{t}_0$ be a Cartan subalgebra of $\mathfrak{c}_0(\mathfrak{a})$ (in other words, it is a maximal toral subalgebra), and let $\mathfrak{t}$ be the toral subalgebra of $\mathfrak{g}_0$ spanned by h and $\mathfrak{t}_0$. Define $\lambda \in \mathfrak{t}^*$ by $[t, e] = \lambda(t)e$, for $t \in \mathfrak{t}$ (so $\lambda(h) = 2$, $\lambda(t) = 0$ for $t \in \mathfrak{t}_0$). Set $\mathfrak{s}(e, \mathfrak{t}) = \mathfrak{s}(\mathfrak{t}, \lambda)$. Then $\mathfrak{s}(e, \mathfrak{t})$ is called a *carrier algebra* of e. Note that $e \in \mathfrak{s}_1$.

Example 8.3.13 Consider the θ-group of Example 8.3.3. Let $e = x_{\alpha_1+\alpha_2+\alpha_3} + x_{\alpha_2+\alpha_3+\alpha_4} + x_{-\alpha_2-\alpha_3}$. This is a nilpotent element in $\mathfrak{g}_1$ lying in a homogeneous $\mathfrak{sl}_2$-triple (h, e, f) with $h = 3h_1 + 2h_2 + 2h_3 + 3h_4$. Letting $\mathfrak{h}$ be the Cartan subalgebra spanned by $h_1, \dots, h_4$, we see that $\mathfrak{c}_0(h) = \mathfrak{h}$. The subalgebra of $\mathfrak{h}$ consisting of the elements commuting with e is spanned by $\hat{h} = h_1 + 2h_2 + h_4$ which generates an irreducible $\mathfrak{sl}_2$-module with highest weight 0, and $[f, \hat{h}] = 0$ as well. We conclude that $\mathfrak{c}_0(\mathfrak{a})$ is spanned by $\hat{h}$. Therefore $\mathfrak{t}$ is spanned by $h, \hat{h}$. By evaluating $\alpha(h)$, $\alpha(\hat{h})$ for $\alpha \in \Phi$, it is clear that $\mathfrak{g}_1(\mathfrak{t}, \lambda)$ is spanned by $x_{\alpha_1+\alpha_2+\alpha_3}, x_{\alpha_2+\alpha_3+\alpha_4}, x_{-\alpha_2-\alpha_3}$. Hence $\mathfrak{g}_{-1}(\mathfrak{t}, \lambda)$ is spanned by $x_{-\alpha_1-\alpha_2-\alpha_3}, x_{-\alpha_2-\alpha_3-\alpha_4}, x_{\alpha_2+\alpha_3}$. From Lemma 8.4.3 it follows that $\mathfrak{s}(e, \mathfrak{t})$ is generated by $\mathfrak{s}_{-1}(e, \mathfrak{t})$, $\mathfrak{s}_0(e, \mathfrak{t})$, $\mathfrak{s}_1(e, \mathfrak{t})$. Therefore bases of the other components of $\mathfrak{g}(\mathfrak{t}, \lambda)$ are easily written and $\mathfrak{s}(e, \mathfrak{t})$ is simple of type A_3.

A $\mathbb{Z}$-graded Lie algebra $\mathfrak{s}$ with $\dim \mathfrak{s}_0 = \dim \mathfrak{s}_1$ is called *locally flat.*

Proposition 8.3.14 *Let $\mathfrak{s} = \mathfrak{s}(e, \mathfrak{t})$. Then $\mathfrak{s}$ is locally flat, and e lies in the open orbit in $\mathfrak{s}_1$.*

Proof. First we show that $\mathfrak{c}_0(\mathfrak{a}) = \mathfrak{c}_0(h) \cap \mathfrak{c}_0(e)$. The left-hand side is obviously contained in the right-hand side. Let $x \in \mathfrak{g}_0$ be such that $[e, x] = [h, x] = 0$. Then x generates an irreducible $\mathfrak{a}$-module of highest weight 0. So its dimension must be 1 and $[f, x] = 0$ as well.

Set $\mathfrak{u} = \mathfrak{c}_0(\mathfrak{t}_0 \oplus \mathfrak{a})$. Then $\mathfrak{u} = \mathfrak{c}_0(\mathfrak{t}_0) \cap \mathfrak{c}_0(\mathfrak{a}) = \mathfrak{c}_0(\mathfrak{t}_0) \cap \mathfrak{c}_0(h) \cap \mathfrak{c}_0(e) = \mathfrak{c}_0(\mathfrak{t}) \cap \mathfrak{c}_0(e) = \mathfrak{g}_0(\mathfrak{t}, \lambda) \cap \mathfrak{c}_0(e)$. But $\mathfrak{t}_0$ is a Cartan subalgebra of $\mathfrak{c}_0(\mathfrak{a})$, so that $\mathfrak{c}_0(\mathfrak{t}_0) \cap \mathfrak{c}_0(\mathfrak{a}) = \mathfrak{t}_0$. Furthermore, as $\lambda(\mathfrak{t}_0) = 0$, the centre of $\mathfrak{g}(\mathfrak{t}, \lambda)$ contains $\mathfrak{t}_0$. Hence $\mathfrak{t}_0 \cap \mathfrak{s} = 0$. It follows that $x \in \mathfrak{s}_0$, $[x, e] = 0$ imply $x = 0$.

Let S_0 be the connected subgroup of G_0 with Lie algebra $\mathfrak{s}_0$. The tangent space of $S_0 e$ at e is $[\mathfrak{s}_0, e]$ (Lemma 8.2.1), and has dimension $\dim \mathfrak{s}_0$ because $\mathfrak{c}_{\mathfrak{s}_0}(e) = 0$. This also shows that the dimension of the orbit of e is the highest possible. Therefore, e lies in the open orbit, which has dimension $\dim \mathfrak{s}_1$. We conclude that $\dim \mathfrak{s}_0 = \dim \mathfrak{s}_1$. □

Lemma 8.3.15 *Let $\mathfrak{s}$ be a semisimple locally flat $\mathbb{Z}$-graded subalgebra of $\mathfrak{g}$. Let $e \in \mathfrak{s}_1$ lie in the open orbit and (h, e, f) be an $\mathfrak{sl}_2$-triple with $h \in \mathfrak{s}_0$, $f \in \mathfrak{s}_{-1}$ (Lemma 8.3.5). Then $\frac{1}{2}h$ is the defining element of the $\mathbb{Z}$-grading of $\mathfrak{s}$.*

Proof. Let $h_0 \in \mathfrak{s}_0$ be the defining element of the $\mathbb{Z}$-grading of $\mathfrak{s}$. Then $[h_0 - \frac{1}{2}h, e] = 0$. But $\mathfrak{c}_0(e) = 0$ because $\mathfrak{s}$ is locally flat and e lies in the open orbit. □

Proposition 8.3.16 *Let (h', e', f') be a second homogeneous $\mathfrak{sl}_2$-triple, and $\mathfrak{a}'$ the subalgebra spanned by it. Let $\mathfrak{s} = \mathfrak{s}(e, \mathfrak{t})$, $\mathfrak{s}' = \mathfrak{s}(e', \mathfrak{t}')$ where $\mathfrak{t}' = \langle h' \rangle \oplus \mathfrak{t}_0'$ and $\mathfrak{t}_0'$ is a Cartan subalgebra of $\mathfrak{c}_0(\mathfrak{a}')$. Then $\mathfrak{s}$ and $\mathfrak{s}'$ are G_0-conjugate if and only if e and e' are G_0-conjugate.*

Proof. Let $g \in G_0$ be such that $g \cdot \mathfrak{s}_i = \mathfrak{s}_i'$. Since e and e' lie in the open orbits of, respectively, $\mathfrak{s}_1$ and $\mathfrak{s}_1'$ (Proposition 8.3.14), it follows that $\frac{1}{2}h$ and $\frac{1}{2}h'$ are the defining elements of the $\mathbb{Z}$-gradings of $\mathfrak{s}$ and $\mathfrak{s}'$ (Lemma 8.3.15). Hence $g \cdot h = h'$, and by Theorem 8.3.6, e and e' are G_0-conjugate.

Now suppose that e and e' are G_0-conjugate. By Proposition 8.3.6, there is a $g \in G_0$ with $g \cdot h = h'$, $g \cdot e = e'$, $g \cdot f = f'$. Then $g(\mathfrak{c}_0(\mathfrak{a})) = \mathfrak{c}_0(\mathfrak{a}')$, and $g \cdot \mathfrak{t}_0$ is a Cartan subalgebra of $\mathfrak{c}_0(\mathfrak{a}')$. Set $Z_{G_0}(\mathfrak{a}') = \{g \in G_0 \mid g(x) = x \text{ for all } x \in \mathfrak{a}'\}$. By Corollary 4.2.8 it follows that $\mathrm{Lie}(Z_{G_0}(\mathfrak{a}')) = \mathfrak{c}_0(\mathfrak{a}')$. So by Proposition 4.3.2, there is a $g' \in Z_{G_0}(\mathfrak{a}')$ with $g'g \cdot \mathfrak{t}_0 = \mathfrak{t}_0'$. Define the linear map $\lambda' : \mathfrak{t}' \to K$ by $\lambda'(h') = 2$, $\lambda'(x) = 0$ for $x \in \mathfrak{t}_0'$. Then, for $t \in \mathfrak{t}$ we have $\lambda'(g'g \cdot t) = \lambda(t)$. So $g'g$ maps $\mathfrak{g}_i(\mathfrak{t}, \lambda)$ to $\mathfrak{g}_i(\mathfrak{t}', \lambda')$. Hence $g'g$ also maps $\mathfrak{s}_i$ to $\mathfrak{s}_i'$. □

Remark 8.3.17 Note that this proposition also implies that $\mathfrak{s}(e, \mathfrak{t})$ only depends on e up to G_0-conjugacy.

Lemma 8.3.18 *Let $\mathfrak{s}$ be a regular $\mathbb{Z}$-graded semisimple subalgebra of $\mathfrak{g}$ and $h_0 \in \mathfrak{s}_0$ be its defining element. Let $\hat{\mathfrak{h}}$ be a Cartan subalgebra of $\mathfrak{c}_0(\mathfrak{s})$ and $\tilde{\mathfrak{h}}$ be the subalgebra spanned by h_0 and $\hat{\mathfrak{h}}$. Define $\mu \in \tilde{\mathfrak{h}}^*$ by $\mu(h_0) = 1$ and $\mu(h) = 0$ for $h \in \hat{\mathfrak{h}}$. Then $\mathfrak{g}(\tilde{\mathfrak{h}}, \mu) = \mathfrak{s}' \oplus \hat{\mathfrak{h}}$ where $\mathfrak{s}'$ is a complete regular semisimple $\mathbb{Z}$-graded subalgebra containing $\mathfrak{s}$ and of the same rank as $\mathfrak{s}$.*

Proof. Since $[\hat{\mathfrak{h}}, \mathfrak{s}] = 0$ and h_0 is the defining element of $\mathfrak{s}$, it is obvious that $\mathfrak{s} \oplus \hat{\mathfrak{h}} \subset \mathfrak{g}(\tilde{\mathfrak{h}}, \mu)$. Furthermore, as the latter algebra is reductive in $\mathfrak{g}$ (Lemma 8.3.11), we have $\mathfrak{g}(\tilde{\mathfrak{h}}, \mu) = \mathfrak{s}' \oplus \mathfrak{d}$ where $\mathfrak{s}'$ is semisimple and $\mathfrak{d}$ is the centre, which is toral. As seen in the proof of Lemma 8.3.11, $\mathfrak{d} \subset \mathfrak{g}_0(\tilde{\mathfrak{h}}, \mu)$. As $\mathfrak{s}'$ is $\mathbb{Z}$-graded, $\mathfrak{s}'_0$ contains a Cartan subalgebra of $\mathfrak{s}'$. It follows that $\mathrm{rank}(\mathfrak{g}(\tilde{\mathfrak{h}}, \mu)) = \mathrm{rank}(\mathfrak{g}_0(\tilde{\mathfrak{h}}, \mu)) = \mathrm{rank}(\mathfrak{g}_0)$. So by Lemma 8.3.10, $\mathrm{rank}(\mathfrak{g}(\tilde{\mathfrak{h}}, \mu)) = \mathrm{rank}(\mathfrak{s}) + \dim \hat{\mathfrak{h}}$. Therefore $\hat{\mathfrak{h}} = \mathfrak{d}$ is the centre of $\mathfrak{g}(\tilde{\mathfrak{h}}, \mu)$ and $\mathfrak{s}$ is a semisimple subalgebra of $\mathfrak{s}'$ of maximal rank. Finally, $\mathfrak{s}'$ is complete by Lemma 8.3.12. □

Proposition 8.3.19 *Let $\mathfrak{s}$ be a semisimple, regular, locally flat, complete, $\mathbb{Z}$-graded subalgebra of $\mathfrak{g}$. Let $e \in \mathfrak{s}_1$ lie in the open orbit. Then there is a homogeneous $\mathfrak{sl}_2$-triple (h, e, f) and a Cartan subalgebra $\mathfrak{t}_0 \subset \mathfrak{c}_0(\mathfrak{a})$ (where $\mathfrak{a}$ is spanned by h, e, f) such that $\mathfrak{s} = \mathfrak{s}(e, \mathfrak{t})$ where $\mathfrak{t} = \langle h \rangle \oplus \mathfrak{t}_0$.*

Proof. Let (h, e, f) be a homogeneous $\mathfrak{sl}_2$-triple with $h, e, f \in \mathfrak{s}$. Let $\mathfrak{a} \subset \mathfrak{g}$ be the subalgebra spanned by h, e, f. Define h_0, $\hat{\mathfrak{h}}$, $\tilde{\mathfrak{h}}$, $\mathfrak{g}(\tilde{\mathfrak{h}}, \mu)$ as in Lemma 8.3.18. By that lemma, as $\mathfrak{s}$ is complete, $\mathfrak{g}(\tilde{\mathfrak{h}}, \mu) = \mathfrak{s} \oplus \hat{\mathfrak{h}}$. Now $\mathfrak{c}_0(\mathfrak{a}) \supset \mathfrak{c}_0(\mathfrak{s})$. Let $\mathfrak{t}_0$ be a Cartan subalgebra of $\mathfrak{c}_0(\mathfrak{a})$ with $\hat{\mathfrak{h}} \subset \mathfrak{t}_0$ and $\mathfrak{t}$ be the subalgebra spanned by h and $\mathfrak{t}_0$. By Lemma 8.3.15, $h = 2h_0$. Therefore, $\tilde{\mathfrak{h}} \subset \mathfrak{t}$ and $\mathfrak{t} \subset \mathfrak{c}_0(\hat{\mathfrak{h}}) = \mathfrak{g}_0(\tilde{\mathfrak{h}}, \mu) = \mathfrak{s}_0 \oplus \hat{\mathfrak{h}}$. Let $t \in \mathfrak{t}_0$ and write $t = x + \hat{t}$ where $x \in \mathfrak{s}_0$ and $\hat{t} \in \hat{\mathfrak{h}}$. Because $[t, e] = [\hat{t}, e] = 0$, also $[x, e] = 0$. But that implies $x = 0$, as $\dim \mathfrak{s}_0 = \dim \mathfrak{s}_1$ and e lies in the open orbit. We conclude that $\tilde{\mathfrak{h}} = \mathfrak{t}$ and $\mathfrak{s} = \mathfrak{s}(e, \mathfrak{t})$. □

Because of this proposition, we say that a subalgebra $\mathfrak{s} \subset \mathfrak{g}$ is a *carrier algebra* if it is semisimple, regular, locally flat, complete, and $\mathbb{Z}$-graded.

Fix a Cartan subalgebra $\mathfrak{h}_0$ of $\mathfrak{g}_0$. A $\mathbb{Z}$-graded subalgebra of $\mathfrak{g}$ which is normalized by $\mathfrak{h}_0$ is said to be $\mathfrak{h}_0$-regular. Let $\mathfrak{s}$ be such a subalgebra. By $R(\mathfrak{s})$ we denote the set of $\bar{\alpha} \in \overline{\Phi}$ (notation as in Section 8.3.1), such that $\mathfrak{g}_{\bar{\alpha}} \subset \mathfrak{s}$. As $\mathfrak{s}$ is $\mathfrak{h}_0$-regular, it is spanned by $\mathfrak{h}_0 \cap \mathfrak{s}$, along with the $\mathfrak{g}_{\bar{\alpha}}$ for $\bar{\alpha} \in R(\mathfrak{s})$. For $i \in \mathbb{Z}$ set $R_i(\mathfrak{s}) = \{\bar{\alpha} \in \overline{\Phi} \mid \mathfrak{g}_{\bar{\alpha}} \subset \mathfrak{s}_i\}$.

Proposition 8.3.20 *Let $\mathfrak{s}$, $\mathfrak{s}'$ be two $\mathfrak{h}_0$-regular carrier algebras. Then $\mathfrak{s}$, $\mathfrak{s}'$ are G_0-conjugate if and only if there is a $w \in W_0$ with $w(R_i(\mathfrak{s})) = R_i(\mathfrak{s}')$ for all $i \in \mathbb{Z}$.*

Proof. Let $g \in G_0$ be such that $g \cdot \mathfrak{s}_i = \mathfrak{s}'_i$ for all i. Set

$$N = \{\bar{g} \in G_0 \mid \bar{g} \cdot \mathfrak{s}'_i = \mathfrak{s}'_i \text{ for all } i\}.$$

By Corollary 4.2.8, its Lie algebra is $\mathfrak{n} = \{x \in \mathfrak{g}_0 \mid [x, \mathfrak{s}'_i] \subset \mathfrak{s}'_i \text{ for all } i\}$. Now $\mathfrak{h}_0$ and $g \cdot \mathfrak{h}_0$ are contained in $\mathfrak{n}$, and moreover they are Cartan subalgebras of it. So, by Proposition 4.3.2, there is a $\bar{g} \in N$ with $\bar{g}g \cdot \mathfrak{h}_0 = \mathfrak{h}_0$. Set $g' = \bar{g}g$. Then $g' \in G_0$ has the properties $g' \cdot \mathfrak{s}_i = \mathfrak{s}'_i$ for all i, and $g' \cdot \mathfrak{h}_0 = \mathfrak{h}_0$. As in Section 8.3.1 we consider the homomorphism $\eta : N_{G_0}(\mathfrak{h}_0) \to W_0$. Let $w' = \eta(g')$. Let $\bar{\alpha}$ be such that $\mathfrak{g}_{\bar{\alpha}} \subset \mathfrak{s}_i$. As seen in Section 8.3.1, $g' \cdot \mathfrak{g}_{\bar{\alpha}} = \mathfrak{g}_{w'(\bar{\alpha})}$. Furthermore, $g' \cdot \mathfrak{g}_{\bar{\alpha}} \subset \mathfrak{s}'_i$. It follows that $w'(R_i(\mathfrak{s})) = R_i(\mathfrak{s}')$ for all i.

For the converse we let $w' \in W_0$ be such that $w'(R_i(\mathfrak{s})) = R_i(\mathfrak{s}')$ for all i, and let $g' \in N_{G_0}(\mathfrak{h}_0)$ be such that $\eta(g') = w'$. Then $g' \cdot \mathfrak{g}_{\bar{\alpha}} = \mathfrak{g}_{w'(\bar{\alpha})} \subset \mathfrak{s}'_i$ for all $\bar{\alpha}$ such that $\mathfrak{g}_{\bar{\alpha}} \subset \mathfrak{s}_i$. Hence $g' \cdot \mathfrak{s}_i = \mathfrak{s}'_i$. □

The results of this section show that mapping a nilpotent orbit $G_0 e$ to the G_0-conjugacy class of the carrier algebra $\mathfrak{s}(e, \mathfrak{t})$, yields a bijective map from the set of nilpotent G_0-orbits to the set of G_0-conjugacy classes of carrier algebras. Indeed, by Proposition 8.3.16 (see also Remark 8.3.17), this map is well-defined and injective. By Proposition 8.3.19 it is surjective.

Furthermore, Proposition 8.3.20 reduces the problem of listing the carrier algebras up to G_0-conjugacy to a combinatorial problem involving the root system $\overline{\Phi}$ and the Weyl group W_0.

8.4 Listing the nilpotent orbits of a θ-group

Let $\mathfrak{g} = \oplus_{i \in \mathcal{C}_m} \mathfrak{g}_i$ be a $\mathcal{C}_m$-graded semisimple Lie algebra, with corresponding θ-group G_0. By Corollary 8.3.8, $\mathfrak{g}_1$ has a finite number of nilpotent G_0-orbits. Here we discuss algorithms to list them. The input to these algorithms is the Lie algebra $\mathfrak{g}$, with bases of the components $\mathfrak{g}_i$. The output is a set of homogeneous $\mathfrak{sl}_2$-triples $\{(h_i, e_i, f_i)\}$ such that the $G_0 e_i$ are exactly the nilpotent G_0-orbits.

One possibility is to use the algorithm for listing the strata of the nullcone (Section 7.4.3) in view of Proposition 8.3.7. However, by making use of the results of the previous section, it is possible to devise much more efficient algorithms. Here we look at two approaches. The first uses $\mathfrak{sl}_2$-triples and the second is based on the theory of carrier algebras.

Throughout we fix a Cartan subalgebra $\mathfrak{h}_0$ of $\mathfrak{g}_0$ lying in the Cartan subalgebra $\mathfrak{h} = \mathfrak{c}_{\mathfrak{g}}(\mathfrak{h}_0)$ (Proposition 8.3.1(iii)). Again W_0 denotes the Weyl group of the root system of $\mathfrak{g}_0$ with respect to $\mathfrak{h}_0$.

8.4.1 Using $\mathfrak{sl}_2$-triples

Here we look at an algorithm that is directly based on the argument in the proof of Corollary 8.3.8. Let $e_1, \ldots, e_r$ be representatives of the nilpotent G-orbits in $\mathfrak{g}$, lying in $\mathfrak{sl}_2$-triples (h_i, e_i, f_i) where $h_i \in \mathfrak{h}$. Section 8.2 shows how these can be obtained. Set

$$\mathcal{H} = \bigcup_{i=1}^{r} W \cdot h_i.$$

Then by Theorem 8.1.4, Corollary 5.8.5, $\mathcal{H}$ is precisely the set of elements of $\mathfrak{h}$ lying in an $\mathfrak{sl}_2$-triple.

Proposition 8.4.1 *Let $h \in \mathfrak{g}_0$ be such that $\mathrm{ad}_{\mathfrak{g}} h$ is semisimple with integral eigenvalues. For $t, i \in \mathbb{Z}$ define $\mathfrak{g}_i(t) = \{x \in \mathfrak{g}_i \mid [h, x] = tx\}$ and set*

$$U = \{u \in \mathfrak{g}_1(2) \mid [\mathfrak{g}_0(0), u] = \mathfrak{g}_1(2)\}.$$

Then U is non-empty and open in $\mathfrak{g}_1(2)$. Moreover, every $e \in U$ lies in a homogeneous $\mathfrak{sl}_2$-triple (h, e, f) or there is no such triple containing h.

Proof. Set $\mathfrak{a} = \oplus_{i \in \mathbb{Z}} \mathfrak{g}_i(2i)$. Let $\mathfrak{t}$ be the subalgebra spanned by h, and define $\lambda \in \mathfrak{t}^*$ by $\lambda(h) = 2$. Then $\mathfrak{t}$ is a toral subalgebra and $\mathfrak{a} = \mathfrak{g}(\mathfrak{t}, \lambda)$ (notation as in Section 8.3.3). Hence by Lemma 8.3.11, $\mathfrak{a}$ is reductive in $\mathfrak{g}$.

Set $\mathfrak{s} = [\mathfrak{a}, \mathfrak{a}]$, which is semisimple and hence algebraic. We have that $\mathfrak{s}$ inherits the $\mathbb{Z}$-grading from $\mathfrak{a}$. One the one hand, all elements of $\mathfrak{s}_1$ are nilpotent. On the other hand, S_0 (the algebraic subgroup of G_0 with Lie algebra $\mathfrak{s}_0$) has a finite number of nilpotent orbits in $\mathfrak{s}_1$ by Corollary 8.3.8. It follows that S_0 has a finite number of orbits in $\mathfrak{s}_1$. Now define $G_{0,h} = \{g \in G_0 \mid g \cdot h = h\}$. By Corollary 4.2.11, $\mathrm{Lie}(G_{0,h}) = \mathfrak{g}_0(0) = \mathfrak{a}_0$. As seen in the proof of Lemma 8.3.11, the centre of $\mathfrak{a}$ is contained in $\mathfrak{a}_0$. So $\mathfrak{s}_0 \subset \mathfrak{a}_0$ and $\mathfrak{s}_1 = \mathfrak{a}_1$. Therefore, $G_{0,h}$ has a finite number of orbits in $\mathfrak{a}_1 = \mathfrak{g}_1(2)$, and by Lemma 3.12.1, there is an open orbit, which we denote O.

Let $x \in O$. Then by Lemma 8.2.1, $[\mathfrak{g}_0(0), x] = \mathfrak{g}_1(2)$. It follows that $O \subset U$. Let $u \in U$. Again using Lemma 8.2.1 we see that the dimension of the closure of $G_{0,h} \cdot u$ is $\dim \mathfrak{g}_1(2)$. Therefore, $G_{0,h} \cdot u$ is the open orbit O. We conclude that $U = O$.

Suppose we have a homogeneous $\mathfrak{sl}_2$-triple (h, x, y). Then $x \in \mathfrak{g}_1(2)$, and by writing $\mathfrak{g}$ as direct sum of $\mathfrak{sl}_2$-modules and using Theorem 2.9.1, we see that $\mathrm{ad}x : \mathfrak{g}_0(0) \to \mathfrak{g}_1(2)$ is surjective. Now $-\mathrm{ad}x$ is the differential of the map $\sigma : G_{0,h} \to \mathfrak{g}_1(2)$, $\sigma(g) = g \cdot x$ (Proposition 3.8.2). So by Proposition 1.4.2, σ is dominant. Theorem 1.4.3 indicates that the image of σ contains an open set of $\mathfrak{g}_1(2)$. It follows that the orbit of x equals O. Hence $x \in U$ and all elements of U lie in a homogeneous $\mathfrak{sl}_2$-triple involving h. □

Now let $h \in \mathcal{H} \cap \mathfrak{h}_0$, and let U be as in the previous proposition. By a few random tries, we find an $x \in U$. By solving a set of linear equations we can decide whether there is a $y \in \mathfrak{g}_{-1}$ such that (h, x, y) is an $\mathfrak{sl}_2$-triple. If there is such a y, we find one and add (h, x, y) to our set of triples. After having done this for each $h \in \mathcal{H} \cap \mathfrak{h}_0$, we obtain a set of homogeneous $\mathfrak{sl}_2$-triples. If there are triples (h_1, e_1, f_1) and (h_2, e_2, f_2) in this set such that h_1 and h_2 are W_0-conjugate, we remove one of them. In view of Theorem 8.1.4, the result of this procedure is exactly the list of triples that we are seeking.

The main practical problem with this algorithm lies in the need to compute the set $\mathcal{H}$: if W is a large group (for example, if $\mathfrak{g}$ is of type E_8); then the method becomes very difficult. However, if $\mathfrak{h}_0$ is a Cartan subalgebra of $\mathfrak{g}$ (this happens if $\mathcal{C}_m = \mathbb{Z}$, or if θ is an inner automorphism), we can make this algorithm much more efficient. We note that the root system of $\mathfrak{g}_0$ with respect to $\mathfrak{h}_0 = \mathfrak{h}$, is naturally a subsystem of the root system Φ of $\mathfrak{g}$. So W_0 is a reflection subgroup of W. Furthermore, all elements of $W_0 w(h)$ are conjugate under W_0. So we compute a set of representatives $w_1, \dots, w_s$ of the left cosets of W_0 in W (Section 2.8.3). Then it suffices to work with the set $\mathcal{H}' = \{w_j \cdot h_i \mid 1 \le i \le r, 1 \le j \le s\}$. This set is usually much smaller than $\mathcal{H}$ so that we obtain a practical algorithm.

Example 8.4.2 Let $\mathfrak{g}$, G_0 be as in Example 8.3.3. Computing the nilpotent orbits with the algorithm of this section takes 0.2 seconds. The set $\mathcal{H}'$ has size 62. There are 30 non-zero nilpotent orbits.

8.4.2 Using carrier algebras

In this section we look at a way to list the nilpotent orbits using the carrier algebras of Section 8.3.3. Because the method of the previous section works quite well when W or the index of W_0 in W is small, we devise a method that does not involve looping over W (for example when computing an orbit of W) or over W/W_0. We use the notation of Section 8.3. In particular we recall the root system $\overline{\Phi}$ of Section 8.3.1. In the following we will use a fixed Chevalley basis of $\mathfrak{g}$ consisting of $x_{\bar{\alpha}}$, for $\bar{\alpha} \in \overline{\Phi}$, and $h_1, \dots, h_\ell$.

Let $\mathfrak{s}$ be a $\mathbb{Z}$-graded $\mathfrak{h}_0$-regular semisimple subalgebra of $\mathfrak{g}$. Then $\mathfrak{s}_0$ contains a Cartan subalgebra of $\mathfrak{s}$, and we consider the corresponding root system Ψ of $\mathfrak{s}$, which is a subset of $\overline{\Phi}$. (We have that Ψ consists of those $\bar{\alpha} \in \overline{\Phi}$ such that $\mathfrak{g}_{\bar{\alpha}} \subset \mathfrak{s}$.) Let Ψ^+ be the union of a set of positive roots of the root system of $\mathfrak{s}_0$, together with all $\bar{\alpha} \in \Psi$ such that $\mathfrak{s}_{\bar{\alpha}} \subset \mathfrak{s}_i$, for $i > 0$. Then Ψ^+ is a set of positive roots of Ψ, and we let Π denote the corresponding set of simple roots. Then $\Pi = \cup_{i \ge 0} \Pi_i$, where $\Pi_i = \{\bar{\alpha} \in \Pi \mid \mathfrak{s}_{\bar{\alpha}} \in \mathfrak{s}_i\}$.

Lemma 8.4.3 *Suppose $\mathfrak{s}$ is locally flat. Then $\Pi = \Pi_0 \cup \Pi_1$.*

Proof. Let $e \in \mathfrak{s}_1$ be an element of the open orbit lying in the homogeneous $\mathfrak{sl}_2$-triple (h, e, f). By Lemma 8.3.15, $\frac{1}{2}h$ is the defining element of the $\mathbb{Z}$-grading of $\mathfrak{s}$. Hence, if $(i, \alpha) \in \Pi_j$, then $\alpha(h) = 2j$. So $\alpha(h)$ is even and non-negative for all $(i,\alpha) \in \Pi$. By Proposition 8.1.6, $\alpha(h) \in \{0,1,2\}$ for $(i,\alpha) \in \Pi$ and the value 1 cannot occur. □

Also we note that Π is a π-system (see Section 5.9) in $\overline{\Phi}$ and Π_0 is a π-system in the root system $\overline{\Phi}_0 = \{\bar{\alpha} \in \overline{\Phi} \mid \mathfrak{g}_{\bar{\alpha}} \subset \mathfrak{g}_0\}$.

We proceed as follows. Using the algorithm of Section 5.9 we obtain a list of all π-systems in $\overline{\Phi}_0$ up to W_0-conjugacy. For each such π-system Π_0, we find, by brute force search, all subsets Π_1 of $\overline{\Phi}_1 = \{\bar{\alpha} \in \overline{\Phi} \mid \mathfrak{g}_{\bar{\alpha}} \subset \mathfrak{g}_1\}$ such that $\Pi_0 \cup \Pi_1$ is a π-system (here we do not care about W_0-conjugacy).

The next step is to check, for each set $\Pi = \Pi_0 \cup \Pi_1$ that we find, whether the $\mathbb{Z}$-graded semisimple subalgebra $\mathfrak{s}^\Pi$ generated by the root spaces $\mathfrak{g}_{\pm\bar{\alpha}}$, $\bar{\alpha} \in \Pi$, is locally flat. There are several ways to do this. The dimension of $\mathfrak{s}_0^\Pi$ follows from the Dynkin diagram of Π_0 and the known dimensions of the simple Lie algebras. Furthermore, $\mathfrak{s}_1^\Pi$ is an $\mathfrak{s}_0^\Pi$-module splitting as a direct sum of irreducible modules of which the $x_{-\bar{\alpha}}$ for $\bar{\alpha} \in \Pi_1$ are the highest weight vectors and their dimensions can be computed with the methods of Section 2.11.3. However, possibly the simplest approach is the following. Let Ψ be the root system of $\mathfrak{s}^\Pi$. After computing the Cartan matrix of Ψ, we apply Algorithm 2.8.14 to list the elements of Ψ^+, writing each root as a linear combination of the elements of Π. Let $c_{\bar{\alpha}}$ be the sum of the coefficients of the elements of Π_1 in the expression for $\bar{\alpha} \in \Psi^+$. Then $\mathfrak{g}_{\bar{\alpha}} \subset \mathfrak{s}_i^\Pi$ if and only if $c_{\bar{\alpha}} = i$, for $i \geq 0$. So if m_i denotes the number of $\bar{\alpha} \in \Psi^+$ with $c_{\bar{\alpha}} = i$, we have $\dim \mathfrak{s}_0^\Pi = 2m_0 + |\Pi|$ and $\dim \mathfrak{s}_1^\Pi = m_1$. By performing this check for all Π found in the first step, we obtain a list L of the π-systems Π such that $\mathfrak{s}^\Pi$ is locally flat.

For each $\Pi \in L$ we compute the elements $h_{\bar{\alpha}} = [x_{\bar{\alpha}}, x_{-\bar{\alpha}}]$ (this follows directly from Theorem 2.9.13) for $\bar{\alpha} \in \Pi$. We note that $\bar{\beta}(h_{\bar{\alpha}}) = \langle \bar{\beta}, \bar{\alpha}^\vee \rangle$, so we can compute $h_\Pi = \sum_{\bar{\alpha}\in\Pi} a_{\bar{\alpha}} h_{\bar{\alpha}}$ such that $\bar{\beta}(h) = j$ for $\bar{\beta} \in \Pi_j$, $j = 1, 2$. Then h_Π is the defining element of the grading of $\mathfrak{s}^\Pi$. Furthermore, by a few random tries we quickly find an $e \in \mathfrak{s}_1^\Pi$ lying in the open orbit. By Lemma 8.3.15, e lies in an $\mathfrak{sl}_2$-triple $(2h_\Pi, e, f_e)$. Let M' be the list of triples $(2h_\Pi, e, f_e)$ for $\Pi \in L$. From this list we take a maximal sublist M such that for each pair of elements $(2h_\Pi, e, f_e)$, $(2h_{\Pi'}, e', f'_e)$ we have that h_Π and $h_{\Pi'}$ are not W_0-conjugate.

By the results of Section 8.3.3, M is an irredundant list of representatives of the nilpotent orbits of G_0. Indeed, we represent a carrier algebra by a basis of its simple roots. By Lemma 8.4.3 these are of the form $\Pi_0 \cup \Pi_1$. By Proposition 8.3.20 it suffices to consider the sets Π up to W_0-conjugacy. So we can limit the possibilities for Π_0 to the π-systems in $\overline{\Phi}_0$ up to W_0-conjugacy. Therefore the list L contains bases of simple roots of all carrier algebras in $\mathfrak{g}$ up to W_0-conjugacy. (In general, the list will also contain bases of simple roots of $\mathbb{Z}$-graded subalgebras which are not complete, so we could make it smaller by adding a completeness test, but it is not necessary to do so.) So the list M' contains at least one $\mathfrak{sl}_2$-triple for each nilpotent orbit, and we conclude by Theorem 8.3.6.

In this algorithm, the most time-consuming step is the brute force enumeration of the possible sets Π_1. It will be increasingly difficult to carry it out with increasing $|\overline{\Phi}_1|$. However, in many cases, when this number is high, so is $|\overline{\Phi}_0|$ and the index of W_0 in W is small (assuming that W_0 is a subgroup of W), making it possible to use the algorithm of the previous section.

Example 8.4.4 Let $\mathfrak{g}$, G_0 be as in Example 8.3.3. Computing the nilpotent orbits with the algorithm of this section takes 0.9 seconds. In the first step of the algorithm 173 π-systems are found.

Remark 8.4.5 If $\mathfrak{h}_0$ is a Cartan subalgebra of $\mathfrak{g}$ so that W_0 is a subgroup of W, we can also formulate an algorithm which uses carrier algebras along with a set of representatives of the left cosets of W_0 in W. This works best when the grading of $\mathfrak{g}$ is a $\mathbb{Z}$-grading, so we focus on that case. Because $\mathfrak{h}_0$ is a Cartan subalgebra of $\mathfrak{g}$, we can work with the root system Φ instead of $\overline{\Phi}$. Let $h_0 \in \mathfrak{h}_0$ be the defining element of the $\mathbb{Z}$-grading of $\mathfrak{g}$ and (h, e, f) be a homogeneous $\mathfrak{sl}_2$-triple in $\mathfrak{g}$, with $h \in \mathfrak{h}_0$. Let $\mathfrak{a}$, $\mathfrak{t}_0$, $\mathfrak{t}$ be as in Section 8.3.3. Then the carrier algebra $\mathfrak{s}(e, \mathfrak{t})$ is the derived algebra of $\mathfrak{g}(\mathfrak{t}, \lambda)$. We claim that $\mathfrak{g}(\mathfrak{t}, \lambda) = \mathfrak{c}_{\mathfrak{g}}(\mathfrak{t}_0)$. For this we first remark that $h_0 - \frac{1}{2}h$ lies in the centre of $\mathfrak{c}_0(\mathfrak{a})$, and thus lies in $\mathfrak{t}_0$. So $h = 2h_0 + t$ for a certain $t \in \mathfrak{t}_0$. But that implies that for an $x \in \mathfrak{g}_i \cap \mathfrak{c}_{\mathfrak{g}}(\mathfrak{t}_0)$ we have $[h, x] = [2h_0 + t, x] = 2ix$, so that $x \in \mathfrak{g}_i(\mathfrak{t}, \lambda)$. The claim follows.

After replacing (h, e, f) by a G_0-conjugate, we may also assume that $\mathfrak{t}_0 \subset \mathfrak{h}_0$. Then $\mathfrak{c}_{\mathfrak{g}}(\mathfrak{t}_0)$ is the sum of $\mathfrak{h}_0$ and the $\mathfrak{g}_\alpha$ where $\alpha \in \Phi$ is such that $\alpha(t) = 0$ for all $t \in \mathfrak{t}_0$. There is a $t_0 \in \mathfrak{t}_0$ such that $\mathfrak{c}_{\mathfrak{g}}(\mathfrak{t}_0) = \mathfrak{c}_{\mathfrak{g}}(t_0)$. Choose a set of positive roots in Φ such that $\alpha(t_0) \geq 0$ for all positive roots α. Then we see that $\mathfrak{s}(e, \mathfrak{t})$ has a basis of simple roots which is a subset of a basis of simple roots of Φ. This means that we can proceed as follows. Let Δ be a fixed basis of simple roots of Φ and $w_1, \ldots, w_s$ be a set of representatives of the left cosets of W_0 in W. Set $\Delta_i = w_i(\Delta)$. For each Δ_i we list its subsets $\Pi = \Pi_0 \cup \Pi_1$ where Π_j consists of $\alpha \in \Delta_i$ such that $\mathfrak{g}_\alpha \subset \mathfrak{g}_j$. For all the sets Π thus obtained, we perform the same steps as in the algorithm above.

If the grading is not a $\mathbb{Z}$-grading, and $\mathfrak{h}_0$ is a Cartan subalgebra of $\mathfrak{g}$ (this means that θ is an inner automorphism), we can perform a similar algorithm. But in this case, instead of just the set Δ we have to consider all π-systems in Φ of rank ℓ (up to W-conjugacy) which can be obtained using the algorithm in Section 5.9.

8.5 Closures of nilpotent orbits

We let $\mathfrak{g} = \oplus_{i \in C_m} \mathfrak{g}_i$ and G_0 be as in Section 8.3. Consider two nilpotent G_0-orbits $\mathcal{O}_1$ and $\mathcal{O}_2$ in $\mathfrak{g}_1$. In this section we give an algorithm for deciding whether $\mathcal{O}_1$ is contained in the closure $\overline{\mathcal{O}}_2$ of $\mathcal{O}_2$. By Proposition 8.3.7, the nilpotent G_0-orbits in $\mathfrak{g}_1$ coincide with the strata of the nullcone $\mathcal{N}_{G_0}(\mathfrak{g}_1)$. More precisely, if (h, e, f) is a homogeneous $\mathfrak{sl}_2$-triple, h is a characteristic of the stratum which coincides with $G_0 e$. So we can use the results of Section 7.4. In particular, Proposition 7.4.24(i) gives a description of the closure of a stratum which will underpin our algorithm.

As before, we let $\mathfrak{h}_0$ denote a fixed Cartan subalgebra of $\mathfrak{g}_0$. Throughout this section, when we speak of a homogeneous $\mathfrak{sl}_2$-triple (h, e, f), it is assumed that $h \in \mathfrak{h}_0$.

Let $\overline{\Phi}$ be as in Section 8.3.1. For $j = 0, 1$ we set $\overline{\Phi}_j = \{\bar{\alpha} \in \overline{\Phi} \mid \mathfrak{g}_{\bar{\alpha}} \subset \mathfrak{g}_j\}$. So $\overline{\Phi}_0$ is the root system of $\mathfrak{g}_0$ with respect to $\mathfrak{h}_0$.

In the following we will write $V = \mathfrak{g}_1$. We start by recalling some notation from Section 7.4.1. For a rational $h \in \mathfrak{h}_0$ and $\tau \in \mathbb{Q}$ we set

$$V_\tau(h) = \{v \in V \mid hv = \tau v\} \text{ and } V_{\geq\tau}(h) = \bigoplus_{\tau' \geq \tau} V_{\tau'}(h).$$

Furthermore, $\mathfrak{c}_0(h) = \{x \in \mathfrak{g}_0 \mid [h, x] = 0\}$, which is the Lie algebra of

$$Z_0(h) = \{g \in G_0 \mid g \cdot h = h\}^\circ.$$

Also, for $\tau \in \mathbb{Q}$ we let $\mathfrak{g}_0^\tau(h) = \{x \in \mathfrak{g}_0 \mid [h, x] = \tau x\}$.

Let (h, e, f) be a homogeneous $\mathfrak{sl}_2$-triple. As seen in the proof of Proposition 8.4.1, $V_2(h)$ has an open $Z_0(h)$-orbit, which is precisely the set of all $v \in V_2(h)$ with $[\mathfrak{c}_0(h), v] = V_2(h)$. Moreover, e lies in this orbit.

Lemma 8.5.1 *$V_2(h) \cap G_0 e$ is equal to the open $Z_0(h)$-orbit in $V_2(h)$.*

Proof. For $\tau \in \mathbb{Q}$ and $v \in V_2(h)$ set

$$\mathfrak{g}_0^v = \{x \in \mathfrak{g}_0 \mid [x, v] = 0\}, \text{ and } \mathfrak{g}_0^{\tau,v} = \mathfrak{g}_0^v \cap \mathfrak{g}_0^\tau(h).$$

Let $e' \in V_2(h) \cap G_0 e$. Since e' lies in the closure of $Z_0(h)e$, we have $\dim \mathfrak{g}_0^{\tau,e'} \geq \dim \mathfrak{g}_0^{\tau,e}$ for all $\tau \in \mathbb{Q}$. As e' and e lie in the same G_0-orbit we see that $\dim \mathfrak{g}_0^{e'} = \dim \mathfrak{g}_0^e$. But $\mathfrak{g}_0^{e'}$ is the direct sum of the various $\mathfrak{g}_0^{\tau,e'}$ and similarly for $\mathfrak{g}_0^e$. It follows that $\dim \mathfrak{g}_0^{\tau,e'} = \dim \mathfrak{g}_0^{\tau,e}$ for all τ. But

$$\mathfrak{g}_0^{0,e'} = \{x \in \mathfrak{c}_0(h) \mid [x, e'] = 0\}$$

and similarly for $\mathfrak{g}_0^{0,e}$. This implies that $\dim[\mathfrak{c}_0(h), e'] = \dim[\mathfrak{c}_0(h), e]$ and e' lies in the open $Z_0(h)$-orbit in $V_2(h)$. □

Proposition 8.5.2 *Let (h, e, f), (h', e', f') be two homogeneous $\mathfrak{sl}_2$-triples. Then $G_0 e'$ is contained in the closure of $G_0 e$ if and only if there is a $w \in W_0$ such that $U_w = V_2(h') \cap V_{\geq 2}(wh)$ contains a point of $G_0 e'$. Furthermore, the intersection of U_w and $G_0 e'$ is open in U_w.*

Proof. By Proposition 7.4.24(i) we have $\overline{G_0 e} = G_0 V_{\geq 2}(h)$, implying the "if" part.

As in Section 7.4.1, we set $\mathfrak{p}_0(h) = \oplus_{\tau \geq 0} \mathfrak{g}_0^\tau(h)$, and denote by $P_0(h)$ the connected algebraic subgroup of G_0 with Lie algebra $\mathfrak{p}_0(h)$. Let $\overline{\Delta}_0$ denote a set of simple roots of $\overline{\Phi}_0$ and $w_i \in W_0$, $i = 1, 2$; then there is a $w \in W_0$ such that U_w contains a point of $G_0 e'$ if and only if there is a $w \in W_0$ such that $V_2(w_1 h') \cap V_{\geq 2}(w w_2 h)$ contains a point of $G_0 e'$. Therefore we may assume that $\alpha(h), \alpha(h') \geq 0$ for all $\alpha \in \overline{\Delta}_0$. Then $\mathfrak{p}_0(h)$ and $\mathfrak{p}_0(h')$ contain the same Borel subalgebra $\mathfrak{b}_0$ (which is the direct sum of $\mathfrak{h}_0$ and the positive root spaces of $\mathfrak{g}_0$), so that $P(h)$ and $P(h')$ contain the same Borel subgroup B_0 of G_0. Let H_0 be the torus of G_0 with Lie algebra $\mathfrak{h}_0$ and $\eta : N_{G_0}(H_0) \to W_0$ be the homomorphism of Theorem 5.8.3. For $w \in W_0$ fix a $\dot{w} \in N_{G_0}(H_0)$ with $\eta(\dot{w}) = w$. Then by Theorem 5.8.3, G_0 is the union of the sets $P_0(h')\dot{w}P_0(h)$. Again using Proposition 7.4.24(i), we see that

$$\overline{G_0 e} = \bigcup_{w \in W_0} P_0(h')\dot{w}P_0(h)(V_{\geq 2}(h)) = \bigcup_{w \in W} P_0(h')\dot{w}(V_{\geq 2}(h)).$$

Suppose $G_0 e' \subset \overline{G_0 e}$. Let $v' \in V_2(h') \cap G_0 e'$. Then there are $p \in P_0(h')$, $w \in W_0$, $v \in V_{\geq 2}(h)$ with $v' = p\dot{w} \cdot v$, or $p^{-1} \cdot v' = \dot{w} \cdot v$. We have $P_0(h') = Z_0(h')N$ where N is the unipotent subgroup of G_0 with Lie algebra $\oplus_{\tau > 0} \mathfrak{g}_0^\tau(h')$. So $p^{-1} = zn$ with $z \in Z_0(h')$, $n \in N$. As $v' \in V_2(h')$, we see that $nv' = v' + v''$ with $v'' \in V_{>2}(h')$. So $p^{-1} \cdot v' = zv' + zv''$ with $zv' \in V_2(h')$, $zv'' \in V_{>2}(h')$. In particular, $p^{-1} \cdot v' \in V_{\geq 2}(h')$ but $\dot{w} \cdot v \in V_{\geq 2}(wh)$. So $p^{-1} \cdot v' \in V_{\geq 2}(h') \cap V_{\geq 2}(wh)$. Denote the latter space by $\widetilde{U}$. Since h' and wh commute, $\widetilde{U}$ is stable under h'. Therefore $\widetilde{U}$ is the direct sum of h'-eigenspaces. Hence $zv' \in \widetilde{U}$. So, in fact, $zv' \in U_w$, and obviously, $zv' \in G_0 e'$.

Finally, by Lemma 8.5.1, $G_0 e' \cap V_2(h')$ is open in $V_2(h')$. Therefore, $G_0 e' \cap U_w$ is open in U_w as well. □

This yields a direct method for checking whether $G_0 e' \subset \overline{G_0 e}$. For each $w \in W_0$ we:

1. Compute the space $U_w = V_2(h') \cap V_{\geq 2}(wh)$.

2. Decide whether there are $u \in U_w$ lying in $G_0 e'$. If this is the case, conclude that $G_0 e' \subset \overline{G_0 e}$.

If the second step never has a positive result, we conclude that $G_0 e'$ is not contained in $\overline{G_0 e}$.

There are several problems with this algorithm. First, we need a method for solving the decision problem in the second step. There is a straightforward randomized algorithm that does the following. Take a random point $u \in U_w$. If $\dim[\mathfrak{c}_0(h'), u] = \dim V_2(h')$, then $u \in G_0 e' \cap U_w$. If not, the intersection is probably empty. By Lemma 8.5.1, $V_2(h') \cap G_0 e'$ is the open $Z_0(h')$-orbit in $V_2(h')$. Therefore, if this gives a positive answer, then the answer is correct.

Furthermore, by Proposition 8.5.2, $G_0e' \cap U_w$ is open in U_w, so if the answer is negative, then the answer is correct with high probability, but not with certainty. This a somewhat unsatisfactory situation. In Section 8.5.1 we will show methods for proving that the intersection is empty.

The second problem is practical. The group W_0 can be large, so we need an efficient algorithm to loop over the orbit W_0h. This is the subject of Section 8.5.2, which will also provide a criterion allowing us to occasionally skip large parts of the orbit.

Remark 8.5.3 The spaces U_w are sums of root spaces $\mathfrak{g}_{\bar\alpha}$ for certain $\bar\alpha \in \overline{\Phi}_1$. Therefore, there are only a finite number of possibilities for these spaces. Hence it is a good idea to keep track of the subspaces U_w for which we have already shown that $U_w \cap G_0e'$ is empty. If the same subspace occurs again, no further work is required.

8.5.1 Deciding emptiness

Here we look at the problem of deciding whether a given space U_w contains a point of G_0e'. We describe two approaches. The first can only give a negative answer, that is, if it returns "no", we proved that there is no such point in U_w. The second gives a complete answer but is computationally much more difficult.

For $x, y \in \mathfrak{g}$ we write $(x, y) = \kappa_{\mathfrak{g}}(x, y)$ (where the latter is the Killing form of $\mathfrak{g}$). Since the restriction of $(\ ,\ )$ to $\mathfrak{h}_0$ is non-degenerate (this follows from Theorem 8.3.1), we get a bijective linear map $\nu : \mathfrak{h}_0 \to \mathfrak{h}_0^*$ by $\nu(h_1)(h_2) = (h_1, h_2)$. Let $\mathfrak{h}_{0,\mathbb{Q}}$ be the vector space over $\mathbb{Q}$ of all $h \in \mathfrak{h}_0$ such that $\mathrm{ad}_{\mathfrak{g}}h$ has rational eigenvalues and set $\mathfrak{h}_{0,\mathbb{R}} = \mathbb{R} \otimes \mathfrak{h}_{0,\mathbb{Q}}$. The form $(\ ,\ )$ restricts to a form on $\mathfrak{h}_{0,\mathbb{Q}}$, which then extends to an inner product on $\mathfrak{h}_{0,\mathbb{R}}$.

We note that the subspaces U_w are stable under $\mathfrak{h}_0$ and h' acts with the single eigenvalue 2. Therefore, U_w is a sum of root spaces $\mathfrak{g}_{\bar\alpha}$, with $\bar\alpha = (1, \alpha)$ and $\alpha(h') = 2$. Let $\bar\alpha_i = (1, \alpha_i)$, $1 \le i \le s$, be the roots such that U_w is the sum of the $\mathfrak{g}_{\bar\alpha_i}$. Let $\hat{h}_i = \nu^{-1}(\alpha_i)$. Then $\hat{h}_i \in \mathfrak{h}_{0,\mathbb{Q}}$ and they lie in the hyperplane in $\mathfrak{h}_{0,\mathbb{R}}$ defined by $(\cdot, h') = 2$. Let $S \subset \mathfrak{h}_{0,\mathbb{R}}$ be the convex hull of the $\hat{h}_i$ and $\tau \in \mathbb{Q}_{>0}$ be such that $(h', \tau h') = 2$.

Lemma 8.5.4 *If $\tau h'$ lies outside S then U_w has no point of G_0e'.*

Proof. Let $u \in U_w \cap G_0e'$. Write u as a linear combination of root vectors, and let S' be the convex hull of the $\hat{h}_i$ where i is such that $\bar\alpha_i$ is involved in the expression for u. Then $S' \subset S$. By Lemma 8.5.1 we have a homogeneous $\mathfrak{sl}_2$-triple (h', u, f'_u). By Proposition 8.3.7, h' is a characteristic of u. By Lemma 7.4.16 there is no other characteristic of u in $\mathfrak{h}_0$. Lemma 7.4.10 shows that the element of minimum norm in S' is $\tau h'$. □

The algorithm now decides whether $\tau h'$ lies outside S (for example by using Kalantari's algorithm; see Remark 7.4.27). If it does, then $U_w \cap G_0 e'$ is empty.

The second method involves much more brute force. Let $u_1, \dots, u_t$, $x_1, \dots, x_m$ and $y_1, \dots, y_r$ be bases of U_w, $\mathfrak{c}_0(h')$ and $V_2(h')$ respectively. Write $[x_i, u_j] = \sum_{k=1}^r c_{ij}^k y_k$. Let $u = z_1 u_1 + \cdots + z_t u_t$, then

$$[x_i, u] = \sum_{k=1}^{r} (\sum_{j=1}^{t} c_{ij}^k z_j) y_k.$$

Let A be the $m \times r$ matrix with $A(i,k) = \sum_j c_{ij}^k z_j$. Then $u \in G_0 e'$ if and only if $[\mathfrak{c}_0(h'), u] = V_2(h')$ (Lemma 8.5.1) if and only if $\mathrm{rank}(A) = r$. Now let $z_1, \dots, z_t$ be the indeterminates of a rational function field F over K, and compute the rank of A over F. Then $U_w \cap G_0 e'$ is empty if and only if this rank is less than r.

Remark 8.5.5 The problem to compute the rank of a matrix over a rational function field is known as Edmonds' problem (see [Lov79]). It is a difficult computation, but can in principle be carried out. For our purposes, however, it is often easier to compute all $r \times r$ minors of A. They are all zero if and only if the rank of A is less than r.

In [GVY12] a second method is given for deciding emptiness, using also the matrix A with entries in F. On some occasions it works better than computing the $r \times r$ minors of A.

8.5.2 Traversing a W_0-orbit

Here we write α instead of $(0, \alpha) \in \overline{\Phi}_0$. Let $\overline{\Delta}_0 = \{\alpha_1, \dots, \alpha_t\}$ be a fixed basis of simple roots of $\overline{\Phi}_0$. For $h \in \mathfrak{h}_0$ we set $v(h) = (\alpha_1(h), \dots, \alpha_t(h))$. Also set $\mathfrak{h}_{0,\mathbb{Z}} = \{h \in \mathfrak{h}_0 \mid v(h) \in \mathbb{Z}^t\}$ and $C_0 = \{h \in \mathfrak{h}_{0,\mathbb{Z}} \mid v(h) \in \mathbb{Z}^t_{\geq 0}\}$. Every W_0-orbit in $\mathfrak{h}_{0,\mathbb{Z}}$ has a unique point in C_0 (this is shown in the same way as Theorem 2.8.29). For $h_0 \in C_0$ we consider the problem of running over the orbit $W_0 \cdot h_0$.

We define the *length* of an $h \in W_0 \cdot h_0$ denoted $\mathcal{L}(h)$ as the length of a shortest $w \in W_0$ such that $w(h_0) = h$.

Proposition 8.5.6 *Let $h \in W_0 \cdot h_0$ with $v(h) = (a_1, \dots, a_m)$. Then*

(i) $\mathcal{L}(h) = |\{\alpha \in \overline{\Phi}_0^+ \mid \alpha(h) < 0\}|$.

(ii) $\mathcal{L}(s_{\alpha_i}(h)) > \mathcal{L}(h)$ *if and only if* $a_i > 0$.

Proof. Let $w \in W_0$ be of shortest length with $w(h_0) = h$. Let $\alpha \in \overline{\Phi}_0^+$ be such that $\alpha(h) = 0$. We claim that $w^{-1}(\alpha) > 0$. Suppose $w^{-1}(\alpha) < 0$ and write $w^{-1}(\alpha) = \sum_j s_j \alpha_j$ with $s_j \in \mathbb{Z}$, $s_j \leq 0$. Then $\alpha(h) = \alpha(w(h_0)) =$

$w^{-1}(\alpha)(h_0) = \sum_j s_j\alpha_j(h_0)$. It follows that for all j with $s_j \neq 0$ we have $\alpha_j(h_0) = 0$. Furthermore, there has to be a j_0 with $s_{j_0} \neq 0$ and $w(\alpha_{j_0}) < 0$. Then $\mathcal{L}(ws_{\alpha_{j_0}}) < \mathcal{L}(w)$ (Corollary 2.8.25) and $s_{\alpha_{j_0}}(h_0) = h_0$. It follows that $ws_{\alpha_{j_0}}$ is a shorter element sending h_0 to h, which is a contradiction. We conclude that for $\alpha \in \overline{\Phi}_0^+$ we have $w^{-1}(\alpha) < 0$ if and only if $\alpha(h) < 0$ which, in view of Lemma 2.8.24, implies (i).

Note that $\alpha(s_{\alpha_i}(h)) = s_{\alpha_i}(\alpha)(h)$. This, together with Lemma 2.8.18, proves (ii). □

Let $h \in W_0 \cdot h_0$ be of length r and write $v(h) = (a_1, \dots, a_t)$. Suppose $a_j > 0$ and write $v(s_{\alpha_j}(h)) = (b_1, \dots, b_t)$. Then $b_j = -a_j < 0$. If j is the minimal index with $b_j < 0$, we say that $s_{\alpha_j}(h)$ is a *successor* of h and h is a *predecessor* of $s_{\alpha_j(h)}$. It is clear that every element of $W_0 \cdot h_0$ has a unique predecessor. We now construct a tree on the points of $W_0 \cdot h_0$ with h_0 as its root. Two elements h and h' are connected if h' is a successor of h. By running through this tree we can efficiently enumerate the orbit $W_0 \cdot h_0$. By the next proposition we may be able to skip large parts of the tree.

Proposition 8.5.7 *Let h, h' and U_w be as in Proposition 8.5.2. Suppose $\kappa_{\mathfrak{g}}(h', w(h)) < \kappa_{\mathfrak{g}}(h', h')$. Then U_w contains no point of G_0e'. Moreover, the same holds for all $U_{w'}$ where $w' \in W_0$ is such that $w'(h)$ occurs in the subtree below $w(h)$.*

Proof. Write $\tilde{\mathfrak{c}}_0(h') = \{x \in \mathfrak{g}_0 \mid [x, h'] = 0 \text{ and } \kappa_{\mathfrak{g}}(x, h') = 0\}$. Then $\mathfrak{c}_0(h') = \langle h' \rangle \oplus \tilde{\mathfrak{c}}_0(h')$ and we have $a \in K$ and $t \in \tilde{\mathfrak{c}}_0(h')$ with $w(h) = ah' + t$. Furthermore, $a = \kappa_{\mathfrak{g}}(h', w(h))/\kappa_{\mathfrak{g}}(h', h')$ so that $a \in \mathbb{Q}$ and $a < 1$. Therefore t has only positive eigenvalues on U_w. Let T, $\widetilde{Z}(h')$ be the connected subgroups of G_0 whose Lie algebras are spanned by t and $\tilde{\mathfrak{c}}_0(h')$ respectively. Then all elements of U_w are unstable with respect to T (see Definition 7.4.2) and specifically with respect to $\widetilde{Z}(h')$. By Theorem 7.4.22, along with the observation that h' is a characteristic of all elements of $G_0e' \cap V_2(h')$ (Lemma 8.5.1), U_w contains no point of G_0e'.

Write $v(w(h)) = (a_1, \dots, a_t)$ and let $s_{\alpha_j}w(h)$ be a successor of $w(h)$. Then $a_j > 0$ and $s_{\alpha_j}w(h) = w(h) - a_j h_{\alpha_j}$ so that $\kappa_{\mathfrak{g}}(h', s_{\alpha_j}w(h)) = \kappa_{\mathfrak{g}}(h', w(h)) - a_j\kappa_{\mathfrak{g}}(h', h_{\alpha_j})$. But $\kappa_{\mathfrak{g}}(h', h_{\alpha_j})$ is a positive rational multiple of $\alpha_j(h') = a_j$. It follows that $\kappa_{\mathfrak{g}}(h', s_{\alpha_j}w(h)) \leq \kappa_{\mathfrak{g}}(h', w(h))$, implying the second statement. □

Example 8.5.8 The algorithm of this section can also be applied to the nilpotent G-orbits in $\mathfrak{g}$ (then $\mathfrak{g}$ is graded by the trivial group). For the group of type E_8, the total computation (determining all orbit inclusions) took 954 seconds. One inclusion had to be decided with the brute force method of Section 8.5.1 and took 156 seconds. We see that the brute force method is not often needed, and that the criterion of Proposition 8.5.7 makes it possible to process examples with very large Weyl groups (in this case the Weyl group has 696,729,600 elements).

For the nilpotent orbits of the θ-group of Example 8.3.3 the entire computation took 0.7 seconds. The brute force decision method was not needed.

8.6 Notes

The algorithm for computing representatives in Section 8.2 is contained in a somewhat weaker form in [Gra08]. In [Cla12], Clarke proposes an attractive algorithm for computing representatives that works for the classical types and yields representatives that are in some sense canonical. The algorithm for deciding whether two nilpotent elements are conjugate appeared (for the exceptional types) in [GE09].

The theory of the carrier algebras is developed by Vinberg in [Vin79]. The algorithms of Section 8.4 appear (in a slightly different form) in [Gra11]. Djokovic ([Djo88]) devised an approach with some similarities to the one in Section 8.4.1. An algorithm roughly along the lines of Remark 8.4.5 has been developed by Littelmann ([Lit96]). The last section is based on [GVY12].

Bibliography

[AL88] J. Apel and W. Lassner. An extension of Buchberger's algorithm and calculations in enveloping fields of Lie algebras. *J. Symbolic Comput.*, 6(2-3):361–370, 1988. Computational aspects of commutative algebra.

[BC76a] P. Bala and R. W. Carter. Classes of unipotent elements in simple algebraic groups. I. *Math. Proc. Cambridge Philos. Soc.*, 79(3):401–425, 1976.

[BC76b] P. Bala and R. W. Carter. Classes of unipotent elements in simple algebraic groups. II. *Math. Proc. Cambridge Philos. Soc.*, 80(1):1–17, 1976.

[BCNS15] Oliver Braun, Renaud Coulangeon, Gabriele Nebe, and Sebastian Schönnenbeck. Computing in arithmetic groups with Voronoï's algorithm. *J. Algebra*, 435:263–285, 2015.

[BCP97] Wieb Bosma, John Cannon, and Catherine Playoust. The Magma algebra system. I. The user language. *J. Symbolic Comput.*, 24(3-4):235–265, 1997. Computational algebra and number theory (London, 1993).

[BDD+10] L. Borsten, D. Dahanayake, M. J. Duff, A. Marrani, and W. Rubens. Four-qubit entanglement classification from string theory. *Phys. Rev. Lett.*, 105(10):100507, 4, 2010.

[Beh75] Helmut Behr. Explizite Präsentation von Chevalley-gruppen über Z. *Math. Z.*, 141:235–241, 1975.

[Ber78] G. M. Bergman. The diamond lemma for ring theory. *Adv. in Math.*, 29(2):178–218, 1978.

[BG06] Enrico Bombieri and Walter Gubler. *Heights in diophantine geometry*, volume 4 of *New Mathematical Monographs*. Cambridge University Press, 2006.

[BHC62] Armand Borel and Harish-Chandra. Arithmetic subgroups of algebraic groups. *Ann. Math. (2)*, 75:485–535, 1962.

[BK89] G. W. Bluman and S. Kumei. *Symmetries and differential equations.* Springer Verlag, Berlin, 1989.

[Bor69] Armand Borel. *Introduction aux groupes arithmétiques.* Publications de l'Institut de Mathématique de l'Université de Strasbourg, XV. Actualités Scientifiques et Industrielles, No. 1341. Hermann, Paris, 1969.

[Bor91] Armand Borel. *Linear algebraic groups.* Springer Verlag, Berlin, second edition, 1991.

[BP10] Andries E. Brouwer and Mihaela Popoviciu. The invariants of the binary decimic. *J. Symbolic Comput.*, 45(8):837–843, 2010.

[Buc90] Johannes Buchmann. A subexponential algorithm for the determination of class groups and regulators of algebraic number fields. In *Séminaire de Théorie des Nombres, Paris 1988–1989*, volume 91 of *Progr. Math.*, pages 27–41. Birkhäuser, Boston, 1990.

[Bur71] N. Burgoyne. Modular representations of some finite groups. In *Representation theory of finite groups and related topics (Proc. Sympos. Pure Math., Vol. XXI, Univ. Wisconsin, Madison, Wis., 1970)*, pages 13–17. American Mathematical Society, Providence, RI, 1971.

[BW71] N. Burgoyne and C. Williamson. Some computations involving simple Lie algebras. In S. R. Petrick, editor, *SYMSAC '71 Proceedings of the second ACM symposium on Symbolic and algebraic manipulation*, pages 162–171, New York, 1971. ACM.

[BW93] Thomas Becker and Volker Weispfenning. *Gröbner bases*, volume 141 of *Graduate Texts in Mathematics.* Springer Verlag, New York, 1993. A computational approach to commutative algebra, In cooperation with Heinz Kredel.

[Car72] R. W. Carter. *Simple groups of Lie type.* John Wiley & Sons, New York, 1972. Pure and Applied Mathematics series, Vol. 28.

[Cas14] Bill Casselman. Structure constants of Kac-Moody Lie algebras. In *Symmetry: representation theory and its applications, Progr. Math.* 257:55–83, 2014.

[CdG09] Serena Cicalò and Willem A. de Graaf. Non-associative Gröbner bases, finitely-presented Lie rings and the Engel condition II. *J. Symbolic Comput.*, 44:786–800, 2009.

[Che51] Claude Chevalley. *Théorie des groupes de Lie. Tome II. Groupes algébriques.* Actualités Sci. Ind. no. 1152. Hermann & Cie., Paris, 1951.

[Che55a] Claude Chevalley. Sur certains groupes simples. *Tôhoku Math. J. (2)*, 7:14–66, 1955.

[Che55b] Claude Chevalley. *Théorie des Groupes de Lie, Tome III. Théorèmes généraux sur les algèbres de Lie.* Hermann, Paris, 1955.

[Che58] Claude Chevalley. *Fondements de la géométrie algébrique.* Secrétariat Mathématique, Paris, 1958.

[Che05] Claude Chevalley. *Classification des groupes algébriques semi-simples.* Springer Verlag, Berlin, 2005. Collected works. Vol. 3, Edited and with a preface by P. Cartier, with the collaboration of Cartier, A. Grothendieck and M. Lazard.

[CHM08] Arjeh M. Cohen, Sergei Haller, and Scott H. Murray. Computing in unipotent and reductive algebraic groups. *LMS J. Comput. Math.*, 11:343–366, 2008.

[Cla12] Matthew C. Clarke. Computing nilpotent and unipotent canonical forms: a symmetric approach. *Math. Proc. Cambridge Philos. Soc.*, 152(1):35–53, 2012.

[CLO15] David A. Cox, John Little, and Donal O'Shea. *Ideals, varieties, and algorithms.* Undergraduate Texts in Mathematics. Springer, Cham, fourth edition, 2015. An introduction to computational algebraic geometry and commutative algebra.

[CM93] David H. Collingwood and William M. McGovern. *Nilpotent orbits in semisimple Lie algebras.* Van Nostrand Reinhold Co., New York, 1993.

[CMT04] Arjeh M. Cohen, Scott H. Murray, and D. E. Taylor. Computing in groups of Lie type. *Math. Comp.*, 73(247):1477–1498, 2004.

[Coh91] P. M. Cohn. *Algebra. Vol. 3.* John Wiley & Sons, Chichester, second edition, 1991.

[Der99] Harm Derksen. Computation of reductive group invariants. *Adv. Math.*, 141:366–384, 1999.

[DFdG15] A. S. Detinko, D. L. Flannery, and W. A. de Graaf. Integrality and arithmeticity of solvable linear groups. *J. Symbolic Comput.*, 68(part 1):138–145, 2015.

[DFH15] A. S. Detinko, D. L. Flannery, and A. Hulpke. Algorithms for arithmetic groups with the congruence subgroup property. *J. Algebra*, 421:234–259, 2015.

[DGPS15] Wolfram Decker, Gert-Martin Greuel, Gerhard Pfister, and Hans Schönemann. Singular 4-0-2: a computer algebra system for polynomial computations, 2015.

[DJK05] Harm Derksen, Emmanuel Jeandel, and Pascal Koiran. Quantum automata and algebraic groups. *J. Symbolic Comput.*, 39(3-4):357–371, 2005.

[Djo88] Dragomir Ž. Djokovic. Classification of nilpotent elements in simple exceptional real Lie algebras of inner type and description of their centralizers. *J. Algebra*, 112(2):503–524, 1988.

[DK02] Harm Derksen and Gregor Kemper. *Computational invariant theory.* Springer Verlag, Berlin, 2002.

[DK08] Harm Derksen and Gregor Kemper. Computing invariants of algebraic groups in arbitrary characteristic. *Adv. Math.*, 217(5):2089–2129, 2008.

[Dyn52] E. B. Dynkin. Semisimple subalgebras of semisimple Lie algebras. *Mat. Sbornik N.S.*, 30(72):349–462 (3 plates), 1952. English translation in: *Amer. Math. Soc. Transl.* (6), (1957), 111–244.

[Ela75] Alexander G. Elashvili. Centralizers of nilpotent elements in Lie algebras. *Sakharth. SSR Mecn. Akad. Math. Inst. Srom.*, 46:109–132, 1975.

[FG07] Claus Fieker and Willem A. de Graaf. Finding integral linear dependencies of algebraic numbers and algebraic Lie algebras. *LMS J. Comput. Math.*, 10:271–287 (electronic), 2007.

[FGP13] Paolo Faccin, Willem A. de Graaf, and Wilhelm Plesken. Computing generators of the unit group of an integral abelian group ring. *J. Algebra*, 373:441–452, 2013.

[For25] L. R. Ford. The fundamental region for a Fuchsian group. *Bull. Amer. Math. Soc.*, 31(9-10):531–539, 1925.

[GAP16] The GAP Group. *GAP – Groups, Algorithms, and Programming, Version 4.8.5*, 2016.

[Ge93] G. Ge. Algorithms related to multiplicative representations of algebraic numbers. PhD thesis, University of California, Berkeley, 1993.

[GE09] Willem A. de Graaf and Alexander G. Elashvili. Induced nilpotent orbits of the simple Lie algebras of exceptional type. *Georgian Math. J.*, 16(2):257–278, 2009.

[GP09] Willem A. de Graaf and Andrea Pavan. Constructing arithmetic subgroups of unipotent groups. *J. Algebra*, 322(11):3950–3970, 2009.

[Gra00] Willem A. de Graaf. *Lie algebras: theory and algorithms*, volume 56 of North-Holland Mathematical Library. Elsevier Science, 2000.

[Gra01] Willem A. de Graaf. Constructing representations of split semisimple Lie algebras. *J. Pure Appl. Algebra*, 164(1-2):87–107, 2001. Effective methods in algebraic geometry (Bath, 2000).

[Gra02] Willem A. de Graaf. Constructing canonical bases of quantized enveloping algebras. *Exp. Math.*, 11(2):161–170, 2002.

[Gra08] Willem A. de Graaf. Computing with nilpotent orbits in simple Lie algebras of exceptional type. *LMS J. Comput. Math.*, 11:280–297 (electronic), 2008.

[Gra09] Willem A. de Graaf. Constructing algebraic groups from their Lie algebras. *J. Symbolic Comput.*, 44:1223–1233, 2009.

[Gra11] Willem A. de Graaf. Computing representatives of nilpotent orbits of θ-groups. *J. Symbolic Comput.*, 46:438–458, 2011.

[GS80a] Fritz Grunewald and Daniel Segal. Some general algorithms. I. Arithmetic groups. *Ann. Math. (2)*, 112(3):531–583, 1980.

[GS80b] Fritz Grunewald and Daniel Segal. Some general algorithms. II. Nilpotent groups. *Ann. Math. (2)*, 112(3):585–617, 1980.

[GS88] P. B. Gilkey and G. M. Seitz. Some representations of exceptional Lie algebras. *Geom. Dedicata*, 25(1-3):407–416, 1988. Geometries and groups (Noordwijkerhout, 1986).

[GT99] L. Yu. Galitski and D. A. Timashev. On classification of metabelian Lie algebras. *J. Lie Theory*, 9:125–156, 1999.

[GVY12] W.A. de Graaf, È.B. Vinberg, and O.S. Yakimova. An effective method to compute closure ordering for nilpotent orbits of θ-representations. *J. Algebra*, 371:38–62, 2012.

[GW09] Roe Goodman and Nolan R. Wallach. *Symmetry, representations, and invariants*, volume 255 of Graduate Texts in Mathematics. Springer Verlag, Dordrecht, 2009.

[Hel78] Sigurdur Helgason. *Differential geometry, Lie groups, and symmetric spaces*, volume 80 of Pure and Applied Mathematics. Academic Press, New York, 1978.

[HEO05] Derek F. Holt, Bettina Eick, and Eamonn A. O'Brien. *Handbook of computational group theory.* Discrete Mathematics and its Applications, Chapman & Hall/CRC, Boca Raton, FL, 2005.

[Hes79] Wim H. Hesselink. Desingularizations of varieties of nullforms. *Invent. Math.*, 55(2):141–163, 1979.

[Hoc81] G. Hochschild. *Basic Theory of algebraic groups and Lie algebras*, volume 75 of Graduate texts in mathematics. Springer Verlag, New York, 1981.

[Hou75] Alston S. Householder. *The theory of matrices in numerical analysis*. Dover Publications, New York, 1975. Reprint of 1964 edition.

[HPP03] Florian Hess, Sebastian Pauli, and Michael E. Pohst. Computing the multiplicative group of residue class rings. *Math. Comp.*, 72(243):1531–1548, 2003.

[Hum75] James E. Humphreys. *Linear algebraic groups*. Springer Verlag, New York, 1975.

[Hum78] James E. Humphreys. *Introduction to Lie algebras and representation theory*, volume 9 of Graduate Texts in Mathematics. Springer Verlag, New York, 1978. Second printing, revised.

[Hun80] Thomas W. Hungerford. *Algebra*, volume 73 of Graduate Texts in Mathematics. Springer Verlag, New York, 1980. Reprint of 1974 original.

[Jac75] Nathan Jacobson. *Lectures in abstract algebra*. Springer Verlag, New York, 1975. Volume II: Linear algebra, Reprint of 1953 edition.

[Jac79] Nathan Jacobson. *Lie algebras*. Interscience Tracts in Pure and Applied Mathematics, No. 10. John Wiley & Sons, New York, 1962.

[Jan96] Jens Carsten Jantzen. *Lectures on quantum groups*, volume 6 of Graduate Studies in Mathematics. American Mathematical Society, 1996.

[Jan03] Jens Carsten Jantzen. *Representations of algebraic groups*, volume 107 of Mathematical Surveys and Monographs. American Mathematical Society, second edition, 2003.

[Kac69] V. G. Kac. Automorphisms of finite order of semisimple Lie algebras. *Funkcional. Anal. i Priložen.*, 3(3):94–96, 1969.

[Kac80] V. G. Kac. Some remarks on nilpotent orbits. *J. Algebra*, 64:190–213, 1980.

[Kac90] V. G. Kac. *Infinite dimensional Lie algebras*. Cambridge University Press, Cambridge, third edition, 1990.

[Kal15] Bahman Kalantari. A characterization theorem and an algorithm for a convex hull problem. *Ann. Oper. Res.*, 226:301–349, 2015.

[Kap95] Irving Kaplansky. *Fields and rings.* Chicago Lectures in Mathematics. University of Chicago Press, 1995. Reprint of second (1972) edition.

[Kas91] M. Kashiwara. On crystal bases of the q-analogue of universal enveloping algebras. *Duke Math. J.*, 63(2):465–516, 1991.

[Kas96] M. Kashiwara. Similarity of crystal bases. In *Lie algebras and their representations (Seoul, 1995)*, pages 177–186. American Mathematical Society, 1996.

[Kat92] Svetlana Katok. *Fuchsian groups.* Chicago Lectures in Mathematics. University of Chicago Press, 1992.

[Kem78] George R. Kempf. Instability in invariant theory. *Ann. Math. (2)*, 108(2):299–316, 1978.

[Kem02] Gregor Kemper. The calculation of radical ideals in positive characteristic. *J. Symbolic Comput.*, 34(3):229–238, 2002.

[Kem07] Gregor Kemper. The computation of invariant fields and a constructive version of a theorem by Rosenlicht. *Transform. Groups*, 12(4):657–670, 2007.

[Kir84] Frances Clare Kirwan. *Cohomology of quotients in symplectic and algebraic geometry*, volume 31 of Mathematical Notes. Princeton University Press, 1984.

[KN79] George Kempf and Linda Ness. The length of vectors in representation spaces. In *Algebraic geometry*, volume 732 of Lecture Notes in Mathematics, pages 233–243. Springer-Verlag, Berlin, 1979.

[Kos66] Bertram Kostant. Groups over Z. In *Algebraic groups and discontinuous subgroups*, pages 90–98. American Mathematical Society, 1966.

[KP05] Jürgen Klüners and Sebastian Pauli. Computing residue class rings and Picard groups of orders. *J. Algebra*, 292(1):47–64, 2005.

[Kra84] Hanspeter Kraft. *Geometrische Methoden in der Invariantentheorie.* Aspects of Mathematics, D1. Friedr. Vieweg & Sohn, Braunschweig, 1984.

[KRW90] A. Kandri-Rody and V. Weispfenning. Noncommutative Gröbner bases in algebras of solvable type. *J. Symbolic Comput.*, 9(1):1–26, 1990.

[Lan02] Serge Lang. *Algebra*, volume 211 of Graduate Texts in Mathematics. Springer Verlag, New York, third edition, 2002.

[Lit94] P. Littelmann. A Littlewood-Richardson rule for symmetrizable Kac-Moody algebras. *Invent. Math.*, 116(1-3):329–346, 1994.

[Lit96] P. Littelmann. An effective method to classify nilpotent orbits. In *Algorithms in algebraic geometry and applications*, volume 143 of *Progr. Math.*, pages 255–269. Birkhäuser, Basel, 1996.

[Lit97] P. Littelmann. Characters of representations and paths in $\mathfrak{H}^*_{\mathbb{R}}$. In *Representation theory and automorphic forms*, volume 61 of *Proc. Sympos. Pure Math.*, pages 29–49. American Mathematical Society, 1997.

[Lit98] P. Littelmann. Cones, crystals, and patterns. *Transform. Groups*, 3(2):145–179, 1998.

[Los06] Ivan V. Losev. The Kempf-Ness theorem and invariant theory. preprint, 2006.

[Lov79] L. Lovász. On determinants, matchings, and random algorithms. In *Fundamentals of computation theory*, volume 2 of *Math. Res.*, pages 565–574. Akademie Verlag, Berlin, 1979.

[Lüb01] Frank Lübeck. Small degree representations of finite Chevalley groups in defining characteristic. *LMS J. Comput. Math.*, 4:135–169 (electronic), 2001.

[Lus90] G. Lusztig. Canonical bases arising from quantized enveloping algebras. *J. Amer. Math. Soc.*, 3(2):447–498, 1990.

[Lus92] G. Lusztig. Introduction to quantized enveloping algebras. In *New developments in Lie theory and their applications*, pages 49–65. Birkhäuser, Boston, 1992.

[Lus93] G. Lusztig. *Introduction to quantum groups.* Birkhäuser, Boston, 1993.

[Mal49] A. I. Mal′cev. On a class of homogeneous spaces. *Izvestiya Akad. Nauk. SSSR. Ser. Mat.*, 13:9–32, 1949.

[Mos56] G. D. Mostow. Fully reducible subgroups of algebraic groups. *Amer. J. Math.*, 78:200–221, 1956.

[MP82] R. V. Moody and J. Patera. Fast recursion formula for weight multiplicities. *Bull. Amer. Math. Soc. (N.S.)*, 7(1):237–242, 1982.

[MQB99] Jörn Müller-Quade and Thomas Beth. Calculating generators for invariant fields of linear algebraic groups. In *Applied algebra, algebraic algorithms and error-correcting codes*, volume 1719 of *Lecture Notes in Computer Science*, pages 392–403. Springer, Berlin, 1999.

[MT11] Gunter Malle and Donna Testerman. *Linear algebraic groups and finite groups of Lie type.* Cambridge Studies in Advanced Mathematics. Cambridge University Press, 2011.

[Mum65] David Mumford. *Geometric invariant theory.* Ergebnisse der Mathematik und ihrer Grenzgebiete, Neue Folge, Band 34. Springer Verlag, New York, 1965.

[Nes84] Linda Ness. A stratification of the null cone via the moment map. *Amer. J. Math.*, 106(6):1281–1329, 1984. With appendix by David Mumford.

[Oni04] Arkady L. Onishchik. *Lectures on Real Semisimple Lie Algebras and Their Representations.* European Mathematical Society, Zürich, 2004.

[OV90] A. L. Onishchik and È. B. Vinberg. *Lie groups and algebraic groups.* Springer Series in Soviet Mathematics. Springer Verlag, Berlin, 1990. Translated from Russian with preface by D. A. Leites.

[Pav09] Andrea Pavan. Computing Arithmetic Subgroups of Affine Algebraic Groups. PhD thesis, Dipartimento di Matematica Pura e Applicata, University of Padova, 2009.

[Pop03] V. L. Popov. The cone of Hilbert null forms. *Tr. Mat. Inst. Steklova*, 241 (Teor. Chisel, Algebra i Algebr. Geom.):192–209, 2003. English translation in *Proc. Steklov Inst. Math.* 241 (2003), no. 1, 177–194.

[Pop09] V. L. Popov. Two orbits: when is one in the closure of the other? *Tr. Mat. Inst. Steklova*, 264 (Mnogomernaya Algebraicheskaya Geometriya):152–164, 2009. English translation in *Proc. Steklov Inst. Math.* 264 (2009), no. 1, 146–158.

[PP80] Wilhelm Plesken and Michael Pohst. On maximal finite irreducible subgroups of $\mathrm{GL}(n, \mathbf{Z})$. V. The eight-dimensional case and a complete description of dimensions less than ten. *Math. Comp.*, 34(149):277–301, loose microfiche suppl, 1980.

[PR94] Vladimir Platonov and Andrei Rapinchuk. *Algebraic groups and number theory*, volume 139 of Pure and Applied Mathematics. Academic Press, Boston, 1994. Translated from 1991 Russian original by Rachel Rowen.

[Rag72] M. S. Raghunathan. *Discrete subgroups of Lie groups.* Springer Verlag, New York, 1972. Ergebnisse der Mathematik und ihrer Grenzgebiete, Band 68.

[Ree10] Mark Reeder. Torsion automorphisms of simple Lie algebras. *L'Enseignement Mathematique*, 56:3–47, 2010.

[Rob96] Derek J.S. Robinson. *A Course in the theory of groups*, volume 80 of Graduate Texts in Mathematics. Springer Verlag, New York, second edition, 1996.

[Rón90] L. Rónyai. Computing the structure of finite algebras. *J. Symbolic Comput.*, 9:355–373, 1990.

[Rou78] Guy Rousseau. Immeubles sphériques et théorie des invariants. *C. R. Acad. Sci. Paris Sér. A-B*, 286(5):A247–A250, 1978.

[Seh93] Sudarshan K. Sehgal. *Units in integral group rings*, volume 69 of Pitman Monographs and Surveys in Pure and Applied Mathematics. Longman, Harlow, 1993. With appendix by Al Weiss.

[Ser66] J.-P. Serre. *Algèbres de Lie semi-simples complexes.* W. A. Benjamin, Amsterdam, 1966.

[Sha94] Igor R. Shafarevich. *Basic algebraic geometry 1.* Springer Verlag, Heidelberg, 1994.

[Sim94] C. C. Sims. *Computation with finitely presented groups.* Cambridge University Press, Cambridge, 1994.

[Slo89] Peter Slodowy. Die Theorie der optimalen Einparameteruntergruppen für instabile Vektoren. In *Algebraische Transformationsgruppen und Invariantentheorie*, volume 13 of *DMV Sem.*, pages 115–131. Birkhäuser, Basel, 1989.

[Spr98] T. A. Springer. *Linear algebraic groups*, volume 9 of Progress in Mathematics. Birkhäuser, Boston, second edition, 1998.

[Ste67] R. Steinberg. *Lectures on Chevalley groups.* Yale University, New Haven, 1967. Notes prepared by John Faulkner and Robert Wilson.

[Tit66] J. Tits. Sur les constantes de structure et le théorème d'existence des algèbres de Lie semi-simples. *Publ. Math. IHES*, 31:21–58, 1966.

[TY05] Patrice Tauvel and Rupert W. T. Yu. *Lie algebras and algebraic groups.* Springer Verlag, New York, 2005.

[VE78] È. B. Vinberg and A. G. Elashvili. A classification of the three-vectors of nine-dimensional space. *Trudy Sem. Vektor. Tenzor. Anal.*, 18:197–233, 1978. English translation in *Selecta Math. Sov.*, 7, 63-98, (1988).

[Vin76] È. B. Vinberg. The Weyl group of a graded Lie algebra. *Izv. Akad. Nauk SSSR Ser. Mat.*, 40(3):488–526, 1976. English translation in *Math. USSR-Izv.* 10, 463–495 (1976).

[Vin79] È. B. Vinberg. Classification of homogeneous nilpotent elements of a semisimple graded Lie algebra. *Trudy Sem. Vektor. Tenzor. Anal.*, (19):155–177, 1979. English translation in *Selecta Math. Sov.* 6, 15-35 (1987).

[VP89] È. B. Vinberg and V. L. Popov. Invariant theory. In *Algebraic geometry, 4 (Russian)*, Itogi Nauki i Tekhniki, pages 137–314. Akad. Nauk SSSR Vsesoyuz. Inst. Nauchn. i Tekhn. Inform., Moscow, 1989. English translation in V. L. Popov and È. B. Vinberg, *Invariant Theory*, in: *Algebraic Geometry IV*, Encyclopedia of Mathematical Sciences, Vol. 55, Springer Verlag, *Proc. Steklov Inst. Math.* 264 (2009), no. 1, 146–158.

[Win74] David Winter. *The structure of fields.* Springer Verlag, New York, 1974. Graduate Texts in Mathematics, No. 16.

Index of Symbols

$<_{\mathrm{dlex}}$, 17
$<_{\mathrm{lex}}$, 17

Ad, 110
$\mathrm{ad}_{\mathfrak{g}}$, 29
$\mathbb{A}^n$, 1

$\mathfrak{c}(\mathfrak{g})$, 30
$\mathfrak{c}_{\mathfrak{g}}(E)$, 30

$\mathrm{Der}(A)$, 33
$\mathrm{D}(n, K)$, 112

$\mathrm{End}(V)$, 28

$\mathfrak{g}^{(j)}$, 40
$\mathbb{G}_{\mathrm{a}}$, 105
$\mathfrak{g}(\mathfrak{a})$, 127
$\mathfrak{g}_\alpha$, 40
G°, 98
$\mathfrak{g}^j$, 33
G_L, 224
$\mathrm{GL}(n, K)$, 96
$\mathfrak{gl}(n, k)$, 28
$\mathrm{GL}(V)$, 95
$\mathfrak{gl}(V)$, 28
$\mathbb{G}_{\mathrm{m}}$, 105
$\mathfrak{g}_\tau(h)$, 267

$\mathrm{ht}(\alpha)$, 54

Int, 110
ι, 95
$\sqrt{I}$, 3
$\mathcal{I}(U)$, 2

$k[a]$, 31

$\kappa_{\mathfrak{g}}$, 46
$K[X]$, 4
$k\langle Z\rangle$, 71

$\mathcal{L}(w)$, 57
$\mathrm{Lie}(G)$, 101
$L(\lambda)$, 189

$M_n(k)$, 28
$\mathfrak{m}_v(X)$, 6

$\mathcal{N}_G(V)$, 259
$N_G(H)$, 157
$\mathfrak{n}_{\mathfrak{g}}(\mathfrak{u})$, 31
$\mathfrak{nr}(\mathfrak{g})$, 35

P, 59
$\mathfrak{p}(h)$, 267
Φ_w, 58
P_+, 59

$\mathrm{Rad}(A)$, 44
$R(G)$, 100
$\mathrm{rank}(\mathfrak{u})$, 295
$R_u(G)$, 118

$\mathrm{SL}(n, K)$, 107
$\mathfrak{sl}(n, k)$, 107
$\mathfrak{sr}(\mathfrak{g})$, 41

$\mathrm{trdeg}(E/F)$, 11
$T_v(X)$, 7

$\overline{U}$, 2
$\mathcal{U}(\mathfrak{g})$, 71
$[\mathfrak{u}, \mathfrak{v}]$, 33

$V_0(\mathfrak{a})$, 36
$V_0(b)$, 35

$V_1(\mathfrak{a})$, 36
$V_{\geq\tau}(h)$, 265
$\mathcal{V}(I)$, 2
$V_\tau(h)$, 265
$\langle v, w^\vee \rangle$, 49

$X^*(G)$, 112
$X_*(T)$, 115

$Z_G(H)$, 155
$\mathfrak{z}(h)$, 266

Index of Terminology

π-system, 217
θ-group, 291

adapted string, 169
additive group, 105
adjoint group, 182
adjoint representation
  of algebraic group, 110
  of Lie algebra, 29
affine space, 1
algebraic group, 95
algebraic hull, 127
algebraic set, 2
arithmetic group, 225
associative algebra, 28

based root datum, 190
  semisimple, 190
Borel
  subalgebra, 214
  subgroup, 214
Bruhat decomposition, 178

C-generating set, 162
C-group, 162
canonical generating set, 67
carrier algebra, 299
  of e, 297
Cartan matrix, 50
  indecomposable, 50
Cartan's theorem, 42
centralizer
  in Lie algebra, 30
  of algebraic groups, 155
centre, 30
character, 112
Chevalley basis, 70
Chevalley commutator formula, 170
Chevalley group, 173
  adjoint, 182
  simply connected, 182
  universal, 182
closed set, 2
  connected, 5
closure, 2
cocharacter, 115
collection algorithms, 201
commensurable, 224
comorphism, 4
composition series, 29
comultiplication, 101
congruence subgroup, 224
connected component of identity, 98
contravariant bilinear form, 193
coordinate ring, 4
crystal graph, 86

defining element of $\mathbb{Z}$-grading, 289
defining lattice, 115
dense set, 2
derivation, 33
  inner, 33
derived series, 40
derived subalgebra, 40
Derksen ideal, 250
diagonalizable algebraic group, 112
differential, 8
dimension of closed set, 12

Dynkin diagram, 51
    weighted, 282

Engel's theorem, 34
exchange condition, 58

Fitting one component, 36
free associative algebra, 71

Gröbner basis, 18, 73
grading, 289

height, 54
    of weight, 89
highest weight module, 79
highest weight vector, 79, 189
Hilbert Mumford criterion, 261
Hirsch length, 233

identity component, 98
integral monomial, 77
invariant field, 252
invariant form, 29
invariant ring, 249
irreducible closed set, 4
irreducible components, 5
irreducible root system, 52

Jacobi identity, 28
Jordan decomposition, 32
    multiplicative, 117

Killing form, 46

lattice, 223
    admissible, 167
    in $\mathbb{Z}^n$, 229
Levi decomposition, 48
Levi's theorem, 48
Lie algebra, 28
    algebraic, 127
    of algebraic group, 101
Lie's theorem, 42
linear algebraic group, 95
locally flat, 298
lower central series, 33

module, 29
    completely reducible, 29
    irreducible, 29
monomial order, 17
    l-elimination, 20
    degree compatible, 17
    degree lexicographical, 17
    lexicographic, 17
morphism of algebraic groups, 97
multiplicative group, 105

nilpotent
    Lie algebra, 33
    linear transformation, 31
nilpotent orbit, 293
nilpotent part, 48
nilradical, 35
normalizer
    in Lie algebra, 31
    of algebraic groups, 157
nullcone, 259

open set, 2

point
    generalized, 13
    generic, 13
    singular, 12
polynomial function, 1
primary decomposition, 26, 36
product space, 33
purification, 230

radical
    of algebraic group, 100
    of associative algebra, 44
    of ideal, 3
    unipotent, 118
rank
    of reductive Lie algebra, 295
    of root system, 52
rational module, 109
rational representation, 109
rational semisimple element, 262
reduced expression, 57
reductive

algebraic group, 118
Lie algebra, 90
linearly, 249
reflection subgroup, 60
regular map, 4
dominant, 14
regular subalgebra, 216
representation
of algebraic group, 109
of Lie algebra, 28
Reynolds operator, 255
root, 39, 52
root lattice, 191
root space, 39
root space decomposition, 40
root subsystem, 60
root system, 52
roots
positive, 53
simple, 53

semidirect product, 100
semisimple
algebraic group, 100
Lie algebra, 46
linear transformation, 31
semisimple part, 48
semistable, 259
simple Lie algebra, 46
simple reflection, 57
$\mathfrak{sl}_2$-triple, 92
homogeneous, 293
solvable Lie algebra, 40
solvable radical, 41
specialization, 13
splitting element, 145
standard $\mathbb{Z}$-grading, 289
stratum, 272
structure constants, 29

$\mathfrak{T}$-group, 232
tangent space, 7
toral algebra, 241
torus, 112
trace form, 29
transcendence degree, 11

unipotent
group, 131
linear transformation, 117
universal enveloping algebra, 71
unstable, 259

vanishing ideal, 2
Verma module, 80

weight
dominant, 59
fundamental, 59, 191
highest, 79, 189
integral, 59
of $\mathfrak{g}$-module, 39
weight lattice, 59, 191
weight space, 39, 188
weight vector, 79, 188
Weyl group, 56
longest element, 58
Weyl's theorem, 47

Zariski topology, 2